Richly Parameterized Linear Models

Additive, Time Series, and Spatial Models Using Random Effects

CHAPMAN & HALL/CRC
Texts in Statistical Science Series

Series Editors
Francesca Dominici, *Harvard School of Public Health, USA*
Julian J. Faraway, *University of Bath, UK*
Martin Tanner, *Northwestern University, USA*
Jim Zidek, *University of British Columbia, Canada*

Analysis of Failure and Survival Data
P. J. Smith

The Analysis of Time Series: An Introduction, Sixth Edition
C. Chatfield

Applied Bayesian Forecasting and Time Series Analysis
A. Pole, M. West, and J. Harrison

Applied Categorical and Count Data Analysis
W. Tang, H. He, and X.M. Tu

Applied Nonparametric Statistical Methods, Fourth Edition
P. Sprent and N.C. Smeeton

Applied Statistics: Handbook of GENSTAT Analyses
E.J. Snell and H. Simpson

Applied Statistics: Principles and Examples
D.R. Cox and E.J. Snell

Applied Stochastic Modelling, Second Edition
B.J.T. Morgan

Bayesian Data Analysis, Third Edition
A. Gelman, J.B. Carlin, H.S. Stern, D.B. Dunson, A. Vehtari, and D.B. Rubin

Bayesian Ideas and Data Analysis: An Introduction for Scientists and Statisticians
R. Christensen, W. Johnson, A. Branscum, and T.E. Hanson

Bayesian Methods for Data Analysis, Third Edition
B.P. Carlin and T.A. Louis

Beyond ANOVA: Basics of Applied Statistics
R.G. Miller, Jr.

The BUGS Book: A Practical Introduction to Bayesian Analysis
D. Lunn, C. Jackson, N. Best, A. Thomas, and D. Spiegelhalter

A Course in Categorical Data Analysis
T. Leonard

A Course in Large Sample Theory
T.S. Ferguson

Data Driven Statistical Methods
P. Sprent

Decision Analysis: A Bayesian Approach
J.Q. Smith

Design and Analysis of Experiments with SAS
J. Lawson

Elementary Applications of Probability Theory, Second Edition
H.C. Tuckwell

Elements of Simulation
B.J.T. Morgan

Epidemiology: Study Design and Data Analysis, Third Edition
M. Woodward

Essential Statistics, Fourth Edition
D.A.G. Rees

Exercises and Solutions in Statistical Theory
L.L. Kupper, B.H. Neelon, and S.M. O'Brien

Exercises and Solutions in Biostatistical Theory
L.L. Kupper, B.H. Neelon, and S.M. O'Brien

Extending the Linear Model with R: Generalized Linear, Mixed Effects and Nonparametric Regression Models
J.J. Faraway

A First Course in Linear Model Theory
N. Ravishanker and D.K. Dey

Generalized Additive Models: An Introduction with R
S. Wood

Generalized Linear Mixed Models: Modern Concepts, Methods and Applications
W. W. Stroup

Graphics for Statistics and Data Analysis with R
K.J. Keen

Interpreting Data: A First Course in Statistics
A.J.B. Anderson

Introduction to General and Generalized Linear Models
H. Madsen and P. Thyregod

An Introduction to Generalized Linear Models, Third Edition
A.J. Dobson and A.G. Barnett

Introduction to Multivariate Analysis
C. Chatfield and A.J. Collins

Introduction to Optimization Methods and Their Applications in Statistics
B.S. Everitt

Introduction to Probability with R
K. Baclawski

Introduction to Randomized Controlled Clinical Trials, Second Edition
J.N.S. Matthews

Introduction to Statistical Inference and Its Applications with R
M.W. Trosset

Introduction to Statistical Limit Theory
A.M. Polansky

Introduction to Statistical Methods for Clinical Trials
T.D. Cook and D.L. DeMets

Introduction to Statistical Process Control
P. Qiu

Introduction to the Theory of Statistical Inference
H. Liero and S. Zwanzig

Large Sample Methods in Statistics
P.K. Sen and J. da Motta Singer

Linear Algebra and Matrix Analysis for Statistics
S. Banerjee and A. Roy

Logistic Regression Models
J.M. Hilbe

Markov Chain Monte Carlo: Stochastic Simulation for Bayesian Inference, Second Edition
D. Gamerman and H.F. Lopes

Mathematical Statistics
K. Knight

Modeling and Analysis of Stochastic Systems, Second Edition
V.G. Kulkarni

Modelling Binary Data, Second Edition
D. Collett

Modelling Survival Data in Medical Research, Second Edition
D. Collett

Multivariate Analysis of Variance and Repeated Measures: A Practical Approach for Behavioural Scientists
D.J. Hand and C.C. Taylor

Multivariate Statistics: A Practical Approach
B. Flury and H. Riedwyl

Multivariate Survival Analysis and Competing Risks
M. Crowder

Nonparametric Methods in Statistics with SAS Applications
O. Korosteleva

Pólya Urn Models
H. Mahmoud

Practical Data Analysis for Designed Experiments
B.S. Yandell

Practical Longitudinal Data Analysis
D.J. Hand and M. Crowder

Practical Multivariate Analysis, Fifth Edition
A. Afifi, S. May, and V.A. Clark

Practical Statistics for Medical Research
D.G. Altman

A Primer on Linear Models
J.F. Monahan

Principles of Uncertainty
J.B. Kadane

Probability: Methods and Measurement
A. O'Hagan

Problem Solving: A Statistician's Guide, Second Edition
C. Chatfield

Randomization, Bootstrap and Monte Carlo Methods in Biology, Third Edition
B.F.J. Manly

Readings in Decision Analysis
S. French

Richly Parameterized Linear Models: Additive, Time Series, and Spatial Models Using Random Effects
J.S. Hodges

Sampling Methodologies with Applications
P.S.R.S. Rao

Stationary Stochastic Processes: Theory and Applications
G. Lindgren

Statistical Analysis of Reliability Data
M.J. Crowder, A.C. Kimber,
T.J. Sweeting, and R.L. Smith

Statistical Methods for Spatial Data Analysis
O. Schabenberger and C.A. Gotway

Statistical Methods for SPC and TQM
D. Bissell

Statistical Methods in Agriculture and Experimental Biology, Second Edition
R. Mead, R.N. Curnow, and A.M. Hasted

Statistical Process Control: Theory and Practice, Third Edition
G.B. Wetherill and D.W. Brown

Statistical Theory: A Concise Introduction
F. Abramovich and Y. Ritov

Statistical Theory, Fourth Edition
B.W. Lindgren

Statistics for Accountants
S. Letchford

Statistics for Epidemiology
N.P. Jewell

Statistics for Technology: A Course in Applied Statistics, Third Edition
C. Chatfield

Statistics in Engineering: A Practical Approach
A.V. Metcalfe

Statistics in Research and Development, Second Edition
R. Caulcutt

Stochastic Processes: An Introduction, Second Edition
P.W. Jones and P. Smith

Survival Analysis Using S: Analysis of Time-to-Event Data
M. Tableman and J.S. Kim

The Theory of Linear Models
B. Jørgensen

Time Series Analysis
H. Madsen

Time Series: Modeling, Computation, and Inference
R. Prado and M. West

Understanding Advanced Statistical Methods
P.H. Westfall and K.S.S. Henning

Texts in Statistical Science

Richly Parameterized Linear Models

Additive, Time Series, and Spatial Models Using Random Effects

James S. Hodges
University of Minnesota
Minneapolis, USA

CRC Press
Taylor & Francis Group
Boca Raton London New York

CRC Press is an imprint of the
Taylor & Francis Group an **informa** business
A CHAPMAN & HALL BOOK

CRC Press
Taylor & Francis Group
6000 Broken Sound Parkway NW, Suite 300
Boca Raton, FL 33487-2742

Printed on acid-free paper
Version Date: 20130730

International Standard Book Number-13: 978-1-4398-6683-2 (Hardback)

Library of Congress Cataloging-in-Publication Data

Hodges, James S., author.
Richly parameterized linear models : additive, time series, and spatial models using random effects / James S. Hodges.
pages cm. -- (Chapman & Hall/CRC texts in statistical science series)
Includes bibliographical references and index.
ISBN 978-1-4398-6683-2 (hardback)
1. Regression analysis--Textbooks. 2. Linear models (Statistics)--Textbooks. I. Title.

QA278.2.H635 2013
519.5'36--dc23 2013026053

Visit the Taylor & Francis Web site at
http://www.taylorandfrancis.com

and the CRC Press Web site at
http://www.crcpress.com

Contents

List of Examples

For each major example, the table below includes the example's number ("#"), the usual label or labels used to refer to it, whether the book's web site has a dataset for the example ("Dataset?"), and the sections in which the example is used.

Table 1: Examples in This Book

#	Label	Dataset?	Appears in ...
1	Molecular structure of a virus	Y	Sections 1.1, 1.4.2
2	Imaging vocal folds	Y	Section 1.1
3	Estimating glomerular filtration rate	N	Section 1.1
4	Experimental plots in a long row	N	Section 2.1
5	Global mean surface temperature	Y	Sections 2.2.2.4, 2.2.3.3, 3.2.1, 3.3.1, 6.1, 13.2, 13.3, 14.3, 15.2, 16.1, 16.2, 17.1
6	Pig jawbone properties	Y	Sections 4.1, 4.2
7	Soft liner gaps	Y	Section 4.2.2
8	Localizing epileptic activity (optical-imaging data)	Y	Sections 6.3, 9.1.4, 12.2, 15.2, 17.2
9	HMO premiums	Y	Chapter 8, Sections 14.5, 19.2
10	Stomach cancer in Slovenia	Y	Sections 5.2.2, 9.1.1, 10.1, 13.2, 15.2
11	Kids and crowns	Y	Sections 9.1.2, 10.2
12	Colon cancer treatment effect	N	Sections 9.1.3, 11
13	Epidermal nerve density	Y	Sections 13.1, 14.4, 17.1, 18.1
14	Health effects of pollutants	N	Section 13.3
15	Periodontal measurements	Y	Sections 5.2, 5.2.2, 12.1.1, 14.1, 14.2, 15.2, 16.1, 16.2, 17.2

List of Figures

List of Tables

Preface

If you believe in things that you don't understand, then you suffer (Wonder 1973).

When I was in graduate school in the early 1980s, linear model theory had just been perfected and I studied with some of the people who had perfected it. With its combination of simplicity and near-total explanatory power, linear model theory is like nothing else in statistics and by the early 1980s, this theory had been honed to the point where it could be taught almost whole to people taking their first serious regression course (e.g., Weisberg 1980).

At about the same time, however, the main thrust of statistical research turned in a different direction, emphasizing breadth over depth by producing methods for specifying and fitting models of greater generality and with weaker assumptions: generalized linear models, which dropped normality; additive models, which dropped linearity; generalized estimating equations (GEE), which dropped independence; hierarchical models, which added layers of structure; mixed models, which added random effects; Markov chain Monte Carlo, with the great flowering of Bayesian methods it enabled; the modeling syntax of the S and then R systems; structural equations models; spatial models and smoothers; dynamic linear models (state-space models); and penalized fits, among many others. Any statistician with a pulse has to love and be impressed by this explosion of sheer modeling power. I do, and I am. If you read something in this book that seems to contradict that statement, go back and read it again.

My admiration notwithstanding, I am mostly an applied statistician and when I use these new methods to analyze my collaborators' data, I routinely get results that are mysterious, inconvenient, or plainly wrong. For mixed linear models, these unhappy results include zero variance estimates, multiple maxima, counterintuitive outlier effects, odd fits (e.g., a wiggly smooth with one smoother but not with another apparently similar smoother), big changes in fit from apparently modest changes in the model or data, and unpredictable convergence of numerical routines, among other things. When my collaborators' datasets produce such puzzles, I urgently need something like the powerful theory of linear models so I can explain them and figure out what to do. As far as I can tell, however, little if anything is known about most of these puzzles. I see very few mentions of them in the listservs I peruse, in the statistical literature, or in talks. There's no reason to think I am a magnet for freak problems, so I suspect that many, perhaps most, working statisticians encounter the same puzzles. When I publish a paper or do a talk about some of these things, I usually hear from people who have had the same problem and are relieved that it wasn't just a programming error, as a referee or their thesis advisor had insisted. For some reason, however, we don't talk or write much about these puzzles.

It seems as if research in statistics has come to mean promoting new methods, as opposed to understanding methods, old or new. Obviously, it would be inaccurate and unfair to say that *nobody* tries to understand existing or new methods; counterexamples include the trace plots used to describe LASSO results or the separation plots used to explain the support-vector machine. But when a classmate and I tried to assemble a catalog of the mysteries and puzzles we've found as a rationale for more investment in understanding these new methods, we found it impossible to formulate anything that looked like a contemporary journal article. These days, a journal article needs to end in a triumph — "Behold! We have conquered this messy dataset/new class of models/previously intractable computing problem/[etc.]!" — and it is hard to make a mystery or puzzle look like a triumph. The triumphant narrative style is so embedded in today's conception of a journal article that even not-too-thoughtful extensions of linear model theory are cast as new methods. If you doubt this, try doing a literature search for outlier-detection methods for hierarchical or other random-effect models. With two exceptions that I know of (one that I wrote and one that I recently refereed), every such paper ends with a standard triumph in which the new method is shown to identify outliers, but none of these papers is based on a systematic understanding of models with random effects.

It's not hard to imagine why the literature looks this way. There are so many new models, where do you begin? With the present-day emphasis on generality, how can you do anything general enough to interest a good journal? The new methods are so complex! And so on. The solution to this quandary, it seems to me, is to stop trying to learn something about all or even a large fraction of the wonderful new methods developed in the last 30 years. They are too disparate and they are developing too quickly.

Having made that unsexy concession, there does seem to be a good place to start. Many of the new methods are now undergoing a process of unification analogous to the unifications that produced generalized linear models and, even earlier, the projection theory of linear models. The unification of models with random effects — so far — consists of a few competing syntaxes for expressing a large class of models and a method for fitting models expressed in each syntax. Parts I and II of this book emphasize one such syntax, mixed linear models using normal distributions, and some of the great variety of models, which I call richly parameterized models, that can be expressed this way and analyzed using conventional and Bayesian methods for mixed linear models. Examples of this unification include Robinson (1991) and Ruppert et al. (2003). This class of models is rich enough to be interesting and close enough to single-error-term linear models to allow many insights and methods to be borrowed or adapted.

A theory of richly parameterized linear models needs more than a syntax and a computing method. It needs to explain things that happen when these models are used to analyze data, to provide ways to detect problems, and when possible, to show how to mitigate or avoid those problems. Part III takes a step from the theory of ordinary linear models toward a theory of richly parameterized models by adapting ideas central to linear model theory. Part IV then takes a step beyond linear model theory by examining the information in the data about the mixed linear model's covariance matrices, which are the difference between ordinary and mixed linear models.

Parts III and IV are founded on two key premises. The first premise comes from linear model theory: Writing down a model and using it to analyze a dataset is equivalent to specifying a function from the data to the inferential or predictive summaries. However you rationalize or interpret that model, it is essential to understand, in a purely mechanical sense, the function from data to summaries that is implied by the model. Often when I present material from this book, people respond with things that are *non sequiturs* in these terms, for example: This model estimates a causal effect while this other model does not; or a Gaussian process probability model with such-and-such covariance function has realizations with such-and-such properties; or this posterior distribution or likelihood is by definition what the data have to say about the model's unknown parameters. I don't dispute such assertions but for the purposes of this book, they are irrelevant. The question asked here is: When I fit *this* model to *this* dataset, why do I get *this* result, and how much would the result change if I made *this* change to the data or model?

The second premise is that we must distinguish between the model *we choose* to analyze a given dataset and the process that *we imagine* produced the data or that, in rare cases, we know actually did produce the data. Our choice to use a model with random effects does not imply that those random effects correspond to any random mechanism out there in the world, and that fact has practical implications. These implications are a recurring theme of this book's first three parts and are summarized in Chapter 13.

Building on these two premises, Parts III and IV are organized around mysterious, inconvenient, or plainly wrong results that turned up in real problems. Most of these are from my collaborative work but some have come from colleagues, for example Michael Lavine's dynamic linear model puzzle (Chapters 6, 9, 12, and 17). Some of these puzzles are now understood to a greater or lesser extent, while others are barely understood at all. In that sense, Parts III and IV are not quite a catalog of unsolved problems — who knows how many puzzles are as yet unreported or undiscovered? — and their theory is grossly incomplete. I will apologize for that once, now. I hope Parts III and IV stimulate enough research so that the second edition of this book, if there is one, can report fewer mysteries and, yes, more triumphs.

Most statistics texts emphasize what we know. This book emphasizes what we don't know. Judging from reviews of my proposals, a lot of academics think mixed linear models are completely understood, when in fact they are still largely not understood. But this is good news: What a bounty of unsolved problems, and for a heavily used class of models! Graduate students, start your engines!

In emphasizing what we don't know, Parts III and IV consider open problems, which leads to a stylistic quandary. Traditionally, statistical theory follows a mathematical style emphasizing results that can be packaged as theorems. I have observed this tradition whenever possible, because you just can't do better than a relevant theorem. Unfortunately, I could not always produce theorem-like material in a reasonable amount of time. In situations like this, statistical theorists usually present nothing at all or present something they *can* package as a theorem, most often some kind of large-sample result. That seemed unproductive, so when I haven't been able to produce relevant theorem-like results, I have instead followed a style used by my scientific (as opposed to statistical) colleagues. They work by posing hypotheses and gathering a variety of evidence, often indirect, so that their hypotheses are either refuted or accumulate credibility while becoming more refined. Obviously accumulated credibility can't replace the iron-clad certainty of a theorem when a relevant theorem can be proven, but the new methods of the last three decades are so complex that it may never be possible to prove relevant theorems about them. We can, however, make progress by approaching our black-box methods in the same way our scientific colleagues approach nature's black boxes, by prying them open gradually and indirectly if necessary. Chapters 11, 12, and 17 are examples of this style, with Chapter 11 refuting a hypothesis and Chapters 12 and 17 developing some hypotheses and producing a first increment of credibility for each.

Along with a sometimes non-mathematical style of inquiry, I've written in a relatively informal narrative style. I've done so because it's friendlier in two senses: It's easier to understand on a first reading and it doesn't hide my opinions and ignorance behind the passive voice and calculated omissions. Students and other readers *should* find fault with the current state of this field, including the things I've contributed to it, and I want them to see those faults and decide they can do better. I will be delighted if this book attracts the interest of young people with better math and computing skills than I have, who can change this area of study from the backwater it is into the thriving area it can and should be.

Some Guidance about Using This Book

The object of Part I is to present a survey of essentials and a particular point of view about them. The object of Parts II, III, and IV is to present the beginnings of a theory of richly parameterized linear models. This book is not intended to be a magisterial overview of everything known about mixed linear models. It is rather intended to present a point of view about what we do and do not understand about mixed linear models and to identify research opportunities. Overviews of mixed linear models include Searle et al. (1992); Ruppert et al. (2003), which focuses on penalized splines represented as mixed linear models; Verbeke & Molenberghs (1997), which focuses on SAS's MIXED procedure; Diggle et al. (1994), which focuses on longitudinal models; Snijders & Bosker (2012), a thorough treatment of hierarchical (multi-level) models obviously based on a lot of experience fitting them and explaining the fits; and Fahrmeir & Tutz (2001), which catalogs models with exponential-family error distributions and linearity in the mean structure.

The hazard of writing a book like this is that I have to write short chapters about sub-fields of statistics with huge literatures. Even D.R. Cox might not be able to master all those sub-fields; I certainly haven't. Academics tend to be territorial and to view the world through a microscope, so whatever I write is guaranteed to offend specialists in each sub-field even if I say nothing that is factually incorrect. Also, the literature and folklore of mixed linear models are gigantic and I know less than

I probably should about them. Therefore I ask your indulgence: I have tried to be nice to everybody and in return, if I've said something factually incorrect or wrong-headed, please tell me and provide detailed citations. If I decide you're right, I'll post a suitable piece on the book's web site with credit to you and replace the relevant passages in the next edition, if there is one.

I've tried simultaneously to give results about both Bayesian and conventional (non-Bayesian) analyses which, these days, mostly revolve around the restricted or residual likelihood. I've done this because Leo Breiman (2001) was right: The two approaches really aren't much different in practice, at least in this area. Chapter 1 is my argument for that claim. I had the good fortune to study in a department where I could become fluent in both languages but most people aren't so fortunate and thus might find it difficult to switch back and forth between Bayesian and conventional language. I've tried to make this as clear as I can and I apologize for any failures of clarity.

I wrote this book from classroom overheads for a one-semester course that I teach for advanced PhD students in the Division of Biostatistics at the University of Minnesota. I use *Semiparametric Regression* by Ruppert, Wand, & Carroll (2003) as a textbook for that course and Parts I and II of the present book refer to it frequently. *Semiparametric Regression* is simply lovely. Among statistical books with hard technical content, it is the friendliest I've ever read and I only disagree with three or four things in the whole book. I recommend it without reservation. If the present book is written half as well, I will be happy.

Each of the present book's chapters ends with exercises, which are of two types. The first type is standard results that PhD students should be able to derive, which are intended to provide practice with the mostly algebraic methods used in this book. Most chapters also include exercises that are, as far as I know, open research questions. Often these include suggestions about where to start, but of course you should feel free to ignore my suggestions.

The book's web site includes datasets analyzed as examples, when I could get permission to include them. I will be happy if you find these datasets useful and horrified if any of them ends up being pawed over eternally like the stack-loss data or the Scottish lip-cancer data. Publish your own datasets! The world will be richer for it.

When I have used one of my published analyses as an example, I have presented it the way it was published. I figured it would be both dishonest and hazardous to make myself look smarter than I actually was and hope nobody checked the original papers. This also gave me an incentive to be nicer to other researchers than I might otherwise be. In each such case I point out what I now believe is wrong with the analysis I published and when it seems worth the effort and space I give a better analysis. If you identify blunders I haven't mentioned and I agree they are blunders, I will post your attempts to alleviate my ignorance on the book's web site.

Finally, Part I refers to SAS frequently because I am in a biostatistics department and even though cognoscenti are obligated to sneer at SAS, we teach it to our students (along with R and WinBUGS) because it helps them find jobs. I don't mean to single out SAS for criticism but it *is* the Microsoft Office of statistical software and that makes it a good example for many purposes.

May 2013

Acknowledgments

Several colleagues read large parts of this book and gave helpful comments and encouragement, including John Adams, Sudipto Banerjee, Yue Cui, Lionel Galway, Yi (Philip) He, Lisa Henn, Galin Jones, Håvard Rue, and Melanie Wall. Adam Szpiro and Jonathan Rougier read the first nine chapters and gave extensive comments that really improved the book, and Adam provided an excellent example for Chapter 13. Galin Jones was especially helpful with the Markov chain Monte Carlo material in Chapter 1, as was Håvard Rue with Sections 7.1 and 7.2. The 20 students who have taken my course were unusually good sports about serving as guinea pigs while I developed this material and they thoughtfully pointed out my errors, inconsistencies, prejudices, and glib assertions. These generous people are not responsible for the result.

Much of the work presented here was done by my PhD students Dan Sargent, Jiannong Liu, Haolan Lu, Brian Reich, Yi (Philip) He, Yue Cui, Yufen Zhang, and Lisa Henn. I have learned a lot from them. Yue Cui, Brian Reich, and Lisa Henn generously gave me permission to use computer code and figures that they created.

Two people deserve particular thanks. Michael Lavine read almost all of the book in one form or other, provided an extremely interesting puzzle that gets (and deserves) more space than any other puzzle in the book, and gave me frequent strong encouragement. Murray Clayton, besides reading much of the book and encouraging me, showed rare generosity in allowing me to write Chapter 13 using a manuscript that we co-authored (Hodges & Clayton 2010), thereby killing it as a separate publication. I am immensely grateful to these old friends.

Several current and former colleagues kindly gave permission to use their datasets in this book and to include them on the book's web site. In order of their dataset's appearance, they are:

- Dwight Anderson, viral structure data;
- Katherine Kendall, vocal folds data;
- Igor Pesun, soft-liner material polishability data;
- Ching-chang Ko, pig jawbone data;
- Michael Lavine, Daryl Hochman, and Michael Haglund, localizing epileptic activity data;
- Vesna Zadnik, Slovenian stomach cancer data;
- Fadi Kass, kids'n'crowns data; and
- William Kennedy, epidermal nerve density data.

Other datasets on the book's web site either are in the public domain or are so old that their creators don't care any more. Dan Sargent arranged for my PhD student Yufen Zhang to analyze the colon cancer dataset for her PhD dissertation and I report excerpts from that work but consortium agreements made it impossible to include the dataset on the book's web site.

Rob Calver of Chapman & Hall gently talked me into writing this book and provided encouragement and good advice along the way. I appreciate it very much.

Finally, I give endless thanks to my wife, Li Chi-ping, and dedicate this book to her. She truly is the most wonderful woman on earth.

Part I

Mixed Linear Models: Syntax, Theory, and Methods

Part I Introduction

When I was in graduate school, a linear regression with 20 or 30 unknown parameters was considered fairly large. Today, it is common to see mixed linear models having random effects with hundreds or thousands of levels; additive models with several predictors and 25 to 50 or more distinct parameters per predictor; spatial random effects with one level for each distinct measurement location, on top of regression coefficients and unknowns in covariance matrices; state-space time series models with at least one unknown parameter per observation and sometimes several; models where each observation has its own error variance; and regression models with hundreds of observations and tens of thousands of predictors. The common feature of these examples is that the model is rich in parameters, so I use the term *richly parameterized models* for these models and others like them. Besides having many unknown parameters, richly parameterized models share a second feature: All such models place constraints on their numerous unknown parameters because otherwise they would be intractable. The variety of structures and constraints is so great and growing so quickly that I won't try to define this class of models more precisely.

In recent years some researchers have begun to unify large classes of richly parameterized models by defining or adopting a syntax in which many such models can be expressed and an approach to analysis and computing applicable to all models expressed in the syntax. As noted, this has at least two precedents, the projection theory of linear models, which unified ANOVA, regression, and many other previously disparate analyses, and generalized linear models, which built on the previous unification by unifying linear models with exponential-family error distributions. The milestone of the latter unification was the first edition of McCullagh & Nelder's *Generalised Linear Models*, which came out in the early 1980s when I was in graduate school. The unification of richly parameterized models has not matured to a point analogous to that represented by McCullagh & Nelder's first edition, but the class of richly parameterized models is vastly larger than the class of generalized linear models and far less well-defined, so the task seems greater.

At least two quite distinct syntaxes have been proposed for expressing large subclasses of richly parameterized models. So far the predominant syntax is based on mixed linear models, with its generalization to non-normal error and random-effect distributions. Part I gives a survey of mixed-linear-model theory and methods, including a closely related alternative formulation that simplifies some derivations. Later parts of the book use mixed linear models to express many richly parameterized models and explore puzzles arising in their application, as a step toward a more complete theory of richly parameterized models. Chapter 7 at the end of Part II briefly discusses two other syntaxes and the associated theory and computational methods. One syntax, based on Gaussian Markov random fields (Rue & Held 2005), is quite distinct from mixed linear models, while the other, Lee & Nelder's extension of generalized linear models (Lee, Nelder, & Pawitan 2006), is arguably an extension of mixed linear models but deserves separate mention. These three syntaxes capture somewhat different classes of richly parameterized models, but many and perhaps all of the phenomena of mixed linear models are relevant to these other syntaxes.

This book generally and Part I in particular will be much easier to read if you are quite familiar with the theory of linear models with one error term. I have not summarized this theory because others have already done it so well. The linear model books that I use most heavily are Weisberg (1980) and Cook & Weisberg (1982). Weisberg's book has had two subsequent editions and a fourth is in the works; the first edition, which I still use regularly, is a gem of concision and clarity. Many other fine presentations of linear model theory are available, including Atkinson (1985) and Christensen (2011), now in its fourth edition. Use the one you find most congenial.

Chapter 1

An Opinionated Survey of Methods for Mixed Linear Models

This chapter introduces the mixed-linear-model form using examples that would have been familiar in, say, the 1960s. Later chapters show some of the vast range of models that can be written in this form, few if any of which would have been recognizable in the 1960s. After the mixed-linear-model form is introduced in Section 1.1, Section 1.2 surveys the conventional (non-Bayesian) approach to analyzing these models, Section 1.3 surveys the Bayesian approach, and Section 1.4 summarizes differences between the two approaches and applies both to an example. Section 1.5 concludes with some comments on computing.

I sometimes refer to the times before and after the revolution in Bayesian computing created by and still consisting mostly of Markov chain Monte Carlo (MCMC) methods. The landmark papers in this great breakthrough were Geman & Geman (1984), Gelfand & Smith (1990), and Tierney (1994); for brevity I use 1990 as the time when Bayesian computing became broadly practicable.

1.1 Mixed Linear Models in the Standard Formulation

The notation defined here will be used throughout the book. Mixed linear models can be written in the form

$$\mathbf{y} = \mathbf{X}\beta + \mathbf{Z}\mathbf{u} + \varepsilon, \text{ where} \tag{1.1}$$

- the observation $\mathbf{y}$ is an n-vector;
- $\mathbf{X}$ is a known design matrix of size $n \times p$, for the so-called fixed effects;
- β, containing the fixed effects, is $p \times 1$ and unknown;
- $\mathbf{Z}$ is a known design matrix of size $n \times q$, for the so-called random effects;
- $\mathbf{u}$, containing the random effects, is $q \times 1$ and unknown;
- $\mathbf{u}$ is modeled as q-variate normal with mean zero and covariance $\mathbf{G}$, which is a function of unknowns in the vector ϕ_G, so $\mathbf{G} = \mathbf{G}(\phi_G)$;
- ε is an unobserved normally distributed error term with mean zero and covariance $\mathbf{R}$, which is a function of unknowns in the vector ϕ_R, so $\mathbf{R} = \mathbf{R}(\phi_R)$; and
- The unknowns in $\mathbf{G}$ and $\mathbf{R}$ are denoted $\phi = (\phi_G, \phi_R)$.

Most commonly, the errors in ε are independent and identically distributed so $\mathbf{R} = \sigma_e^2\mathbf{I}_n$ for the unknown error variance $\phi_R = \sigma_e^2$ ($\mathbf{I}_n$ being the $n \times n$ identity matrix), but this is not always the case.

If $\mathbf{Zu}$ is eliminated from (1.1) or equivalently if $\mathbf{u}$'s covariance matrix $\mathbf{G}$ is set to zero, then (1.1) is the familiar linear model with a single error term, ε. Thus, the novelty in mixed linear models arises from $\mathbf{Zu}$ and the unknown ϕ_G. Writing a model, such as a penalized spline, in this form mostly involves specifying $\mathbf{Z}$ and $\mathbf{G}(\phi_G)$ and, as far as I know, *all* of the oddities and inconveniences examined in this book arise because ϕ_G is unknown.

The simplest useful example of a mixed linear model is the balanced one-way random-effects model. In this model, observations y_{ij} come in clusters indexed by $i = 1,\ldots,q$ with observations in each cluster indexed by $j = 1,\ldots,m$. One way to write this model is

$$y_{ij} = \mu + u_i + \varepsilon_{ij}, \tag{1.2}$$

where the u_i are independent and identically distributed (iid) $N(0,\sigma_s^2)$ and the ε_{ij} are iid $N(0,\sigma_e^2)$ and independent of the u_i. The y_{ij} have two components of variance, to use an old jargon term, with u_i capturing variation between clusters that affects all the observations in cluster i, and the ε_{ij} representing variation specific to observation (i,j). In the standard notation, $n = qm$, $\mathbf{y} = (y_{11},\ldots,y_{qm})'$, $\mathbf{X} = \mathbf{1}_{qm}$, a qm-vector of 1's, $\beta = \mu$, $\mathbf{Z} = \mathbf{I}_q \otimes \mathbf{1}_m$, $\mathbf{u} = (u_1,\ldots,u_q)'$, $\varepsilon = (\varepsilon_{11},\ldots,\varepsilon_{qm})'$, $\mathbf{G} = \sigma_s^2\mathbf{I}_q$, and $\mathbf{R} = \sigma_e^2\mathbf{I}_{qm}$, where $\otimes$ is the Kronecker product, defined for matrices $\mathbf{A} = (a_{ij})$ and $\mathbf{B}$ as $\mathbf{A}\otimes\mathbf{B} = (a_{ij}\mathbf{B})$.

The definitions above referred to "so-called fixed effects" and "so-called random effects" because the term *random effect* is no longer very well defined, to the point that at least one prominent statistician has advocated abolishing it (Gelman 2005a, Section 6). This issue is discussed at length in Chapter 13, which relies on material in the intervening chapters. For now, note that the term *random effect* originally referred to a very specific thing but now it includes a much larger and more miscellaneous collection of things:

- *Original meaning (old-style random effects)*: The levels of a random effect (in analysis-of-variance jargon) are draws from a population, and the draws are not of interest in themselves but only as samples from the larger population, which *is* of interest. In this original meaning, the random effects $\mathbf{Zu}$ provide a way to model sources of variation that affect several observations in a common way as in the one-way random effect model, where the cluster-specific random effect u_1 affects all the y_{1j}.

- *Current meaning (new-style random effects)*: In addition to the above, a random effect may have levels that are not draws from any population, or that are the entire population, or that may be a sample but a new draw from the random effect could not conceivably be drawn, and in all these cases the levels themselves are of interest. In this extended usage, the random effect $\mathbf{Zu}$ provides a model that is flexible because it is richly parameterized but that avoids overfitting because it constrains $\mathbf{u}$ by means of its covariance $\mathbf{G}$.

The original meaning given above is much like the definition in Scheffé (1959, p. 238) but differs in that Scheffé required a random sample from a population, not merely the more vague "draws" used above. (I use "draws" because outside of survey-sampling contexts, I have never seen anyone draw a genuinely random sample for analysis with a mixed linear model.) The current meaning of *random effects* has come into being implicitly as more statisticians notice that the mathematical *form* of a random effect can be used to model situations that do not fit the original meaning of random effect. Throughout this book, I will call these old-style and new-style random effects.

Later chapters show many examples of new-style random effects. The following three examples include old-style random effects; all arose in projects on which I collaborated. Besides exemplifying mixed-linear-model notation, they give a sense of the variety of models that fit in this framework even if we restrict ourselves to old-style random effects. Also, each example's analysis produced an inconvenient, mysterious, or plainly wrong result that no current theory of linear mixed models can explain or remedy. The second and third examples are dead ends in that as yet nobody can explain why these problems occur or offer advice more specific than "try something else." They suggest how readily such problems occur — and how many research opportunities await.

Example 1. Molecular structure of a virus. Dwight Anderson's lab at the University of Minnesota School of Dentistry used cryoelectron microscopy and other methods to develop a hypothesized molecular description of the outer shell (prohead) of the bacteriophage virus $\phi 29$ (Peterson et al. 2001). To test this hypothesized model, they estimated the count of each kind of protein in the prohead, to compare to counts they had hypothesized. They did so by breaking the prohead or phage into its constituent proteins, weighing those proteins, and converting the weight into a count of copies. I use as an example the major capsid protein, gp8. Counting copies of gp8 had four major steps:

- Selecting a *parent*; there were 2 prohead parents and 2 phage parents;
- Preparing a *batch* of the parent;
- On a given gel date, creating electrophoretic *gels*, separating the different proteins on the gels, and cutting out the piece of the gel relevant to gp8; and
- Burning several such gel pieces in an *oxidizer run* to get a gp8 weight from each gel, which was then converted into a count of gp8 copies.

For the gp8 counts, there were 4 parents, 9 batches, 11 gels, and 7 oxidizer runs for a total of 98 measurements. In the analyses published in Peterson et al. (2001), I treated each of these measurement steps as an old-style random effect, though now that I am older I recognize this is debatable at best for the four parents. The batches, gels, and oxidizer runs, however, are clearly old-style random effects: They can be viewed as draws from hypothetical infinite populations of batches, gels, and oxidizer runs; these 9, 11, and 7 draws, respectively, are of interest only for what they tell us about variation between batches, gels, and runs; and they have no interest in themselves. Table 1.1 shows the first 33 measurements of the gp8 count, each with its

Table 1.1: Molecular Structure of a Virus: First 33 Rows of Data for gp8 Weight

Parent	Batch	Gel Date	Oxidizer Run	gp8 Weight
1	1	1	1	244
1	1	2	1	267
1	1	2	1	259
1	1	2	1	286
1	3	1	1	218
1	3	2	1	249
1	3	2	1	266
1	3	2	1	259
1	7	4	3	293
1	7	4	3	277
1	7	4	3	286
1	7	4	3	297
1	7	5	4	315
1	7	5	4	302
1	7	5	4	312
1	7	5	4	319
1	7	5	4	316
1	7	5	4	321
1	7	5	4	293
1	7	5	4	283
1	7	7	4	311
1	7	7	4	282
1	7	7	4	283
1	7	7	4	276
1	7	7	4	331
1	7	7	4	252
1	7	7	4	326
1	7	7	4	334
2	2	1	1	272
2	2	2	1	223
2	4	1	1	208
2	4	2	1	226
2	4	2	1	223

parent, batch, gel date, and oxidizer run. (This design is far from ideal, but this is what was presented to me. In view of the ingenuity and labor that went into gathering these data, I waited until the paper was accepted and then gently advised my collaborators to talk to me before they did something like this again.)

To analyze these data, I used the oldest and simplest mixed linear model, the variance component model, in which the ith measurement of the gp8 count, y_i, is modeled as

$$
\begin{aligned}
y_i &= \mu + \text{parent}_{j(i)} + \text{batch}_{k(i)} + \text{gel}_{l(i)} + \text{run}_{m(i)} + \varepsilon_i \\
\text{parent}_{j(i)} &\overset{iid}{\sim} N(0, \sigma_p^2), j = 1, \ldots, 4 \\
\text{batch}_{k(i)} &\overset{iid}{\sim} N(0, \sigma_b^2), k = 1, \ldots, 9 \\
\text{gel}_{l(i)} &\overset{iid}{\sim} N(0, \sigma_g^2), l = 1, \ldots, 11 \\
\text{run}_{m(i)} &\overset{iid}{\sim} N(0, \sigma_r^2), m = 1, \ldots, 7 \\
\varepsilon_i &\overset{iid}{\sim} N(0, \sigma_e^2), i = 1, \ldots, 98, \qquad (1.3)
\end{aligned}
$$

where the deterministic functions $j(i)$, $k(i)$, $l(i)$, and $m(i)$ map i into the parent, batch, gel, and run indices, respectively.

In the standard formulation of the mixed linear model, equation (1.1), $\mathbf{X} = \mathbf{1}_{98}$, $\beta = \mu$,

$$
\mathbf{Z} = \begin{matrix}
\overbrace{1000}^{\text{parent (4 cols)}} & \overbrace{10\ldots0}^{\text{batch (9 cols)}} & \overbrace{10\ldots0}^{\text{gel (11 cols)}} & \overbrace{10\ldots0}^{\text{run (7 cols)}} \\
1000 & 10\ldots0 & 01\ldots0 & 10\ldots0 \\
\vdots & \vdots & \vdots & \vdots \\
0001 & 00\ldots1 & 00\ldots1 & 00\ldots1
\end{matrix},
$$

$$
\mathbf{u} = [\text{parent}_1, \ldots, \text{parent}_4, \text{batch}_1, \ldots, \text{batch}_9, \text{gel}_1, \ldots, \text{gel}_{11}, \text{run}_1, \ldots, \text{run}_7]',
$$

where $\mathbf{Z}$ is 98×31 and $\mathbf{u}$ is 31×1. $\mathbf{G}$, the covariance matrix of $\mathbf{u}$, is 31×31 block-diagonal with blocks $\sigma_p^2\mathbf{I}_4, \sigma_b^2\mathbf{I}_9, \sigma_g^2\mathbf{I}_{11}$, and $\sigma_r^2\mathbf{I}_7$, so that the unknowns in $\mathbf{G}$ are $\phi_G = (\sigma_p^2, \sigma_b^2, \sigma_g^2, \sigma_r^2)$. Finally, $\mathbf{R}$, the covariance matrix of the errors ε, is $\sigma_e^2\mathbf{I}_{98}$, so $\phi_R = \sigma_e^2$.

Similar models were used to analyze counts of the other proteins in the ϕ29 bacteriophage virus. In Section 1.4.2, we will see that the conventional analysis, maximizing the restricted likelihood, produces zero estimates of several of these variance components, which is obviously wrong and about which the theory of the conventional analysis is silent. Chapter 18 discusses zero variance estimates and associated practical questions, such as determining whether the restricted likelihood for ϕ or the data's contribution to the marginal posterior of ϕ is flat (contains little information) near zero.

Example 2. Testing a new system for imaging vocal folds. Ear-nose-and-throat doctors evaluate certain speech or larynx problems by taking video images of the inside of the larynx during speech and having trained raters assess the resulting images. As of 2008, the standard method used strobe lighting with a slightly longer period than the period of the vocal folds' vibration, a clever trick that produces an artificially slowed-down image of the folds' vibration. The new method under study used high-speed video (HSV) to allow very slow-motion replay of the video and thus a direct

Table 1.2: Imaging Vocal Folds: First 9 Subjects and 26 Rows of Data for "Percent Open Phase"

Subject ID	Imaging Method	Rater	% Open Phase
1	strobe	CR	56
1	strobe	KK	70
1	strobe	KK	70
1	HSV	KUC	70
1	HSV	KK	70
1	HSV	KK	60
2	strobe	KUC	50
2	HSV	CR	54
3	strobe	KUC	60
3	strobe	KUC	70
3	HSV	KK	56
4	strobe	CR	65
4	HSV	KK	56
5	strobe	KK	50
5	HSV	KUC	55
5	HSV	KUC	67
6	strobe	KUC	50
6	strobe	KUC	50
6	HSV	KUC	50
7	strobe	KK	50
7	HSV	KUC	57
8	strobe	CR	56
8	strobe	KUC	60
8	HSV	CR	50
9	strobe	CR	92
9	HSV	KUC	78

view of the folds' vibration. Katherine Kendall (2009) applied both imaging methods to each of 50 subjects, some healthy and some not. Each of the resulting 100 images was assessed by raters, labeled CR, KK, and KUC. Some images were rated by more than one rater and/or by a single rater more than once, in a haphazard design having 154 total ratings. (Again, this unfortunate design was presented to me as a *fait accompli.*) For each image, a rating consisted of 5 quantities on continuous scales. Dr. Kendall's interest was in differences between the new and old imaging methods and between raters, and in their interaction. Table 1.2 shows the design variables for the first 9 subjects (26 ratings), with the outcome "percent open phase."

This is a generalization (or corruption, if you prefer) of a repeated-measures design, where the subject effect is "subject ID" and the within-subject fixed effects are method, rater, and their interaction. I analyzed these data using a mixed linear model including those fixed effects and random effects for a subject's overall level, for the interaction of subject and method (describing how the difference between methods varies between subjects), for the interaction of subject and rater (describing how the difference between raters varies between subjects), and a residual. The last could also be called the three-way interaction of subject, method, and rater (describing how the method-by-rater interaction varies between subjects) but conventionally it is called the residual and treated as an error term. As for Example 1, these three subject effects are old-style random effects that would have been recognized by Scheffé's contemporaries: They all arise from drawing subjects from the population of potential subjects, though not by a formal randomization, and the subjects themselves have no inherent interest.

To fit this analysis into the mixed-linear-model framework (1.1), the residual covariance matrix is $\mathbf{R} = \sigma_e^2 \mathbf{I}_{154}$ with unknown σ_e^2. The fixed-effect design matrix $\mathbf{X}$ and parameter β encode imaging method and rater and could use any of several parameterizations, each with its own specification of $\mathbf{X}$ and β. For the indicator parameterization (sometimes called "treatment contrasts") with strobe and KK as the reference (base) levels for method and rater respectively, the first 26 rows of $\mathbf{X}$ are given in Table 1.3. The values of β corresponding to these columns are, from left, the average for strobe rated by KK; HSV minus strobe for KK; CR minus KK for strobe; KUC minus KK for strobe; and two βs for interactions.

Regarding the random effects, with a design this messy it is easiest to think first of the elements of $\mathbf{u}$ and then construct the corresponding $\mathbf{Z}$. Here is the random-effect specification used by the JMP software in which I did the computations (which is the same specification as in SAS's MIXED procedure). The subject effect has 50 elements, $u_{s1}, \ldots, u_{s50}$; the subject-by-method effect has 100 elements $u_{t1}, u_{H1}, \ldots, u_{t50}, u_{H50}$, where u_{ti}, u_{Hi} are subject i's random effects for s<u>t</u>robe and <u>H</u>SV respectively. For the interaction of subject and rater, we have one random effect (one u) for each unique combination of a rater and a subject ID appearing in the dataset, giving $u_{1,CR}, u_{1,KK}, u_{1,KUC}, u_{2,CR}, u_{2,KUK}, u_{3,KK}, u_{3,KUC}, u_{4,CR}, u_{4,KK}, \ldots, u_{49,CR}, u_{50,CR}$. Table 1.4 gives the columns of $\mathbf{Z}$ corresponding to the first three subjects in the dataset (the left three columns are labels, the right three clusters of columns are the columns of $\mathbf{Z}$).

This specification of the random effects differs from the specification of a balanced mixed model given by Scheffé (1959, e.g., Section 8.1), in which the random effects are defined to sum to zero across subscripts referring to fixed effects. The specification given here avoids quandaries arising from empty design cells, which the present problem has.

When I applied this analysis to the five outcome measures, the numerical routine performing the standard analysis (maximizing restricted likelihood) converged for four outcomes but not for the fifth. The designs were identical for the 5 outcomes $\mathbf{y}$ so the differences in convergence must arise from differences in the outcomes, not the design.[1] However, it is unknown *how* differences in the outcomes produce differences in convergence. Unfortunately, I have made zero progress on the latter

[1]Actually, they weren't quite identical. One of the outcomes for which the algorithm converged had a missing observation.

Table 1.3: Imaging Vocal Folds: First 26 Rows of Fixed-Effect Design Matrix **X**

1	0	1	0	0	0
1	0	0	0	0	0
1	0	0	0	0	0
1	1	0	1	0	1
1	1	0	0	0	0
1	1	0	0	0	0
1	0	0	1	0	0
1	1	1	0	1	0
1	0	0	1	0	0
1	0	0	1	0	0
1	1	0	0	0	0
1	0	1	0	0	0
1	1	0	0	0	0
1	0	0	0	0	0
1	1	0	1	0	1
1	1	0	1	0	1
1	0	0	1	0	0
1	0	0	1	0	0
1	1	0	1	0	1
1	0	0	0	0	0
1	1	0	1	0	1
1	0	1	0	0	0
1	0	0	1	0	0
1	1	1	0	1	0
1	0	1	0	0	0
1	1	0	1	0	1

Table 1.4: Imaging Vocal Folds: Rows and Non-zero Columns of the Random-Effect Design Matrix **Z** for the First Three Subjects

Subject ID	Method	Rater	Subject	Subject-by-Method	Subject-by-Rater
1	strobe	CR	100	100000	1000000
1	strobe	KK	100	100000	0100000
1	strobe	KK	100	100000	0100000
1	HSV	KUC	100	010000	0010000
1	HSV	KK	100	010000	0100000
1	HSV	KK	100	010000	0100000
2	strobe	KUC	010	001000	0001000
2	HSV	CR	010	000100	0000100
3	strobe	KUC	001	000010	0000010
3	strobe	KUC	001	000010	0000010
3	HSV	KK	001	000001	0000001

question, so I have nothing more to say about this example. The dataset is on the book's web site; perhaps you can figure it out.

Example 3. Estimating glomerular filtration rate in kidney-transplant recipients. Glomerular filtration rate (GFR) describes the flow rate of filtered fluids through a person's kidneys. Direct measurements of GFR, such as iothalamate or iohexol GFR, are considered ideal but are very expensive and time-consuming. Outside of research settings, GFR is estimated by plugging standard clinical lab measurements into an equation that predicts GFR. Many such equations have been proposed, most commonly using serum creatinine. However, serum creatinine can be influenced by changes in muscle mass from steroid use, which may give misleading GFR estimates for the large number of kidney-transplant recipients whose immunosuppressive regimens include steroids.

Ward (2010) tested the accuracy and precision of 11 GFR prediction equations against iothalamate GFR (iGFR) using data from 153 subjects in a randomized trial enrolling kidney-transplant recipients. Of the 11 equations, 7 were functions of cystatin C only while 4 were functions of serum creatinine. I describe Ward's project in terms of a generic estimated GFR (eGFR). The study protocol specified that subjects would be randomized about a month after their transplant and have iGFR measured then and annually for the next 5 years. At these annual visits, standard labs were taken and eGFR was computed so each subject had up to 6 pairs of eGFR and the gold-standard iGFR, though many subjects had missed visits or had not finished follow-up at the time of these analyses. One question was: Does eGFR capture the trend over time in iGFR? Even if a given eGFR is biased high (say), if it is consistently biased high, it might accurately capture the trend over time in iGFR. To answer this question, each eGFR was compared separately to iGFR, using the following analysis. I describe the simplest analysis; variants included a covariate describing steroid use and analyses of subsets of subjects.

The dataset included two observations ("cases") per subject per annual visit, one for iGFR and one for eGFR. The mixed linear model fit a straight line in time (visit number) for each combination of a subject and a GFR method (iGFR or eGFR). For each subject these four quantities — a slope and intercept for each method — were treated as an iid draw from a 4-variate normal distribution. Thus the fixed effects design matrix **X** had one row per measurement and 4 columns corresponding to population-average values of the intercept for iGFR, the slope in visits for iGFR, the intercept for eGFR, and the slope in visits for eGFR. Table 1.5 shows the rows of **X** for two subjects, the first having all 6 visits and the second having only 3 visits. The corresponding fixed effects are $\beta' = (\beta_{0I}, \beta_{1I}, \beta_{0e}, \beta_{1e})$, with subscripts I and e indicating iGFR and eGFR, and subscripts 0 and 1 indicating slope and intercept. The substantive question is whether β_{1I} and β_{1e} differ.

The random-effects design matrix **Z** had 4 columns per subject, analogous to the columns for fixed effects. Each subject's columns in **Z** were the same as his/her columns in **X** for rows corresponding to his/her observations, and were zeros in rows corresponding to other subjects' observations. Table 1.5 shows the rows and columns of **Z** corresponding to the rows of **X** shown in the same table. Subject i's random

Table 1.5: Glomerular Filtration Rate: Fixed-Effect Design Matrix **X** and Random-Effect Design Matrix **Z** for the First Two Subjects

$$
\mathbf{X} = \left[\begin{array}{cccc}
1 & 0 & 0 & 0 \\
1 & 1 & 0 & 0 \\
1 & 2 & 0 & 0 \\
1 & 3 & 0 & 0 \\
1 & 4 & 0 & 0 \\
1 & 5 & 0 & 0 \\
0 & 0 & 1 & 0 \\
0 & 0 & 1 & 1 \\
0 & 0 & 1 & 2 \\
0 & 0 & 1 & 3 \\
0 & 0 & 1 & 4 \\
0 & 0 & 1 & 5 \\
\hline
1 & 0 & 0 & 0 \\
1 & 1 & 0 & 0 \\
1 & 2 & 0 & 0 \\
0 & 0 & 1 & 0 \\
0 & 0 & 1 & 1 \\
0 & 0 & 1 & 2 \\
\vdots & \vdots & \vdots & \vdots
\end{array}\right]
\quad
\mathbf{Z} = \left[\begin{array}{cccc|cccc}
1 & 0 & 0 & 0 & 0 & 0 & 0 & 0\ldots \\
1 & 1 & 0 & 0 & 0 & 0 & 0 & 0\ldots \\
1 & 2 & 0 & 0 & 0 & 0 & 0 & 0\ldots \\
1 & 3 & 0 & 0 & 0 & 0 & 0 & 0\ldots \\
1 & 4 & 0 & 0 & 0 & 0 & 0 & 0\ldots \\
1 & 5 & 0 & 0 & 0 & 0 & 0 & 0\ldots \\
0 & 0 & 1 & 0 & 0 & 0 & 0 & 0\ldots \\
0 & 0 & 1 & 1 & 0 & 0 & 0 & 0\ldots \\
0 & 0 & 1 & 2 & 0 & 0 & 0 & 0\ldots \\
0 & 0 & 1 & 3 & 0 & 0 & 0 & 0\ldots \\
0 & 0 & 1 & 4 & 0 & 0 & 0 & 0\ldots \\
0 & 0 & 1 & 5 & 0 & 0 & 0 & 0\ldots \\
\hline
0 & 0 & 0 & 0 & 1 & 0 & 0 & 0\ldots \\
0 & 0 & 0 & 0 & 1 & 1 & 0 & 0\ldots \\
0 & 0 & 0 & 0 & 1 & 2 & 0 & 0\ldots \\
0 & 0 & 0 & 0 & 0 & 0 & 1 & 0\ldots \\
0 & 0 & 0 & 0 & 0 & 0 & 1 & 1\ldots \\
0 & 0 & 0 & 0 & 0 & 0 & 1 & 2\ldots \\
\vdots & \vdots & \vdots & \vdots & \vdots & \vdots & \vdots & \vdots
\end{array}\right]
$$

effects $\mathbf{u}_i' = (u_{0I}, u_{1I}, u_{0e}, u_{1e})$ are, respectively, that subject's deviations from the population averages of the intercept for iGFR, the slope in visits for iGFR, the intercept for eGFR, and the slope in visits for eGFR. These analyses assumed **u** was iid normal with mean zero and covariance matrix **G** that was a general 4×4 covariance matrix.

For this analysis, we judged that the errors ε might show two kinds of correlation: Between iGFR and eGFR at a given visit, and serially across visits for a given measurement method. Thus, the errors were modeled as independent between persons, and within persons were autocorrelated to order 1 (AR(1)) within each method and correlated between methods. In the syntax of SAS's MIXED procedure, the REPEATED command had TYPE = UN@AR(1). Thus **R** was block-diagonal with blocks corresponding to subjects and block size two times the number of visits a subject attended. Table 1.6 shows the block in **R** for the second subject, who attended the first 3 visits only.

I will not try to rationalize this analysis beyond the following. The iGFR outcomes were quite variable and a person-specific straight-line trend with serial correlation fit as well as anything could; the investigators did specifically ask about time trends, which clinicians usually interpret as rates, i.e., slopes; and as it turned out, all seven cystatin C equations gave as estimates nearly identical increasing trends for eGFR while iGFR showed a flat or slightly declining estimated trend (depending on

Table 1.6: Glomerular Filtration Rate: Block of Error Covariance Matrix **R** for the Second Subject, with 3 Visits

σ_1^2	$\sigma_1^2\rho$	$\sigma_1^2\rho^2$	σ_{12}	$\sigma_{12}\rho$	$\sigma_{12}\rho^2$
$\sigma_1^2\rho$	σ_1^2	$\sigma_1^2\rho$	$\sigma_{12}\rho$	σ_{12}	$\sigma_{12}\rho$
$\sigma_1^2\rho^2$	$\sigma_1^2\rho$	σ_1^2	$\sigma_{12}\rho^2$	$\sigma_{12}\rho$	σ_{12}
σ_{12}	$\sigma_{12}\rho$	$\sigma_{12}\rho^2$	σ_2^2	$\sigma_2^2\rho$	$\sigma_2^2\rho^2$
$\sigma_{12}\rho$	σ_{12}	$\sigma_{12}\rho$	$\sigma_2^2\rho$	σ_2^2	$\sigma_2^2\rho$
$\sigma_{12}\rho^2$	$\sigma_{12}\rho$	σ_{12}	$\sigma_2^2\rho^2$	$\sigma_2^2\rho$	σ_2^2

the group of subjects), while all four serum creatinine equations captured the iGFR time trend reasonably well.

For the present purpose, the analysis had a few interesting features. Although the 11 eGFRs were fairly highly correlated with each other for this group of measurements, they differed considerably in the ease with which restricted likelihood (as implemented in MIXED) fit the model to them, particularly for subsets of the data, e.g., excluding the 27 people who received continuous steroid treatment. For some eGFRs, the model fit happily with MIXED's defaults, while for other eGFRs the maximizing routine would not converge unless we specified particular starting values and allowed a large number of iterations, which makes me wonder whether there were multiple maxima. (We didn't find any.) Obvious tricks like centering covariates helped for some analyses but for some analyses that converged readily without centering, centering made it harder to get convergence.

Anybody with any experience knows that fitting this kind of model is a fussy business, but I am not aware of any detailed attempt to explain that fussiness. It is well-known that in the simplest version of this model — clustered data with a random effect for a cluster and compound symmetry correlation structure for the errors — the random effect variance and error correlation are not identified. The present model should be identified because we used AR(1) errors instead of compound-symmetric errors, but with relatively short series for each person, no one can say how close this model is to being non-identified. One sign of difficulty is that many of these fits gave estimates of **G** with correlations of 1. In fact, the TYPE=UN@AR1 specification in MIXED's RANDOM statement, which specifies the model for **G**, gave correlations greater than 1 in absolute value — before these analyses, I hadn't known MIXED would allow that — so we had to use the TYPE=UNR@AR1 specification to avoid illegal estimates for random-effect correlations.

Unfortunately, I cannot make Ward's dataset available. However, I get estimates on the boundary of legal values almost every time I fit a model in which two random effects are correlated, so you should have little trouble finding examples of your own. This is akin to the problem of zero variance estimates but it turns out to be harder to think about. I have made no progress on this example either and will not return to it.

The point of these examples — and of this book as a whole — is that we now have a tremendous ability to fit models of the form (1.1), but we have little understanding of the resulting fits. Parts III and IV of this book are one attempt to begin developing a theory that would provide the necessary understanding, but it is only the barest beginning.

1.2 Conventional Analysis of the Mixed Linear Model

This section is not intended to be an exhaustive or even particularly full presentation of conventional analyses of the mixed linear model, which center on the restricted likelihood. More detailed treatments are available in many books, for example Searle et al. (1992); Snijders & Bosker (2012), which focuses on hierarchical (multi-level) models; Verbeke & Molenberghs (1997), which focuses on SAS's MIXED procedure; Ruppert et al. (2003), which focuses on penalized splines; and Diggle et al. (1994), which focuses on longitudinal analyses. Fahrmeier & Tutz (2010) consider a variety of analysis methods for a broad range of models that all have error distributions in the exponential family.

This section's purpose, besides defining notation and other things, is to emphasize three points:

- The theory and methods of mixed linear models are strongly connected to the theory and methods of linear models, though the differences are important;
- The restricted likelihood is the posterior distribution from a particular Bayesian analysis; and
- Conventional and Bayesian analyses are incomplete or problematic in many respects.

Regarding the first of these, we have an immense collection of tools and intuition from linear models and we should use them as much as possible in developing a theory of mixed linear models. Currently researchers appear to do this inconsistently; for example, ideas of outliers and influence have been carried over to mixed linear models but not collinearity, which features prominently in the present book's Part III.

Regarding incomplete or problematic features of the approaches, I emphasize these because so many people seem either unaware of or complacent about them. For the conventional analysis, Ruppert et al. (2003) give the best treatment of these problems that I know of, while most textbooks seem to ignore them, Snijders & Bosker (2012) being an exception. I often summarize and refer to Ruppert et al. (2003) rather than repeat them, and otherwise I emphasize problems that seem to receive little or no attention elsewhere.

1.2.1 Overview

If an ordinary linear model with a single homoscedastic error term (constant variance) is fit using ordinary least squares, the unknown error variance has no effect on estimates of the mean structure, i.e., the regression coefficients. Thus, the linear model literature emphasizes estimation, testing, etc., for the mean structure with the error variance being a secondary matter. This is not the case in mixed linear models. Estimates of the unknowns in the mean structure — and I include in this the random effect $\mathbf{u}$ — depend on the unknowns ϕ in the random-effect covariance $\mathbf{G}$ and the error covariance $\mathbf{R}$.

The conventional analysis usually proceeds in three steps: Estimate ϕ; treating that estimate as if it is known to be true, estimate β and $\mathbf{u}$; and then compute tests, confidence intervals, etc., again treating the estimate of ϕ as if it is known to be true. This section begins with estimation of β and $\mathbf{u}$, then moves to estimating ϕ, and then on to tests, etc. The obvious problem here — treating the estimate of ϕ as if it is known to be true — is well known and is discussed briefly below.

1.2.2 Mean Structure Estimates

Mixed linear models have a long history and have accumulated a variety of jargon reflecting that history. In the traditional usage — which is still widely used (e.g., Ruppert et al. 2003) — one *estimates* the fixed effects β but *predicts* the random effects $\mathbf{u}$. This distinction is absent in Bayesian analyses of mixed linear models and appears to be losing ground in conventional non-Bayesian analyses. I maintain this distinction when it seems helpful to do so.

Contemporary conventional analyses can be interpreted (and I do so) as using a unified approach to estimating fixed effects and predicting random effects based on the likelihood function that results from treating ϕ as if it is known (except that, as we will see, this is not really a likelihood, at least by today's definition). The presentation here assumes normal distributions for errors and random effects but the approach is the same when other distributions are used.

To begin, then, we have the mixed linear model as in (1.1),

$$\mathbf{y} = \mathbf{X}\beta + \mathbf{Z}\mathbf{u} + \varepsilon, \qquad \mathbf{u} \sim N_q(0, \mathbf{G}(\phi_G)), \qquad \varepsilon \sim N_n(0, \mathbf{R}(\phi_R)). \tag{1.4}$$

In the conventional view, the random variables here are $\mathbf{u}$ and ε, but by implication $\mathbf{y}$ is also a random variable. Customarily, then, the analysis proceeds by writing the joint density of $\mathbf{y}$ and $\mathbf{u}$ as

$$f(\mathbf{y}, \mathbf{u}|\beta, \phi) = f(\mathbf{y}|\mathbf{u}, \beta, \phi_R) f(\mathbf{u}|\phi_G), \tag{1.5}$$

where f will be used generically to represent a probability density, in this case Gaussian densities. Taking the log of both sides gives

$$\begin{aligned} \log f(\mathbf{y}, \mathbf{u}|\beta, \phi) &= K - \frac{1}{2}\log|\mathbf{R}(\phi_R)| - \frac{1}{2}\log|\mathbf{G}(\phi_G)| \\ &\quad -\frac{1}{2}\left\{(\mathbf{y} - \mathbf{X}\beta - \mathbf{Z}\mathbf{u})'\mathbf{R}(\phi_R)^{-1}(\mathbf{y} - \mathbf{X}\beta - \mathbf{Z}\mathbf{u}) + \mathbf{u}'\mathbf{G}(\phi_G)\mathbf{u}\right\}, \end{aligned} \tag{1.6}$$

where K is an unimportant constant.

Readers may note the awkwardness inherent in (1.6). It is tempting to call this a likelihood. However, the usual likelihood arises as the probability or probability density of an observed random variable treated as a function of unknown parameters, while in (1.6), the random variable $\mathbf{u}$ is unobserved. Lee et al. (2006, Chapters 1 to 4) try valiantly to resolve this difficulty in a way that allows them to say they do not use prior distributions. However, I am persuaded by other work (Bayarri et al. 1988, Bayarri & DeGroot 1992) that the likelihood and prior distributions cannot be cleanly distinguished so that this difficulty is inherent and cannot be resolved tidily without adopting a Bayesian approach, in which the distinction is not important; see, e.g., Rappold et al. (2007) for a *bona fide* scientific application. Perhaps in the long run statisticians will not be so fussy about labeling things as likelihoods or priors. (This is arguably true already for conventional likelihood-based analyses of state-space models; see Chapter 6.) The conventional analysis, as practiced today at least, is already one step in this direction: In effect, it ignores this difficulty and treats (1.6) as if it were a likelihood function for the unknowns β and $\mathbf{u}$, with ϕ treated as if it were known. However, it is worth remembering that as of today, (1.6) is not considered a likelihood.

If we now maximize this not-really-a-likelihood ("quasi-likelihood" and "pseudo-likelihood" having already been claimed), then β and $\mathbf{u}$ are estimated by minimizing

$$\underbrace{\left[\mathbf{y}-\mathbf{C}\begin{pmatrix}\beta\\ \mathbf{u}\end{pmatrix}\right]'\mathbf{R}^{-1}\left[\mathbf{y}-\mathbf{C}\begin{pmatrix}\beta\\ \mathbf{u}\end{pmatrix}\right]}_{\text{"likelihood"}}+\underbrace{[\beta|\mathbf{u}]\begin{pmatrix}\mathbf{0} & \mathbf{0}\\ \mathbf{0} & \mathbf{G}^{-1}\end{pmatrix}\left[\begin{array}{c}\beta\\ \mathbf{u}\end{array}\right]}_{\text{"penalty"}} \tag{1.7}$$

where the $\mathbf{0}$ are matrices of zeroes of the appropriate dimensions, I've suppressed the dependence on ϕ for simplicity, and $\mathbf{C} = [\mathbf{X}|\mathbf{Z}]$, where this notation denotes concatenation of the columns of $\mathbf{X}$ and $\mathbf{Z}$.

Equation (1.7) introduces two jargon terms commonly used in the conventional theory of mixed linear models and generalizations of it. Equation (1.7) is sometimes called a "penalized likelihood," where the first term fills the role of the likelihood — and it *would* be a likelihood, if $\mathbf{u}$ were simply a vector of unknown regression coefficients — and the second term, the "penalty," changes the likelihood by adding a penalty for values of $\mathbf{u}$ that are large relative to $\mathbf{G}$. The idea of a penalized likelihood is to allow a rich, flexible parameterization of the mean structure through a high-dimensional $\mathbf{u}$, but to avoid overfitting by pushing $\mathbf{u}$'s elements toward zero by means of the penalty. Here is more jargon: Pushing $\mathbf{u}$'s elements toward zero in this manner is called "shrinkage," a term that originated with Charles Stein's work in the late 1950s. Shrinkage is a key aspect of analyses of richly parameterized models and often recurs in later chapters.

It is easy to show that this maximization problem produces these point estimates for β and $\mathbf{u}$:

$$\left[\begin{array}{c}\tilde{\beta}\\ \tilde{\mathbf{u}}\end{array}\right]_{\phi} = \left[\mathbf{C}'\mathbf{R}^{-1}\mathbf{C}+\begin{pmatrix}\mathbf{0} & \mathbf{0}\\ \mathbf{0} & \mathbf{G}^{-1}\end{pmatrix}\right]^{-1}\mathbf{C}'\mathbf{R}^{-1}\mathbf{y} \tag{1.8}$$

where the tildes above β and $\mathbf{u}$ mean these are estimates and the subscript ϕ indicates that the estimates depend on ϕ. (Derivation of (1.8) is an exercise at the end of the chapter.)

If the term arising from (1.7)'s penalty,

$$\begin{bmatrix} \mathbf{0} & \mathbf{0} \\ \mathbf{0} & \mathbf{G}^{-1} \end{bmatrix}, \tag{1.9}$$

were omitted from (1.8), $\tilde{\beta}$ and $\tilde{\mathbf{u}}$ would simply be the generalized least squares (GLS) estimates for the assumed value of ϕ_R. The penalty alters the GLS estimates of β and $\mathbf{u}$ by adding $\mathbf{G}^{-1}$ to the precision matrix for $\mathbf{u}$ from an ordinary regression, $\mathbf{Z}'\mathbf{R}^{-1}\mathbf{Z}$. In this sense, the penalty term provides an extra piece of information about $\mathbf{u}$, namely $\mathbf{u} \sim N_q(\mathbf{0}, \mathbf{G})$.

The estimates (1.8) give fitted values:

$$\tilde{\mathbf{y}}_\phi = \mathbf{C} \begin{bmatrix} \tilde{\beta} \\ \tilde{\mathbf{u}} \end{bmatrix}_\phi \tag{1.10}$$

$$= \mathbf{C} \left[\mathbf{C}'\mathbf{R}^{-1}\mathbf{C} + \begin{pmatrix} \mathbf{0} & \mathbf{0} \\ \mathbf{0} & \mathbf{G}^{-1} \end{pmatrix} \right]^{-1} \mathbf{C}'\mathbf{R}^{-1}\mathbf{y}. \tag{1.11}$$

For a given value of ϕ, the fitted values $\tilde{\mathbf{y}}_\phi$ are simply the observations $\mathbf{y}$ pre-multiplied by a known square matrix.

In traditional usage, the estimates (1.8) of $\mathbf{u}$ are called the best linear unbiased predictors or predictions (BLUPs). In this usage, "bias" refers to hypothetical repeated sampling from the distributions of both $\mathbf{u}$ and ε. For new-style random effects, hypothetical repeated sampling over the distribution of $\mathbf{u}$ is often plainly meaningless, and BLUPs are in fact biased over repeated sampling from ε. In such cases, the term BLUP is somewhat misleading. This issue is treated at length in Chapters 3 and 13.

The original derivation of the estimates (1.8) used a two-step procedure, first estimating the fixed effects β and then using those estimates to "predict" the random effects $\mathbf{u}$. Specifically, the mixed linear model (1.1) was re-written by making $\mathbf{Zu}$ part of the error, so that $\mathbf{y} = \mathbf{X}\beta + \varepsilon^*$, where $\varepsilon^* = \mathbf{Zu} + \varepsilon$, so that $\mathbf{y} \sim N_n(\mathbf{X}\beta, \mathbf{V})$ for $\mathbf{V} = \text{cov}(\varepsilon^*) = \mathbf{ZGZ}' + \mathbf{R}$. For given $\mathbf{V}$, i.e., for given ϕ, the standard textbook estimator of β is the GLS estimator $\tilde{\beta}$. The so-called best linear prediction of $\mathbf{u}$ is then $\mathbf{GZ}'\mathbf{V}^{-1}(\mathbf{y} - \mathbf{X}\beta)$ (Ruppert et al. 2003, Sections 4.4, 4.5), and when the unknown β is replaced by its estimate $\tilde{\beta}$, the resulting "prediction" of $\mathbf{u}$ is $\tilde{\mathbf{u}}$ as in (1.8).

More traditional jargon: The foregoing takes $\mathbf{G}(\phi_G)$ and $\mathbf{R}(\phi_R)$ as given. If an estimate of ϕ is plugged into (1.8), the resulting estimates of β and $\mathbf{u}$ are called "estimated BLUPs" or EBLUPs.

1.2.3 Estimating ϕ, the Unknowns in G *and* R

1.2.3.1 Maximizing the, or Rather a, Likelihood

Perhaps the first thing that would come to most people's minds — at least, those who don't consider themselves Bayesians — would be to write down the likelihood and maximize it. In simpler problems, the likelihood is a function of all the unknowns in the model, but as noted in the previous section, if $\mathbf{u}$ is considered among the unknowns, then it is unclear how the likelihood is defined for the mixed linear model or indeed whether it can be.

One pragmatic way to avoid this quandary is to get rid of the random effects $\mathbf{u}$ as in the traditional derivation of the point estimate $\tilde{\beta}$ shown in the previous section. With this approach, the model for the data $\mathbf{y}$ is

$$\mathbf{y} \sim N_n(\mathbf{X}\beta, \mathbf{V}(\phi)) \text{ where } \mathbf{V}(\phi) = \mathbf{Z}\mathbf{G}(\phi_G)\mathbf{Z}' + \mathbf{R}(\phi_R). \tag{1.12}$$

Having disposed of $\mathbf{u}$, (1.12) provides a well-defined likelihood that is a function of (β, ϕ) and which can be maximized to give estimates of β and ϕ.

Unfortunately, maximizing this likelihood produces a point estimate of ϕ with known flaws that can be quite severe. The simplest case is familiar from a first course in statistics. If $X_1, \ldots, X_n$ are independent and identically distributed (iid) as $N(\mu, \sigma_e^2)$ with unknown μ and σ_e^2, σ_e^2 has maximum likelihood estimate (MLE) $\sum(X_i - \bar{X})^2/n$, where $\bar{X}$ is the average of the X_i. This MLE is biased as an estimate of σ_e^2, with expected value $(n-1)\sigma_e^2/n$. This bias becomes trivial as n grows, but the same is not true for some simple elaborations of this problem, as Neyman and Scott showed in 1948 with the following example. If X_{i1} and X_{i2} are iid $N(\alpha_i, \sigma_e^2)$ for $i = 1, \ldots, n$, the MLE for σ_e^2 is $\sum_i (X_{i1} - X_{i2})^2/4n$ and has expected value $\sigma_e^2/2$ for all n. If α_i is treated as a random effect with variance σ_s^2, making this problem a mixed linear model, then the MLE of σ_e^2 becomes unbiased but now the MLE of σ_s^2 is biased. (Proofs are given as exercises.) In general, the problem is that the MLE of ϕ does not account for degrees of freedom — in this usage, linearly independent functions of the data $\mathbf{y}$ — that are used to estimate fixed effects.

1.2.3.2 Maximizing the Restricted (Residual) Likelihood

Dissatisfaction with the MLE's bias, among other things, prompted a search for alternatives that were unbiased, at least for the problems that were within reach at the time. This work, beginning in the 1950s, led to the present-day theory of the restricted likelihood, also called the residual likelihood. Various rationales have been given for the restricted likelihood, each providing some insight into how it differs from the likelihood in (1.12). Those seeking more detail about the rationales summarized here could start with Searle et al. (1992), Section 6.6, and references given there.

One rationale is that the restricted likelihood summarizes information about ϕ using the residuals from fitting the model $\mathbf{y} = \mathbf{X}\beta +$ [iid error] by ordinary least squares (hence "residual likelihood"). A closely related definition of the restricted likelihood is the likelihood arising from particular linear combinations of the data, $\mathbf{w} = \mathbf{K}'\mathbf{y}$, where $\mathbf{K}$ is a known, full-rank $n \times (n - \text{rank}(\mathbf{X}))$ matrix chosen so that $\mathbf{w}$'s distribution is independent of β. One such $\mathbf{K}$ has columns forming an orthonormal basis for the orthogonal complement of the column space of $\mathbf{X}$, i.e., the space of residuals from a least squares fit of $\mathbf{y}$ on $\mathbf{X}$. In general many such $\mathbf{K}$ exist but the likelihood of $\mathbf{K}'\mathbf{y}$ ("restricted likelihood") is invariant to the choice of $\mathbf{K}$ (Searle et al. 1992, Section 6.6). If the model for $\mathbf{y}$ is written as in equation (1.12), then for any $\mathbf{K}$, $\mathbf{w} = \mathbf{K}'\mathbf{y}$ is distributed as

$$\mathbf{w} = \mathbf{K}'\mathbf{y} \sim N\left(\mathbf{0}, \mathbf{K}'\mathbf{V}(\phi)\mathbf{K}\right). \tag{1.13}$$

where the random variable $\mathbf{w}$ and thus the covariance matrix $\mathbf{K}'\mathbf{V}(\phi)\mathbf{K}$ have dimension $n - \text{rank}(\mathbf{X})$.

Finally, the restricted likelihood can be derived as a particular "marginal likelihood." Begin with the likelihood arising from (1.12). This likelihood is a function of (β, ϕ); integrate β out of this likelihood to give a function of ϕ. The result, which we now explore in some detail, is the restricted likelihood. This integral has a Bayesian interpretation, to which I return below.

The likelihood arising from (1.12) is

$$L(\beta, \mathbf{V}) = K|\mathbf{V}|^{-\frac{1}{2}} \exp\left(-\frac{1}{2}(\mathbf{y} - \mathbf{X}\beta)'\mathbf{V}^{-1}(\mathbf{y} - \mathbf{X}\beta)\right). \tag{1.14}$$

To integrate out β, expand the quadratic form in the exponent and complete the square (I am showing this instead of leaving it for an exercise because this technique is so useful):

$$\begin{aligned}(\mathbf{y} - \mathbf{X}\beta)'\mathbf{V}^{-1}(\mathbf{y} - \mathbf{X}\beta) &= \mathbf{y}'\mathbf{V}^{-1}\mathbf{y} + \beta'\mathbf{X}'\mathbf{V}^{-1}\mathbf{X}\beta - 2\beta'\mathbf{X}'\mathbf{V}^{-1}\mathbf{y} \\ &= \mathbf{y}'\mathbf{V}^{-1}\mathbf{y} - \tilde{\beta}'\mathbf{X}'\mathbf{V}^{-1}\mathbf{X}\tilde{\beta} + (\beta - \tilde{\beta})'\mathbf{X}'\mathbf{V}^{-1}\mathbf{X}(\beta - \tilde{\beta}),\end{aligned}$$

where $\tilde{\beta} = (\mathbf{X}'\mathbf{V}^{-1}\mathbf{X})^{-1}\mathbf{X}'\mathbf{V}^{-1}\mathbf{y}$, the generalized least-squares estimate given $\mathbf{V}$. Integrating β out of (1.14) is then just the integral of a multivariate normal density, so

$$\int L(\beta, \mathbf{V})d\beta = K|\mathbf{V}|^{-\frac{1}{2}}|\mathbf{X}'\mathbf{V}^{-1}\mathbf{X}|^{-\frac{1}{2}} \exp\left(-\frac{1}{2}\left[\mathbf{y}'\mathbf{V}^{-1}\mathbf{y} - \tilde{\beta}'\mathbf{X}'\mathbf{V}^{-1}\mathbf{X}\tilde{\beta}\right]\right). \tag{1.15}$$

The natural logarithm of (1.15) is almost always used. Taking the log of (1.15) and substituting the expression for $\tilde{\beta}$ gives the log restricted likelihood

$$RL(\phi|\mathbf{y}) = K - 0.5\left(\log|\mathbf{V}| + \log|\mathbf{X}'\mathbf{V}^{-1}\mathbf{X}| + \mathbf{y}'\left[\mathbf{V}^{-1} - \mathbf{V}^{-1}\mathbf{X}(\mathbf{X}'\mathbf{V}^{-1}\mathbf{X})^{-1}\mathbf{X}\mathbf{V}^{-1}\right]\mathbf{y}\right), \tag{1.16}$$

where $\mathbf{V} = \mathbf{Z}\mathbf{G}(\phi_G)\mathbf{Z}' + \mathbf{R}(\phi_R)$ is a function of the unknown ϕ in the covariance matrices $\mathbf{G}$ and $\mathbf{R}$. (See, for example, Searle et al. 1992, Sections 8.3f and M.f.)

Within likelihood theory, this marginalizing integral is quite *ad hoc* and ultimately is justified only by the desirable properties of the resulting point estimates. This integral has a much more natural Bayesian interpretation: If the fixed effects β have an improper flat prior $\pi(\beta) \propto 1$, then (1.15), multiplied by a prior distribution for ϕ, is just the marginal posterior distribution of ϕ. Within Bayesian theory, there is nothing inherently unclean about this marginalizing integral although the improper flat prior on β can have striking consequences, some of which are pertinent to matters discussed in this book and will be noted when they arise.

1.2.4 Other Machinery of Conventional Statistical Inference

Now we have point estimates $\hat{\phi}$ of ϕ from maximizing the restricted likelihood; these are plugged into (1.8) to give estimates $(\hat{\beta}, \hat{\mathbf{u}})$ of $(\beta, \mathbf{u})$ which in turn are plugged into (1.10) to give fitted values $\hat{\mathbf{y}}$. I use hats $\hat{\bullet}$ instead of tildes $\tilde{\bullet}$ here to indicate that these estimates and fitted values are computed using $\hat{\phi}$. These $(\hat{\beta},\hat{\mathbf{u}})$ and $\hat{\mathbf{y}}$ were derived from linear model theory by treating $\hat{\phi}$ as if it were the true value of ϕ and the rest of the conventional theory is derived from linear model theory the same way. These derivations are straightforward, so they have been left as exercises.

I believe it is not controversial to say that the conventional approach to analyzing mixed linear models has an unsatisfactory feature that is probably impossible to fix. Except for some special cases, there is no feasible alternative to treating the estimates $\hat{\phi}$ as if they were the true values of the unknowns in **G** and **R**. This assumes away the usually large variation in $\hat{\phi}_G$ (or uncertainty about ϕ_G, if you prefer) and the often non-trivial variation in $\hat{\phi}_R$. Among the dissatisfied, some consider Bayesian analyses a promising alternative now that they can be computed readily (e.g., Ruppert et al. 2003, Section 4.7, p. 102). Thus, I give just a brief overview of the conventional machinery as implemented in widely used software, leaning heavily on Ruppert et al. (2003).

1.2.4.1 Standard Errors for Fixed and Random Effects

Standard errors for the fixed-effect estimates $\hat{\beta}$ can be derived from the model (1.12) in which the random effects **u** have been incorporated into the error term. Under that model, $\mathbf{y} \sim N_n(\mathbf{X}\beta, \mathbf{V}(\phi))$ where $\mathbf{V}(\phi) = \mathbf{Z}\mathbf{G}(\phi_G)\mathbf{Z}' + \mathbf{R}(\phi_R)$. If ϕ is set to $\hat{\phi}$ and treated as known, giving $\hat{\mathbf{V}}$, then $\hat{\beta}$ is the familiar GLS estimator $\hat{\beta} = (\mathbf{X}'\hat{\mathbf{V}}^{-1}\mathbf{X})^{-1}\mathbf{X}'\hat{\mathbf{V}}^{-1}\mathbf{y}$, which has the familiar covariance matrix $\text{cov}(\hat{\beta}) = (\mathbf{X}'\hat{\mathbf{V}}^{-1}\mathbf{X})^{-1}$. This covariance — which is exact if **V** is known, and approximate otherwise (Section 1.2.4.2 below has more to say on this) — provides the standard errors for $\hat{\beta}$ given in, for example, SAS's MIXED procedure: $SE(\hat{\beta}_i) = \text{cov}(\hat{\beta})_{ii}^{0.5}$, where the subscript *ii* indicates the ith diagonal element of $\text{cov}(\hat{\beta})$. (The same standard errors are obtained from (1.19) below.)

To see the deficiency of this usual standard error, note that

$$\text{cov}(\hat{\beta}) = E(\text{cov}\{\hat{\beta}|\phi\}) + \text{cov}(E\{\hat{\beta}|\phi\}), \tag{1.17}$$

where the outer expectation and covariance are with respect to the distribution of $\hat{\phi}$. By the familiar fact that generalized least squares estimates are unbiased even if the covariance matrix is wrong, $E\{\hat{\beta}|\phi\} = \beta$, so $\text{cov}(E\{\hat{\beta}|\phi\}) = 0$. The usual standard error for $\hat{\beta}$, given above, amounts to approximating the first term in (1.17) by $\text{cov}\{\hat{\beta}|\hat{\phi}\}$. This deficiency is well known, e.g., it is noted in the documentation for the MIXED procedure.

Considering both the fixed and random effects simultaneously, recall from (1.8) that the estimates (EBLUPs) of the fixed and random effects are

$$\begin{bmatrix} \hat{\beta} \\ \hat{\mathbf{u}} \end{bmatrix} = \left[\mathbf{C}'\mathbf{R}(\hat{\phi}_R)^{-1}\mathbf{C} + \begin{pmatrix} \mathbf{0} & \mathbf{0} \\ \mathbf{0} & \mathbf{G}(\hat{\phi}_G)^{-1} \end{pmatrix}\right]^{-1} \mathbf{C}'\mathbf{R}(\hat{\phi}_R)^{-1}\mathbf{y}, \tag{1.18}$$

where, as before, $\mathbf{C} = [\mathbf{X}|\mathbf{Z}]$ and the estimates now have hats instead of tildes because they depend on $\hat{\phi}$. It will be useful to have two different covariances for $(\hat{\beta},\hat{\mathbf{u}})$. The first involves the unconditional covariance of $\hat{\mathbf{u}}-\mathbf{u}$:

$$\mathrm{cov}\begin{pmatrix}\hat{\beta}\\ \hat{\mathbf{u}}-\mathbf{u}\end{pmatrix} = \left[\mathbf{C}'\mathbf{R}(\hat{\phi}_R)^{-1}\mathbf{C}+\begin{pmatrix}\mathbf{0} & \mathbf{0}\\ \mathbf{0} & \mathbf{G}(\hat{\phi}_G)^{-1}\end{pmatrix}\right]^{-1}, \tag{1.19}$$

This covariance is with respect to the distributions of both ε and $\mathbf{u}$ and is used to give standard errors for the random effects in SAS's MIXED procedure. (The derivation is left as an exercise.)

The second covariance conditions on $\mathbf{u}$. At this point in the book, it may seem odd to condition on $\mathbf{u}$ but it will seem more natural when we consider methods in which $\mathbf{u}$ is a new-style random effect, that is, an unknown parameter that happens to be constrained (by $\mathbf{G}$). This covariance has a key role in Chapter 3's discussion of confidence bands for penalized-spline fits. The covariance conditional on $\mathbf{u}$ is

$$\begin{aligned}\mathrm{cov}\left(\begin{matrix}\hat{\beta}\\ \hat{\mathbf{u}}\end{matrix}\;\middle|\;\mathbf{u}\right) &= \left[\mathbf{C}'\mathbf{R}(\hat{\phi}_R)^{-1}\mathbf{C}+\begin{pmatrix}\mathbf{0} & \mathbf{0}\\ \mathbf{0} & \mathbf{G}(\hat{\phi}_G)^{-1}\end{pmatrix}\right]^{-1}\mathbf{C}'\mathbf{R}^{-1}\mathbf{C}\\ &\quad\times\left[\mathbf{C}'\mathbf{R}(\hat{\phi}_R)^{-1}\mathbf{C}+\begin{pmatrix}\mathbf{0} & \mathbf{0}\\ \mathbf{0} & \mathbf{G}(\hat{\phi}_G)^{-1}\end{pmatrix}\right]^{-1}.\end{aligned} \tag{1.20}$$

This expression is derived trivially by noting that conditional on $\mathbf{u}$, $\mathrm{cov}(\mathbf{y}) = \mathbf{R}$.

1.2.4.2 Testing and Intervals for Fixed and Random Effects

More complex tests involving the fixed and random effects use the covariances just given in Section 1.2.4.1 and again, the standard tests fix ϕ at its estimate $\hat{\phi}$ and treat it as if known. Thus, for example, SAS's test for whether $l'\beta$, a linear combination of β, is zero is the Wald test based on

$$z = \frac{l'\hat{\beta}}{(l'\mathrm{cov}(\hat{\beta})l)^{0.5}} \tag{1.21}$$

which is treated as if it is approximately distributed as $N(0,1)$. The 95% Wald confidence interval is constructed by replacing the numerator of z in (1.21) with $l'\hat{\beta}-l'\beta$.

The distribution in (1.21) is exact if ϕ is known, which, of course, it never is. On this point, Ruppert et al. (2003, p. 104) issue a sober warning:

> [T]he theoretical justification of [(1.21)] for general mixed models is somewhat elusive owing to the dependence in **y** imposed by the random effects. Theoretical back-up for [(1.21)] exists in certain special cases, such as those arising in analysis of variance and longitudinal data analysis. ... For some mixed models, including many used in the subsequent chapters of this book [and the present book as well], justification of [(1.21)] remains an open problem.

In other words, use this at your own risk. Ruppert et al. (2003) note that as in simpler problems, the likelihood ratio test provides an alternative to the Wald test (1.21), but they also note that as for the Wald test, "its justification ... is dependent on the type of correlation structure induced by the **G** and **R** matrices" (Ruppert et al. 2003, p. 105), i.e., this doesn't solve the problem.

For linear functions of both β and $\mathbf{u}$, similar Wald tests can be constructed using the joint covariance of β and $\mathbf{u}$ given in Section 1.2.4.1. Similarly, tests for entire effects, e.g., for the 4 degrees of freedom corresponding to a 5-category grouping variable, can be done as Wald tests (above) or as likelihood ratio tests. However, these tests have no better theoretical rationale than the test above that only involves β.

As an alternative to relying on large-sample approximations, Ruppert et al. (2003) propose a parametric bootstrap to produce reasonably accurate small-sample confidence intervals and P-values (e.g., p. 144 for fitted values).

1.2.4.3 Testing and Intervals for ϕ

In many applications of mixed linear models, the unknowns in ϕ are nuisances and there is little interest in estimates or intervals for them. In such cases, the only interesting question is whether the variance of a random effect, σ_s^2, is zero because if so, it can be omitted from the model. For example, in Example 2 about imaging vocal folds, it would be convenient if one of the variance components was zero, because removing that component would make the analysis simpler and better-behaved. Taken literally, of course, it seems obvious that none of these variance components could be exactly zero, but the same objection can be made about any sharp null hypothesis.

For testing whether $\sigma_s^2 = 0$, the conventional theory offers little beside likelihood ratio or restricted likelihood ratio tests, in which the maximized (restricted) likelihoods are compared for an unrestricted fit and a fit with σ_s^2 fixed at zero (e.g., Ruppert et al. 2003, Section 4.8). The usual large-sample approximate chi-square distribution for the (restricted) likelihood ratio test does not apply because the null hypothesis $\sigma_s^2 = 0$ is on the boundary of the parameter space. An asymptotic argument appropriate to the situation shows that in large samples, the (restricted) likelihood ratio statistic has a null distribution that is approximately an equal mixture of a point mass at zero and the chi-square distribution from the usual asymptotic theory. Unfortunately, this is well-known to be a poor approximation in mixed linear models for the same reasons discussed in the preceding section.[2] Thus Ruppert et al. (2003, p. 107), among others, recommend against using this asymptotic test and suggest instead a parametric bootstrap P-value.

Sometimes, however, the unknowns in ϕ are interesting in themselves. In Example 1, measuring the virus's structure, the elements of ϕ describe different components of measurement error and information about the variances of those components might guide design of future experiments or an effort to improve the measurement process. Again, the conventional theory offers little more than standard large-sample theory for maximum likelihood and its extension to restricted likelihood: The approximate covariance matrix for $\hat{\phi}$ is the inverse of -1 times the matrix of second derivatives of the log likelihood or restricted likelihood, evaluated at $\hat{\phi}$. See, for ex-

[2]However, references given in Snijders & Bosker (2012, Section 6.2.1) show that this approximate test has reasonably good properties for the special case of variances of intercepts and slopes in multi-level models, sometimes called random regressions.

ample, Searle et al. (1992), Section 6.3 for maximum likelihood and Section 6.6e for restricted likelihood. SAS's MIXED procedure uses this approach to provide standard errors for estimates of the unknowns in ϕ, and uses it to compute a z-statistic and P-value for testing whether each element of ϕ is zero.

The resulting covariance matrix and standard errors are, as far as I can tell, universally regarded as poor approximations; Ruppert et al. (2003) do not consider them worth mentioning. Even SAS's own documentation for MIXED says "tests and confidence intervals based on asymptotic normality can be obtained. However, these can be unreliable in small samples [and, of course, nobody knows what "small sample" means, though SAS is hardly alone in giving this sort of useless advice], especially for parameters such as variance components which have sampling distributions that tend to be skewed to the right" (v. 9.1.3 on-line manual). Other authors are less restrained: Verbeke & Molenberghs (1997) devote their Appendix B to criticizing MIXED's z-statistic and P-value for non-negative elements of ϕ.

One glaring problem with the large-sample approximation on the original scale is that it frequently gives standard errors for variances that are larger than the variance estimates, making Wald-style confidence intervals — estimate plus-or-minus 2 standard errors — generally inappropriate. A common alternative confidence interval uses a Satterthwaite approximation. As implemented in SAS's MIXED procedure (Milliken & Johnson 1992, p. 348), the Satterthwaite approximate confidence interval for a variance σ_s^2 is

$$\frac{\nu\hat{\sigma}_s^2}{\chi^2_{\nu,1-\alpha/2}} \leq \sigma_s^2 \leq \frac{\nu\hat{\sigma}_s^2}{\chi^2_{\nu,\alpha/2}}, \tag{1.22}$$

where the denominators are quantiles of a chi-squared distribution with degrees of freedom $\nu = 2z^2$ for $z = \hat{\sigma}_s^2/(\text{standard error } \hat{\sigma}_s^2)$. In MIXED, $\hat{\sigma}_s^2$ is obtained by maximizing the likelihood or restricted likelihood and its standard error is from the large-sample approximation. The Satterthwaite interval is derived by assuming z^2 is approximately distributed as chi-squared with r degrees of freedom, matching the variances of z^2 and the chi-squared distribution, and solving for r.

In some simple cases this distribution is exact and for some cases where it is approximate, Milliken & Johnson (1992) describe simulation experiments in which it performs well. However, they note (p. 349) "when there are very few levels [in the sense of ANOVA] associated with a random effect and consequently very few degrees of freedom [in the sense of independent pieces of information, not ν], the resulting confidence intervals are going to be very wide." Indeed, in my experience this approximation routinely gives intervals ranging effectively from zero to infinity, particularly when the variance estimate is barely positive. In fairness, such nonsensical results should probably be taken to mean "the approximation is not appropriate in this case."

The mixed linear model can be parameterized using log variances instead of variances and the large-sample approximation will undoubtedly work better with this parameterization. As far as I know, however, no statistical package does this.

1.3 Bayesian Analysis of the Mixed Linear Model

Deriving the conventional analysis involved various *ad hoc* acts. Such *ad hockery* is entirely consistent with the frequentist approach of deriving procedures by any convenient method and justifying them based on their operating characteristics, e.g., by the reduced bias of maximum restricted likelihood estimates compared to maximum likelihood estimates. As a pragmatist, I have no problem with this approach in general. For mixed linear models, however, we have already seen that the conventional analysis has several deficiencies and we will see more. These problems are bad enough that people who had no apparent interest in Bayesian methods before 1990 now see Bayesian methods as an elegant way to alleviate some problems of the conventional approach (e.g., Ruppert et al. 2003, Chapter 16). In this sense, those who are unconvinced by Bayesian ideology may view Bayesian methods as simply procedures derived using a particular general approach, which must be justified by their operating characteristics like any other procedure. This section presents Bayesian methods for mixed linear models from this viewpoint. Like the preceding section, this section is not intended to be exhaustive but is rather intended to define things and to emphasize aspects in which Bayesian analysis is incomplete, incompletely understood, or problematic.

1.3.1 A Very Brief Review of Bayesian Analysis in General

The Bayesian approach was for a long time considered revolutionary and thus disreputable to those opposed to it, and it still carries the lingering odors of both revolutionary zeal and disrepute. Thus, any treatment of it must begin with some kind of manifesto; this is mine. If you are unfamiliar with Bayesian analysis, you will almost certainly need to spend some time with a more general introduction, e.g., the early chapters of Carlin & Louis (2008) or Gelman et al. (2004), where you can read those authors' manifestos.

The fundamental idea of Bayesian analysis is to describe all uncertainty and variation using probability distributions. This contrasts with the conventional approach in the preceding section: There, for example, we did not know the values of the fixed effects β and although we produced probability statements about an *estimator* of β, $\hat{\beta}$, we made no probability statements about β itself. By contrast, a Bayesian analysis assigns probability distributions to entities like β that are conceived to be fixed in value but unknown. A Bayesian analysis also assigns probability distributions to observable quantities like $\mathbf{y}$ conditional on unknowns such as β, $\mathbf{u}$, and ϕ_R, just as in the conventional approach. Assigning a joint probability distribution to all observed and unobserved quantities allows a Bayesian analysis to proceed using *only* the probability calculus, with particular reliance on one theorem, Bayes's Theorem.

Using probability this way brings an undeniable theoretical tidiness and, with modern Bayesian computing, real practical advantages in certain problems. It has also made possible a great deal of confusion, some of which is especially relevant to contemporary uses of mixed linear models. To a person trained in conventional statistics, probability describes one type of thing — variation between replicate observations of a random mechanism (often hypothetical), like a coin flip — and "the probability of the event A" describes the long-run frequency of the occurence of event A in a sequence of replicate observations from the random mechanism. In addition to this, Bayesian analysis uses probability to describe uncertainty about unknown and unobservable things which, like β, may have no real existence and which cannot in any sense be conceived as arising from a random mechanism, let alone one that can produce replicate observations. The random effects $\mathbf{u}$ have an ambiguous status between these two extremes: The traditional definition conceives of them as arising from a random mechanism like coin flips, but they are inherently unobservable. The newer varieties of random effect add to this difficulty because, as Chapter 13 argues in detail, their probability distributions cannot meaningfully be understood as describing variation between replicate realizations of a random mechanism.

One school of Bayesian thinking avoids this conceptual difficulty by arguing that the conventional notion of probability, as a long-run frequency, is so inconsistent with itself and with reality that it must be discarded. The leading figure in this school was Bruno de Finetti, whose magnum opus, *Theory of Probability* (1974–75, English translation by Antonio Machi and Adrian Smith), develops the view that probability can only be meaningfully interpreted as a quantification of an individual's personal uncertainty. From this beginning, statistical analysis becomes the process of updating personal beliefs expressed as probability distributions using the probability calculus, Bayes's Theorem in particular. For a long time, de Finetti's view was known to few people because his writing is — deliberately, it seems — nearly impenetrable even in a sympathetic translation and because the necessary calculations were intractable outside of a few simple problems. With the advent of modern Bayesian computing, far more people find Bayesian analyses worth pursuing because of their practical advantages but de Finetti has gained few if any adherents. To compress a complex subject perhaps excessively, it seems that few people find de Finetti's subjective approach relevant to their problems let alone convincing, though his devotees certainly remain devoted. (If you'd like to meet one, just say my previous sentence to a large mixed audience of academic statisticians.)

Indeed, as Kass (2011) argues, these days fewer and fewer people identify themselves as either Bayesian or frequentist but rather identify themselves as practicing statisticians who use tools from both approaches. This arises from the convergence of several trends: The increasing power and ease of data analysis using Bayesian methods; the growing belief that academic statisticians should become deeply involved in applications; the realization that the old Bayes-frequentist dispute was conducted in a theoretical world with no connection to the reality of statistical applications; and, as a consequence, growing acceptance of the view that "[statistical] procedures should be judged according to their performance under theoretical conditions thought to capture relevant real-world variation in a particular applied setting" (Kass 2011, p. 7) and not by the abstract properties with which advocates of the two schools have flogged each other. In particular, "it makes more sense to place in the center of our logical framework the match or mis-match of theoretical assumptions with the real world of data." Kass (2011) is, to my knowledge, the first explicit articulation of a view that he calls "statistical pragmatism" and while his view is, of course, not yet completely fleshed out, I endorse it heartily.

Having disposed of the obligatory manifesto, I now proceed in a manner I consider pragmatic.

Let $\mathbf{w}$ represent observed outcomes and let θ represent any unknown quantities. Very generally, Bayesian analysis uses the following items.

- *The Likelihood.* Denote the probability model for the data $\mathbf{w}$ given the unknowns θ as $\pi(\mathbf{w}|\theta)$, a function of $\mathbf{w}$ with θ treated as fixed (though unknown). The likelihood $L(\theta;\mathbf{w})$ is $\pi(\mathbf{w}|\theta)$ treated as a function of θ; $\mathbf{w}$ is treated as fixed once the data have been observed.
- *The Prior Distribution.* Information about θ external to $\mathbf{w}$ is represented by a probability distribution $\pi(\theta)$. This is called "the prior distribution" in the sense that it represents information or beliefs about θ *prior to* updating those beliefs in light of $\mathbf{w}$ using Bayes's Theorem (see the next item). However, the information summarized in $\pi(\theta)$ need not precede $\mathbf{w}$ in any temporal sense. For example, $\mathbf{w}$ could describe an epidemic that happened in the 1700s and $\pi(\theta)$ might capture present-day knowledge about the vectors that spread this particular disease.
- *Bayes's Theorem.* This standard and uncontroversial theorem states that if A and B are events in the sense of probability theory, $P(A|B) = P(B|A)P(A)/P(B)$, where "$P(A|B)$" means the conditional probability of event A given that the event B is supposed to have happened, while $P(A)$ is the unconditional probability of event A. Another version of this theorem applies to situations in which the events A and B refer to quantities that take values on a continuous scale, where probability densities for those continuous variates are denoted by π: $\pi(\theta|\mathbf{w}) = \pi(\mathbf{w}|\theta)\pi(\theta)/\pi(\mathbf{w})$, where $\pi(\mathbf{w}) = \int \pi(\mathbf{w}|\theta)\pi(\theta)d\theta$.
- *The Posterior Distribution.* Suppose the observable data $\mathbf{w}$ has probability density $\pi(\mathbf{w}|\theta)$ depending on the unknown θ. Suppose further that θ has prior distribution $\pi(\theta)$. (This common wording glosses over the miraculous origin of $\pi(\theta)$, to which we will return.) Then we can use Bayes's theorem to update the information in the prior distribution, giving the posterior distribution $\pi(\theta|\mathbf{w}) \propto L(\theta;\mathbf{w})\pi(\theta)$, where the likelihood $L(\theta;\mathbf{w})$ fills the role of $\pi(\mathbf{w}|\theta)$. Again, "posterior" need not have a temporal meaning; it merely means *after* accounting for the data $\mathbf{w}$.
- *The Fundamental Principle.* In a Bayesian analysis, *all* inferential or predictive statements about θ depend on the data $\mathbf{w}$ *only* through the posterior distribution.

A few comments are in order.

For people trained in conventional statistics, The Fundamental Principle seems like a needless constraint. In conventional statistics, you can use any method you like to devise procedures that give inferential or predictive statements about θ; all that matters is their performance. From a Bayesian viewpoint, however, procedures that do not conform to The Fundamental Principle should be avoided because they are subject to technical defects, called *incoherence* in aggregate, while Bayesian procedures are not (if they use proper priors, and in special cases if they use improper priors). In practice, even people who call themselves Bayesian use all manner of non-conforming methods for model building, i.e., for selecting $\pi(\mathbf{w}|\theta)$, and leading advocates of Bayesian methods have sanctioned this (e.g., Smith 1986) even though in theory one could use only Bayesian methods even for model-building. One reason they sanction non-Bayesian methods for model building is that using Bayesian methods for this purpose is like playing the piano with your feet, in most if not all cases. Since the advent of modern Bayesian computing, one rarely hears objections to using non-Bayesian methods to specify $\pi(\mathbf{w}|\theta)$.

For decades, while the Foundations of Statistics were a live topic of debate, anti-Bayesians criticized the prior distribution as introducing a needless element of subjectivity into the analysis. Those toeing de Finetti's line countered that the prior *adds* no subjectivity because objectivity in statistical analysis is a delusion. More conciliatory pro-Bayesians argued that it is usually possible to pick a prior that adds little or no information to the analysis, as in Jimmy Savage's principle of stable estimation. By the time Bayesian analyses became practical and widespread in the early 1990s, this argument had mostly died along with the disputants. (But not completely; see Lee et al. 2006, Lee & Nelder 2009.) In practice, it is rare to see prior distributions that describe subjective personal belief. Sometimes the prior represents *bona fide* external information, e.g., information about a medical device based on clinical trials of similar devices. Most often, the prior simply conditions the likelihood $L(\theta;\mathbf{w}) = \pi(\theta|\mathbf{w})$ by restricting the range of θ or by shortening the tails of $L(\theta;\mathbf{w})$. In either role, a prior can improve a procedure's performance markedly. However, most often people would like to use a minimally informative prior and unfortunately no consensus exists on what such priors might be. If you want to do a Bayesian analysis, you must supply a prior distribution and selection of priors is a live issue in analysis of mixed linear models.

1.3.2 Bayesian Analysis of Mixed Linear Models

Applying the preceding section's terminology to the mixed linear model as in (1.1):

- The unknown θ includes
  - unknowns in the mean structure, β and $\mathbf{u}$, and
  - unknowns in the variance structure, $\phi = (\phi_G, \phi_R)$.
- The data $\mathbf{w}$ are, by convention, the outcome $\mathbf{y}$.
- The likelihood is specified by (1.1) and the lines immediately following it; $\pi(\mathbf{y}|\beta, \mathbf{u}, \phi_R)$ is Gaussian (normal) with mean $\mathbf{X}\beta + \mathbf{Z}\mathbf{u}$ and covariance $\mathbf{R}(\phi_R)$.
- The prior for $\mathbf{u}$ given ϕ_G is $N(0, \mathbf{G}(\phi_G))$; β is generally given a multivariate normal prior, most commonly an improper flat prior $\pi(\beta) \propto 1$. Usually β and $\mathbf{u}$ are independent conditional on ϕ though this is not necessary. Finally, (ϕ_G, ϕ_R) are given a prior $\pi(\phi_G, \phi_R)$ which will be discussed at length below.

Although in common usage $\mathbf{X}$ and $\mathbf{Z}$ are considered part of the *dataset* for a mixed-linear-model analysis, for historical reasons they are not part of the *data* $\mathbf{y}$. Rather, they are simply treated as known quantities. The foregoing specifications permit a Bayesian analysis.

1.3.2.1 Tests and Intervals for Unknowns

By The Fundamental Principle, inferential statements about any unknowns depend on the data only through their posterior distributions. Thus, all inferential statements about β depend on the data only through the marginal posterior distribution of β, $\pi(\beta|\mathbf{y})$; similarly for individual β_j, for $\mathbf{u}$ or individual u_j, and for ϕ or individual elements of it. Posterior intervals or regions analogous to confidence intervals are obtained from these marginal distributions. Any interval or region containing the desired probability can be used. Conventionally this is 95%, although in the dismal Bayesian tradition of offering naïve consumers something for nothing, narrower 90% intervals are becoming more common. These intervals naturally account for uncertainty about the unknowns ϕ, in distinct contrast to the conventional approach, which treats $\hat{\phi}$ as if it were the true ϕ.

Point estimates for elements of β and $\mathbf{u}$ are generally posterior means or medians, which are usually quite similar if not identical. Selection of point estimates is much more complicated for elements of ϕ, because their posteriors are not always unimodal and rarely if ever symmetric. The most commonly used point estimates are the mean or median of marginal distributions of individual elements of ϕ. The simulation study in He & Hodges (2008) strongly suggests that the posterior mean should not be used as a point estimate because it is extremely sensitive to the tails of the posterior distribution, which for elements of ϕ are often long and poorly determined by the data. The posterior median is less sensitive than the mean to the distribution's tails but can still be sensitive if the upper tail is long. The simulation study in He & Hodges (2008) reached the surprising conclusion that the posterior mode is a better point estimate for standard deviations, variances, and precisions than the posterior mean or median because the posterior mode is quite insensitive to tail behavior. This grates against the standard intuition that modes are unstable but that intuition derives from histograms of data, not from posterior distributions. A more serious objection to the posterior mode as a point estimate is that the posterior for ϕ can be multimodal (Chapter 14 gives an example and Chapter 19 explores this further). Given our generally poor understanding of ϕ's posterior distribution, perhaps this odd finding is best interpreted as showing the complexity of that posterior distribution.

In the Bayesian approach, tests for entire effects, e.g., testing a 5-category grouping variable, can be specified but are somewhat more problematic. The conventional analysis allows Wald or likelihood ratio tests for entire effects; the Bayesian analog to Wald tests is to compute a 95% posterior region for the vector of coefficients in β and/or $\mathbf{u}$ corresponding to the effect and reject the null hypothesis of no effect if the region excludes the origin.

More commonly these days, an effect is tested implicitly by doing the Bayesian analysis with and without the effect and comparing the two model fits using the Deviance Information Criterion (DIC, Spiegelhalter et al. 2002). DIC is a Bayesian analog to Akaike's Information Criterion (AIC) and the Schwarz Criterion (also known as the Bayesian Information Criterion or BIC) in that each has the form of a goodness-of-fit measure penalized by a measure of model complexity. Although the DIC became popular very quickly after its introduction as an all-purpose method for comparing models, it is rather *ad hoc* as its originators readily admit, bless their hearts (see especially the discussion and rejoinder following Spiegelhalter et al. 2002). The DIC's measure of model complexity in particular has been criticized for its *ad hoc* definition and some odd behavior such as occasional negative values

(Celeux et al. 2006, Plummer 2008, Cui et al. 2010). Thus, although DIC is widely used, in my judgment it is of questionable value. Leaving aside such qualms, DIC, like AIC and its many variants, has no associated formal testing machinery so there is no calibration for differences between models in DIC analogous to P-values or chi-squared statistics.

Finally, Bayes factors provide a more traditional sort of Bayesian test. The Bayes factor comparing two hypotheses, say the null hypothesis that an effect is zero versus the alternative hypothesis that it is not, is the ratio of the marginal probability of the data under the null divided by the marginal probability of the data under the alternative (see, e.g., Bernardo & Smith 1994, pp. 389–395; Kass & Raftery 1995 give a review). If the Bayes factor is multiplied by the ratio of the prior probability of the null hypothesis divided by the prior probability of the alternative hypothesis, the result is the ratio of the posterior probability of the null and alternative; the Bayes factor allows a user to avoid putting prior probabilities on the competing models. Although these Bayesian entities have a long history, they have not come into widespread use because they depend on the prior distributions placed on the unknowns in the null and alternative models and the extent of that dependence is rather generally unknown (though Kass & Raftery 1995 take a more sanguine view). One approximate Bayes factor that turns up in a number of applications is computed from the model-selection criterion called the Schwarz Criterion or Bayesian Information Criterion (BIC); see Kass & Raftery (1995, p. 778). Raftery and his collaborators use this approximation in several applications, e.g., the R package BMA, which does Bayesian model averaging for linear and generalized linear models and Cox regressions. As far as I know, no software package uses this approximation for testing effects in mixed linear models; an exercise involves deriving such Bayes factors.

1.3.2.2 Some Comments on the Bayesian Analysis

In constrast to the conventional analysis, a Bayesian analysis makes no particular distinction between β and $\mathbf{u}$: Each is just a vector of unknown constants. Customarily, β and $\mathbf{u}$ are independent *a priori* so the prior distribution for $(\beta, \mathbf{u})$ factors as $\pi(\beta, \mathbf{u}) = \pi(\beta)\pi(\mathbf{u})$. This book follows that custom, but it is not necessary.

The list above described $\mathbf{u}$'s distribution as part of the prior, a label used by many people who call themselves Bayesian. However, as noted, Bayarri et al. (1988) and Bayarri & DeGroot (1992) have argued convincingly, using mixed linear models as an example, that it is not possible to distinguish cleanly the likelihood and the prior. I prefer to call $\mathbf{u}$'s distribution part of the model for the data and to reserve the term "prior" for $\pi(\beta, \phi)$. This choice is admittedly arbitrary but makes it easier to communicate with the large majority of statisticians who were primarily trained in conventional statistics, to whom Bayesian jargon is like a foreign language. In particular, this usage makes it possible to communicate with the declining but still non-trivial group of statisticians who consider Bayesian methods to be somewhere between unnecessary and loathsome. In my first job after graduate school, I had an older colleague who became visibly upset when the conversation turned to Bayesian methods, as if he was in the presence of an abomination. By using the term *model* for assumptions like "$\mathbf{u} \sim N(0, \mathbf{G})$," I was able to maintain a conversation with this well-read colleague and learn many useful things.

Fortunately, this labeling quandary creates no problem for a Bayesian analysis: No matter what you call **u**'s probability distribution, the Bayesian calculations are the same. Even if you eliminate **u** as in (1.12), no problems arise within the Bayesian framework because **u** is eliminated simply by marginalizing with respect to it, i.e., integrating **u** out of the model or the posterior, either of which gives the same marginal posterior distribution of (β, ϕ). This is an example of the coherence of Bayesian analysis.

Before discussing specific prior distributions for ϕ, I offer some tentative generalizations about their importance. In analyses of mixed linear models with $\mathbf{R} = \sigma_e^2 \mathbf{I}$, the data generally provide fairly strong information about σ_e^2. The theory in Part IV of this book gives good reasons to expect this and except for tiny datasets, I have never seen a case in which the posterior for σ_e^2 depended much on its prior — that is, when the prior was varied moderately within the classes described below, because obviously you can make the posterior for σ_e^2 do almost anything if you make the prior sufficiently extreme.

However, for normal-error mixed linear models at least, the data only provide weak information about ϕ_G. Indeed, for these models ϕ_G is Bayesianly unidentified by the definition of Gelfand and Sahu (1999) because $\pi(\phi_G | \mathbf{u}, \mathbf{y}) = \pi(\phi_G | \mathbf{u})$, i.e., "observing the data **y** does not increase our prior knowledge about $[\phi_G]$ given $[\mathbf{u}]$" (Gelfand & Sahu 1999, p. 248). This differs from the conventional notion of identification, which Gelfand and Sahu call "likelihood identification," and apart from exceptions arising from, e.g., extremely small datasets, the *marginal* posterior of ϕ_G differs from the *marginal* prior of ϕ_G, so the data do provide information about ϕ_G. Generally speaking, however, the posterior of ϕ_G — and thus also the restricted likelihood — is not very well determined by the data. The theory in Part IV of this book gives some explanation for this empirical observation for an interesting subset of mixed linear models. Thus the prior for ϕ_G seems to be important in considerable generality and the consequences of different choices are poorly understood. Although some relevant literature exists, it is not large and "what is a good prior for general-purpose use?" is very much an open question. This is given as an exercise, of doctoral-dissertation scale, at the end of the chapter.

(A reader commented that the posterior predictive distribution of **y** or a new $\mathbf{y}_0$ is probably much less sensitive to ϕ_G's prior than the marginal posterior of β or **u**, so the importance of ϕ_G's prior may depend on whether the goal is parameter estimation or prediction. This is certainly plausible but I would say unproven; it may depend on specifics of the model or data.)

Part of the problem of specifying priors for ϕ is the tremendous variety of covariance structures included in the class of mixed linear models. Such literature as exists has been almost exclusively about priors for individual variances or for covariance matrices.

1.3.2.3 Prior Distributions for Variances

For mixed linear models, the gamma prior for the precision — the reciprocal of the variance — is conditionally conjugate: It gives a gamma posterior for that precision conditional on all other unknowns. This prior became popular before 1990 when computation was rarely possible with non-conjugate priors but it continues to be popular even though modern Bayesian computing has made conjugacy unnecessary.

Such gamma priors can insert specific information into the analysis by specifying the distribution's mean and variance or by specifying a gamma distribution with particular percentiles, e.g., the 2.5^{th} and 97.5^{th} percentiles. However, most commonly gamma priors for the precision are intended to insert minimal information into the analysis. The most frequently used prior of this sort is a gamma distribution with parameters $(\varepsilon, \varepsilon)$, which has mean 1 and variance $1/\varepsilon$. (This distribution transforms to an inverse gamma prior on the variance, although the inverse gamma distribution has no mean or higher moments when $\varepsilon \leq 1$.) Small ε gives this gamma prior a large variance, which is generally taken to mean that the prior inserts little information into the analysis. One of this chapter's exercises shows another sense in which this prior inserts little information into the analysis, by interpreting it as being worth ε observations.

Common values for ε in practice are 0.01 and 0.001, which give gamma distributions with mean 1 and variance 100 and 1000, respectively. However, for a positive random variable with a fixed mean, increasing the variance does not give the distribution a roughly constant probability density, as it does for random variables taking values on the real line. Instead, the variance of a positive random variable can be increased with fixed mean only by concentrating probability near zero and in an extremely long right tail. This results in weird distributions: The Gamma(0.001,0.001) distribution has mean 1, variance 1000, and 95th percentile 3×10^{-20}, which seems anything but diffuse. When first presented with this 95th percentile, most people refuse to believe it. See for yourself: The R command is "qgamma(0.95, shape=0.001, rate=0.001)."

The Jeffreys prior for a variance σ^2 is sometimes used. It is a special case of the gamma prior on the precision with $\varepsilon = 0$, which is equivalent to an improper flat distribution on the real line for $\log \sigma^2$. As we will see, for some models the likelihood's contribution to the posterior distribution is a strange function of variances in ϕ, so this impropriety can cause problems.

Gelman has proposed at least two alternatives to gamma priors; the following two were proposed in Gelman (2006), which gives a nice survey of priors for variance parameters. The first idea is a flat prior on the standard deviation, the rationale being the natural desire to let the posterior reflect the likelihood's information about the standard deviation, which has the same units as the random variable it describes. This prior's weakness is that a user must specify upper and lower limits for its sample space. The lower limit of this interval is usually zero or close to it (sometimes the prior is bounded away from zero to avoid computing problems) and the upper limit is usually somewhat arbitrary. Estimates and intervals for variances based on this prior can be sensitive to the upper limit when the likelihood has a flat upper tail, as it often does. This prior performed reasonably well in the simulation studies in He et al. (2007) and He & Hodges (2008) but otherwise little systematic information is available about its performance.

Gelman (2006) also proposed a half-t prior for the standard deviation of a random effect; this can also be described as the distribution of the absolute value of a central t variate. The half-Cauchy distribution is a special case. The full conditional posterior implied by this prior is a folded noncentral t distribution, so if the half-t is understood as a particular folded noncentral t, then it is conditionally conjugate for a standard deviation. This prior has largely been studied as a way to improve the mixing behavior in MCMC routines (see references in Gelman 2006, p. 519) but little is known about its statistical performance.

He et al. (2007) proposed a rather different prior for the common case in which $\mathbf{G}$ is diagonal with two or more unknown variances on its diagonal and $\mathbf{R} = \sigma_e^2\mathbf{I}$. The prior is defined in terms of the precisions, i.e., the reciprocals of the variances. Suppose the vector of unknown precisions is $(\tau_e, \tau_1, \ldots, \tau_m)$, where $\tau_e = 1/\sigma_e^2$ is the error precision and $\tau_1, \ldots, \tau_m$ are the precisions of the m random effects. Reparameterize the m random effect precisions as follows. First, define the total relative precision $\lambda = \sum_{k=1}^m \tau_k/\tau_e$, where "relative" means "precision of the random effects relative to the error precision." Now define the allocation of total relative precision among the m random effects as $\gamma = (\gamma_1, \ldots, \gamma_m)$, where

$$\gamma_k = \frac{\tau_k/\tau_e}{\lambda} = \frac{\tau_k}{\sum_{j=1}^m \tau_j}, \tag{1.23}$$

so $\tau_k = \tau_e\gamma_k\lambda$. Note that $\sum_{k=1}^m \gamma_k = 1$, and $\gamma = (\gamma_1, \ldots, \gamma_m)$ takes values in the m-dimensional simplex. This parameterization has two nice features: γ is invariant to changes in the data's scale and it takes values in a compact set. Thus a flat prior on γ is proper and treats the random-effect variances as exchangeable. Total relative precision, a positive real number, is better determined by the data than are the individual random-effect precisions, so its prior has relatively little impact. The simulation studies by He et al. (2007) and He & Hodges (2008) suggest this prior has good properties but it too has received little study.

A third approach was proposed by Hodges & Sargent (2001) and extended in Cui et al. (2010). This approach uses a measure of the complexity of the model's fit, degrees of freedom, which Section 2.2 discusses in detail so for now we'll use a simplified description and defer a detailed discussion. For the case in which $\mathbf{R} = \sigma_e^2\mathbf{I}$ and $\mathbf{G}$ is a diagonal matrix with L unknown variances on its diagonal, so $\phi_G = (\sigma_1^2, \ldots, \sigma_L^2)$, the degrees of freedom in the mixed linear model's fit is the sum of the degrees of freedom in the L components of the fit corresponding to the L variances in ϕ_G. The vector of component-specific degrees of freedom, $(\nu_1, \ldots, \nu_L)$, is a function of the L ratios $(\sigma_1^2/\sigma_e^2, \ldots, \sigma_L^2/\sigma_e^2)$; thus a prior distribution on $(\nu_1, \ldots, \nu_L)$ induces a prior distribution on $(\sigma_1^2/\sigma_e^2, \ldots, \sigma_L^2/\sigma_e^2)$. (All of this is well-defined under conditions given in Section 2.2.2.3.) Specifying a prior distribution on the error variance σ_e^2 completes the specification of a prior on ϕ. Published examples include Hodges et al. (2007), Cui et al. (2010) and, for a spatial model, Paciorek & Schervish (2006).

Putting a prior distribution on the degrees-of-freedom scale has two main advantages. First, for models with new-style random effects such as penalized splines, the variances in ϕ_G are merely a device to implement smoothing and it is nearly impossible to develop intuition or external information about them. By contrast, the degrees of freedom in a fit is subject to much more direct interpretation and thus intuition. In such uses, a prior on the degrees of freedom allows a user to directly condition the smoothness of the fit. Chapter 2 gives an example. Second, the prior on degrees of freedom is invariant to a large range of scale changes and re-parameterizations of the mixed linear model (Section 2.2.2.3), which is generally not true for priors specified directly on variances in ϕ. For the balanced one-way ANOVA model, a flat prior on the degrees of freedom in the fit is equivalent to the uniform-shrinkage prior (Daniels 1999).

The performance of this prior was examined in a simulation study in Hodges et al. (2007). Once again, however, little is known about this prior's properties outside of very simple cases.

1.3.2.4 Prior Distributions for Covariance Matrices

For unstructured $\mathbf{G}$, in practice it is rare to see a prior distribution other than the Wishart distribution for $\mathbf{G}^{-1}$; see e.g., Bernardo & Smith (1994, pp. 138–140). The Wishart is a generalization of the gamma distribution and is conditionally conjugate for $\mathbf{G}^{-1}$. This distribution has two parameters, degrees of freedom α and a symmetric non-singular matrix $\mathbf{B}$, where $\alpha > (q-1)/2$, q being the dimension of $\mathbf{G}$, and where the Wishart's expectation is $\alpha\mathbf{B}^{-1}$. Most commonly, these are specified as $\mathbf{B} = \mathbf{I}_q$ and α the smallest integer that specifies a proper Wishart. The Wishart's popularity appears to be a relic of the time before modern Bayesian computing because it resists intuition. Under the Wishart, the marginal distributions of individual precisions are chi-squared, and the marginal distributions of individual correlations are beta-distributed on the interval $[-1,1]$ (Barnard et al. 2000, Section 2), so the correlations can be given marginal uniform priors by setting the Wishart's parameters a certain way. However, I know of nothing that provides intuition about the correlations between unknowns implied by the Wishart. Little is known about the performance of posterior distributions obtained using this prior, though it performs poorly in some cases in the simulation studies of Daniels & Kass (1999, 2001).

There is a modest literature of alternatives to the Wishart. Much of this work was motivated by simple problems, e.g., X_i iid $N(\mu,\Sigma)$ with unknown μ and Σ, but it is applicable to mixed linear models. Daniels & Kass (1999, 2001) review the literature up to 2001. They discuss the Jeffreys prior and Berger-Bernardo priors for a covariance matrix, but these priors perform poorly in their simulations. The poor performace of the Jeffreys prior has some theoretical basis. Eaton & Freedman (2004) showed that it gives incoherent predictive distributions in the iid $N(\mu,\Sigma)$ case and surveyed literature with similar results. Eaton & Sudderth (2004) showed the same thing as a by-product of a more general argument about priors for covariance matrices. Incoherence also holds for the estimation problem by a similar argument (M.L. Eaton, personal communication). I do not know whether the Berger-Bernardo prior has a similar flaw. In any case, the Jeffreys prior and Berger-Bernardo priors will not be discussed further here.

Three alternatives to the Wishart were proposed by Daniels & Kass (1999), Barnard, McCulloch, & Meng (2000), and Chen & Dunson (2003). Each depends on a decomposition of the covariance matrix; the first two are non-conjugate while the third is partially conjugate.

Daniels & Kass (1999, 2001) intended their prior to shrink the covariance toward a diagonal matrix and to shrink the eigenvalues of the covariance toward each other. It uses the spectral decomposition $\mathbf{G} = \Gamma \mathbf{D} \Gamma'$, where the orthogonal matrix Γ has $\mathbf{G}$'s eigenvectors as its columns, and the diagonal matrix $\mathbf{D}$ has $\mathbf{G}$'s eigenvalues as its diagonal entries. Daniels & Kass (1999) propose a flat prior on the eigenvalues, which has the effect of shrinking them toward each other (p. 1256). The $q \times q$ orthogonal matrix Γ is written as the product of $q(q-1)$ simple matrices, each of which is a function of a so-called Givens angle θ_{ij} which "may be considered a rotation in the plane spanned by [the] i and j components of the basis defining the matrix [Γ]." Each $\theta_{ij} \in (-\pi/2, \pi/2)$; the logit of each θ_{ij} receives a normal prior centered at zero.

Barnard et al. (2000) write $\mathbf{G} = \text{diag}(\mathbf{S})\ \mathbf{B}\ \text{diag}(\mathbf{S})$, where $\mathbf{S}$ is a q-vector of standard deviations, the square roots of the diagonals of $\mathbf{G}$, and $\mathbf{B}$ is the correlation matrix corresponding to $\mathbf{G}$. The motivation is that practitioners are trained to think in terms of standard deviations, which are on the same scale as the data, and correlations. Although these authors discuss a variety of possibilities within this framework, they favor making $\mathbf{S}$ and $\mathbf{B}$ independent *a priori*. For $\log(\mathbf{S})$, they favor a multivariate normal with a diagonal covariance. For $\mathbf{B}$, they focus on two priors: The marginal distribution of the correlation matrix $\mathbf{B}$ implied by an inverse-Wishart on $\mathbf{G}$, which for a specific setting of the Wishart parameters gives a uniform marginal distribution for each correlation in $\mathbf{B}$; and a flat proper prior on the space of legal correlation matrices $\mathbf{B}$, which is uniform on the correlations jointly but not uniform marginally for individual correlations.

The prior of Chen & Dunson (2003) is somewhat like that of Barnard et al. (2000) but uses a Cholesky decomposition. For the covariance matrix $\mathbf{G}$ of the random effects in the mixed linear model (1.1), let $\mathbf{G} = \mathbf{LL}'$ be the unique Cholesky decomposition, where $\mathbf{L}$ is lower triangular with nonnegative diagonal elements. Let $\mathbf{L} = \Lambda\Gamma$, where Λ is diagonal with non-negative elements and Γ is lower triangular with 1's in the diagonal entries. Then the random effects of (1.1) can be rewritten as $\mathbf{Zu} = \mathbf{Z}\Lambda\Gamma\mathbf{b}$, where $\mathbf{b}$ is a vector of iid standard normal latent variables. This re-expression of the random effects is handy because Λ contains the standard deviations of the random effects $\mathbf{u}$ and the lower-triangular Γ, which carries information about correlations between random effects, can be interpreted as regression coefficients. Because of the latter, Chen & Dunson (2003) give it a multivariate normal prior conditional on Λ, which is conditionally conjugate. It would then be possible to follow Barnard et al. (2000) by putting a multivariate normal prior on $\log(\Lambda)$, but Chen & Dunson's purpose is to do variable selection for random effects, so instead their prior treats the elements of Λ as independent and gives each element a mixture distribution with positive prior probability on zero. Pourahmadi proposed a non-Bayesian method based on a Cholesky decomposition with 1's on the diagonal, though it treats the standard deviations differently than Chen & Dunson (2003). Pourahmadi (2007) gives a brief review and contrasts his approach with the approach of Chen & Dunson (2003).

Simulation studies in Daniels & Kass (1999, 2001) show that the Givens-angle prior and Barnard et al.'s decomposition perform well in a range of cases and perform much better than the Wishart prior in some cases. Pourahmadi (2007, p. 1006) claims the Cholesky-based approach "has been used effectively" in several areas but none of the citations gives information about its performance apart from illustrative examples.

1.3.3 Computing for Bayesian Analyses

This section is not intended to be a survey of an area, Bayesian computing, that has simply exploded in the last two decades. Rather, it emphasizes one method, Markov chain Monte Carlo, which brought Bayes to the masses and dominates Bayesian applications (which was considered an oxymoron when I was in graduate school).

Before Gelfand and Smith (1990), explicit expressions for $\pi(\beta,\mathbf{u}|\mathbf{y})$, $\pi(\beta|\mathbf{y})$, $\pi(\mathbf{u}|\mathbf{y})$, or $\pi(\mathbf{X}\beta+\mathbf{Z}\mathbf{u}|\mathbf{y})$ were only tractable in simple or unrealistic special cases. It is easy to derive $\pi(\phi_G,\phi_R|\mathbf{y})$, the marginal posterior of the unknowns in the covariances $\mathbf{G}$ and $\mathbf{R}$; this is given as an exercise. Similarly, if $\mathbf{R}=\sigma_e^2\mathbf{I}$ and $1/\sigma_e^2$ has a gamma prior, it is easy to derive the marginal posterior of the unknowns in $\frac{1}{\sigma_e^2}\mathbf{G}$; this is also given as an exercise. Unfortunately, these marginal posteriors have nonstandard forms so it is hard to compute posterior intervals with pre-1990 methods except when ϕ_G has a single dimension. Thus while these results are sometimes useful in developing theory and methods, otherwise they have little practical utility.

In practice, Bayesian analyses since 1990 have almost always been performed by drawing samples from the posterior distribution of the unknowns and summarizing those draws as (estimates of) posterior percentiles, means, etc. By far the most common method of sampling from the posterior is Markov chain Monte Carlo (MCMC). If $\theta=(\beta,\mathbf{u},\phi)$ is the vector of unknowns in a mixed linear model, an MCMC method produces a sequence of draws from the joint posterior distribution of θ, $\theta_{(b)}, b=1,\ldots,B$. The sequence is a Markov chain because the distribution of $\theta_{(b)}$ is a function only of the preceding draw $\theta_{(b-1)}$ (and $\mathbf{y}$, $\mathbf{X}$, and $\mathbf{Z}$, of course) but not of any earlier draws in the chain $\theta_{(j)}, j\le b-2$. With a few exceptions, it is usually much easier to make draws from certain Markov chains than to make iid draws from the posterior; the price is that the sequence of MCMC draws is serially correlated. One extremely useful feature of MCMC is that once you have drawn a sequence $\theta_{(1)},\theta_{(2)},\ldots,\theta_{(B)}$, then for *any* function $g(\theta)$, the sequence $g(\theta_{(1)}),g(\theta_{(2)}),\ldots,g(\theta_{(B)})$ is a sequence of MCMC draws from the posterior of $g(\theta)$.

The MCMC literature is enormous and I will not try to summarize it even for mixed linear models. I give an example below of an MCMC algorithm for mixed linear models which is as well understood as any MCMC algorithm for an interesting class of models but which nobody claims is optimal. Interested readers looking for more background on MCMC can consult the relatively friendly introductions in Carlin & Louis (2008) or Gilks et al. (1996) although new texts are published routinely and you might find it worthwhile to shop around.

Here is an MCMC algorithm for mixed linear models; in MCMC jargon, this is a Gibbs sampler that blocks on $(\beta, \mathbf{u})$. Given a starting value for θ, $\theta_{(0)}$, one draw of the full vector θ is made by drawing, in order, subsets of θ conditional on the most recent draws for the rest of θ. In the Gibbs sampler shown here, a draw is first made from the so-called full conditional of $(\beta, \mathbf{u})$, i.e., the posterior distribution of $(\beta, \mathbf{u})$ conditional on the previous draw of ϕ (the full conditionals shown here are straightforward to derive and given as an exercise):

$$
\begin{aligned}
(\beta, \mathbf{u})|\mathbf{y}, \phi_{(b-1)} \quad &\sim \quad \text{normal with mean} \\
& \left[\mathbf{C}'\mathbf{R}_{(b-1)}^{-1}\mathbf{C} + \begin{pmatrix} \mathbf{0} & \mathbf{0} \\ \mathbf{0} & \mathbf{G}_{(b-1)}^{-1} \end{pmatrix}\right]^{-1} \mathbf{C}'\mathbf{R}_{(b-1)}^{-1}\mathbf{y} \\
\text{and covariance} \quad & \left[\mathbf{C}'\mathbf{R}_{(b-1)}^{-1}\mathbf{C} + \begin{pmatrix} \mathbf{0} & \mathbf{0} \\ \mathbf{0} & \mathbf{G}_{(b-1)}^{-1} \end{pmatrix}\right]^{-1}, \qquad (1.24)
\end{aligned}
$$

where, as before, $\mathbf{C} = [\mathbf{X}|\mathbf{Z}]$, the $\mathbf{0}$ are appropriate-sized matrices of zeroes, and the subscripts on $\mathbf{R}$ and $\mathbf{G}$ indicate that they are computed using the $(b-1)^{\text{th}}$ draw of ϕ. The conditional posterior mean is formally identical to Section 1.2.2's equation (1.8), the BLUPs, which is not a coincidence. Having drawn $(\beta_{(b)}, \mathbf{u}_{(b)})$ from this conditional distribution, $\phi_{(b)}$ is now drawn from its full conditional distribution given $(\beta_{(b)}, \mathbf{u}_{(b)})$:

$$
\begin{aligned}
& \pi(\phi_G, \phi_R | \mathbf{y}, \beta_{(b)}, \mathbf{u}_{(b)}) \\
\propto \; & |\mathbf{R}(\phi_R)|^{-\frac{1}{2}} \exp\left(-\frac{1}{2}(\mathbf{y} - \mathbf{X}\beta_{(b)} - \mathbf{Z}\mathbf{u}_{(b)})'\mathbf{R}(\phi_R)^{-1}(\mathbf{y} - \mathbf{X}\beta_{(b)} - \mathbf{Z}\mathbf{u}_{(b)})\right) \\
& \times |\mathbf{G}(\phi_G)|^{-\frac{1}{2}} \exp\left(-\frac{1}{2}\mathbf{u}'_{(b)}\mathbf{G}(\phi_G)^{-1}\mathbf{u}_{(b)}\right) \pi(\phi_G, \phi_R), \qquad (1.25)
\end{aligned}
$$

where $\pi(\phi_G, \phi_R)$ is the prior of (ϕ_G, ϕ_R). Note that if ϕ_G and ϕ_R are independent *a priori*, so $\pi(\phi_G, \phi_R) = \pi(\phi_G)\pi(\phi_R)$, then ϕ_G and ϕ_R are conditionally independent *a posteriori*, which is convenient computationally. For many problems (1.25) is a well-known distributional form. For example, if $\mathbf{G} = \sigma_s^2\mathbf{I}_q$ and $\mathbf{R} = \sigma_e^2\mathbf{I}_n$ so $\phi_G = \sigma_s^2$ and $\phi_R = \sigma_e^2$, then the full conditional distribution for (σ_s^2, σ_e^2) is

$$
\begin{aligned}
& \pi(\sigma_s^2, \sigma_e^2 | \mathbf{y}, \beta_{(b)}, \mathbf{u}_{(b)}) \\
\propto \; & (\sigma_e^2)^{-\frac{n}{2}} \exp\left(-\frac{1}{2\sigma_e^2}(\mathbf{y} - \mathbf{X}\beta_{(b)} - \mathbf{Z}\mathbf{u}_{(b)})'(\mathbf{y} - \mathbf{X}\beta_{(b)} - \mathbf{Z}\mathbf{u}_{(b)})\right) \\
& \times (\sigma_s^2)^{-\frac{q}{2}} \exp\left(-\frac{1}{2\sigma_s^2}\mathbf{u}'_{(b)}\mathbf{u}_{(b)}\right) \pi(\sigma_s^2, \sigma_e^2), \qquad (1.26)
\end{aligned}
$$

where $\pi(\sigma_s^2, \sigma_e^2)$ is the prior of (σ_s^2, σ_e^2). If the precisions $1/\sigma_s^2$ and $1/\sigma_e^2$ have the conventional gamma prior, then their full conditionals are also gamma distributions.

Ruppert et al. (2003, Section 16.6) give another MCMC algorithm for the simple case just above, which generalizes straightforwardly to all mixed linear models. In this so-called Rao-Blackwellized algorithm, MCMC draws are first made from the *marginal* posterior of (ϕ_G, ϕ_R), i.e., the posterior of (ϕ_G, ϕ_R) after having integrated $(\beta, \mathbf{u})$ out of the joint posterior. Then the marginal posterior density of $(\beta, \mathbf{u})$ is estimated by averaging the full-conditional density (1.24), for each value of $(\beta, \mathbf{u})$, over the MCMC draws of (ϕ_G, ϕ_R). This gives an efficient estimate of the marginal posterior density of $(\beta, \mathbf{u})$. Taking advantage of normal theory applied to (1.24), Rao-Blackwellization also allows similarly efficient estimates of posterior moments and marginal densities of linear functions of $(\beta, \mathbf{u})$.

That's the good news: In the glorious new world of Bayesian computing, there are algorithms for pretty much any problem and many possible algorithms for mixed linear models. Now for the not-so-good news: For innocent-looking mixed linear models, it is easy to write down legal MCMC samplers that have extremely poor properties. For example, Chapter 17 examines in detail a simple model yielding very strange marginal posteriors for ϕ, which was a nasty but productive surprise (Reich et al. 2007). "Poor properties" refers to several related things: The sequence of MCMC draws has high serial autocorrelation; the series of draws moves slowly around the space in which θ takes values; the sampler can "get stuck" in a small region of values of θ for many draws; and so on. Presence of any of these problems means the sampler provides little information about the posterior distribution for a given amount of computing. These problems afflict MCMC generally and many diagnostics have been proposed for detecting them. As early as 1996, Cowles & Carlin (1996) gave detailed reviews of 13 such diagnostics but concluded that "all ... can fail to detect the sorts of convergence failure that they were designed to identify."

Recently, encouraging progress has been made in an approach pioneered by Jim Hobert, Galin Jones, and their students. For models with $\mathbf{G} = \sigma_s^2\mathbf{I}$ and $\mathbf{R} = \sigma_e^2\mathbf{I}$ and inverse gamma priors on the two variances, Johnson & Jones (2010) gave sufficent conditions on the gamma priors under which the blocked Gibbs sampler described above is geometrically ergodic. Roman & Hobert (2012) extended those results to include models with multiple random effects and a wider class of priors, including a flat (improper) prior on the fixed effects (β) and inverse gammas on the random-effect variances including hyperparameter values that give improper priors. Johnson & Geyer (2012) extended this approach in a different direction by showing how to change variables in the posterior to obtain a posterior density that satisfies the conditions for geometric ergodicity, so that running an MCMC on the transformed variable and back-transforming gives a geometrically ergodic sampler for the original variable. This approach has very wide applicability. These geometric ergodicity results permit derivation of asymptotic MCMC standard errors for expectations of functions of the unknown parameters, which gives a way to determine the number of MCMC iterations B needed to obtain an estimate of the posterior mean of a scalar function $g(\theta)$ with a given (small) MCMC error (Flegel et al. 2008). One of these methods, the consistent batch means method (Jones et al. 2006), is extremely easy to use and is available in the R package mcmcse, written by James Flegal and John Hughes.

The results just described establish asymptotic properties of the MCMC sequence and there is no guarantee that in "small samples" (i.e., any practical B), a given chain has converged. However, this work is progressing quickly and I think it is not unrealistic to look forward to Bayesian software for a respectable class of mixed linear models that is simple for users and as foolproof as statistical software can be.

A final comment on MCMC: Proponents of Bayesian methods sometimes claim that analyses using MCMC or another sampling method are exact in that they do not rely on large-sample approximations, as conventional analyses do for mixed linear models. This is true: If you run a well-functioning MCMC long enough, you can obtain an arbitrarily precise estimate of any posterior quantity. However, nobody knows how long is "long enough" and there are examples in which "long enough" is on the order of 10^{20} iterations. In other words, this claim should be taken as hype, not fact, at least for the near future.

As noted at the beginning of this section, Bayesian computing continues to develop quickly and now includes other sampling methods and powerful approximations such as integrated nested Laplace approximations (Rue et al. 2009). Interested readers looking for a place to start might consider the latest edition of Carlin & Louis's text.

1.4 Conventional and Bayesian Approaches Compared

This section summarizes the advantages and disadvantages of the two approaches, then illustrates them using analyses I did for Example 1, the molecular structure of a virus (Peterson et al. 2001).

1.4.1 Advantages and Disadvantages of the Two Approaches

Some key advantages of the conventional approach, based on maximizing the restricted likelihood, are as follows:

- Maximizing is usually simpler than integrating; this is the case for mixed models.
- As a result, flexible mixed-linear-model analyses using the conventional approach are available in every major statistical package.
- Maximizing the restricted likelihood is often much faster than Bayesian analyses using MCMC, if it's done right.

The following are some key disadvantages:

- It is somewhere between difficult and impossible to account for variability in the estimate $\hat{\phi}$ (or account for uncertainty about ϕ, if you prefer) in intervals and tests for β and $\mathbf{u}$. To my knowledge, no software package does this.
- Maximizing the restricted likelihood routinely produces zero estimates for variances in $\mathbf{G}$. When this happens, conventional theory offers no way to construct an interval describing likely values of the variance that was estimated to be zero.

The statement "Maximizing the restricted likelihood is [very fast] if it's done right" requires some comment. In maximizing the restricted likelihood, some software packages use brute force to invert the matrices involved, which can be prohibitively large. The MIXED procedure in SAS does this and it is not hard to specify relatively simple problems that MIXED cannot practically compute. By contrast, for mixed linear models with $\mathbf{R} = \sigma_e^2 \mathbf{I}_n$, the algorithm in the lme4 package in R (Bates & DebRoy 2004) uses a clever matrix decomposition to compute extremely quickly, even for some problems that are effectively impossible to compute in MIXED. Unfortunately, the restriction to $\mathbf{R} = \sigma_e^2 \mathbf{I}_n$ excludes many models that can be specified in MIXED.

Regarding zero variance estimates, I use mixed linear models to analyze perhaps 25 datasets a year, fitting maybe 200 mixed linear models, and I get zero variance estimates routinely. Most of these analyses are some version of repeated-measures or variance-components analysis. In some cases, the random effects are present merely to account for clustering in computing standard errors for β, so the difference between a zero variance and a small variance has little effect and the zero variance estimate seems innocuous. In many cases, however, the variance component itself is of interest and in such cases a zero estimate is simply wrong — the variance component is not zero, though it may be small — and the absence of an interval of plausible values is worse than useless.

To me, the single most striking thing about text and reference books for mixed linear models is their silence on this subject. I have read Ruppert et al. (2003) very carefully and found exactly one mention of zero variance estimates, on p. 177, which "we cannot explain."[3] Perhaps these authors rarely obtain zero estimates in the problems they work, although my students get zero variance estimates often enough when they fit the same models (penalized splines). Ruppert et al. (2003) do discuss testing whether a variance component is zero (e.g., pp. 106–107), with laudable warnings about relying on large-sample approximations. I have found no mention of zero variance estimates in Lee et al. (2006), Fahrmeir & Tutz (2010), Searle et al. (1992, which does discuss negative variance estimates produced by obsolete methods), Verbeke & Molenberghs (1997), or Diggle et al. (1994), though I might have missed some brief mentions. Snijders & Bosker (2012) is alone in forthrightly stating that zero variance estimates are common and that standard software gives a misleading standard error of zero for such zero variance estimates (Sections 4.7, 5.4). Otherwise, no text or monograph, as far as I know, gives prominent treatment to this routine annoyance and research in the conventional analysis has produced no methods to handle it. In fairness to these authors and others, most books in this field focus on specifying and fitting models and interpreting the results of tidy examples, which may be the right emphasis given that this class of models has not yet penetrated the general statistics user market as much as, say, ordinary linear models or Cox regression. Nonetheless, it is an odd hole in a field for which model-fitting methods have been intensively developed for at least 40 years. Zero variance estimates are present in the example in Section 1.4.2 below, while Chapter 18 of the present book explores this topic in detail.

Some key advantages of the Bayesian approach are as follows:

[3] Actually, I missed this lone mention but my student Lisa Henn didn't.

- A Bayesian analysis naturally accounts for uncertainty about ϕ in tests and intervals for β and $\mathbf{u}$.
- The Bayesian analyses specified above do not give zero point estimates for variances in ϕ_G (but see the related comment below) and naturally provide intervals of plausible values in cases in which the conventional analysis would give a zero point estimate and no interval.

Some key disadvantages are as follows:

- The results depend on prior distributions, particularly for ϕ_G, to an extent that is not close to being well understood. Sensitivity analyses can address this but I've never seen a sensitivity analysis for priors that really explored a range of alternative priors (including, alas, my own published analyses, such as the example given below).
- With existing software, a Bayesian analysis is harder to compute than a conventional analysis, although presumably this will improve as time passes.
- MCMC routines for mixed linear models are not well understood in any generality and in a strict view would be considered experimental for all cases not meeting the sufficient conditions described above.

It is not difficult to specify Bayesian analyses that allow a variance's posterior distribution to have positive mass on zero; this is one Bayesian way to test whether that variance is zero. (The method of Chen & Dunson 2003, discussed above in Section 1.3.2.4, is an example.) Such specifications are not exotic today but they are not widely used and little is known about how they perform.

Saying that a Bayesian analysis "naturally gives intervals of plausible values" glosses over the question of how much you should trust those intervals. In problems where maximizing the restricted likelihood gives a zero estimate for a variance, the restricted likelihood can decline very slowly from zero as that variance is increased. I would like to say this happens frequently, but no one knows whether that is true. Chapters 14 and 18 explore an example of a zero variance estimate in which the restricted likelihood is quite flat near zero. In this example, I only discovered that the restricted likelihood was flat near zero because I did Bayesian analyses with many different priors and happened to notice that for each prior, the prior and posterior had very similar percentiles up to about the median. Thus, while a Bayesian analysis avoids the obviously wrong zero estimates for variances and provides a plausible interval, in cases where the restricted likelihood — the data contribution to the marginal distribution of ϕ — is flat near that peak at zero, the posterior is largely determined by the prior. If you begin with a Bayesian analysis and are not alerted to the possibility of this problem, there is no way to know that the data contribution to the marginal distribution of ϕ (the restricted likelihood) is quite flat unless you run many analyses with different priors. In other words, the ability to do a Bayesian analysis here makes it easy for a user to fool himself.

1.4.2 Conventional and Bayesian Analyses for the Viral-Structure Example

To be fair, I chose this example to make the conventional approach look bad: It emphasizes neglected consequences of the zero variance estimates routinely produced by maximizing the restricted likelihood. I also chose it because the analysis appeared in a good non-statistical journal. Be aware that in 2000 the Bayesian analyses presented here took me about 10 times as long to execute as the conventional analyses and that I have no compelling rationale for the prior distributions.

Peterson et al. (2001) collected data to estimate numbers of each of six proteins in the prohead of the bacteriophage virus $\phi 29$. The model I used to analyze these data was described in Section 1.1 and given in equation (1.3). Among the six proteins, gp8 had the most measurements with 98, while the other proteins had fewer than half as many measurements, so that parts of model (1.3) were omitted in analyzing them.

The conventional analyses were done in SAS's VARCOMP procedure with no fixed effects. The 95% interval in Table 1.7 is the estimate for the intercept plus and minus two times the standard error given by VARCOMP. (I should have used VARCOMP's confidence interval, which is based on the t distribution and thus wider.)

For the Bayesian analyses, the prior distributions were an improper flat prior for the intercept (β); for $1/\sigma_e^2$, gamma with mean 0.025 and standard deviation 0.05 (i.e., shape 0.25, scale 10); for $1/\sigma_j^2$, for $j = p,b,g,r$, gamma with mean 0.5 and standard deviation 1 (i.e., shape 0.25, scale 0.5). Peterson et al. (2001) gave this rationale for the prior on ϕ: "The prior distributions for the variances provide information equivalent to a quarter of a measurement [i.e., shape 0.25] and, apart from pure measurement error [σ_e^2], treat the components of variation equally." Subject to treating $1/\sigma_j^2, j = p,b,g,r$ equally, I chose this prior because it was consistent with what I knew about σ_e^2 and induced a nearly flat prior distribution on the degrees of freedom in the model fit. (Chapter 2 discusses degrees of freedom in the fitted values of mixed linear models and priors based on them.) For computations, I used the Gibbs sampler described in Section 1.3.3 implemented in an S+ program that I wrote myself. I ran the sampler for 5100 cycles and discarded the first 100 for burn-in. The point estimate and 95% interval are the average and the 2.5th and 97.5th percentiles of the 5000 retained Gibbs draws, respectively. WARNING: *Do not* mimic this MCMC sampler! I did this a long time ago when I was so ignorant about MCMC that I didn't know how ignorant I was. There was no need to write my own code; the number of iterations should be determined using objective methods like consistent batch means (Jones et al. 2006); discarding iterations for burn-in is unnecessary (Flegal et al. 2008, Flegal & Jones 2011); and the posterior mean is a worse point estimator than the posterior median for a variance, standard deviation, or precision (Section 1.3.2.1).

Table 1.7 gives the estimates and nominal 95% intervals for numbers ("Copies") of each molecule, as published in Peterson et al. (2001, Table II). The Bayesian interval is wider for all six molecules. The difference between the Bayesian and conventional ("REML") analyses is negligible for gp8 but quite large for the other five molecules. It is not a coincidence that gp8 has the largest sample size by quite a bit, but this difference between the two approaches is also affected by the efficiency of the design, in this case by the crossing and nesting of random effects.

Table 1.7: Molecular Structure of a Virus: Estimates and Intervals for Copy Numbers

Molecule	N Measurements	Method	Copies	95% Interval
gp7	31	REML	25.8	(19.0,32.6)
		Bayesian	26.9	(13.6,40.9)
gp8	98	REML	231.9	(200.5,263.3)
		Bayesian	232.4	(199.8,266.5)
gp8.5	46	REML	53.5	(37.7,69.3)
		Bayesian	52.6	(29.1,74.4)
gp9	40	REML	9.1	(7.9,10.3)
		Bayesian	9.1	(3.3,15.2)
gp11	40	REML	11.2	(10.3,12.1)
		Bayesian	11.2	(5.6,16.6)
gp12	40	REML	58.6	(51.9,65.3)
		Bayesian	58.2	(44.7,71.2)

The difference between the Bayesian and conventional intervals in Table 1.7 can be understood by considering the estimates of the variance components, which Table 1.8 gives for three molecules. For gp8, where the conventional and Bayesian intervals for "Copies" are quite similar, the REML estimate for a variance component's standard deviation is larger for some components while the Bayesian estimate is larger for others. The same is true for gp8.5, where the Bayesian interval for copies (Table 1.7) is 43% wider than the conventional interval and REML gives a zero estimate for the variance of the oxidizer-run random effect. For gp9, where the Bayesian interval for copies is nearly five times as wide as the conventional interval, the posterior medians are much larger than the REML estimates for the four variance components that are present. For all three molecules, the Bayesian 95% interval is extremely wide for all variance components except error (σ_e), reflecting a great deal of uncertainty. This uncertainty is ignored in the conventional analysis; this tends to make the conventional intervals narrower than the Bayesian intervals.

Table 1.8 shows some other things that, in my experience, are common in analyses of mixed linear models. For gp8 and gp8.5, the REML estimate and the two Bayesian estimates of σ_e are quite similar, while for gp9 the estimates differ less than they do for other variance components. The Bayesian interval for σ_e is also fairly narrow (the conventional interval would be, too, if it were shown). These observations are, in my experience, true of the error variance quite generally and Chapters 15 and 16 gives some theory suggesting — I won't go so far as to say "explaining" — why the data provide better information about the error variance than about random-

Table 1.8: Molecular Structure of a Virus: Estimates and Intervals for Variance Components (Expressed as Standard Deviations)

	σ_e	σ_p	σ_b	σ_g	σ_r
gp8					
REML estimate	20.3	23.8	20.6	0	15.1
Posterior mean	20.6	15.5	26.4	5.3	11.9
Posterior median	20.6	9.7	25.5	3.3	10.4
Bayes 95% interval	(17.7,24.1)	(0.6,63)	(4.0,53)	(0.5,20)	(0.7,34)
gp8.5					
REML estimate	4.3	9.1	7.6	3.6	0
Posterior mean	4.5	8.3	9.7	3.6	2.3
Posterior median	4.4	3.2	8.1	3.1	1.6
Bayes 95% interval	(3.6,5.6)	(0.5,49)	(2.6,27)	(0.8,9.1)	(0.5,9.3)
gp9					
REML estimate	0.96	—	0.85	0	0
Posterior mean	1.2	—	2.7	1.3	1.4
Posterior median	1.2	—	1.4	0.95	1.1
Bayes 95% interval	(0.99,1.6)	—	(0.5,12)	(0.4,4.1)	(0.4,4.9)

effect variances, even in mixed linear models that have no replication within design cells.

On the other hand, the intervals for variance components other than error are all quite wide. They are especially wide here because the sample sizes are small, but in my experience this is true in considerable generality. The theory in Chapters 15 and 16 also suggests why this happens.

Note also the similarity of the point estimates of copies produced by the conventional and Bayesian methods (Table 1.7). In this simple case, differences arise because the two approaches tend to give somewhat different estimates of variance components and thus to weight individual observations differently. Fixed effect estimates from the two approaches are not this similar in general. I have noticed, though, that even in cases where two variance-structure models fit quite differently according to standard criteria like a restricted likelihood ratio test or the Bayesian Information (Schwarz) criterion, that rarely has much effect on tests of fixed effects. There must be instances in which this fails, that is, apart from cases in which one of the models is grossly inappropriate, but I don't know what conditions are required.

Finally, the posterior means and medians differ a great deal for some of the random effects' standard deviations. This happens when the marginal posterior has an extremely long upper tail. The simulation study by He & Hodges (2008) suggests that in some generality the posterior mean is unstable as an estimate of a variance, standard error, or precision so it should not be used as a point estimate.

1.5 A Few Words about Computing

I am nowhere near expert on computing for mixed linear models or anything else. I have included this section, weak as it is, for the benefit of readers who know even less than I do and who are looking for a place to start. I say a bit about packages I have actually used and briefly mention other packages that I have not used but which have a following. Let the reader beware. Snijders & Bosker (2012, Chapter 18) give a survey of software for hierarchical (multi-level) models that includes almost all software mentioned here and quite a few packages not mentioned here.

The MIXED program in SAS performs the conventional analysis for a large range of mixed linear models using the methods and approximations discussed in previous sections. The REPEATED command specifies $\mathbf{R}$, the covariance of the errors ε, and RANDOM commands specify $\mathbf{G}$, the covariance of $\mathbf{u}$. It has taken me many years to become comfortable with the syntax of these two commands but MIXED can specify a very large collection of models. SAS's documentation for MIXED does not include many of the models discussed in this book; Ruppert et al. (2003, Appendix B.3) gives a SAS macro calling MIXED that fits many of the penalized splines in their book and their book's web site gives more SAS code. No doubt a web search would turn up a lot more SAS code for models in the present book. MIXED does not do tests comparing models with different variance structures but provides the maximized log restricted likelihood from which such tests can be computed. Computing for the conventional approach is by brute force and can be very slow. MIXED does Bayesian analyses using MCMC but only for the small subset of its models in which $\mathbf{R} = \sigma_e^2\mathbf{I}$ and $\mathbf{G}$ is diagonal (the documentation calls these "variance components models"). The easiest way to access documentation about MIXED is at SAS's web site, www.sas.com.

The R system has a few contributed packages that do computations for mixed linear models. The two that I have used are nlme and lme4, which also do computations for some generalized mixed linear models. Doug Bates was a co-author of both packages but only lme4, the later of the two, uses the matrix decompositions that speed up the necessary matrix inversions. Also, the two packages have very different syntaxes for specifying random effects. Although neither package has built-in capability for the same range of models as SAS's MIXED procedure, each allows user-specified variance functions. The lme4 package includes some Bayesian capability but I have never used it. Pinheiro & Bates (2000) gives a very friendly introduction to an earlier version of nlme implemented in the S-plus system. Documentation for the R version of nlme and for lme4 are available at the CRAN web site, under contributed packages (http://cran.r-project.org/web/packages/). The web site for Ruppert et al. (2003) has some code suitable for R or S-plus. I am advised that as of 2010 the SemiPar R package associated with their book is no longer actively maintained so you should probably not use it, although it was available on the CRAN web site the last time I checked. Finally, I am advised that the R package mgcv is popular and powerful but I have never done more than peruse its documentation.

Although MCMC methods seem quite simple for many problems, if it's practicable you are better off using existing MCMC engines instead of writing your own code because they're far less likely to have undiscovered bugs. For larger problems, unfortunately, there may be no practicable alternative to writing your own code but given how little we understand posterior distributions even for mixed linear models, this is a hazardous undertaking.

The best-known Bayesian software is produced by the BUGS project and available in many forms. The Windows version is WinBUGS, OpenBUGS works on many platforms, and the BRugs package allows WinBUGS to be called from within the R system. Current information is available on the web site of The BUGS Project, http://www.mrc-bsu.cam.ac.uk/bugs/winbugs/contents.shtml. Unlike SAS or R, BUGS does not have packages for particular classes of commonly used models analogous to MIXED or lme4, but rather provides a flexible programming environment in which models can be specified. BUGS's internals determine an MCMC routine and then run it. The latter is no small feat and the BUGS team has been extremely open about their blunders, for all of which they deserve immense credit. However, this general design means that you have to write a new program for every new model you want to fit instead of being able, as in SAS or lme4, to specify entirely new models very quickly using an interface specialized for mixed linear models. No doubt if I were a better programmer I would find this aspect of BUGS less onerous but for me it is a real disincentive to doing Bayesian analyses. I only make the effort of doing a Bayesian analysis in collaborative work when I think it likely to give a substantially different result from the conventional analysis which could affect the answer to the subject-matter question, as in the viral-structure example in Section 1.4.2. BUGS uses Gibbs or Metropolis-Hastings sampling, generally sampling one unknown at a time although it has a limited facility for sampling multivariate normal vectors. In this sense, BUGS's great strength, its generality, is also its great weakness because it is unlikely to give an efficient MCMC algorithm for any given problem. However, it is usually much easier than writing your own MCMC sampler in R or C++ and far less likely to suffer from coding errors. The R system also has a contributed MCMC package called JAGS, written by Martyn Plummer, which has nearly the same syntax as BUGS but has very different internals so it can be used as a check on BUGS. JAGS has the advantage that it will run on any platform that R runs on, so a computer ignoramus like me can run it on the Macintosh operating system without having to figure out how to use the Mac's Windows emulator, which I would need to use the BRugs package to access WinBUGS. JAGS documentation is available from the CRAN R web site.

Those are the software systems I have used. Other popular packages with extensive facilities for doing the conventional analysis include the general-purpose packages SPSS and STATA. Two popular specialized packages are HLM and MLwiN. These were originally designed to do analyses for hierarchical or multilevel models, which is a special case of mixed linear models for normal errors, though both systems handle some families of non-normal errors. Both systems do some form of both conventional as well as Bayesian analyses though neither does Bayesian analyses for the range of models that BUGS handles. Snijders & Bosker (2012, Chapter 18) discuss these packages in detail. Also, Rue and his collaborators (Rue et al. 2009) have developed an R package for doing approximate Bayesian calculations using integrated nested Laplace approximations (INLA); Section 7.2 gives a bit more information.

If you want to do conventional analyses and are already comfortable in SPSS and STATA you are probably better off learning their mixed linear model functionality than switching to a new system. If you are keen to do Bayesian analyses, you probably need to learn BUGS or JAGS if you want to do Bayesian analyses for classes of models besides mixed linear models (unless you are a good or keen programmer, I suppose).

Exercises

Regular Exercises

1. (Section 1.2.2) Derive the conventional point estimate (1.8) of $(\beta, \mathbf{u})$ given $\mathbf{G}$ and $\mathbf{R}$.
2. (Section 1.2.3) Neyman-Scott paradox: If Y_{i1} and Y_{i2} are iid $N(\alpha_i, \sigma_e^2)$ for each of $i = 1, \ldots, n$, show that the MLE for σ_e^2 is $\sum_i (Y_{i1} - Y_{i2})^2/4n$ and show that it has expected value $\sigma_e^2/2$ for all n. Show that the maximum restricted likelihood estimate for σ_e^2 is unbiased. Now turn this into a mixed linear model by adding $\alpha_i \sim$ iid $N(0, \sigma_s^2)$. Show that the MLE for σ_e^2 in this mixed linear model is unbiased — using the expectation over the distributions of $\mathbf{u}$ and ε — but the MLE for σ_s^2 has a bias of order $1/n$. Finally, show that maximizing the restricted likelihood for this mixed linear model gives unbiased estimates for both variances. In deriving these results, allow the variance estimates to take negative values. (Hint: Re-parameterize (σ_e^2, σ_s^2) as (σ_e^2, γ), where $\gamma = \sigma_e^2 + 2\sigma_s^2$.)
3. (Section 1.2.4.1) Derive cov $\begin{pmatrix} \hat{\beta} \\ \hat{\mathbf{u}} - \mathbf{u} \end{pmatrix}$ in (1.19).
4. (Section 1.3.2.3) For the simple problem where $X_i, i = 1, \ldots, n$ are iid $N(0, 1/\tau)$, τ being the precision, derive the posterior distribution of τ for the prior $\tau \sim$ *Gamma*(a, b) with density $\pi(\tau) \propto \tau^{a-1} e^{-b\tau}$, so τ has prior mean a/b. Note the role n plays in τ's posterior and use that to interpret the parameter a of the gamma prior as the "value" of the prior measured in terms of observations. Note also how b enters into the posterior. I think this, more than anything else, accounts for the popularity of the gamma$(\varepsilon,\varepsilon)$ prior for precisions: It appears to contribute almost no information to the posterior. But it is a *very* weird distribution.
5. (Section 1.3.3) For the Bayesian analysis, derive the marginal posterior of ϕ for a general prior on ϕ, i.e., the result should have the form (function of ϕ) $\times$ (prior on ϕ). Hint: Integrate out $(\beta, \mathbf{u})$ with a single integral.
6. (Following the preceding exercise.) Assume $\mathbf{R} = \sigma_e^2 \mathbf{I}$. Re-parameterize ϕ from (σ_e^2, ϕ_G) to (σ_e^2, ϕ_G^*), where ϕ_G^* is the vector of unknowns in $\mathbf{G}/\sigma_e^2$. Assume $1/\sigma_e^2$ has a gamma prior distribution, independent of the prior for ϕ_G^*. Derive the marginal posterior of ϕ_G^*. Hint: If this seems hard, you're doing it the wrong way.
7. (Section 1.3.3) For the blocked Gibbs sampler given in this section, derive the full conditional posteriors of $(\beta, \mathbf{u})$, ϕ_G, ϕ_R. This should be easy, but if you're getting stuck, Ruppert et al. (2003) Section 16.3 shows the full conditionals for a special case.

Open Questions

1. (Section 1.2.4.2) For an interesting class of models, e.g., in one of Chapters 3 to 6, derive better large-sample approximations for the various test statistics and pivotal quantities used in the conventional theory.
2. (Sections 1.4.1 and 1.4.2) For cases in which maximizing the restricted likelihood gives a zero estimate for a variance, derive a simple diagnostic test that detects when the restricted likelihood is quite flat near zero, and a simple one-sided interval for that variance with reasonably accurate coverage.
3. (Sections 1.3.2.3 and 1.3.2.4) For an interesting class of models, find a prior for ϕ that gives point estimates with good properties and posterior intervals with at worst near-nominal frequentist coverage. For an example of what such a project might look like, see He et al. (2007) and He & Hodges (2008). I consider these papers *components* of an examination of a prior's performance but by no means complete.
4. (Section 1.3.2.1) For an interesting class of models, examine the effect of priors for (β, ϕ) on Bayes factors for testing effects with multiple degrees of freedom, and if possible, specify tests using these priors with more or less reliable frequentist properties.
5. (Section 1.3.2.1) Derive an approximate Bayes factor for an effect in a mixed linear model based on the Schwarz Criterion/BIC approximation in Kass & Raftery (1995, p. 778). See if you can extend this to testing elements of ϕ. Be careful: The asymptotic theory that rationalizes this approximation may not apply for some tests of interest.
6. (Section 1.3.2.3 and 1.4.1) For an interesting class of models, examine the effect of priors for (β, ϕ) on Bayes factors for testing whether a variance component in ϕ_G is zero, and if possible, specify tests with more or less reliable frequentist properties.

Chapter 2

Two More Tools: Alternative Formulation, Measures of Complexity

Chapter 1 surveyed standard tools for analyzing mixed linear models. This chapter gives two more tools that later chapters will use.

2.1 Alternative Formulation: The "Constraint-Case" Formulation

Although this book emphasizes the mixed-linear-model formulation (1.1), another closely related formulation is sometimes useful. I call this alternative formulation the "constraint-case" formulation, though other names have been used. This section presents the constraint case formulation of mixed linear models and gives a brief history of its use.

2.1.1 The Constraint-Case Formulation

Consider the balanced one-way random effect model, which can be written as follows:

$$\begin{aligned} y_{ij} &= \theta_i + \varepsilon_{ij}, \text{ where } \varepsilon_{ij} \overset{iid}{\sim} N(0, \sigma_e^2) && (2.1) \\ \theta_i &= \mu + \delta_i, \text{ where } \delta_i \overset{iid}{\sim} N(0, \sigma_s^2) && (2.2) \\ \mu &= M + \xi, \text{ where } \xi \sim N(0, \sigma_p^2) && (2.3) \\ & \text{for } i = 1, \ldots, q \text{ and } j = 1, \ldots, m. \end{aligned}$$

The last row is a normal prior distribution for μ, so M and σ_p^2 are known while σ_e^2 and σ_s^2 are unknown, as in the mixed linear model. As usual, $\theta_i, \varepsilon_{ij}, \mu$, and δ_i are unknown. Equations (2.1), (2.2), and (2.3) could be rewritten to fit into the mixed-linear-model formulation (1.1) and will be below. For now, however, rewrite (2.2) and (2.3) as, respectively,

$$\begin{aligned} 0 &= -\theta_i + \mu + \delta_i && (2.4) \\ M &= \mu - \xi. && (2.5) \end{aligned}$$

(Some people find this odd, but it is just the obvious symbol manipulation.) Now "stack" equations (2.1), (2.4), and (2.5) as if they are simply different groups of "cases" or observations:

$$\left[\begin{array}{c} \mathbf{y} \\ \hline \mathbf{0}_q \\ \hline M \end{array}\right] = \left[\begin{array}{c|c} \mathbf{I}_q \otimes \mathbf{1}_m & \mathbf{0}_{qm} \\ \hline -\mathbf{I}_q & \mathbf{1}_q \\ \hline \mathbf{0}'_q & 1 \end{array}\right] \left[\begin{array}{c} \theta_1 \\ \vdots \\ \theta_q \\ \hline \mu \end{array}\right] + \left[\begin{array}{c} \varepsilon \\ \hline \delta \\ \hline -\xi \end{array}\right], \qquad (2.6)$$

where $\otimes$ is the Kronecker product, defined for matrices $\mathbf{A} = (a_{ij})$ and $\mathbf{B}$ as $\mathbf{A} \otimes \mathbf{B} = (a_{ij}\mathbf{B})$. Equation (2.6) has the form of a linear model. The left side of the equals sign is a known vector of length $qm+q+1$. On the right side of the equals sign is a known design matrix of dimension $(qm+q+1) \times (q+1)$ multiplied by a $(q+1)$-vector of unknown coefficients, plus an unknown error vector of length $qm+q+1$. The vector of errors has a diagonal covariance matrix and is heteroscedastic, that is, the $qm+q+1$ entries in the error term do not all have the same variance. The first qm "observations" or "cases" have error variance σ_e^2, the next q "cases" have error variance σ_s^2, and the last "case" has (known) error variance σ_p^2.

If you take the model in (2.6), write down the corresponding probability model, and multiply by a prior distribution for (σ_e^2, σ_s^2), the result is identical to the joint posterior distribution you obtain from writing this model as a mixed linear model and specifying a prior for (σ_e^2, σ_s^2). The proof is an exercise. If you omit the prior for (σ_e^2, σ_s^2) and let σ_p^2 go to infinity, this formulation also produces the restricted likelihood. In other words, nothing deep is happening here; equation (2.6) is just an accounting identity (Whittaker 1998).

Here is some terminology that will be used later. The rows of (2.6) arising from equation (2.1) are called "data cases," where the biostatistical jargon term "cases" is equivalent to the regression literature's "observations." These are the cases in which the outcome data or dependent variable $\mathbf{y}$ appear. The rows of (2.6) arising from equations (2.2) and (2.4) are called "constraint cases"; these artificial cases *constrain* the θ_i in a stochastic manner, with the constraint being more or less binding as σ_s^2 is smaller or larger. Finally, the rows of (2.6) arising from equations (2.3) and (2.5) are called "prior cases," capturing the prior distribution on the fixed effect μ.

There is more than one way to write this model in the constraint-case formulation. Instead of (2.1), (2.2), and (2.3), the balanced one-way random effects model can be written in the form of a mixed linear model (1.1), with $\mathbf{X} = \mathbf{1}_{qm}$, $\beta = \mu$, $\mathbf{Z} = \mathbf{I}_q \otimes \mathbf{1}_m$, $\mathbf{u} = (\delta_1, \ldots, \delta_q)'$, $\mathbf{G} = \sigma_s^2 \mathbf{I}_q$, and $\mathbf{R} = \sigma_e^2 \mathbf{I}_{qm}$. Begin by writing the mixed linear model as three equations:

$$\mathbf{y} = \mathbf{X}\beta + \mathbf{Z}\mathbf{u} + \varepsilon, \text{ where } \varepsilon \sim N_{qm}(0, \sigma_e^2 \mathbf{I}_{qm}) \qquad (2.7)$$

$$\mathbf{u} = \delta, \text{ where } \delta \sim N_q(0, \sigma_s^2 \mathbf{I}_q) \qquad (2.8)$$

$$\mu = M + \xi, \text{ where } \xi \sim N(0, \sigma_p^2), \qquad (2.9)$$

and then rewrite the constraint and prior cases (2.8) and (2.9) in the same manner as above, as

$$\mathbf{0}_q = -\mathbf{u} + \delta \qquad (2.10)$$

$$M = \mu - \xi. \qquad (2.11)$$

Now stack these equations:

$$\left[\begin{array}{c}\mathbf{y}\\ \hline \mathbf{0}_q\\ \hline M\end{array}\right] = \left[\begin{array}{c|c}\mathbf{1}_{qm} & \mathbf{I}_q\otimes\mathbf{1}_m\\ \hline \mathbf{0}_q & -\mathbf{I}_q\\ \hline 1 & \mathbf{0}'_q\end{array}\right]\left[\begin{array}{c}\mu\\ \hline \mathbf{u}\end{array}\right] + \left[\begin{array}{c}\varepsilon\\ \hline \delta\\ \hline -\xi\end{array}\right]. \tag{2.12}$$

Like equation (2.6), equation (2.12) has the form of a linear model with data, constraint, and prior cases and a heteroscedastic error term, and it too gives the same joint posterior for all unknowns as the original mixed-linear-model formulation (the proof is an exercise). Again, nothing deep is happening here, this is simply a different way to write the same accounting identity.

More generally, the mixed linear model (1.1), with a $N_p(M,\Sigma)$ prior on β, can be written in constraint-case form as

$$\left[\begin{array}{c}\mathbf{y}\\ \hline \mathbf{0}_q\\ \hline M\end{array}\right] = \left[\begin{array}{c|c}\mathbf{X} & \mathbf{Z}\\ \hline \mathbf{0}_q & -\mathbf{I}_q\\ \hline \mathbf{I}_p & \mathbf{0}_{p\times q}\end{array}\right]\left[\begin{array}{c}\beta\\ \hline \mathbf{u}\end{array}\right] + \left[\begin{array}{c}\varepsilon\\ \hline \delta\\ \hline -\xi\end{array}\right] \tag{2.13}$$

where the error term has covariance matrix

$$\text{cov}\left(\begin{array}{c}\varepsilon\\ \hline \delta\\ \hline -\xi\end{array}\right) = \left[\begin{array}{ccc}\mathbf{R} & \mathbf{0} & \mathbf{0}\\ \mathbf{0} & \mathbf{G} & \mathbf{0}\\ \mathbf{0} & \mathbf{0} & \Sigma\end{array}\right]. \tag{2.14}$$

Thus, all mixed linear models can be written in the constraint-case form, and as far as I know any model that can be written in the constraint-case form can also be written as a mixed linear model.

Why bother with this reformulation? Lee et al. (2006) use it and its extension to non-normal errors as a vehicle for a variety of analyses. This book uses it for two reasons. Sometimes the constraint-case form gives different intuition than the standard form (Section 2.2.5.1 is an example) or it allows simpler derivations because the covariance matrix of the combined error term is often diagonal. (Chapters 11 and 12 use it to derive restricted likelihoods.) Second, some models are simpler in the constraint-case form than in the standard form. Here is an example from Julian Besag's work, a special case of the dynamic linear models considered in Chapter 6.

Example 4. Experimental plots in a long row. Plots in a field experiment are laid out in one long row, labeled $i = 1,...,n$. (Apparently Besag encountered this design in an actual field trial with a rather large n.) Two treatments are allocated randomly to the plots, indicated by $T_i = 0$ or 1. Suppose F_i is the unobserved fertility of plot i, and that F_i is assumed to change along the row of plots according to the model $F_i = F_{i-1} + \delta_i$, where $\delta_i \overset{iid}{\sim} N(0,\sigma_s^2)$. Model the yield in plot i as

$$y_i = T_i\beta + F_i + \varepsilon_i, \text{ where } \varepsilon_i \overset{iid}{\sim} N(0,\sigma_e^2). \tag{2.15}$$

(The intercept is implicit in the F_i.) Rewrite the model for F_i as

$$0 = -F_i + F_{i-1} + \delta_i, i = 2,\ldots,n, \tag{2.16}$$

which supplies constraint cases for the F_i. Put a $N(M, \sigma_p^2)$ prior on β and write it as

$$M = \beta + \xi, \text{ where } \xi \sim N(0, \sigma_p^2) \tag{2.17}$$

to supply a prior case for β. Stacking these "cases" as before gives the constraint-case formulation of this model:

$$\left[\begin{array}{c} \mathbf{y} \\ \hline \mathbf{0}_{n-1} \\ \hline M \end{array}\right] = \left[\begin{array}{c|cccccc} \mathbf{T} & & & \mathbf{I}_n & & & \\ \hline & 1 & -1 & 0 & \dots & 0 & 0 \\ & 0 & 1 & -1 & \dots & 0 & 0 \\ \mathbf{0}_{n-1} & \vdots & & & \ddots & & \vdots \\ & 0 & 0 & 0 & \dots & 1 & -1 \\ \hline 1 & & & \mathbf{0}_{1\times n} & & & \end{array}\right] \left[\begin{array}{c} \beta \\ \hline F_1 \\ \vdots \\ F_n \end{array}\right] + \left[\begin{array}{c} \varepsilon \\ \hline \delta \\ \hline -\xi \end{array}\right], \tag{2.18}$$

where $\mathbf{T} = (T_1, \dots, T_n)'$.

This is considerably simpler than the mixed-linear-model formulation of the same model, presented below; Chapter 6 derives this for a fairly general dynamic linear model. In dynamic linear model jargon, the observation equation, i.e., the model for the observations y_i, is

$$y_i = \mathbf{H}_i \theta_i + \varepsilon_i, \tag{2.19}$$

for $\mathbf{H}_i = \left[\begin{array}{cc} T_i & 1 \end{array}\right]$ and $\theta_i' = (\beta, F_i)$. The state equation, describing the evolution of the state θ_i, is

$$\theta_i = \theta_{i-1} + \mathbf{w}_i, \tag{2.20}$$

where the state error vector $\mathbf{w}_i$ is 2×1 with covariance matrix

$$\mathbf{G}_i = \left[\begin{array}{cc} 0 & 0 \\ 0 & \sigma_s^2 \end{array}\right]. \tag{2.21}$$

As shown in Chapter 6, this can be written as a mixed linear model with

$$\mathbf{X} = \left[\begin{array}{c} \mathbf{H}_1 \\ \mathbf{H}_2 \\ \vdots \\ \mathbf{H}_n \end{array}\right]; \quad \beta = \left[\begin{array}{c} \beta \\ F_1 \end{array}\right] \tag{2.22}$$

$$\mathbf{Z} = \left[\begin{array}{ccccc} \mathbf{0} & \mathbf{0} & \mathbf{0} & \dots & \mathbf{0} \\ \mathbf{H}_2 & \mathbf{0} & \mathbf{0} & \dots & \mathbf{0} \\ \mathbf{H}_3 & \mathbf{H}_3 & \mathbf{0} & \dots & \mathbf{0} \\ \mathbf{H}_4 & \mathbf{H}_4 & \mathbf{H}_4 & \dots & \mathbf{0} \\ & & \vdots & & \\ \mathbf{H}_n & \mathbf{H}_n & \mathbf{H}_n & \dots & \mathbf{H}_n \end{array}\right]; \quad \mathbf{u}' = (\mathbf{w}_2', \dots, \mathbf{w}_n') \tag{2.23}$$

$$\mathbf{G} = \text{blockdiag}(\mathbf{G}_2, \dots, \mathbf{G}_n). \tag{2.24}$$

(Proof that this is equivalent to (2.15), (2.16) is an exercise.) The only advantage of this mixed-linear-model expression is that it makes explicit something that the constraint-case formulation left implicit, namely the flat (improper) prior on a one-dimensional subspace of the sample space of $(F_1, \ldots, F_n)$.

Note also that in contrast to the earlier examples, this example's random effect $(F_1, \ldots, F_n)$ does not obviously qualify as an old-style random effect. While the levels F_i are arguably draws from a population, they are also arguably the whole population. However, the F_i are not of interest in themselves but only as a means of modeling the distribution of the errors around the treatment-group means.

2.1.2 A Brief History of the Constraint-Case Formulation

Henderson et al. (1959) used data and constraint cases, which they called "mixed model equations," to compute best linear unbiased predictors. Bates & DebRoy (2004) applied a decomposition to these mixed model equations to radically speed up computation of the likelihood and restricted likelihood (see also Searle et al. 1992, section 7.6). Theil (1971) used prior cases to insert prior information about regression coefficients into a non-Bayesian analysis. In linear regression models, Salkever (1976) and Fuller (1980) obtained a frequentist predictive distribution for a new observable y_f by adding y_f to the vector of regression coefficients and adding an artificial case $0 = \mathbf{x}'\beta - y_f + \varepsilon$. Duncan & Horn (1972) re-expressed dynamic linear models as linear models, generalizing Example 4 above, to prove Gauss-Markov-style theorems about Kalman filters. Fellner (1986) used data and constraint cases to construct M-estimators for variance components. Lee & Nelder (1996) used the constraint-case formulation to express a large collection of normal and non-normal hierarchical models. Given the diverse ways the constraint-case formulation has been employed, the literature probably has other uses that I have missed.

2.2 Measuring the Complexity of a Mixed Linear Model Fit

The complexity of a mixed-linear-model fit and of components of the fit are often described using a measure called "degrees of freedom" (DF), which generalizes the familiar notion of DF from ordinary and generalized linear models. For mixed linear models, DF are used to specify F-tests; to describe how large a model is, often as part of the penalty term of model-selection criteria analogous to Akaike's Information Criterion (AIC); and to specify prior distributions on ϕ. Section 2.2.1 gives the generally used measure of complexity for the whole model's fit and some rationales for this measure, while Section 2.2.2 gives a way to partition the DF in the whole model into DF for components of the model. With that background, Section 2.2.3 discusses uses of DF in more detail. Section 2.2.4 compares DF to another widely used measure of model complexity, p_D, which Spiegelhalter et al. (2002) defined as part of their model-choice criterion *DIC*. Finally, Section 2.2.5 uses the constraint-case and mixed model formulations to give some intuition for the notion of a fractional degree of freedom.

2.2.1 DF in the Whole Fit

For initial motivation, consider again the balanced one-way random effects model:

$$y_{ij} = \theta_i + \varepsilon_{ij}, \text{ where } \varepsilon_{ij} \overset{iid}{\sim} N(0, \sigma_e^2) \quad (2.25)$$

$$\theta_i = \mu + \delta_i, \text{ where } \delta_i \overset{iid}{\sim} N(0, \sigma_s^2) \quad (2.26)$$

$$\text{for } i = 1, \ldots, q \text{ and } j = 1, \ldots, m.$$

The fitted values for this model are $\hat{y}_{ij} = \hat{\mu} + \hat{\delta}_i$. For concreteness, suppose these were obtained using conventional (non-Bayesian) methods, though this is not necessary.

What is the complexity of this fit? If $\hat{\sigma}_s^2 \to \infty$ for fixed $\hat{\sigma}_e^2$, then this is just a one-way ANOVA without random effects, so $\hat{y}_{ij} = \bar{y}_{i.}$, the average of $y_{ij}, j = 1, \ldots, m$, and this fit has q degrees of freedom. On the other hand, if $\hat{\sigma}_s^2 \to 0$ for fixed $\hat{\sigma}_e^2$, then the y_{ij} are iid $N(\mu, \sigma_e^2)$, $\hat{y}_{ij} = \bar{y}_{..}$, the average of all the y_{ij}, and this fit has 1 degree of freedom. For any positive finite σ_s^2, this model's mean structure appears to have q freely varying parameters, the θ_i, though as the previous section's jargon term "constraint case" suggests, the θ_i are constrained in a stochastic sense, which complicates interpretation of "freely varying." Moreover, it seems awkward to suggest that the complexity of the fit is q as long as σ_s^2 is positive, but drops discontinuously to 1 as soon as σ_s^2 reaches the limiting value of zero.

This awkwardness can be avoided by treating DF in a mixed-linear-model fit as a real number instead of an integer. As far as I know, every approach proposed for defining or rationalizing DF leads to the same definition of the DF in the whole fit of a mixed linear model. The following develops the definition from linear regression through linear smoothers to mixed linear models, and then gives some formal rationales for this definition.

In an ordinary linear regression $\mathbf{y} = \mathbf{X}\beta + \varepsilon$ fit by ordinary least squares, the fitted values are $\hat{\mathbf{y}} = \mathbf{H}\mathbf{y}$ for $\mathbf{H} = \mathbf{X}(\mathbf{X}'\mathbf{X})^{-1}\mathbf{X}'$. The degrees of freedom in the fitted model is rank($\mathbf{X}$) = trace($\mathbf{H}$). For a linear smoother $\mathbf{S}_\lambda$, where the fitted values are $\hat{\mathbf{y}} = \mathbf{S}_\lambda \mathbf{y}$ — the smoother is a function of a tuning parameter λ which is treated as fixed when computing fitted values — then by analogy with linear models, the DF in the fit is trace($\mathbf{S}_\lambda$).

Mixed linear models can be viewed as a linear smoother, as follows. Given values for ϕ, the unknowns in the covariance matrices $\mathbf{G}$ and $\mathbf{R}$, the fitted values for a mixed linear model are, as in equation (1.10),

$$\hat{\mathbf{y}}_\phi = \mathbf{C}\left[\mathbf{C}'\mathbf{R}^{-1}\mathbf{C} + \begin{pmatrix} \mathbf{0} & \mathbf{0} \\ \mathbf{0} & \mathbf{G}^{-1} \end{pmatrix}\right]^{-1} \mathbf{C}'\mathbf{R}^{-1}\mathbf{y}. \quad (2.27)$$

where $\mathbf{C} = [\mathbf{X}|\mathbf{Z}]$, $\mathbf{G} = \mathbf{G}(\phi_G)$, and $\mathbf{R} = \mathbf{R}(\phi_R)$. This is a linear smoother with

$$\mathbf{S}_\lambda = \mathbf{C}\left[\mathbf{C}'\mathbf{R}^{-1}\mathbf{C} + \begin{pmatrix} \mathbf{0} & \mathbf{0} \\ \mathbf{0} & \mathbf{G}^{-1} \end{pmatrix}\right]^{-1} \mathbf{C}'\mathbf{R}^{-1} \quad (2.28)$$

and $\lambda = \phi$. Thus, the DF in the mixed linear model's fitted values is

$$\text{trace}(\mathbf{S}_\lambda) = \text{trace}\left(\mathbf{C}\left[\mathbf{C}'\mathbf{R}^{-1}\mathbf{C} + \begin{pmatrix} \mathbf{0} & \mathbf{0} \\ \mathbf{0} & \mathbf{G}^{-1} \end{pmatrix}\right]^{-1} \mathbf{C}'\mathbf{R}^{-1}\right) \quad (2.29)$$

(see, e.g., Ruppert et al. 2003, Sections 3.13, 8.3, which are a bit less general).

For example, for the balanced one-way random effects model, specified as a mixed linear model just above equation (2.7), the DF in the fit for given (σ_e^2, σ_s^2) is obtained by plugging the pieces of the mixed-linear-model specification into (2.29). The result is DF = $1 + (q-1)mr/(mr+1)$ for $r = \sigma_s^2/\sigma_e^2$; the proof is an exercise. This result for the simplest interesting model has features that are present quite generally. First, the DF in the fit takes values in the interval $[1, q]$ and increases continuously with σ_s^2 for given σ_e^2, as our initial motivation suggested it should. Second, for models with normal errors and random effects, DF is a function of the *ratio* of variances $r = \sigma_s^2/\sigma_e^2$, not the individual variances. In the common case where $\mathbf{R} = \sigma_e^2 \mathbf{I}_n$, DF is a function of $\mathbf{G}/\sigma_e^2$; if $\mathbf{G}$ is a diagonal matrix with variances on its diagonal, DF is a function of the ratios of these variances to σ_e^2, not of the variances themselves. Finally, for fixed σ_s^2, DF decreases as σ_e^2 increases. The intuition is that as noise variance increases for fixed σ_s^2, the estimates of the random effects $\mathbf{u}$ are shrunk more toward zero, so the complexity of the fit (DF) decreases.

Various rationales have been offered for this definition of DF. Ruppert et al. (2003) simply assert it by analogy with linear models, as did Hodges & Sargent (2001). Hastie & Tibshirani (1990) assert this definition for linear smoothers $\mathbf{S}_\lambda$ but note that for λ fixed, this DF is the correct adjustment for model complexity in Mallows' C_p for smoothers: "In particular, the C_p statistic corrects [the average squared residual] to make it unbiased for [the prediction squared error] by adding a quantity 2 trace$(\mathbf{S}_\lambda)\hat{\sigma}^2/n$, where $\hat{\sigma}^2$ is an estimate of [the error variance σ_e^2]" (p. 53). Vaida and Blanchard (2005) obtained a similar result extending Akaike's information criterion to clustered-data situations; see below in Section 2.2.3.

This definition of DF is also the generalized degrees of freedom defined by Ye (1998). Ye (1998) defined the generalized degrees of freedom of a fit produced by literally any model-fitting procedure as its aggregate sensitivity to the individual observations. Suppose $\mathbf{y}$ has an n-variate normal distribution with mean μ and covariance $\sigma^2 \mathbf{I}_n$, with σ^2 known, and that a procedure gives an estimate $\hat{\mu}(\mathbf{y})$ of μ. Then Ye defined the generalized DF in the fit as

$$\sum_{i=1}^{n} \frac{\partial E_\mu\{\hat{\mu}_i(\mathbf{y})\}}{\partial \mu_i} = \sum_{i=1}^{n} E_\mu\{\hat{\mu}_i(\mathbf{y})(y_i - \mu_i)\}/\sigma^2, \tag{2.30}$$

where E_μ means expectation over the distribution of $\mathbf{y}$ given μ. This is the sum over the observations i of $\hat{\mu}_i(\mathbf{y})$'s sensitivity to a small change in y_i, and thus measures the estimation procedure's flexibility in Ye's terms, or complexity in the terms used here. For mixed linear models with known ϕ, the DF formula (2.29) is the same as Ye's generalized DF (the proof is an exercise).

Two other interpretations of DF in the fit are discussed below in Sections 2.2.2 and 2.2.5. These are, respectively, that the DF in the fit is, loosely speaking, the ratio of the fit's modeled covariance matrix to the total covariance matrix; and that the mixed-linear-model fit shrinks the unconstrained least-squares fit by specific fractions along canonical basis vectors of the fitted-value space, with those fractions adding to the DF in the whole model fit.

2.2.2 Partitioning a Fit's DF into Components

Ruppert et al. (2003, Section 8.3) assert a formula for partitioning the DF of a fit into DF for the components of the fit. Cui et al. (2010) developed a formula from first principles, which agrees with the formula of Ruppert et al. (2003) when both are defined. This section summarizes Cui et al. (2010); those interested in proofs and further details are referred to that paper and its online supplement.

2.2.2.1 Notation, Motivation for the Partition of DF

Consider again the mixed linear model (1.1)

$$\mathbf{y} = \mathbf{X}\beta + \mathbf{Z}\mathbf{u} + \varepsilon. \tag{2.31}$$

Partition $\mathbf{Z}$'s columns as $\mathbf{Z} = [\mathbf{Z}_1, \ldots, \mathbf{Z}_l, \ldots, \mathbf{Z}_L]$ and conformably partition $\mathbf{u} = (\mathbf{u}_1', \ldots, \mathbf{u}_L')'$. Model the lth component of $\mathbf{u}$ as $\mathbf{u}_l | \mathbf{G}_l \sim N(\mathbf{0}, \mathbf{G}_l)$ with the $\mathbf{u}_l | \mathbf{G}_l$ mutually independent. Define $\mathbf{G}$ as the block diagonal matrix with the $\mathbf{G}_l$ as the diagonal blocks. The covariance matrices $\mathbf{R}$ and $\mathbf{G}_l$ are nonnegative definite.

To motivate the partition of DF, consider a simpler model with no fixed effects:

$$\mathbf{y} = \mathbf{Z}\mathbf{u} + \varepsilon, \tag{2.32}$$

where $\mathbf{u} \sim N(\mathbf{0}, \mathbf{G})$ and $\varepsilon \sim N(\mathbf{0}, \mathbf{R})$, and assume $\mathbf{G}$ and $\mathbf{R}$ are positive definite. From Section 2.2.1, the DF in the whole fit is

$$\text{trace}[\mathbf{Z}(\mathbf{Z}'\mathbf{R}^{-1}\mathbf{Z} + \mathbf{G}^{-1})^{-1}\mathbf{Z}'\mathbf{R}^{-1}]. \tag{2.33}$$

Rewrite the matrix inverse in (2.33) as (Schott 1997, Theorem 1.7)

$$(\mathbf{Z}'\mathbf{R}^{-1}\mathbf{Z} + \mathbf{G}^{-1})^{-1} = \mathbf{G} - \mathbf{G}\mathbf{Z}'(\mathbf{Z}\mathbf{G}\mathbf{Z}' + \mathbf{R})^{-1}\mathbf{Z}\mathbf{G}. \tag{2.34}$$

The left-hand side of (2.34) is familiar from Bayesian linear models as the conditional posterior covariance of $\mathbf{u}|\mathbf{y}, \mathbf{R}$ for the prior $\mathbf{u} \sim N(\mathbf{0}, \mathbf{G})$, while the right-hand side is familiar as the conditional covariance of $\mathbf{u}$ given $\mathbf{y}$ from the joint multivariate normal distribution of $\mathbf{u}$ and $\mathbf{y}$. Use (2.34) to rewrite (2.33):

$$\begin{aligned} DF &= \text{trace}[\mathbf{Z}\mathbf{G}\mathbf{Z}'\mathbf{R}^{-1} - \mathbf{Z}\mathbf{G}\mathbf{Z}'(\mathbf{Z}\mathbf{G}\mathbf{Z}' + \mathbf{R})^{-1}\mathbf{Z}\mathbf{G}\mathbf{Z}'\mathbf{R}^{-1}] \\ &= \text{trace}[\mathbf{Z}\mathbf{G}\mathbf{Z}'(\mathbf{Z}\mathbf{G}\mathbf{Z}' + \mathbf{R})^{-1}(\mathbf{Z}\mathbf{G}\mathbf{Z}' + \mathbf{R} - \mathbf{Z}\mathbf{G}\mathbf{Z}')\mathbf{R}^{-1}] \\ &= \text{trace}[\mathbf{Z}\mathbf{G}\mathbf{Z}'(\mathbf{Z}\mathbf{G}\mathbf{Z}' + \mathbf{R})^{-1}], \end{aligned}$$

which has the form: trace[ratio of $\{$modeled variance matrix$\}$ to $\{$total variance matrix$\}$]. (The second line follows from the first by pulling $\mathbf{R}^{-1}$ out to the right, then writing $\mathbf{Z}\mathbf{G}\mathbf{Z}'$ as $\mathbf{Z}\mathbf{G}\mathbf{Z}'(\mathbf{Z}\mathbf{G}\mathbf{Z}' + \mathbf{R})^{-1}(\mathbf{Z}\mathbf{G}\mathbf{Z}' + \mathbf{R})$.) This suggests defining an effect-specific DF as that effect's contribution of variance to total variance. In the full model (2.31), the $\mathbf{u}_l$ are independent of each other conditional on their covariances $\mathbf{G}_l$, and the variance in $\mathbf{y}$ arising from the effect represented by $\mathbf{u}_l$ is $\mathbf{Z}_l\mathbf{G}_l\mathbf{Z}_l'$. Along with variation from the error ε, the total modeled variance for $\mathbf{y}$ is $\sum_1^L \mathbf{Z}_l\mathbf{G}_l\mathbf{Z}_l' + \mathbf{R} = \mathbf{Z}\mathbf{G}\mathbf{Z}' + \mathbf{R}$. Thus, in this special case when all effects are random effects (i.e., all effects are smoothed), β is null and all $\mathbf{G}_l$ are positive definite, the DF for the component represented by the columns in Z_l is $\text{trace}[\mathbf{Z}_l\mathbf{G}_l\mathbf{Z}_l'(\mathbf{Z}\mathbf{G}\mathbf{Z}' + \mathbf{R})^{-1}]$.

2.2.2.2 Partitioning DF

The preceding reformulation of DF for the model without fixed effects (2.32) suggests a way to rewrite DF as a sum of meaningful quantities, but any such partition of DF must accommodate the fixed effects $\mathbf{X}\beta$. A fixed effect can be viewed as the limit of a random effect for which the covariance matrix goes to infinity and imposes no constraint. The definition that follows uses a covariance matrix $\lambda\mathbf{G}^0$ as β's covariance, where $\mathbf{G}^0$ is unspecified but positive definite and λ is a positive scalar. In the limit as λ goes to $+\infty$, the specific $\mathbf{G}^0$ does not matter as long as it is positive definite. In Bayesian terms, this is a flat prior on the fixed effects and it can have nontrivial consequences (see, e.g., Section 10.2). If instead informative normal priors are placed on all fixed effects, this case becomes the simpler one considered just above in Section 2.2.2.1.

Before proceeding, we need more notation. $P_{\mathbf{H}}$ is the orthogonal projection onto the column space of a matrix $\mathbf{H}$, and $\mathbf{H}^+$ is the Moore-Penrose generalized inverse of $\mathbf{H}$.

For the full mixed linear model (2.31), define DF for the fixed effects, denoted by $DF(\mathbf{X})$, as

$$\begin{aligned} DF(\mathbf{X}) &= lim_{\lambda\to+\infty}\text{trace}[\mathbf{X}\lambda\mathbf{G}^0\mathbf{X}'(\mathbf{X}\lambda\mathbf{G}^0\mathbf{X}'+\mathbf{ZGZ}'+\mathbf{R})^+] \\ &= \text{trace}[P_{\mathbf{X}}] = rank(\mathbf{X}). \end{aligned}$$

(The online supplement to Cui et al. 2010 proves the above and the following.) For the random effect $\mathbf{u}_l$ corresponding to the columns $\mathbf{Z}_l$, DF is defined analogously:

$$\begin{aligned} DF(\mathbf{Z}_l) &= lim_{\lambda\to+\infty}\text{trace}[\mathbf{Z}_l\mathbf{G}_l\mathbf{Z}_l'(\mathbf{X}\lambda\mathbf{G}^0\mathbf{X}'+\mathbf{ZGZ}'+\mathbf{R})^+] \\ &= \text{trace}\{\mathbf{Z}_l\mathbf{G}_l\mathbf{Z}_l'[(\mathbf{I}-P_{\mathbf{X}})(\mathbf{ZGZ}'+\mathbf{R})(\mathbf{I}-P_{\mathbf{X}})]^+\}. \end{aligned} \quad (2.35)$$

Similarly, for the error term ε,

$$\begin{aligned} DF(\varepsilon) &= lim_{\lambda\to+\infty}\text{trace}[\mathbf{I}_n\mathbf{R}\mathbf{I}_n'(\mathbf{X}\lambda\mathbf{G}^0\mathbf{X}'+\mathbf{ZGZ}'+\mathbf{R})^+] \\ &= \text{trace}\{\mathbf{R}[(\mathbf{I}-P_{\mathbf{X}})(\mathbf{ZGZ}'+\mathbf{R})(\mathbf{I}-P_{\mathbf{X}})]^+\}. \end{aligned}$$

In general this DF for the error term differs from residual degrees of freedom as defined in Ruppert et al. (2003, Section 3.14) or Hastie & Tibshirani (1990, Section 3.5). We return to this point below, in Section 2.2.3.

By these definitions, the DF of a random effect Z_l or of error ε is the fraction of total variation in $\mathbf{y}$ contributed by Z_l or ε out of variation not accounted for by the fixed effects $\mathbf{X}$. Giving priority to the fixed effects in this manner is a consequence of the flat prior on the fixed effects, the consequences of which are not well understood.

2.2.2.3 Properties of the Definition

This partition has four properties with the not inconsiderable virtue of seeming reasonable.

- The sum of DF over all fixed and random effects equals DF for the whole model as given in Section 2.2.1.

- When $\mathbf{R}$ is positive definite, $DF(\mathbf{X}) + \sum_1^L DF(\mathbf{Z}_l) + DF(\varepsilon) = n$, the number of observations. For DF of error as defined in Ruppert et al. (2003) and Hastie & Tibshirani (1990), $DF(\mathbf{X}) + \sum_1^L DF(\mathbf{Z}_l) + DF(\text{error}) < n$, with the deficiency arising from $DF(\text{error})$. Section 2.2.3 below says more about this.
- The DF of a random effect has a reasonable upper bound: $DF(\mathbf{Z}_l) \leq rank((\mathbf{I} - P_\mathbf{X})\mathbf{Z}_l) = rank([\mathbf{X}, \mathbf{Z}_l]) - rank(\mathbf{X}) \leq rank(\mathbf{Z}_l)$.
- Scale-invariance property. DF and prior distributions on DF avoid problems arising from scaling of columns of the design matrix. (Section 2.2.3 discusses priors on DF.) Suppose that in the model (2.31), the covariance of each random effect $\mathbf{u}_l$ and the covariance of the error term ε are characterized by a single unknown parameter: $\mathbf{G}_l = \sigma_l^2 \mathbf{G}_l^0$ and $\mathbf{R} = \sigma_e^2 \mathbf{R}^0$, where the σ_l^2 are unknown scalars and $\mathbf{G}_l^0, \mathbf{R}^0$ are known and positive definite. Let $\mathbf{B} \in R^{p \times p}$ be nonsingular, and $\mathbf{H}_l$ be a matrix such that $\mathbf{H}_l \mathbf{G}_l^0 \mathbf{H}_l' = \mathbf{G}_l^0$, e.g., $\mathbf{G}_l^0$ is the identity and $\mathbf{H}_l$ is orthogonal. Then the posterior of $\mathbf{X}\beta + \mathbf{Zu}$ arising from independent priors on $(DF(\mathbf{Z}_1)), \cdots, DF(\mathbf{Z}_L))$ and σ_e^2 is the same when $\mathbf{X}$ is transformed to $\mathbf{X}^* = \mathbf{XB}$, and $\mathbf{Z}_l$ is transformed to $\mathbf{Z}_l^* = t_l \mathbf{Z}_l \mathbf{H}_l$ for nonzero scalars t_l.

Because $DF(\mathbf{Z}_l)$ can be laborious to compute, Cui (2008, p. 79) derived the approximation $DF_p(\mathbf{Z}_l) = \text{trace}\{\mathbf{Z}_l \mathbf{G}_l \mathbf{Z}_l' [(\mathbf{I} - P_\mathbf{X})(\mathbf{Z}_l \mathbf{G}_l \mathbf{Z}_l' + \mathbf{R})(\mathbf{I} - P_\mathbf{X})]^+\}$, which is equivalent to the approximation in Ruppert et al. (2003, pp. 175–176) when both are defined.

Regarding this approximation, Ruppert et al. (2003, p. 176) say "for all examples we have observed ... there is practically no difference between the approximate and exact degrees-of-freedom values." Cui et al. (2010) and Cui (2008) give examples where the approximation fails badly. These examples are all models in which the design matrices for different sets of effects are perfectly collinear with each other so that the effects are identified only because of shrinkage. This routinely happens in dynamic linear models (Chapter 6) and some spatial models (Chapter 5).

Regarding penalized splines, with which Ruppert et al. (2003) were mainly concerned, I have managed to produce an example for which the approximation fails. Let X_1 and X_2 be iid bivariate normal, each with mean zero and marginal variance 1 and with correlation 0.95. Consider a model where X_1 enters linearly and X_2 enters as a penalized spline with a truncated quadratic basis; for $n = 200$, I used 35 knots at the $(k+1)/37$ quantiles, $k = 1, \ldots, 35$. For $\sigma_s^2/\sigma_e^2 = 10^6$, the exact DF for X_2 is 30.6 while the approximate DF is 26.5. This is a really extreme fit in which the spline is hardly smoothing at all. If σ_s^2/σ_e^2 is reduced by a factor of just 10, to 10^5, the exact and approximate DF for X_2 agree to at least 7 digits, giving just under 25 DF to the spline for X_2. Even this is far wigglier than any reasonable spline fit.

This fragment of evidence suggests that the approximation can fail for penalized splines when the effects in question are highly collinear but that this may not be a practical concern. Further exploration of this question is given as an exercise.

2.2.2.4 Example of Partitioning DF

Example 5. Global mean surface temperature. Measures of the earth's surface temperature are used to describe global climate. Consider the series y_t, $t = 0, \ldots, 124$,

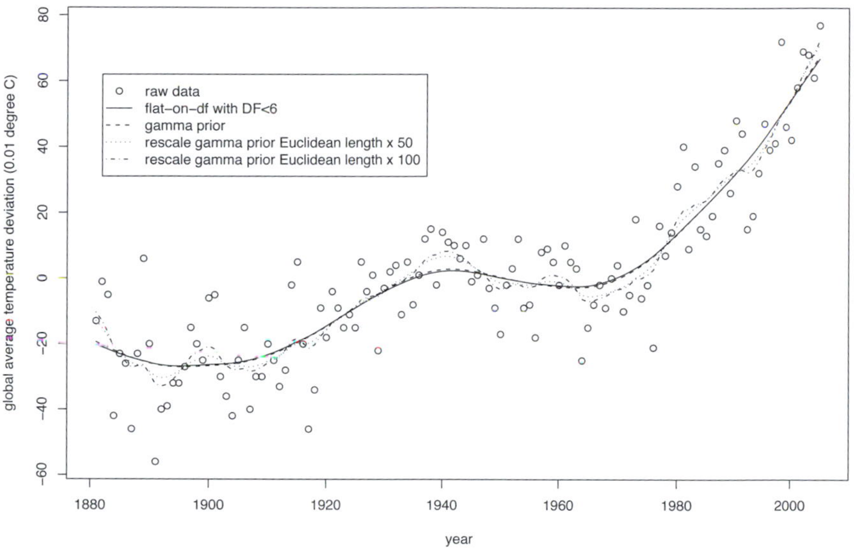

Figure 2.1: Global mean surface data, data and fitted smooths for gamma priors. The various fits are discussed in Section 2.2.3.3.

of global average temperature deviations (units 0.01 degrees Celsius) from 1881 to 2005. Figure 2.1 plots y_t. (This version of the series was downloaded in 2006 and is available on the book's web site. Current versions of this and related series are available at http://data.giss.nasa.gov/gistemp/tabledata/GLB.Ts.txt.) The rest of the plot is explained below in Section 2.2.3.3.

For this example, we smooth y_t using the linear growth model, which captures variation using a time-varying mean and trend and independent noise (West & Harrison 1997, Chapter 2). (Other ways to smooth this series are considered later in this book.) This model has equations for observation error, variation in local mean, and variation in trend respectively:

$$
\begin{aligned}
y_t &= \mu_t + n_t, t = 0, \cdots, T, & (2.36)\\
\mu_t &= \mu_{t-1} + \theta_{t-1} + w_{1,t}, t = 1, \cdots, T, & (2.37)\\
\theta_t &= \theta_{t-1} + w_{2,t}, t = 1, \cdots, T-1. &
\end{aligned}
$$

Let $\mu, \theta, \mathbf{n}, \mathbf{w}_1$ and $\mathbf{w}_2$ be the vectors of $\mu_t, \theta_t, n_t, w_{1t}$, and w_{2t} respectively. Assume $\mathbf{n}$, $\mathbf{w}_1$ and $\mathbf{w}_2$ are mutually independent and $\mathbf{n} \sim N(\mathbf{0}, \sigma_n^2 \mathbf{I}_{T+1}), \mathbf{w}_1 \sim N(\mathbf{0}, \sigma_1^2 \mathbf{I}_T)$, and $\mathbf{w}_2 \sim N(\mathbf{0}, \sigma_2^2 \mathbf{I}_{T-1})$, so σ_n^2, σ_1^2 and σ_2^2 describe respectively the size of observational error n_t and the smoothness of level μ_t and trend θ_t.

Section 6.1 shows that the linear growth model can be written as a mixed linear model with fixed effects μ_0 and θ_0, the initial local mean and trend respectively, and random effects $\mathbf{w}_1$and $\mathbf{w}_2$, the noise in the local mean and local trend respectively:

$$\begin{pmatrix} y_0 \\ y_1 \\ y_2 \\ y_3 \\ \vdots \\ y_T \end{pmatrix} = \begin{pmatrix} 1 & 0 \\ 1 & 1 \\ 1 & 2 \\ 1 & 3 \\ \vdots & \vdots \\ 1 & T \end{pmatrix} \begin{pmatrix} \mu_0 \\ \theta_0 \end{pmatrix} + \begin{pmatrix} 0 & 0 & \cdots & & & 0 \\ 1 & 0 & \cdots & & & 0 \\ 1 & 1 & 0 & \cdots & & 0 \\ 1 & 1 & 1 & 0 & \cdots & 0 \\ \vdots & \vdots & & & \ddots & \vdots \\ 1 & 1 & 1 & 1 & \cdots & 1 \end{pmatrix} \begin{pmatrix} w_{11} \\ w_{12} \\ \vdots \\ w_{1T} \end{pmatrix} \tag{2.38}$$

$$+ \begin{pmatrix} 0 & 0 & \cdots & & & 0 \\ 0 & 0 & \cdots & & & 0 \\ 1 & 0 & 0 & \cdots & & 0 \\ 2 & 1 & 0 & 0 & \cdots & 0 \\ \vdots & \vdots & & & \ddots & \vdots \\ T-1 & T-2 & T-3 & T-4 & \cdots & 1 \end{pmatrix} \begin{pmatrix} w_{21} \\ w_{22} \\ \vdots \\ w_{2,T-1} \end{pmatrix} + \begin{pmatrix} n_0 \\ n_1 \\ \vdots \\ n_T \end{pmatrix}. \tag{2.39}$$

To complete the mixed-linear-model specification, $\mathbf{R} = \sigma_n^2 \mathbf{I}_{T+1}$, $\mathbf{G}_1 = \sigma_1^2 \mathbf{I}_T$, and $\mathbf{G}_2 = \sigma_s^2 \mathbf{I}_{T-1}$. In the dynamic linear model literature, it is customary to add a distribution for the fixed effects (μ_0, θ_0). It is hard to see how to interpret the latter as anything other than a prior distribution, though the non-Bayesian dynamic linear model literature is blasé about this nicety. Commonly (μ_0, θ_0) is given a normal distribution with zero mean and large variance. The analyses to be shown below simply use a flat prior, i.e., with infinite variance, consistent with this book's usual treatment of fixed effects.

$DF(\mathbf{w}_1)$,$DF(\mathbf{w}_2)$, and $DF(\mathbf{n})$ follow straightforwardly using the formulas in Section 2.2.2 but the expressions do not simplify, so I omit them and illustrate their properties with some special cases. When the local mean and trend do not vary ($\sigma_1^2 = \sigma_2^2 = 0$), the model reduces to a linear regression with intercept μ_0 and slope β_0, with 2 DF. Thus when $\sigma_1^2 > 0$ or $\sigma_2^2 > 0$, $DF(\mathbf{w}_1)$ and $DF(\mathbf{w}_2)$ describe complexity in the fit beyond a linear trend attributable to variation in the local mean and the local trend respectively. Note that the matrix $[\mathbf{X}|\mathbf{Z}_1]$ has full column rank, so $\mathbf{Z}_2 = \mathbf{Z}_1\mathbf{H}$ for a specific $\mathbf{H}$, i.e., $\mathbf{Z}_2$ and $\mathbf{Z}_1$ are exactly collinear. Thus if $\sigma_1^2 > 0$ and $\sigma_2^2 > 0$, $\mathbf{w}_1$ and $\mathbf{w}_2$ compete with each other. Table 2.1 shows how they compete for various (σ_1^2, σ_2^2); noise variance σ_n^2 is fixed at 1 and T at 124. When $\sigma_2 = 0$ (Table 2.1A), $DF(\mathbf{w}_1)$ increases as σ_1 grows, with analogous results when $\sigma_1 = 0$ and σ_2 grows. However, for fixed $\sigma_1 > 0$ (Table 2.1B), $DF(\mathbf{w}_1)$ decreases as σ_2 grows because the larger σ_2 allows $\mathbf{w}_2$ to compete more aggressively, so $DF(\mathbf{w}_2)$ grows at the expense of $DF(\mathbf{w}_1)$. The analogous result occurs when σ_1 increases for fixed σ_2. Chapter 12 explores this sort of competition in greater detail.

Note that this example's random effects $\mathbf{w}_1$ and $\mathbf{w}_2$, like Example 4's F_i, do not appear to be old-style random effects. Again, the levels of the random effects — the annual adjustments to the mean and slope — are arguably draws from a population, but again, they are also arguably the whole population (you can't go back and draw

Table 2.1: DF for $\mathbf{w}_1$ and $\mathbf{w}_2$ in the Linear Growth Model in Different Scenarios (σ_n = 1)

A. Assuming $\sigma_j = 0$ for $j = 1$ or 2.

ξ	DF($\mathbf{w}_1$) if $\sigma_1 = \xi, \sigma_2 = 0$	DF($\mathbf{w}_2$) if $\sigma_1 = 0$, $\sigma_2 = \xi$
0.01	0.1	3.4
0.1	4.8	13.1
1	54.3	47.4
10	120.6	116.4

B. Assuming various values of σ_2 for fixed σ_1.

	$\sigma_1 = 0.1$		$\sigma_1 = 1$	
σ_2	DF($\mathbf{w}_1$)	DF($\mathbf{w}_2$)	DF($\mathbf{w}_1$)	DF($\mathbf{w}_2$)
0.01	3.3	2.8	54.1	0.2
0.1	1.3	12.8	49.5	5.2
1	0.4	47.2	27.6	38.1
10	0.02	116.4	2.2	114.4

a new slope for 1943). Further, in this case, the object of this exercise is to get a smooth fit to the series, so the realized values of the random effects are themselves of interest through their effect on $\hat{\mathbf{y}}$.

2.2.3 *Uses of DF for Mixed Linear Models*

DF are used to specify F-tests; to describe how large a model is, often as part of the penalty term of model-selection criteria; and to specify prior distributions on ϕ. The following subsections describe these uses, emphasizing aspects that are less well-known.

2.2.3.1 *Using DF to Specify F-Tests*

A conventional F-test of an effect in a mixed linear model requires degrees of freedom for the effect and for residuals. In ordinary linear models this is straightforward and for normal errors the theory is exact. Such F-tests can be extended to mixed linear models, but "parametric F-tests ... are only approximate" (Ruppert et al. 2003, Section 6.6.2). Hastie & Tibshirani (1990, Section 3.9) derived such two such approximate F-tests. The first was a simple extension of the test for ordinary linear models: If "*RSS*" means "residual sum of squares" and the subscripts "1" and "2" refer to smaller and larger nested models respectively, then this approximate test uses

$$\frac{(RSS_1 - RSS_2)/(\gamma_2 - \gamma_1)}{RSS_2/(n - \gamma_2)} \sim F_{\gamma_2 - \gamma_1, n - \gamma_2} \tag{2.40}$$

as a test statistic, where $n-\gamma_j$ is the degrees of freedom for error, $j=1,2$. The second approximation is somewhat more complicated and has closer to nominal operating characteristics. For the present purpose, the key point is that it also depends on the $n-\gamma_j$.

Hastie & Tibshirani (1990), as well as Ruppert et al. (2003), use this expression for the degrees of freedom for error for the linear smoother $\mathbf{S}_\lambda$ (dropping the subscript j on γ):

$$n-\gamma = n - \text{trace}(2\mathbf{S}_\lambda - \mathbf{S}_\lambda\mathbf{S}'_\lambda). \tag{2.41}$$

The rationale for this definition (Hastie & Tibshirani 1990, Section 3.5) is that for a linear smoother $\mathbf{S}_\lambda$ with λ treated as known,

$$E\{RSS(\lambda)\} = \{n - \text{trace}(2\mathbf{S}_\lambda - \mathbf{S}_\lambda\mathbf{S}'_\lambda)\}\sigma_e^2 + \mathbf{b}'_\lambda\mathbf{b}_\lambda, \tag{2.42}$$

where σ_e^2 is the error variance; $RSS(\lambda)$ is the residual sum of squares of the smooth fit given λ; $\mathbf{b}_\lambda = (\mathbf{S}_\lambda - \mathbf{I}_n)f$ is the bias of the smoother given λ, for $f = E(\mathbf{y})$; and all expectations are with respect to the distribution of the residual error term ε, *not* the distribution of the random effect $\mathbf{u}$, if one is present. To derive an F-test of whether a smooth fit is superior to a linear fit, Hastie & Tibshirani assume that the larger model (with the smoother) is unbiased and that the smaller model (with the linear fit) is unbiased under the null, so the bias term $\mathbf{b}'_\lambda\mathbf{b}_\lambda$ is zero. Then by analogy to linear regression, they define degrees of freedom for error as in (2.41). Ruppert et al. (2003) derive the same definition of degrees of freedom for error and assume somewhat less explicitly that "the bias term ... is negligible" (p. 83).

Hastie & Tibshirani's rationale for γ requires a bias of zero, which is exactly correct under their assumptions. Like most good ideas, however, this idea has been extended beyond its original use to cases where bias is not negligible. Also, as noted above, even if the bias $\mathbf{b}_\lambda$ is zero, this definition has the untidy property that

$$\begin{aligned}(\text{DF in fit}) + (\text{DF for error}) &= \text{trace}(\mathbf{S}_\lambda) + n - \text{trace}(2\mathbf{S}_\lambda - \mathbf{S}_\lambda\mathbf{S}'_\lambda) \\ &= n - \text{trace}(\mathbf{S}_\lambda - \mathbf{S}_\lambda\mathbf{S}'_\lambda) < n.\end{aligned} \tag{2.43}$$

The inequality holds because $\mathbf{S}_\lambda$ is symmetric with eigenvalues in $[0,1]$ and at least one eigenvalue less than 1, so $\text{trace}(\mathbf{S}'_\lambda\mathbf{S}_\lambda) < \text{trace}(\mathbf{S}_\lambda)$.

As noted, the expectation $E\{RSS(\lambda)\}$ in equation (2.42) is with respect to the distribution of the error term ε, not the random effect's distribution. Cui et al. (2010, Section 4) show for mixed linear models that if the expectation is also taken over the distribution of the random effect $\mathbf{u}$, $E\{RSS(\lambda)\} = \sigma_e^2(n - \text{trace}(\mathbf{S}_\lambda))$, which means that the missing degrees of freedom in equation (2.43) are for the bias term. Taking the expectation with respect to the random effects implies a count of degrees of freedom for error equal to DF as defined above in Section 2.2.2. Cui et al. (2010) give an example in which the difference between these two definitions of degrees of freedom for error appears large, 764 versus 700, although this difference would have negligible practical impact on an F-test.

2.2.3.2 Using DF to Describe Model Size, for Model-Selection Criteria

Many model-selection criteria, e.g., C_p, AIC, BIC, DIC, and other ICs, have the form {goodness-of-fit measure} + {penalty}. The various criteria use different penalties but for all these criteria, the penalty increases as the model's complexity increases. The deviance information criterion DIC (Spiegelhalter et al. 2002) uses a measure p_D of model complexity; this is compared to DF below in Section 2.2.4. For Mallows' C_p, DF is the model-complexity measure in the penalty term, interpreting the fitted mixed linear model as a linear smoother and taking ϕ as known (Hastie & Tibshirani 1990, Section 3.5).

More recently, Vaida & Blanchard (2005) derived DF as the model complexity in the penalty for a version of Akaike's Information Criterion (AIC) for clustered observations. They distinguished a *population focus* for the analysis from a *cluster focus*. With a population focus, the random effect $\mathbf{u}$ is simply a device for modeling the correlation of the outcome $\mathbf{y}$ within clusters, and the mixed linear model is equivalent to the model $\mathbf{y} \sim N_n(\mathbf{X}\beta, \mathbf{V})$ for $\mathbf{V} = \mathbf{ZGZ}' + \mathbf{R}$. (In the present book's terms, $\mathbf{u}$ is an old-style random effect.) The random effects $\mathbf{u}$ are incorporated into the error covariance and the complexity of the model is rank($\mathbf{X}$) + {number of free parameters in ϕ}. This is the version of AIC that is usually computed for mixed linear models (e.g., in SAS's MIXED procedure).

By contrast, if the analysis has a *cluster focus*, prediction at the cluster level is conditional on the cluster means, so the random effect $\mathbf{u}$ is no longer a nuisance introduced only to induce correlation within clusters but rather is part of the mean structure used for the prediction. (In the present book's terms, $\mathbf{u}$ is a new-style random effect.) From this beginning, Vaida & Blanchard returned to Akaike's original definition and derived from first principles the conditional Akaike information, conditional on the random effects $\mathbf{u}$, and an unbiased estimator of the conditional Akaike information, cAIC, the conditional AIC.

The penalty term in cAIC is, as in AIC, 2 $\times$ model complexity, but with the cluster focus, the model's complexity depends on the extent of shrinkage of cluster means. If ϕ is known, Vaida & Blanchard (2005) showed that the correct model-complexity measure is DF as defined in Section 2.2.1. For the case in which $\mathbf{R} = \sigma_e^2\mathbf{I}$ with σ_e^2 unknown, but $\mathbf{G}$ is known, Vaida & Blanchard showed that the correct model-complexity measure is $q_1(\text{DF}+1)+q_2$, where q_1 and q_2 are functions of n and p, and as $n \rightarrow \infty$, $q_1 \rightarrow 1$ and $q_2 \rightarrow 0$. The "+1" in the model-complexity term is for σ_e^2; AIC accounts for all unknown parameters, while DF only describes the complexity of the fitted mean structure.

2.2.3.3 Prior Distributions on DF

The most familiar notion of DF is for linear models with one error term, in which a model's DF is the fixed known rank of its design matrix. As extended to scatterplot smoothers (e.g., Hastie & Tibshirani 1990), a fit's DF is not fixed and known before the model is fit, but rather is a function of the tuning constant λ that controls the fit's smoothness. For DF as defined in Section 2.2.2, e.g., in Example 5's model for global mean surface temperature, the vector of effect-specific DF, $(DF(\mathbf{w}_1), DF(\mathbf{w}_2))$, is a one-to-one function of the vector of variance ratios $(\sigma_1^2/\sigma_e^2, \sigma_2^2/\sigma_e^2)$ and again is not fixed or known before the model is fit. Because of this one-to-one function, placing a prior distribution on $(\sigma_1^2/\sigma_e^2, \sigma_2^2/\sigma_e^2)$ induces a prior on $(DF(\mathbf{w}_1), DF(\mathbf{w}_2))$. But it is equally legitimate to place a prior on $(DF(\mathbf{w}_1), DF(\mathbf{w}_2))$ to induce a prior on the unknown variance ratios $(\sigma_1^2/\sigma_e^2, \sigma_2^2/\sigma_e^2)$. For new-style random effects, such a prior on DF has the advantage of being specified on the interpretable complexity measure instead of on the uninterpretable variances or variance ratios.

This can be stated more precisely, using the notation of Section 2.2.2. In the mixed linear model (2.31), suppose each random effect $\mathbf{u}_l$ and the error term ε have covariances characterized by one unknown parameter, respectively $\mathbf{G}_l = \sigma_l^2 \mathbf{G}_l^0$ and $\mathbf{R} = \sigma_e^2 \mathbf{R}^0$, where the σ_j^2 are unknown scalars and $\mathbf{G}_l^0$ and $\mathbf{R}^0$ are known and positive definite. Assume further that the space spanned by $\mathbf{Z}_l$, the design matrix for $\mathbf{u}_l$, is not a subset of the space spanned by $\mathbf{X}$, the design matrix for the fixed effects. Then $DF(\cdot)$ is a 1-1 mapping between $\nu = (DF(\mathbf{Z}_1), \cdots, DF(\mathbf{Z}_L))'$ on ν's range and $\mathbf{s} = (\log(\sigma_1^2/\sigma_e^2), \cdots, \log(\sigma_L^2/\sigma_e^2)) \in R^L$, so a prior on ν induces a unique prior on $\mathbf{s}$. The online supplement for Cui et al. (2010) gives a proof and the Jacobian. Specification of a prior on ϕ is completed by specifying a prior on σ_e^2.

By the scale-invariance property in Section 2.2.2.3, the posterior distribution of $\mathbf{X}\beta + \mathbf{Z}\mathbf{u}$ using a prior on ν as above is invariant under certain transformations of $\mathbf{X}$ and $\mathbf{Z}$. Sometimes it is desirable to put a prior on functions of ν, e.g., for the linear-growth model we could put a prior on $DF(\mathbf{w}_1)/(DF(\mathbf{w}_1) + DF(\mathbf{w}_2))$, the fraction of the fitted DF attributed to changes in the local mean. If an L-variate function of ν, $\mathbf{k} = (k_1(\nu), \cdots, k_L(\nu))$ is a 1-1 mapping, then a prior on $\mathbf{k}$ induces a unique prior on $\mathbf{s}$.

When $\mathbf{G}_l$'s form is more complex than assumed above, a prior on DF partially specifies a prior on the unknowns in $\mathbf{G}_l$ and a complete prior specification requires a prior on other functions of $\mathbf{G}_l$. For example, if $\mathbf{G}_l = diag(\sigma_{l1}^2, \sigma_{l2}^2)$, then a uniform prior on $DF(\mathbf{Z}_l)$ induces a prior on a scalar function of $(\sigma_{l1}^2/\sigma_e^2, \sigma_{l2}^2/\sigma_e^2)$. A complete specification requires a prior on another scalar function of $(\sigma_{l1}^2/\sigma_e^2, \sigma_{l2}^2/\sigma_e^2)$, e.g., on $\sigma_{l1}^2/\sigma_{l2}^2$.

A prior placed on DF can be subjected to conditions, like any other prior. Thus, for example, a prior on $\nu = (DF(\mathbf{Z}_1), \cdots, DF(\mathbf{Z}_L))'$ could be subject to the condition $DF(\mathbf{Z}_1) + \cdots + DF(\mathbf{Z}_L) < K$, for K a constant. Such a prior constrains the total complexity of the fit but allows the Bayesian machinery to allocate that complexity among the various effects.

For a simple example of a prior on DF, consider the balanced one-way random effects model as in Section 2.2.1. For this model, DF = $1 + (q-1)mr/(mr+1)$ for $r = \sigma_s^2/\sigma_e^2$, with DF $\in [1, q]$. Thus, a prior distribution on DF induces a prior distribution on the variance ratio r. An obvious (though sometimes ill-advised) prior on DF is a uniform distribution, which for the present example is indifferent among degrees of shrinkage (and indeed for this problem is equivalent to the uniform-shrinkage prior, Daniels 1999). This prior on DF implies a prior distribution for r with cumulative distribution function $Pr(r \leq k) = mk/(mk+1)$ for $r \in (0, \infty)$. (The proof is an exercise, which also explores an unusual property of this induced prior on r.) For this prior on r, if the number of observations per group, m, is increased, $Pr(r \leq k)$ increases for given k, so prior probability is pushed toward smaller values of r a priori and thus greater shrinkage. In other words, the data are discounted more steeply for larger m. To someone accustomed to Bayesian analyses, it may seem odd to have the prior depend on the sample size, but this is true of some other so-called objective or reference priors and this is legal in Bayesian terms (i.e., it creates no incoherence) as long as the sample size m is fixed (i.e., the study giving rise to the dataset is not analyzed sequentially).

For a more complex example, consider Example 5, the linear growth model applied to the global mean surface temperature data, and consider two types of priors on $(\sigma_n^2, \sigma_1^2, \sigma_2^2)$, an independent Gamma(0.001,0.001) on each $1/\sigma_j^2$, and DF-induced priors. Figure 2.1 in Section 2.2.2.4 plotted the data and three fits (posterior means) arising from gamma priors. For one fit, the analysis used the random-effects design matrix $\mathbf{Z}$ as given in Section 2.2.2.4; for the other two fits, the columns of $\mathbf{Z}$ were multiplied by 50 and 100. The fit with the original scaling smooths the most, with $\text{DF}(\mathbf{w}_1) + \text{DF}(\mathbf{w}_2)$ having posterior mean 5.6. The gamma prior's effect is not invariant to the scaling of $\mathbf{Z}$'s columns, so the posteriors differ noticeably when the same gamma prior is used with re-scaled design matrices. When $\mathbf{Z}$'s columns have been multiplied by 100, the fit is quite wiggly; the posterior means of $\text{DF}(\mathbf{w}_1)$ and $\text{DF}(\mathbf{w}_2)$ sum to 21.5. It was simply luck that in the "natural" scaling, the gamma prior gave a reasonable result.

Priors specified on DF, by contrast, avoid the problem of scaling. Instead, they allow you to control the fit's smoothness directly by constraining DF in the fit. Figure 2.2 plots fits from flat priors on ($\text{DF}(\mathbf{w}_1)$, $\text{DF}(\mathbf{w}_2)$) with five different constraints on $\text{DF}(\mathbf{w}_1) + \text{DF}(\mathbf{w}_2)$ in the smoothed effects. The fit becomes smoother as the constraint on total DF is reduced. Figure 2.1 plots the fit from the prior with total DF constrained to be less than 6 using a solid line; it is nearly indistinguishable from the fit arising from the Gamma(0.001,0.001) priors. Figure 2.3 shows histograms of 10,000 draws from the posterior of $\text{DF}(\mathbf{w}_1)$ and $\text{DF}(\mathbf{w}_2)$ arising from this same fit. Both posteriors are skewed, but in different directions. For the local mean $E(DF(\mathbf{w}_1)|\mathbf{y}) = 0.86$, while for the local slope $E(DF(\mathbf{w}_2)|\mathbf{y}) = 4.58$.

This example also gives an instance in which the approximate DF (Section 2.2.2.3; Ruppert et al. 2003, Section 8.3) performs poorly. For the flat priors on ($\text{DF}(\mathbf{w}_1)$, $\text{DF}(\mathbf{w}_2)$) with total DF constrained to be less than 6, the posterior mean of exact and approximate DF for $\mathbf{w}_1$ are 0.86 and 1.87 respectively, differing by more than 1 DF, while the posterior mean of exact and approximate DF for $\mathbf{w}_2$ are closer, 4.58 and 4.76 respectively. When total DF is constrained to be less than 10, the posterior mean of exact and approximate DF are 2.97 and 5.13 for $\mathbf{w}_1$ — differing by more than 2 DF — and 5.14 and 5.75 for $\mathbf{w}_2$. As noted, this is a case in which the two effects $\mathbf{w}_1$ and $\mathbf{w}_2$ have design matrices that are exactly collinear.

I do not propose an *unconstrained* flat prior on DF as a default prior. Fr this model and dataset, a flat prior on ($\text{DF}(\mathbf{w}_1)$, $\text{DF}(\mathbf{w}_2)$) without a constraint on total DF gives a grossly undersmoothed fit with about 45 DF *a posteriori*. It is still true that overparameterized models often need strong prior information.

Other examples of priors on degrees of freedom are given in Hodges & Sargent (2001), Paciorek & Schervisch (2006), Hodges et al. (2007), and Cui et al. (2010).

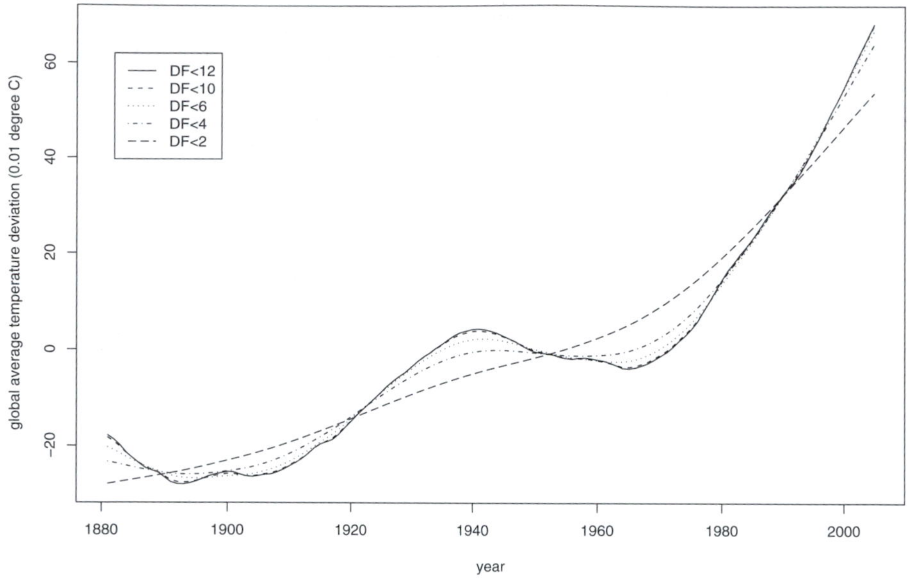

Figure 2.2: Global mean surface data, DF prior with different sum constraints on total smoothed DF.

Hodges et al. (2007) presents a designed simulation experiment comparing three priors for smoothed ANOVA, two priors specified on DF and one specified on the precisions themselves. This experiment's results are summarized in Section 4.2.3.2 below. Hodges et al. (2007) also explored an example of a prior on DF conditioned to fix the total DF in an effect; this is described in detail in Section 4.2.3.3 below.

2.2.4 DF Compared to p_D of Spiegelhalter et al. (2002)

In defining the deviance information criterion (DIC) — which, like the other ICs, has the form {goodness-of-fit measure} + {penalty} — Spiegelhalter et al. (2002) defined their penalty, the effective number of parameters p_D, in terms of a measure of model fit. Specifically, their Sections 2.4 to 2.6 define $p_D = \overline{D(\theta)} + D(\bar{\theta})$ where $D(\theta) = -2\log f(\mathbf{y}|\theta)$, for f the probability or density of $\mathbf{y}$ given the unknown θ, and over-bars indicate the posterior mean. In view of Plummer (2008), it is not clear that the "right" penalty for using the same data to fit and evaluate a model is a simple function of *any* measure of model complexity, so describing model complexity and comparing models should be considered distinct problems (although many people seem determined to use DF to penalize fit in model-comparison criteria.)

DF, as defined in Cui et al. (2010), was intended to describe a model's complexity in terms drawn from simple, well-understood models, making it possible to control complexity in fitting the model to a dataset by means of a prior distribution on complexity. Regarding p_D's interpretability, its definition (Spiegelhalter et al. 2002, Section 2.3) is opaque and the only concrete interpretation Spiegelhalter et al. (2002)

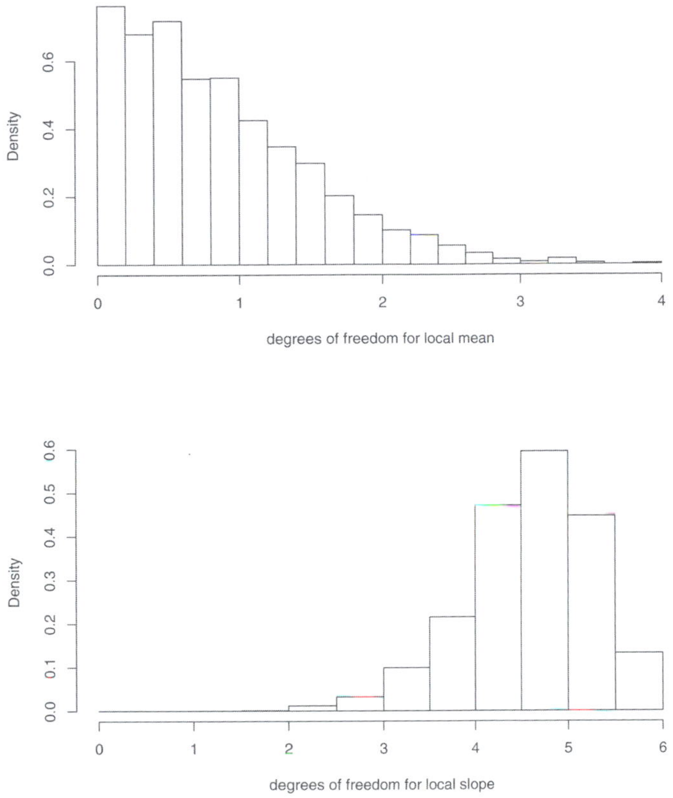

Figure 2.3: Global mean surface data, histograms of posterior DF for local mean (top) and local slope (bottom), for the flat prior on DF with sum of total smoothed DF constrained to < 6.

offer is that in simple cases — where complexity measures already exist — p_D takes values that people generally like. Second, the definition of DF implicitly presumes that a model exists independently of any realized outcome $\mathbf{y}$, so that model's complexity can be defined without a realized $\mathbf{y}$. This is true of DF and of most other complexity measures (e.g., Hastie & Tibshirani 1990, Ye 1998), but not p_D. The definition of p_D *requires* a realized $\mathbf{y}$ through its dependence on the posterior mean of the model's unknowns, $\bar{\theta}$, and of the deviance, $\overline{D(\theta)}$. In special cases, such as normal hierarchical models with all covariance parameters known, p_D does not depend on $\mathbf{y}$ and in these cases p_D agrees with DF (Spiegelhalter et al. 2002, Sections 2, 4; Green 2002). But with unknown variances — i.e., in practical situations — p_D and DF differ.

A complexity measure like DF, defined independently of the outcome $\mathbf{y}$, will in general be a function of unknowns. This is acceptable: Statisticians routinely specify models as functions of unknowns — in Bayesian terms, conditional on unknowns

— and it creates no difficulty to describe a model's complexity as a function of its unknowns. The DF of a *fitted* model is obviously a function of the data, either through plug-in estimates or posterior distributions of the unknown parameters in the covariance matrices. But a complexity measure defined independently of $\mathbf{y}$, like DF, allows a prior distribution on complexity to control softly the complexity of the fit. This is not possible with p_D because in general it is defined in terms of a specific realized $\mathbf{y}$.

It is not clear p_D can be partitioned corresponding to components of the model, as DF can. Spiegelhalter et al. (2002, Sections 6.3, 8.1) do partition p_D corresponding to subdivisions of the deviance to create an outlier diagnostic from DIC. Also, in problems where the likelihood factors, DIC and p_D partition analogously, e.g., for errors-in-covariates models. Green (2002) gives a partition of p_D identical to the partition of DF in Section 2.2.2 for the special case where all covariance matrices are known, but it is not clear this can be extended to the case of unknown covariances.

Thus, while p_D's relatively easy computation is certainly desirable, for purposes such as placing priors on ϕ, this advantage seems outweighed by p_D's conceptual and practical disadvantages.

2.2.5 *Some Intuition for Fractional Degrees of Freedom*

People accustomed to integer degrees of freedom are often puzzled by fractional degrees of freedom. How can fractional degrees of freedom be interpreted? In ordinary linear models, "two degrees of freedom" implies a projection space specified by two basis vectors, in which the fitted values can vary arbitrarily. For a mixed linear model, "two degrees of freedom" means that the projection (fitted-value) space is a subspace of R^n with as many as $p+q$ dimensions, but the fitted values cannot vary arbitrarily in the q dimensions corresponding to the random effects. The distribution placed on the random effects — or the constraint cases, using the constraint-case formulation of Section 2.1 — constrain the fitted values, restricting their variation in certain directions of fitted-value space to an extent determined by ϕ. This constraint reduces the flexibility and thus complexity of the fit in a continuous manner, requiring DF on a continuous scale. The rest of this section uses the constraint-case formulation and then the mixed-model formulation to make this idea somewhat more formal.

2.2.5.1 *Using the Constraint-Case Formulation*

To do this, we first need to recapitulate the definition of DF in the whole model fit using the constraint-case formulation. The mixed linear model in constraint-case form was given in Section 2.1.1's equation (2.13). Here, as in Hodges & Sargent (2001) we use an improper flat prior on β, so the prior cases are omitted:

$$\left[\begin{array}{c}\mathbf{y}\\ \hline \mathbf{0}_q\end{array}\right] = \left[\begin{array}{c|c}\mathbf{X} & \mathbf{Z}\\ \hline \mathbf{0}_q & -\mathbf{I}_q\end{array}\right]\left[\begin{array}{c}\beta\\ \hline \mathbf{u}\end{array}\right] + \left[\begin{array}{c}\varepsilon\\ \hline \delta\end{array}\right] \tag{2.44}$$

where the error term has covariance matrix

$$\Gamma = \mathrm{cov}\left(\frac{\varepsilon}{\delta}\right) = \left[\begin{array}{cc}\mathbf{R} & \mathbf{0}\\ \mathbf{0} & \mathbf{G}\end{array}\right]. \tag{2.45}$$

For the rest of this section, treat ϕ as known. Summarize equation (2.44) as

$$\mathbf{Y} = \mathbf{W}\Theta + \mathbf{e}, \tag{2.46}$$

with $\mathbf{Y} = (\mathbf{y}', \mathbf{0}')'$ and so on. Pre-multiply (2.46) by $\Gamma^{-1/2}$, where $\Gamma = \text{cov}(\mathbf{e})$, giving

$$\mathbf{Y}_\Gamma = \mathbf{W}_\Gamma\Theta + \mathbf{e}_\Gamma, \quad \text{cov}(\mathbf{e}_\Gamma) = \mathbf{I}_{n+q}, \tag{2.47}$$

where the subscript Γ indicates the pre-multiplication, e.g., $\mathbf{Y}_\Gamma = \Gamma^{-1/2}\mathbf{Y}$, etc. This gives a model with homoscedastic errors. Now let $\mathbf{W}_\Gamma$ have singular value decomposition $\mathbf{W}_\Gamma = \mathbf{U}\Delta\mathbf{V}'$, where $\mathbf{U}$ has the same dimension as $\mathbf{W}$ and orthonormal columns, $\mathbf{U}'\mathbf{U} = \mathbf{I}_{p+q}$, Δ is diagonal, and $\mathbf{V}$ is $(p+q) \times (p+q)$ and an orthogonal matrix. The subscript Γ is suppressed in the singular value decomposition, but $\mathbf{U}$, Δ, and $\mathbf{V}$ depend on Γ. Re-parameterize (2.47) by defining $\Xi = \Delta\mathbf{V}'\Theta$, so the model (2.47) becomes

$$\left[\frac{\mathbf{y}_\Gamma}{\mathbf{0}_q}\right] = \mathbf{Y}_\Gamma = \mathbf{U}\Xi + \mathbf{e}_\Gamma \tag{2.48}$$

$$= \left[\frac{\mathbf{U}_1}{\mathbf{U}_2}\right]\Xi + \mathbf{e}_\Gamma, \tag{2.49}$$

where $\mathbf{U}_1$ corresponds to the data cases and has n rows like $\mathbf{y}_\Gamma$; $\mathbf{U}_2$ corresponds to the constraint cases and has q rows; and $\mathbf{I}_{p+q} = \mathbf{U}'\mathbf{U} = \mathbf{U}_1'\mathbf{U}_1 + \mathbf{U}_2'\mathbf{U}_2$, so that $\mathbf{U}_2'\mathbf{U}_2 = \mathbf{I} - \mathbf{U}_1'\mathbf{U}_1$ is non-negative definite and positive definite as long as the (known) value of ϕ does not push either $\mathbf{G}$ or $\mathbf{R}$ to either extreme of zero or infinity.

With this machinery, the DF in the whole model fit are defined as follows. The fitted values (again, treating ϕ as known) are

$$\hat{\mathbf{y}}_\Gamma = \mathbf{U}_1\hat{\Xi} = \mathbf{U}_1(\mathbf{U}'\mathbf{U})^{-1}\mathbf{U}'\mathbf{Y}_\Gamma \tag{2.50}$$

$$= \mathbf{U}_1[\mathbf{U}_1'|\mathbf{U}_2']\left[\frac{\mathbf{y}_\Gamma}{\mathbf{0}_q}\right] \tag{2.51}$$

$$= \mathbf{U}_1\mathbf{U}_1'\mathbf{y}_\Gamma. \tag{2.52}$$

Note that in computing the fitted values in equation (2.50), we have thrown away part of the fit of the pseudo-data vector $\mathbf{Y}_\Gamma$, namely $\mathbf{U}_2(\mathbf{U}'\mathbf{U})^{-1}\mathbf{U}'\mathbf{Y}_\Gamma$.

Equation (2.52) defines a linear smoother with $\mathbf{S}_\lambda = \mathbf{U}_1\mathbf{U}_1'$, where the tuning constant $\lambda = \Gamma$ is implicit in $\mathbf{U}_1$. Thus the DF in the whole fit is

$$DF = \text{trace}(\mathbf{U}_1\mathbf{U}_1') = \text{trace}(\mathbf{U}_1'\mathbf{U}_1) = \sum_{k=1}^{p+q} \mathbf{U}_1^{k\prime}\mathbf{U}_1^k, \tag{2.53}$$

where the superscript k indexes the columns of $\mathbf{U}_1$. Thus $\mathbf{y}_\Gamma$'s projection onto $\mathbf{U}_1^k$ has $\rho_k = \mathbf{U}_1^{k\prime}\mathbf{U}_1^k$ degrees of freedom, where $\rho_k = \mathbf{U}_1^{k\prime}\mathbf{U}_1^k \leq 1$. Note that while $\mathbf{U}_1\mathbf{U}_1'$ is a projection, it is not an orthogonal projection.

We can now interpret the idea, stated at the beginning of this section, that the constraint cases constrain the fitted values, restricting their variation in certain directions to an extent determined by ϕ, the unknowns in Γ. If $\hat{\mathbf{y}}_{\Gamma,k} = \mathbf{U}_1^k\mathbf{U}_1^{k\prime}\mathbf{y}_\Gamma$ is the component of $\hat{\mathbf{y}}_\Gamma$ along $\mathbf{U}_1^k$'s direction, and if $\| \bullet \|$ denotes Euclidean norm, then

$$\|\hat{\mathbf{y}}_{\Gamma,k}\| = \|\mathbf{U}_1^k\mathbf{U}_1^{k\prime}\hat{\mathbf{y}}_\Gamma\| = \|\mathbf{U}_1^k\||\mathbf{U}_1^{k\prime}\hat{\mathbf{y}}_\Gamma| = \|\mathbf{U}_1^k\|^2|\mathbf{U}^{k*\prime}\hat{\mathbf{y}}_\Gamma|, \tag{2.54}$$

where $\mathbf{U}^{k*}$ has unit length and the same direction as $\mathbf{U}_1^k$. The DF along direction $\mathbf{U}_1^k$ is $\rho_k = \|\mathbf{U}_1^k\|^2$; thus, the constraint cases can be understood as restricting $\hat{\mathbf{y}}_{\Gamma,k}$ by multiplying the projection in the direction $\mathbf{U}^{k*}$ by a factor ρ_k, the DF in that direction.

For any $(p+q)$-dimensional orthogonal matrix $\mathbf{P}$, reparameterizing (2.48) again using $\mathbf{U}^\circ = \mathbf{UP}$ and $\Xi^\circ = \mathbf{P}'\Xi$ produces the same DF in the whole fit but a different set of canonical directions $\mathbf{U}_1^{\circ k}$ and corresponding ρ_k°.

2.2.5.2 *Using the Mixed Model Formulation*

As in the preceding subsection, assume ϕ and thus $\mathbf{G}$ and $\mathbf{R}$ are known. Write the mixed linear model as

$$\mathbf{y} = \mathbf{C}\begin{pmatrix}\beta \\ \mathbf{u}\end{pmatrix} + \varepsilon, \tag{2.55}$$

where, as before, $\mathbf{C} = [\mathbf{X}|\mathbf{Z}]$. Pre-multiply (2.55) by $\mathbf{R}^{-1/2}$ to make this a homoscedastic model:

$$\mathbf{y}_* = \mathbf{C}_*\begin{pmatrix}\beta \\ \mathbf{u}\end{pmatrix} + \varepsilon_*, \tag{2.56}$$

where the subscript "*" indicates premultiplication by $\mathbf{R}^{-1/2}$ and $\text{cov}(\varepsilon_*) = \mathbf{I}_n$. For $\mathbf{R} = \sigma_e^2\mathbf{I}_n$, equation (2.56) is simply equation (2.55) with both sides divided by σ_e.

The fitted values arising from (2.56) are

$$\hat{\mathbf{y}}_* = \mathbf{C}_*\left[\mathbf{C}_*'\mathbf{C}_* + \begin{pmatrix}\mathbf{0} & \mathbf{0} \\ \mathbf{0} & \mathbf{G}^{-1}\end{pmatrix}\right]^{-1}\mathbf{C}_*'\mathbf{y}_*. \tag{2.57}$$

Let $\mathbf{C}_*$ have singular value decomposition $\mathbf{C}_* = \mathbf{U}_*\mathbf{DV}'$, where $\mathbf{U}_*$ is $n \times (p+q)$ with orthonormal columns, $\mathbf{D}$ is $(p+q)\times(p+q)$ diagonal with non-negative diagonal entries, and $\mathbf{V}$ is a $(p+q)\times(p+q)$ orthogonal matrix. The fitted values in (2.57) can then be rewritten as

$$\hat{\mathbf{y}}_* = \mathbf{U}_*\mathbf{D}\left[\mathbf{D}^2 + \mathbf{V}_2'\mathbf{G}^{-1}\mathbf{V}_2\right]^{-1}\mathbf{D}\mathbf{U}_*'\mathbf{y}_*, \tag{2.58}$$

where $\mathbf{V}' = [\mathbf{V}_1'|\mathbf{V}_2']$ and $\mathbf{V}_2'$ is $(p+q)\times q$. If we now assume that $\mathbf{D}$ has only positive diagonal elements — i.e., that $\mathbf{X}, \mathbf{Z}$, and $\mathbf{C}$ are full rank — then the fitted values become

$$\hat{\mathbf{y}}_* = \mathbf{U}_*\left[\mathbf{I}_{p+q} + \mathbf{D}^{-1}\mathbf{V}_2'\mathbf{G}^{-1}\mathbf{V}_2\mathbf{D}^{-1}\right]^{-1}\mathbf{U}_*'\mathbf{y}_*. \tag{2.59}$$

Let $\mathbf{D}^{-1}\mathbf{V}_2'\mathbf{G}^{-1}\mathbf{V}_2\mathbf{D}^{-1}$ have spectral decomposition $\mathbf{D}^{-1}\mathbf{V}_2'\mathbf{G}^{-1}\mathbf{V}_2\mathbf{D}^{-1} = \Phi\Delta\Phi'$, where Φ is $(p+q)\times(p+q)$ orthogonal and Δ is diagonal with diagonal entries $\delta_k, k = 1,\ldots,p+q$. For simplicity, assume $\mathbf{G}$ has full rank q; then $\mathbf{D}^{-1}\mathbf{V}_2'\mathbf{G}^{-1}\mathbf{V}_2\mathbf{D}^{-1}$ has rank q, so Δ has q positive diagonal entries and p zero diagonal entries. The fitted values now become

$$\begin{aligned}
\hat{\mathbf{y}}_* &= \mathbf{U}_*\Phi\left[\mathbf{I}_{p+q} + \Delta\right]^{-1}\Phi'\mathbf{U}_*'\mathbf{y}_* && (2.60)\\
&= \mathbf{U}\left[\mathbf{I}_{p+q} + \Delta\right]^{-1}\mathbf{U}'\mathbf{y}_* && (2.61)\\
&= \sum_{k=1}^{p+q}\mathbf{U}^k(1+\delta_k)^{-1}\mathbf{U}^{k\prime}\mathbf{y}_*, && (2.62)
\end{aligned}$$

where $\mathbf{U} = \mathbf{U}_* \Phi$ and $\mathbf{U}^k$ is the kth column of $\mathbf{U}$. This expression for the fitted values corresponds to re-parameterizing (2.56) as $\mathbf{C}_* \binom{\beta}{\mathbf{u}} = \mathbf{U}\xi$ for $\xi = \Phi' \mathbf{D} \mathbf{V} \binom{\beta}{\mathbf{u}}$. Thus the fitted values are the sum of components $\hat{\mathbf{y}}_*^k = \mathbf{U}^k (1+\delta_k)^{-1} \mathbf{U}^{k\prime} \mathbf{y}_*$; the component or projection along $\mathbf{U}^k$ is reduced in length by the factor $(1+\delta_k)^{-1} \in [0,1]$. But the DF in the fit, from (2.60), is trace$(\mathbf{U} [\mathbf{I}_{p+q} + \Delta]^{-1} \mathbf{U}') = \sum (1+\delta_k)^{-1}$, where the component $\hat{\mathbf{y}}_*^k$ of the fit has $(1+\delta_k)^{-1}$ DF. Recall that p of the δ_k are 0 and the other q are positive. Thus, as we found using the constraint-case formulation in Section 2.2.5.1, q components of the fit $\hat{\mathbf{y}}_*^k$ are shrunk by reducing their projection in the direction $\mathbf{U}^k$ by a factor equal to the DF in that direction.

Exercises

Regular Exercises

1. (Section 2.1) Show that the same joint posterior arises from the mixed-linear-model formulation of the balanced one-way random effects model and from the two constraint-case formulations, equations (2.6) and (2.12). Hint: Write $\theta_i = \mu + u_i$.
2. (Section 2.1) Show that the mixed-linear-model expression for the field trial with a long row of plots (2.22) is the same model as the original expression (2.15), (2.16).
3. (Section 2.2.1) Derive the DF in the fit of the balanced one-way random effects model, using the formula (2.29).
4. (Section 2.2.1) Show that the DF for mixed linear models (2.29) is the same as the generalized DF of Ye (1998), defined in (2.30), for known ϕ. (Hint: Hodges & Sargent (2001), Section 5.2 gives a brief proof.)
5. (Section 2.2.3.3) For the balanced one-way random effect model, derive the distribution function of the prior on $r = \sigma_s^2/\sigma_e^2$ that is induced by the flat prior on DF. Show that this prior on r has no mean (i.e., an infinite mean) and thus no higher positive moments.

Open Questions

1. (Section 2.2.2.3) Either prove that the DF approximation works well for reasonable penalized spline fits — i.e., the fitted DF are low relative to the maximum possible DF — or produce a counterexample.

Part II

Richly Parameterized Models as Mixed Linear Models

Part II Introduction

With Part I's theory and methods in hand, each chapter in Part II does three things: introduces a class of richly parameterized models, shows how to express it as a mixed linear model, and uses that expression to make some observations. Each class of models has a large literature and my observations are not intended to be an exhaustive consideration in light of the mixed-linear-model re-expression; for one thing, not enough is known to give an exhaustive consideration. Rather, my observations are intended to present some specific points and to give a sense of the fertility of insight made available by the mixed-linear-model formulation. Each chapter also refers forward to chapters in Parts III and IV in which the mixed-linear-model formulation is used as a vehicle to explore some mysterious, inconvenient, or plainly wrong result obtained using the chapter's class of models to analyze data.

After Chapters 3 to 6 give a sense of the utility of the mixed-linear-model form, Chapter 7 briefly discusses two less well-known syntaxes for unifying theory and computation for richly parameterized models. Each syntax, like mixed linear models, captures a large sub-class of richly parameterized models and each may prove as fertile in insight as the mixed-linear-model syntax although this potential is, in my view, as yet unrealized.

Chapter 3

Penalized Splines as Mixed Linear Models

This chapter's purpose is to introduce penalized splines, show how they can be represented as mixed linear models, and emphasize some features of mixed-linear-model theory as applied to penalized splines. It considers only penalized splines for estimating a one-dimensional function of a single variable, $y = f(x)$, i.e., scatterplot smoothing. Later chapters consider additive models with multiple x variables and spatial and other non-additive models in more than one dimension.

The literature on splines is enormous and this chapter is *not* intended to be a survey of that literature. Ruppert et al. (2003) is a very friendly introduction to penalized splines; this chapter leans on it heavily. Green & Silverman (1994) give a more general and historical introduction to splines that is very accessible.

3.1 Penalized Splines: Basis, Knots, and Penalty

Given n paired observations (x_i, y_i), fitting a penalized spline to estimate $y_i = f(x_i)$ is an optimization problem specified in terms of

- a *basis* for the space in which the fitted values $\hat{y}_i = \hat{f}(x_i)$ lie;
- *knots* in x's range, at which the fitted smooth curve is allowed to change in a specific way determined by the basis; and
- a *penalty* function that controls the fit's smoothness.

Basis, knots, and penalty will be introduced in that order.

The word *basis* has the same meaning as in linear model theory, so we begin there. For a simple linear model $y_i = \beta_0 + \beta_1 x_i + [\text{error}], i = 1, \ldots, n$ with $x_i \in [0, 1]$, the fitted values $\hat{y}_i = \hat{\beta}_0 + \hat{\beta}_1 x_i$ take values in a two-dimensional subspace of real n-space defined by these two basis vectors:

$$\begin{matrix} 1 & x_1 \\ 1 & x_2 \\ 1 & x_3 \\ \vdots & \vdots \\ 1 & x_n \end{matrix} \tag{3.1}$$

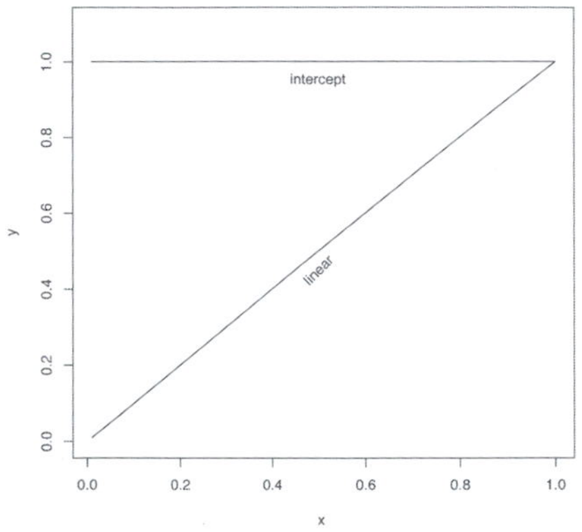

(a) Linear basis

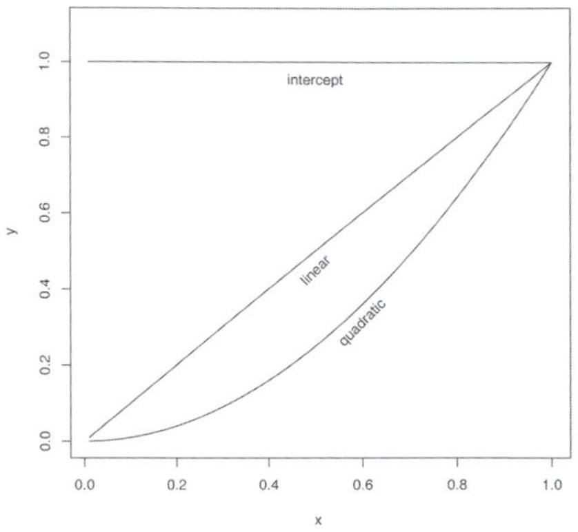

(b) Quadratic basis

Figure 3.1: Bases for linear regressions. Panel (a): linear fit in x; Panel (b): quadratic fit in x.

Figure 3.1a plots these basis vectors, with x_i on the horizontal axis and entries in these vectors on the vertical axis, with plot points connected by solid lines. All possible vectors of fitted values $\hat{\mathbf{y}}$ are linear combinations of these two basis vectors, specified by $\hat{\beta}_0$ for the intercept and $\hat{\beta}_1$ for the slope. Similarly, for a quadratic regression model $y_i = \beta_0 + \beta_1 x_i + \beta_2 x_i^2 + [\text{error}]$, the vector $\hat{\mathbf{y}}$ of fitted values $\hat{y}_i = \hat{\beta}_0 + \hat{\beta}_1 x_i + \hat{\beta}_2 x_i^2$ takes values in a three-dimensional subspace of real n-space defined by these three basis vectors:

$$\begin{array}{ccc} 1 & x_1 & x_1^2 \\ 1 & x_2 & x_2^2 \\ 1 & x_3 & x_3^2 \\ \vdots & \vdots & \vdots \\ 1 & x_n & x_n^2 \end{array} \tag{3.2}$$

Figure 3.1b shows these three basis vectors, again plotted with x_i on the horizontal axis and entries in these vectors on the vertical axis.

Sometimes it is desirable to fit a "broken stick" regression, in which the slope of the fitted line changes at a pre-specified point. Figure 3.2a shows some artificial data with a broken stick fit to it; the slope changes at $x = 0.6$. The linear model representing this fit is $y_i = \beta_0 + \beta_1 x_i + \beta_{11}(x_i - 0.6)_+ + [\text{error}]$, where

$$(z)_+ = \begin{cases} z & \text{if } z \geq 0, \\ 0 & \text{otherwise.} \end{cases} \tag{3.3}$$

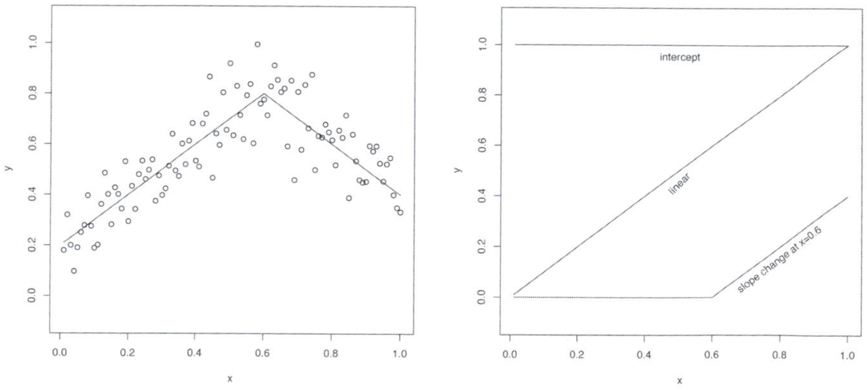

(a) Broken-stick fit

(b) Broken-stick basis

Figure 3.2: "Broken stick" regression. Panel (a): artificial data with slope changing at $x = 0.6$; Panel (b): basis for the broken-stick fit with slope changing at $x = 0.6$.

The fitted curve has slope β_1 for $x_i < 0.6$ and $\beta_1 + \beta_{11}$ for $x_i > 0.6$; the model's fitted-value space has basis

$$\begin{array}{ccc} 1 & x_1 & (x_1 - 0.6)_+ \\ 1 & x_2 & (x_2 - 0.6)_+ \\ 1 & x_3 & (x_3 - 0.6)_+ \\ \vdots & \vdots & \vdots \\ 1 & x_n & (x_n - 0.6)_+ \end{array} \tag{3.4}$$

Figure 3.2b shows this basis.

The obvious generalization of the broken stick model allows the fitted curve's slope to change in several places, which leads to a simple spline (which is also ill-advised, as we'll see). A spline's *knots* are the values of x at which the slope is allowed to change; call them $\kappa_i, i = 1, \ldots, K$. Figure 3.3a shows artificial data with a fit having 5 knots, at $x = 0.18, 0.34, 0.51, 0.67$, and 0.84, and thus 5 changes in slope, drawn with a solid line. (The change in slope at 0.51 is very close to zero.) The linear model representing this fit is $y_i = \beta_0 + \beta_1 x_i + \sum_{j=1}^{K} \beta_{1j}(x_i - \kappa_j)_+ +$ [error]; the slope at x_i is $\beta_1 + \sum_{j=1}^{K} \beta_{1j} I(x_i > \kappa_j)$, where the indicator function $I(z)$ is 1 if the condition z is true and 0 otherwise. The basis of this model's fitted-value space is

$$\begin{array}{ccccc} 1 & x_1 & (x_1 - \kappa_1)_+ & \ldots & (x_1 - \kappa_5)_+ \\ 1 & x_2 & (x_2 - \kappa_1)_+ & \ldots & (x_2 - \kappa_5)_+ \\ 1 & x_3 & (x_3 - \kappa_1)_+ & \ldots & (x_3 - \kappa_5)_+ \\ \vdots & \vdots & \vdots & & \vdots \\ 1 & x_n & (x_n - \kappa_1)_+ & \ldots & (x_n - \kappa_5)_+ \end{array} \tag{3.5}$$

which is shown in Figure 3.3b.

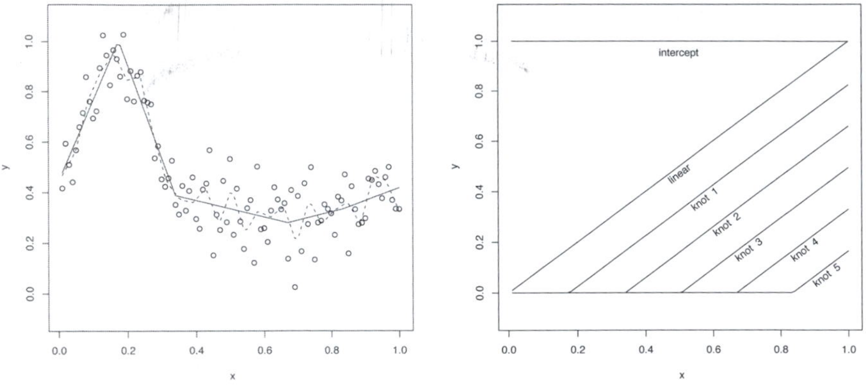

(a) Data and two fits (b) 5-knot basis

Figure 3.3: Fitting data with more knots. Panel (a): Artificial data with two fits: solid line, fit with 5 change-points (knots); dashed line, fit with 25 change-points. Panel (b): Basis for fit with 5 change-points. The line labeled "knot j" is the basis vector $(x_i - \kappa_j)_+$.

I chose five slope changes for an illustration, but that was an arbitrary choice, as were the choices of the five x values where the slope could change. In general, you won't know where the slope *should* change a priori, so for the sake of flexibility it should be allowed to change in many places. But allowing the slope to change at many places will overfit the data, giving fitted values and predictions with a lot of meaningless noise, like the dashed-line fit in Figure 3.3a, which has 25 equally spaced knots for 100 observations.

Penalized splines compromise between these two conflicting goals — allowing flexibility but avoiding overfitting — by using many knots, which allows the slope to change in many places, but favoring smaller slope changes over larger slope changes. The latter is implemented by means of a *penalty* on the slope change at each knot, which constrains the size of the changes. As the vagueness of my last sentence suggests, the penalty can take a great variety of forms. For illustration, the least squares criterion

$$\text{choose } (\beta_0, \beta_1, \{\beta_{1j}\}) \text{ to minimize}$$
$$\sum_i \left(y_i - \beta_0 - \beta_1 x_i - \sum_{j=1}^{K} \beta_{1j}(x_i - \kappa_j)_+ \right)^2 \tag{3.6}$$

can be modified by adding a penalty term $\lambda \sum_{j=1}^{K} \beta_{1j}^2$, for a positive scalar λ, to give a new optimization problem:

$$\text{choose } (\beta_0, \beta_1, \{\beta_{1j}\}) \text{ to minimize}$$
$$\sum_i (y_i - \beta_0 - \beta_1 x_i - \sum_{j=1}^{K} \beta_{1j}(x_i - \kappa_j)_+)^2 + \lambda \sum_{j=1}^{K} \beta_{1j}^2. \quad (3.7)$$

Adding the penalty term to the minimization problem tends to push each individual β_{1j} toward zero, to a greater extent for larger λ. Applying this penalty to the bumpy fit in Figure 3.3a (choosing λ in a manner to be dicussed) produces a smoother fit.

These are the three components of a penalized spline: basis, knots, and penalty. Section 3.2 develops these ideas in somewhat greater generality and then Section 3.3 re-expresses penalized splines as mixed linear models and explores some implications of that re-expression.

3.2 More on Basis, Knots, and Penalty

3.2.1 A Few More Bases

Many spline bases have been proposed. Ruppert et al. (2003) emphasize truncated power (polynomial) bases and I follow their lead here. B-splines and natural splines are two other bases derived from polynomials that feature prominently in the numerical-analysis literature and in easy-to-use functions in the R system. Apart from a brief mention below, I do not discuss them further; interested readers can consult the friendly introduction by Hastie & Tibshirani (1990) and references given there.

The basis used as an example in Section 3.1 is called a truncated-line basis because each basis vector $(x_i - \kappa_k)_+, i = 1, \ldots, n$ defines a line that is truncated (set to zero) for $x_i < \kappa_k$. This basis can be generalized to a truncated polynomial of any order. A truncated quadratic basis with knots κ_j has basis vectors corresponding to $1, x_i, x_i^2, (x_i - \kappa_1)_+^2, \ldots, (x_i - \kappa_K)_+^2$, so the coefficient of the quadratic term changes at each knot. This produces a smoother fit than the truncated-line basis; Figure 3.4 illustrates the two fits using the artificial data from Figure 3.3a. The fit using the truncated-line basis (solid line) has no derivatives at the knots and makes sharp turns, while the fit using the truncated-quadratic basis (dashed line) has one derivative at each knot and makes smoother turns.

The obvious generalization is a truncated p^{th}-degree polynomial, with basis vectors corresponding to $1, x_i, \ldots, x_i^p, (x_i - \kappa_1)_+^p, \ldots, (x_i - \kappa_K)_+^p$; this gives a fit with $p-1$ derivatives at each knot. Cubic splines ($p = 3$) have long been popular among cognoscenti, although in my admittedly modest experience they give fits that differ little from truncated-quadratic splines. For example, in Figure 3.4 compare the quadratic fit (dashed line) with the cubic fit (dotted line). In this case, all three bases do a good job of recovering the true function

$$y = f(x) = 0.4 + 0.5\sin(3\pi x)I(x < 1/3) - 0.125\sin(1.5\pi[x - 1/3])I(x \geq 1/3), \quad (3.8)$$

where the function $I(\textit{condition})$ is 1 if *condition* is true and zero otherwise. The

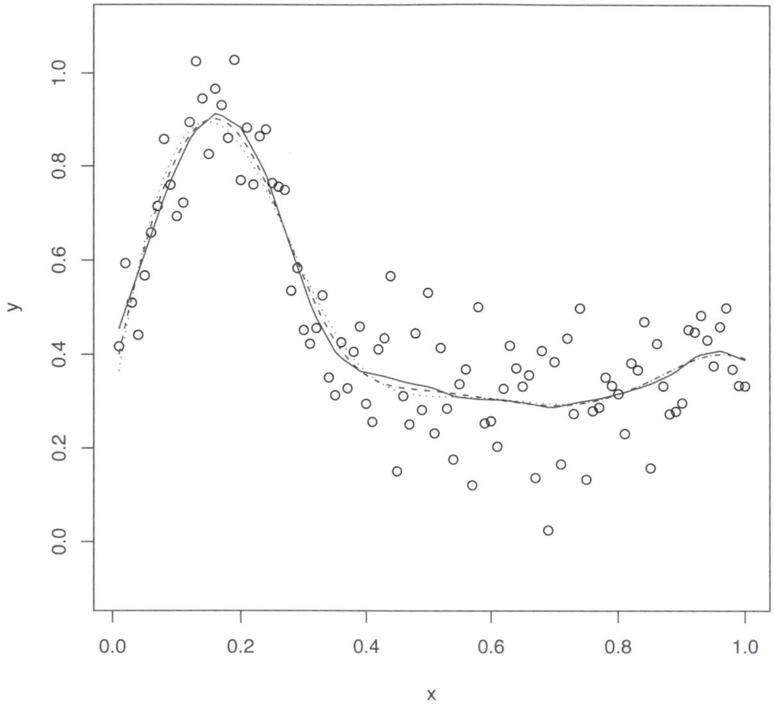

Figure 3.4: Penalized splines with 25 equally spaced knots for three bases: truncated-line (solid line), truncated-quadratic (dashed line), and truncated-cubic (dotted line). The true function is given in the text; errors are iid normal with mean zero and standard deviation 0.1.

errors were iid normal with standard deviation 0.1. Apparently those working in the area find it desirable to have a fitted function with second derivatives everywhere. Note that for the three fits in Figure 3.4, the DF of the fitted spline declines as p increases: DF is 10.8, 8.4, and 7.6 in the truncated-linear, -quadratic, and -cubic fits, and thus the random-effects part of the fitted spline has 8.8, 5.4, and 3.6 DF respectively out of a maximum of 25. (These models were fit by maximizing the restricted likelihood as described below in Section 3.3.) In other words, the higher-order splines permit a fit of lower complexity, as measured by DF. This has been true in the few examples that my students or I have examined: A truncated line apparently needs to change more at knots to turn a corner than does a truncated quadratic, which in turn needs to change more at knots than a truncated cubic.

For a truncated-line basis, it is easy to see how the slope changes at the knots. The effect of the knots is somewhat more obscure for higher-order truncated polynomials; Figure 3.5 makes this concrete using Example 5's global mean surface temperature data. Figure 3.5a shows the data and a penalized spline fit using a truncated quadratic

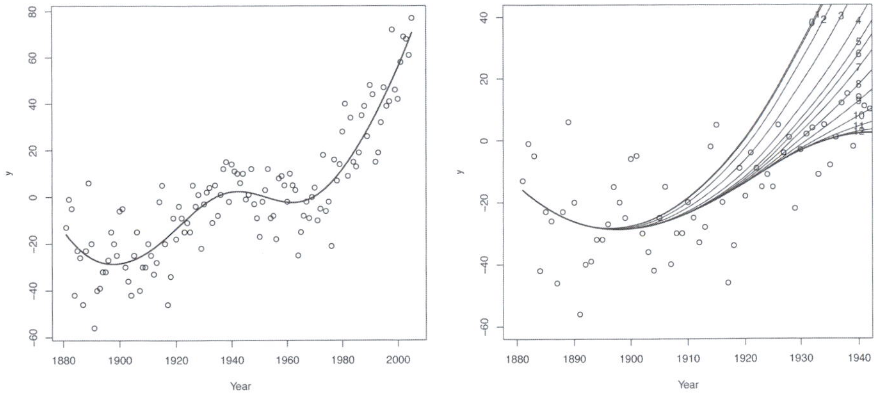

(a) Data and penalized spline fit

(b) Effect of increments to quadratic coefficient

Figure 3.5: Global-mean surface temperature data. Panel (a): Fit with truncated-quadratic basis and 30 knots. Panel (b): Effect of the change at each knot: the line labeled j is the sum of the fitted fixed effects and fitted random effects up to knot j; the line labeled "0" is the fit before the first knot.

basis with 30 equally spaced knots, which is discussed at length below. Figure 3.5b shows how the change in the quadratic coefficient at each knot changes the fitted curve up to 1940. The line labeled with the number j is the sum of the fitted fixed effects and all fitted random effects up to knot j. The changes at the knots simply make the curve follow the data's shape. At a first look, you might wonder (as I did) why the spline fit doesn't make the coefficient of the linear term $\beta_1 x_i$ change at each knot as well, but Figure 3.5b suggests that this would be redundant.

Radial bases are a different kind of truncated-power basis. At first they seem odd, but their value is clearer for multi-dimensional x (Section 5.3). A radial basis of order p and knots κ_j has basis vectors corresponding to $1, x_i, \ldots, x_i^p, |x_i - \kappa_1|_+^p, \ldots, |x_i - \kappa_K|_+^p$. Figure 3.6 shows the basis vectors for a radial basis of order $p = 1$ and the same 5 knots as in Figure 3.3. For a radial basis of order $p = 1$, the slope at x_i is $\beta_1 + \sum_j \left(\beta_{1j} I(x_i > \kappa_j) - \beta_{1j} I(x_i < \kappa_j)\right)$; if the fit is unpenalized, this radial basis will give exactly the same fit as a truncated-line basis with the same knots, though the $\hat{\beta}_{1j}$ are different. The advantage of a radial basis is that the term $|x_i - \kappa_j|$ is a function of x_i's distance from knot κ_j but not its direction from κ_j, so the radial basis generalizes readily to x of more than one dimension.

3.2.2 A Bit More on Penalty Functions

At equation (3.7) above the penalty function was simply asserted. Now we go back a little farther and derive penalty functions, which also makes a nice transition to the mixed-linear-model representation.

To make this concrete, we use the example of a truncated quadratic basis with knots κ_j, basis vectors corresponding to $1, x_i, x_i^2, (x_i - \kappa_1)_+^2, \ldots, (x_i - \kappa_K)_+^2$, and cor-

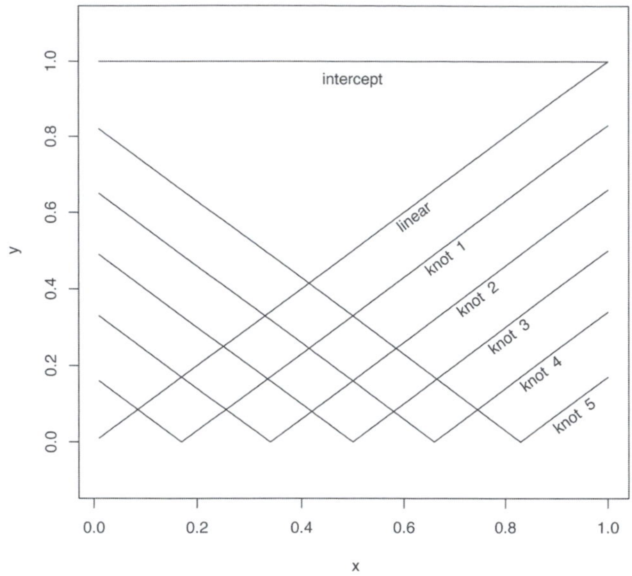

Figure 3.6: Radial basis with 5 knots.

responding coefficients $\beta_0, \beta_1, \beta_2, \beta_{21}, \ldots, \beta_{2K}$. We want to constrain $\beta_{21}, \ldots, \beta_{2K}$ to avoid overfitting while allowing enough flexibility so the fit can follow the data. A penalized spline balances these two goals by solving this optimization problem:

$$\begin{gathered}\text{choose } \beta = (\beta_0, \beta_1, \beta_2, \{\beta_{2j}\}) \text{ to minimize} \\ \{\text{lack-of-fit measure}\} \text{ subject to } \{\text{penalty } \leq M\},\end{gathered} \tag{3.9}$$

for M a positive scalar constant (the penalty is positive by convention). A great variety of lack-of-fit measures have been proposed and rationalized. For the normal-errors models considered here, we use residual sum of squares; for generalized linear models this is usually the deviance.

The penalty term constrains the coefficients compared to a fit relying only on the lack-of-fit measure. The literature provides many penalty functions. Most recently penalties have been crafted to do variable selection in problems with many predictors and relatively few observations, the so-called "large p small n problem." An early landmark in this area was the lasso (Tibshirani 1996); for our problem, the lasso corresponds to the penalty $\lambda \sum_j |\beta_{2j}|$ for a positive λ. In the optimization problem (3.9), this penalty has the effect of forcing individual β_{2j} to be exactly zero, which is equivalent to doing variable selection and thus appropriate for the lasso's purpose, but which may be unnecessary for penalizing spline fits. (I know of no literature using this penalty for spline fits; an exercise explores this penalty.)

Ruppert et al. (2003, Sections 3.5 and 3.8) discuss various penalties and end up using penalties of the form $\beta' \mathbf{D} \beta$ for a symmetric non-negative definite matrix $\mathbf{D}$.

Later still they restrict themselves to diagonal **D** with either 0 or 1 on the diagonal. This chapter follows them in both choices.

Given these choices, a penalized spline solves the optimization problem

$$\begin{array}{c}\text{choose } \beta \text{ to minimize} \\ (\mathbf{y}-\mathbf{X}\beta)'(\mathbf{y}-\mathbf{X}\beta) \text{ subject to } \beta'\mathbf{D}\beta \leq M,\end{array} \tag{3.10}$$

where **X** is a design matrix with columns containing the spline's basis. This turns out to be equivalent to another optimization problem:

$$\begin{array}{c}\text{choose } \beta \text{ to minimize} \\ (\mathbf{y}-\mathbf{X}\beta)'(\mathbf{y}-\mathbf{X}\beta)+\lambda^2\beta'\mathbf{D}\beta,\end{array} \tag{3.11}$$

for a particular λ that is a function of M. The term $\lambda^2\beta'\mathbf{D}\beta$ is sometimes called a "roughness penalty." For given λ, the solution to this optimization problem is

$$\hat{\beta}_\lambda = (\mathbf{X}'\mathbf{X}+\lambda^2 D)^{-1}\mathbf{X}'\mathbf{y}, \tag{3.12}$$

which gives fitted values

$$\hat{\mathbf{y}} = \mathbf{X}\hat{\beta}_\lambda = \mathbf{X}(\mathbf{X}'\mathbf{X}+\lambda^2 D)^{-1}\mathbf{X}'\mathbf{y}, \tag{3.13}$$

with degrees of freedom in the fit $\text{DF}_\lambda = \text{trace}(\mathbf{X}(\mathbf{X}'\mathbf{X}+\lambda^2 D)^{-1}\mathbf{X}')$. These expressions should look familiar from Chapters 1 and 2; they are special cases of results for mixed linear models. Older readers will recognize this as ridge regression, a method intended to stabilize coefficient estimates in linear regressions with collinear predictors, which is in turn equivalent to the point estimate from a particular Bayesian analysis of the same linear regression.

3.2.3 Brief Comments on Some Operational Matters

As noted, splines generally and penalized splines in particular are a large class of procedures and a user must make choices about basis, knots, and penalty. How should you choose a penalized spline or indeed a scatterplot smoother, of which penalized splines are a special case? This question is beyond the scope of the present book, which is about using mixed linear models as a vehicle for developing theory about many different models. Ruppert et al. (2003, Section 3.16) gives a nice brief overview of these issues, adapted from Marron (1996). Here I summarize this discussion very briefly and give some default choices that I use in examples.

Choosing a basis involves a trade-off among smoothness, simplicity, and numerical considerations. Truncated-power bases are simple and provide enough smoothness but imply highly collinear design matrices **X**, **Z**, which can cause numerical problems. Alternatives like B-splines and natural splines preserve smoothness and avoid numerical problems at the cost of much greater obscurity. The latter is unimportant, however, for those focusing on the fitted smooth.

How many knots should you use and where should you put them? Ruppert et al. (2003, Section 5.5, or at least my interpretation of it) suggest that for most problems, if you use many knots their precise locations have little effect on the fit. Thus, they propose a default number of knots K that is the smaller of 35 and {number of unique x_i}/4, and proposed locations κ_j equal to the $[(j+1)/(K+2)]^{\text{th}}$ sample quantiles of the unique x_i. They note (p. 126) that sometimes these defaults are inadequate and both K and the knot locations need to be selected based on features of the smooth curve being estimated, e.g., if $y = f(x)$ "seems to have a lot of fine detail, then K should be increased" (p. 127) and knots should be concentrated in intervals of x where $f(x)$ has fine detail.

Regarding the penalty, Ruppert et al. (2003) discuss various penalties (e.g., Section 3.8) and use as a default the simple quadratic-form penalties described above in Section 3.2.2. I did not find an explicit rationale for this choice but it corresponds to a particular penalty that is popular in the spline literature. It also has the great virtue of allowing penalized splines to be represented as mixed linear models, to which we now turn.

3.3 Mixed Linear Model Representation

Splines developed in their own little universe and have a huge literature, as noted. Unifying penalized splines into the mixed-linear-model framework has a price: This huge literature must be pruned to fit into the unifying framework and if this framework becomes popular, the pruned-off parts could be forgotten. For example, the advent of generalized linear models (GLMs) in the late 1970s unified many previously disparate models and literatures. GLMs are conventionally fit by maximizing the likelihood and the popularity of GLMs killed off alternative model-fitting criteria like Berkson's minimum chi-squared criterion (Berkson 1980). I have not heard a mention of Berkson's criterion since I left graduate school in 1985 and I am not persuaded by arguments in its favor but we should not presume that the pruning needed for unification is always benign.

The penalized spline model with truncated-power basis and penalty $\beta'\mathbf{D}\beta$ can be written as a mixed linear model as follows. The model is

$$\begin{aligned} y_i &= f(x_i) + \text{[error]} \qquad \text{for } i = 1, \ldots, n \\ &= \beta_0 + \beta_1 x_i + \ldots + \beta_p x_i^p + \sum_{j=1}^{K} \beta_{pj}(x_i - \kappa_j)_+^p + \text{[error]}. \end{aligned} \tag{3.14}$$

A squared-error lack-of-fit measure, as in the optimization problem (3.10), corresponds to iid normal errors, so [error] in (3.14) becomes $\varepsilon_i \overset{iid}{\sim} N(0, \sigma_e^2)$. Then equation (3.14)'s elements can be formulated as a mixed linear model with these definitions:

$$\mathbf{X} = \begin{bmatrix} 1 & x_1 & \ldots & x_1^p \\ 1 & x_2 & \ldots & x_2^p \\ & & \vdots & \\ 1 & x_n & \ldots & x_n^p \end{bmatrix}; \quad \beta = \begin{bmatrix} \beta_0 \\ \beta_1 \\ \vdots \\ \beta_p \end{bmatrix} \tag{3.15}$$

$$\mathbf{Z} = \begin{bmatrix} (x_1-\kappa_1)_+^p & \dots & (x_1-\kappa_K)_+^p \\ & \vdots & \\ (x_n-\kappa_1)_+^p & \dots & (x_n-\kappa_K)_+^p \end{bmatrix}; \quad \mathbf{u}' = \begin{bmatrix} \beta_{p1} \\ \vdots \\ \beta_{pK} \end{bmatrix},$$

and $\mathbf{R} = \sigma_e^2 \mathbf{I}_n$. Different choices of a basis and knots correspond to different choices of $\mathbf{X}$ and $\mathbf{Z}$, with corresponding definitions of β and $\mathbf{u}$.

Specification of $\text{cov}(\mathbf{u}) = \mathbf{G}$ is not so obvious. The penalized-spline optimization problem (3.11), in the notation just defined, corresponds to choosing $(\beta, \mathbf{u})$ to minimize

$$\frac{1}{\sigma_e^2}(\mathbf{y}-\mathbf{X}\beta-\mathbf{Z}\mathbf{u})'(\mathbf{y}-\mathbf{X}\beta-\mathbf{Z}\mathbf{u}) + \frac{\lambda^2}{\sigma_e^2}\mathbf{u}'\mathbf{u}, \tag{3.16}$$

where the objective function in (3.11) has been divided by σ_e^2 and the penalty matrix $\mathbf{D}$ is diagonal with $p+1$ zero diagonal elements corresponding to the fixed effects $\beta' = (\beta_0, \dots, \beta_p)$ and K diagonal elements of 1 corresponding to $\mathbf{u}' = (\beta_{p1}, \dots, \beta_{pK})$. This has the same form as equation (1.7) in Section 1.2.2, the objective function that is minimized to give best linear unbiased predictors (BLUPs) in the conventional analysis of mixed linear models. This suggests finishing the mixed-linear-model specification of the penalized spline by setting $\text{cov}(\mathbf{u}) = \mathbf{G} = \sigma_s^2 \mathbf{I}_K$ for $\sigma_s^2 = \frac{\sigma_e^2}{\lambda^2}$.

The penalized spline now has the mathematical *form* of a mixed linear model. But go back to Section 1.1 and consider the original meaning of a "random effect." The penalized spline's $\mathbf{u}$ is not a random effect under the definition used by Scheffé (1959, p. 238). The "levels" of $\mathbf{u}$ are in no sense a draw from a population; there is no population. Even the ANOVA jargon "levels" is strained past its breaking point as a label for the entries in the penalized spline's $\mathbf{u}$. Also, we are interested in these specific values of $\mathbf{u}$ because they determine the fitted spline. The penalized spline's random effect $\mathbf{u}$ is part of the model's mean, not part of its variance. The fitted model is $\hat{y} = \hat{f}(x) = \mathbf{X}\hat{\beta} + \mathbf{Z}\hat{\mathbf{u}}$ with error ε; it is *not* $\hat{y} = \mathbf{X}\hat{\beta}$ with error $\mathbf{Z}\hat{\mathbf{u}} + \varepsilon$, as it would be for an old-style random effect. This has concrete consequences, as we will soon see.

So $\mathbf{u}$ is a clear example of a random effect that meets the current definition of "random effect" but not the original definition. Ruppert et al. (2003) say "we [use] the mixed-linear-model formulation of penalized splines as a convenient fiction" (p. 138), using "fiction" in the sense of "a belief or statement that is false, but that is often held to be true because it is expedient to do so" (*New Oxford American Dictionary*, MacIntosh edition). Ruppert et al. (2003) are making a bow toward the original meaning of "random effect" while acknowledging that penalized splines do not meet it.

Having expressed the penalized spline as a mixed linear model, the fitted values are, from equation (1.10),

$$\tilde{\mathbf{f}}_\phi = \mathbf{X}\tilde{\beta}_\phi + \mathbf{Z}\tilde{\mathbf{u}}_\phi = \mathbf{C}\left[\mathbf{C}'\mathbf{C} + \frac{\sigma_e^2}{\sigma_s^2}\mathbf{D}\right]^{-1}\mathbf{C}'\mathbf{y} \tag{3.17}$$

where, as before, $\mathbf{C} = (\mathbf{X}|\mathbf{Z})$ and $\mathbf{D} = \text{diag}(0, \dots, 0, 1, \dots, 1)$. These are identical to the fitted values in (3.13). The complexity of the whole fit is DF = $\text{trace}(\mathbf{C}\left[\mathbf{C}'\mathbf{C} + \frac{\sigma_e^2}{\sigma_s^2}\mathbf{D}\right]^{-1}\mathbf{C}')$. The fitted values are indicated by tildes because they depend on $\phi = (\sigma_e^2, \sigma_s^2)$; choosing a value of ϕ is discussed just below. Note that the fitted spline does not depend on the value of either variance individually, but only on their ratio σ_e^2/σ_s^2. The predicted or fitted value for a new value of x, x_0, is defined similarly. If $l(x_0) = (1, x_0, \ldots x_0^p, (x_0 - \kappa_1)_+^p, \ldots (x_0 - \kappa_K)_+^p)'$, then

$$\begin{aligned} \tilde{\mathbf{f}}(x_0)_\phi &= l(x_0)' \left[\frac{\tilde{\beta}}{\tilde{\mathbf{u}}} \right] \qquad (3.18) \\ \text{for} \left[\begin{array}{c} \tilde{\beta} \\ \tilde{\mathbf{u}} \end{array} \right] &= \left[\mathbf{C}'\mathbf{C} + \frac{\sigma_e^2}{\sigma_s^2} \mathbf{D} \right]^{-1} \mathbf{C}'\mathbf{y}. \end{aligned}$$

So far we have gained nothing from representing penalized splines as mixed linear models. The payoff comes from applying mixed-linear-model methods to penalized splines, specifically methods for fitting models and making inferential or predictive statements. In applying methods for mixed linear models, ϕ is selected in the conventional analysis by maximizing the restricted likelihood, or in the Bayesian approach it is an unknown parameter with prior and posterior distributions. Similarly, either analytic approach can be used to produce intervals around the spline fit describing uncertainty. Regarding this wholesale importation of methods, Ruppert et al. (2003) carefully note "we used the mixed model formulation of penalized splines as a convenient fiction to estimate smoothing parameters. The mixed model is a reasonable (though not compelling) Bayesian prior for a smooth curve, and ... [maximized restricted-likelihood] estimates of variance components give estimates of the smoothing parameter that generally behave well. ... Although we are attracted by the automatic nature of the mixed model-REML approach to fitting additive models, we discourage blind acceptance of whatever answer it provides and recommend looking at other amounts of smoothing" (pp. 138 and 177). In other words, here is a fertile source of methods that must be justified by their performance, a pragmatic attitude I share.

Of course, splines already have a body of fitting and inferential methods, but to me at least that body seems *ad hoc* and jerry-built while statistical methods for mixed linear models, for all their problems, seem considerably less so. However, adopting mixed linear models as an analytic framework leads to some quandaries. The global-mean surface temperature example will now be used to explore the promise and puzzles of treating penalized splines as a special case of mixed linear models.

3.3.1 Applying the Mixed-Linear-Model Approach

Consider again the global-mean surface temperature example, introduced in Section 2.2.2.4 and plotted in Figure 2.1 with a fit from a different kind of smoother. The series has 125 equally spaced measurements from 1881 to 2005 inclusive. The penalized spline analysis presented here uses 30 equally spaced knots at the years 1880 + {4, 8, ..., 120} and a truncated-quadratic basis, i.e., equation (3.15) with $p = 2$. (My R code to set up the design matrices **X** and **Z** has 10 lines and I am a dismal programmer, so this is really simple.)

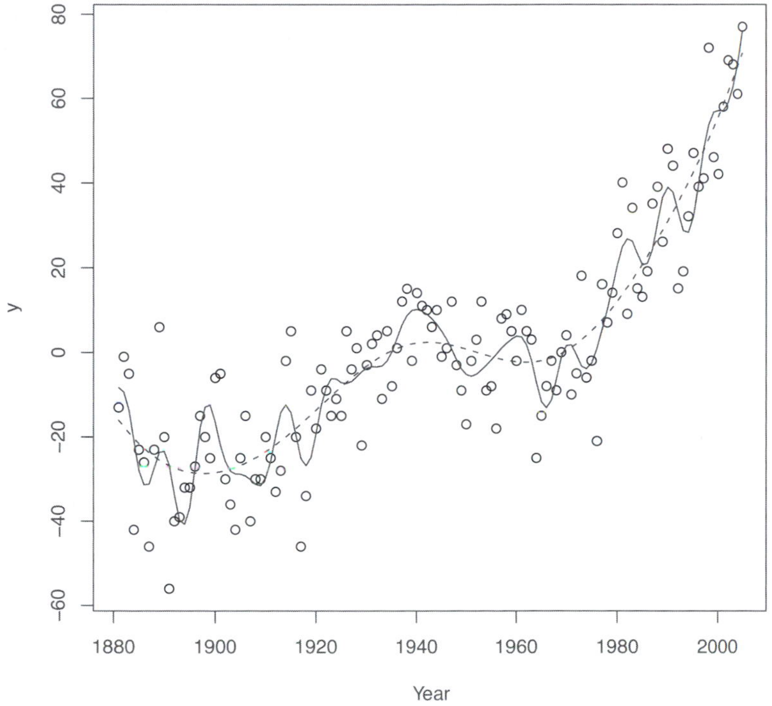

Figure 3.7: Global-mean surface temperature data: unpenalized and penalized fits with the truncated-quadratic basis.

If the data $\mathbf{y}$ are regressed on the resulting design matrix $\mathbf{C} = [\mathbf{X}|\mathbf{Z}]$ without any constraint (penalty) on the coefficients $(\beta, \mathbf{u})$, the resulting fit is the wiggly solid line in Figure 3.7. The smooth dashed line in Figure 3.7 is a penalized spline fit using the conventional analysis of the corresponding mixed linear model. The value of ϕ that maximizes the restricted likelihood is $\hat{\sigma}_e^2 = 145$ (i.e., the error standard deviation is estimated to be about 12) and $\hat{\sigma}_s^2 = 947$. The fit has 6.7 DF including 3 DF for the fixed effects, so the smoothed part of the model has 3.7 DF out of a possible 30 (one per knot), a substantial degree of smoothing. Maximizing the restricted likelihood has, in this case at least, produced a reasonable smooth that goes through the middle of the data while avoiding the gross overfitting of the unpenalized fit.

For a penalized-spline-as-mixed-linear-model, the mean-structure parameters $(\beta, \mathbf{u})$ are interesting only in that they determine the fitted values, but a brief look at them is illuminating. Table 3.1 shows the estimated coefficients and standard errors for the wiggly unpenalized fit (columns on the left) and for the penalized smooth fit (columns on the right). Given that $\mathbf{y}$ ranges from about -60 to +80, the unpenalized coefficients and their standard errors are simply enormous. These reflect the extreme collinearity of the design matrix: In $\mathbf{C} = [\mathbf{X}|\mathbf{Z}]$, 179 of the 496 distinct correlations among columns are greater than 0.95, 80 are greater than 0.99, and **C** has condition number (largest singular value divided by smallest) 202110.1, which is bad according to the numerical-analysis advice I can find on the Internet. Nonetheless, an F-test of the 30 unpenalized truncated-quadratic terms gives $F = 17.3$ on (30,92) degrees of freedom, so the P-value is less than 10^{-15} despite dilution of the test by gross overfitting. By comparison, the penalized coefficients and their standard errors (Table 3.1, columns on the right) are two to three orders of magnitude smaller than their unpenalized counterparts, consistent with the substantial smoothing in the penalized analysis. All the $\hat{u}_j$ are smaller in magnitude than their respective standard errors, but the spline fit tests clearly better than the fixed-effects-only quadratic fit: $P = 2 \times 10^{-18}$ in the naïve asymptotic test; $P = 4 \times 10^{-10}$ in the perhaps less naïve F-test; and in the bootstrapped restricted likelihood ratio test suggested by Ruppert et al. (2003), 1000 bootstrap draws from the null gave a maximum test statistic of 14.5 compared to an observed statistic of 75.3, so P is less than 0.001 by a lot.

Now look more closely at the estimated (or predicted, if you prefer) random effects $\hat{\mathbf{u}}$ from the penalized spline fit. These are in the second column from the right in Table 3.1 and are plotted in Figure 3.8. The $\hat{\mathbf{u}}$ bear no resemblance to an iid sample from a normal or any other distribution. Runs of positive and negative signs as in the table and figure, while necessary to fit a curve to the GMST data, are spectacularly improbable if **u** is truly iid. This grates on our statistical sensibilities, which tell us to condemn this model as a poor fit to the data. However, iid is not the wrong model for **u**; rather, when we use a mixed linear model as a mere device, a convenient fiction, to fit a smooth curve, there is no reason to expect the fitted **u** to look like an iid sample. The distributional assumption for **u** simply constrains the fit. Other parts of this mixed linear model are "wrong" as well, e.g., the GMST series clearly does not fit a model of iid errors around a smooth curve (see, e.g., Adams et al. 2000), and the error variance is plainly larger in the early years. Nonetheless, maximizing the restricted likelihood from this mixed linear model is an effective way to draw a smooth curve through the middle of these data, which was the purpose of fitting a spline.

How should we draw a band around the fitted spline to indicate our uncertainty about the true curve $y = f(x)$? One often-used informal method is to draw B bootstrap samples from the data, fit the model to each bootstrap sample, and draw the bootstrapped fits on the same plot as the fit to the real data. This plot of the bootstrap fits shows a kind of band around the actual fit, suggesting the extent of uncertainty about the fit. (Doing so for the GMST fit is an exercise.)

More formal methods for drawing confidence bands can be derived from mixed-linear-model theory. This leads to a puzzle, the solution of which depends on how the random effect **u** is interpreted. Specifically, the question is how to construct pointwise confidence intervals around the fitted spline. To make the discussion concrete, we follow the development in Ruppert et al. (2003, Section 6.4), changing their notation slightly. We do so not to single out these authors for criticism; rather, the exceptional clarity and care of their development makes it an ideal platform for discussing this puzzle. (Indeed, these pages in Ruppert et al. 2003 are so carefully written that I

Table 3.1: Global-Mean Surface Temperature Data

	Unpenalized fit		Penalized fit	
	Estimate	SE	Estimate	SE
Fixed effects				
Intercept	–5924.8	10661.9	85.33	131.12
Linear	–6930.8	12906.3	176.37	174.76
Quadratic	–2029.7	3903.7	68.45	58.67
Random effects				
u1	4387.7	5258.6	–0.53	30.77
u2	–4688.3	2817.7	–3.48	30.68
u3	5679.9	2519.4	–6.61	30.35
u4	–6310.3	2465.8	–8.41	29.78
u5	4015.4	2455.8	–5.68	29.14
u6	–1305.9	2453.9	–3.93	28.62
u7	1662.1	2453.5	–5.80	28.31
u8	–3786.6	2453.5	–9.19	28.17
u9	4810.5	2453.5	–12.12	28.13
u10	–3868.4	2453.4	–15.85	28.13
u11	1859.5	2453.4	–16.95	28.12
u12	–704.7	2453.4	–15.36	28.12
u13	921.5	2453.4	–11.43	28.11
u14	–1280.9	2453.4	–4.55	28.10
u15	430.3	2453.4	5.42	28.10
u16	19.8	2453.4	14.97	28.10
u17	776.0	2453.4	20.24	28.10
u18	–500.1	2453.4	20.76	28.11
u19	–222.0	2453.4	19.35	28.11
u20	–977.6	2453.4	17.33	28.12
u21	3185.8	2453.4	12.59	28.13
u22	–3738.0	2453.4	4.26	28.13
u23	2975.1	2453.5	–3.45	28.14
u24	–1127.1	2453.5	–8.82	28.22
u25	–1763.1	2453.5	–8.75	28.44
u26	3538.6	2453.6	–4.15	28.86
u27	–4306.1	2454.3	0.20	29.46
u28	5013.2	2458.1	2.68	30.09
u29	–4427.8	2482.3	1.58	30.55
u30	3008.7	2650.0	0.57	30.75

Note: Coefficient estimates with standard errors (SE) from unpenalized fit (left) and penalized fit (right). Figure 3.7 shows fitted values from these fits.

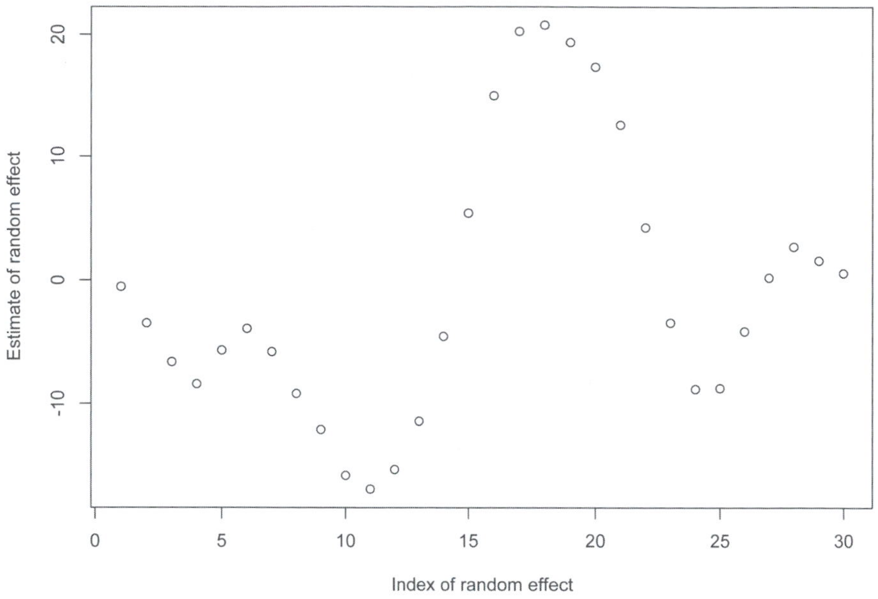

Figure 3.8: Global mean surface data: $\hat{\mathbf{u}}$ from the mixed-linear-model fit.

wonder whether these authors disagree.) This discussion uses the conditional and unconditional covariance matrices for $(\beta, \mathbf{u})$ in Section 1.2.4.1.

Ruppert et al. (2003, p. 137) note that "Variability estimates differ depending on whether randomness in $\mathbf{u}$ is taken into account. One argument ... is that randomness of $\mathbf{u}$ is a device used to model curvature, while ε accounts for variability about the curve. According to this argument, variance calculations should be done with respect to the conditional distribution $\mathbf{y}|\mathbf{u}$ rather than the unconditional distribution of $\mathbf{y}$." In other words, $\mathbf{u}$ is just an unknown and $\mathbf{u} \sim N(\mathbf{0}, \mathbf{G})$ just constrains the estimate of $\mathbf{u}$. Recalling that $\mathbf{C} = [\mathbf{X}|\mathbf{Z}]$, define $\mathbf{C}_i = [\mathbf{X}_i|\mathbf{Z}_i]$, where $\mathbf{X}_i$ and $\mathbf{Z}_i$ are the rows of $\mathbf{X}$ and $\mathbf{Z}$ corresponding to year i. Then $\hat{f}(x_i) = \mathbf{X}_i\hat{\beta} + \mathbf{Z}_i\hat{\mathbf{u}}$ is the fitted value of f at x_i, where the hats indicate use of the maximum restricted likelihood estimates, and

$$\text{var}\{\hat{f}(x_i)|\mathbf{u}\} = \mathbf{C}_i \text{cov}([\hat{\beta}, \hat{\mathbf{u}}]'|\mathbf{u})\mathbf{C}_i', \tag{3.19}$$

where $\text{cov}([\hat{\beta}, \hat{\mathbf{u}}]'|\mathbf{u})$ is a function of σ_e^2 and σ_s^2 but not a function of $\mathbf{u}$ despite the conditioning. (More precisely, for any fixed ϕ, this conditional covariance is not a function of $\mathbf{u}$; see equation (1.20) or Ruppert et al. 2003, p. 139, equation 6.10.) Because $\hat{f}(x)|\mathbf{u}$, for given x, has a normal distribution with mean $E[\hat{f}(x)|\mathbf{u}]$ and variance (3.19), a 95% confidence interval can be constructed from this distribution in the usual way. This interval is centered at $E[\hat{f}(x)|\mathbf{u}]$, but we will call it the conditional confidence interval for $f(x)$. (As Ruppert et al. note, (3.19) ignores variance arising from estimating σ_e^2 and σ_s^2, as usual in conventional analyses of mixed linear models, but this is irrelevant for the present purpose.)

"If there is no appreciable bias [in the spline fit]," Ruppert et al. continue, "then $E[\hat{f}(x)|\mathbf{u}] \approx f(x)$, and this interval can be interpreted as a confidence interval for $f(x)$." But the mixed-model framework allows an estimate of the bias conditional on $\mathbf{u}$, i.e., taking $\mathbf{u}$ as fixed though unknown:

$$E[\hat{f}(x) - f(x)|\mathbf{u}] = -r\mathbf{C}(\mathbf{C}'\mathbf{C} + r\mathbf{I})^{-1}\begin{pmatrix}\mathbf{0}_p \\ \mathbf{u}\end{pmatrix}, \tag{3.20}$$

where $\mathbf{0}_p$ is a p-vector of zeroes (one for each of the fixed effects) and $r = \sigma_e^2/\sigma_s^2$, with larger values implying a smoother fit. (The proof is an exercise.) This bias is a function of x and non-zero in general. Thus, the conditional confidence interval is an interval for $E[\hat{f}(x)|\mathbf{u}]$ but not for $f(x)$. Although Ruppert et al. do not say it in these words, the conditional confidence interval is not a proper confidence interval for $f(x)$ in general because its coverage is too low for certain x, and its coverage is low because the conditional interval has the wrong center for those x.

Up to this point, Ruppert et al.'s development is consistent with $\mathbf{u}$'s nature as a new-style random effect that does not meet the original definition of a random effect, that is, the development treats $\mathbf{u}$ simply as a convenient way to describe an ensemble of fixed but unknown constants. But now Ruppert et al. make a choice that seems odd for a new-style random effect; we follow their development to its conclusion, then give a choice that seems more consistent with $\mathbf{u}$'s nature.

After Ruppert et al. give (3.20), the bias of $\hat{f}(x)$ given $\mathbf{u}$, and note that it is non-zero in general, they observe "But, since $E(\mathbf{u}) = \mathbf{0}$, the unconditional bias [i.e., averaging over $\mathbf{u}$'s distribution] is $E[\hat{f}(x) - f(x)] = 0$. Thus, on average over the distribution of $\mathbf{u}$, $\hat{f}(x)$ is unbiased for $f(x)$. To account for bias in the confidence intervals, the [conditional] variance $\text{var}\{\hat{f}(x)|\mathbf{u}\}$ should be replaced by the conditional mean-squared error $E[\{\hat{f}(x) - f(x)\}^2|\mathbf{u}]$ ... and then averaged over the $\mathbf{u}$ distribution." This gives

$$E[\{\hat{f}(x_i) - f(x_i)\}^2] = \mathbf{C}_i \text{cov}\begin{pmatrix}\hat{\beta} \\ \hat{\mathbf{u}} - \mathbf{u}\end{pmatrix}\mathbf{C}_i'. \tag{3.21}$$

Ruppert et al. suggest using a confidence interval centered at $\hat{f}(x)$ with upper and lower limits determined by (3.21). This interval, which we call the unconditional confidence interval, is wider than the conditional interval arising from (3.19): $\mathbf{C}_i\text{cov}([\hat{\beta}, \hat{\mathbf{u}} - \mathbf{u}]')\mathbf{C}_i' > \mathbf{C}_i\text{cov}([\hat{\beta}, \hat{\mathbf{u}}]'|\mathbf{u})\mathbf{C}_i'$, "because [(3.21)] accounts for both components of error (variance and squared bias) whereas [(3.19)] accounts only for variance and covers $E[\hat{f}(x)]$, not $f(x)$." (Proof of this inequality is an exercise.)

For each year of the GMST fit, Figure 3.9 shows the conditional standard error (square root of (3.19), solid line) and unconditional standard error (square root of (3.21), dashed line), as well as the square root of $\{\text{conditional variance + bias}^2\}$ (dotted line) for each x (year). As noted, the unconditional covariance (3.21) is derived by averaging $\{\text{conditional variance + bias}^2\}$ given $\mathbf{u}$ over the distribution of $\mathbf{u}$. Figure 3.9 shows that the unconditional standard error is too small at the places

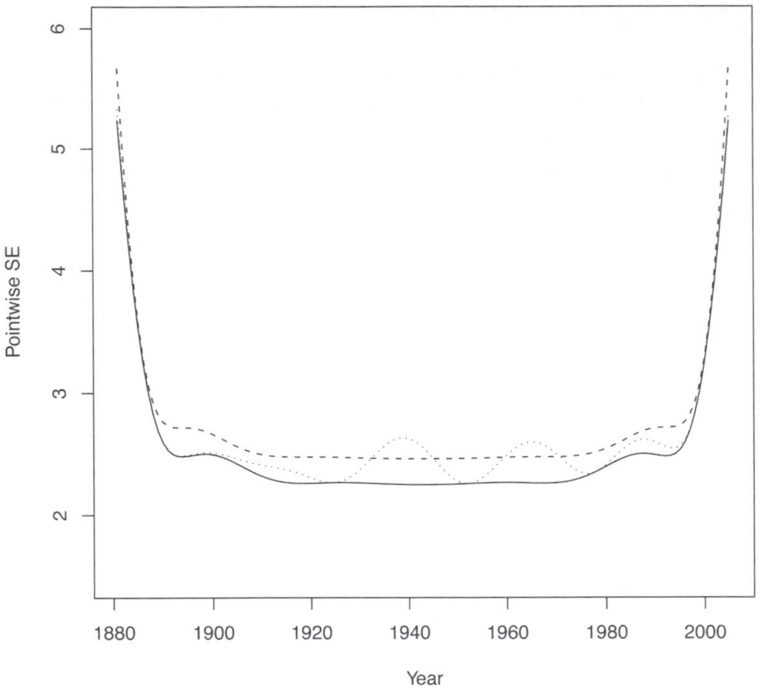

Figure 3.9: Global-mean surface temperature data: Different standard errors for pointwise 95% confidence intervals. Solid line: Conditional standard error (square root of (3.19)). Dashed line: Unconditional standard error (square root of (3.21)). Dotted line: Square root of $\{\text{conditional variance} + \text{bias}^2\}$.

where bias is largest and too large where bias is small. Ruppert et al. cite a result of Nychka (1988), that the unconditional pointwise confidence interval using (3.21) gives 95% coverage *averaging over the values of the predictor* (year, in the GMST example), but note that coverage is too low in areas of high curvature, where bias is greatest, and too high in areas of low curvature, where bias is negligible. It is clear from Figure 3.9 why this happens.

Now let's reconsider Ruppert et al.'s choice to account for bias in $\hat{f}(x)$ given $\mathbf{u}$ by inflating the variance of $\hat{f}(x)$ conditional on $\mathbf{u}$, by adding squared bias and taking the expectation of this inflated variance with respect to $\mathbf{u}$'s distribution. If $\mathbf{u}$ is understood as a new-style random effect — a fixed but unknown quantity, for which the distribution is just a device to induce shrinkage/smoothing — Ruppert et al.'s choice appears to be a *non sequitur*. The problem with the conditional interval for $f(x)$ based on (3.19) is not that it is too narrow, but that because of bias in the spline fit, it is centered at the wrong value for each x. Widening the confidence interval does not solve the problem, as Ruppert et al. note; the unconditional interval's coverage is still too low for precisely those values of x where bias in $\hat{f}(x)$ is largest. Viewing

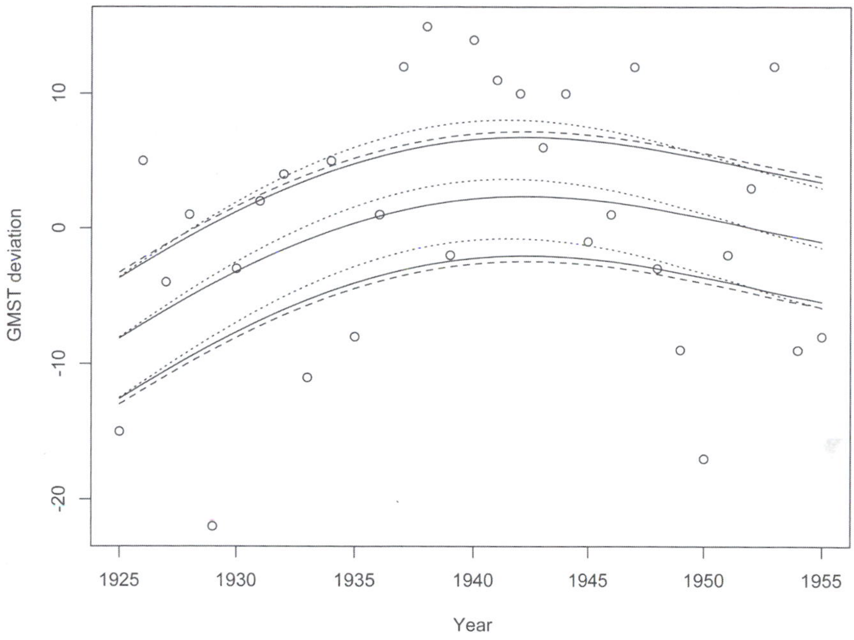

Figure 3.10: Detail of GMST series and fit for the years 1925–1955, with different confidence intervals. Solid lines: Spline fit and simple conditional interval around this (biased) fit $\hat{f}$. Dashed lines: Unconditional interval centered around $\hat{f}$. Dotted lines: Bias-corrected fit and conditional interval around it.

u as a new-style random effect, the obvious way to fix this problem is not to inflate (3.19) but to center the confidence interval in the right place. The obvious (though not necessarily best) way to do this is to subtract the estimated bias — (3.20) evaluated at $\hat{\mathbf{u}}$ — from the fit $\hat{f}(x)$ and construct a confidence interval around this bias-corrected estimate using the conditional standard error derived from (3.19).

Figure 3.10 compares the various intervals using the GMST example. It shows the data (circles), the spline fit (the central solid line), and the three nominally 95% confidence intervals for a period with large curvature, 1925 to 1955. The (biased) fit $\hat{f}$ and the simple conditional interval around it are drawn using solid lines; the unconditional interval centered around $\hat{f}$ is the dashed lines; and the bias-corrected fit and the conditional interval around it are the dotted lines. The bias-corrected interval is shifted up by a maximum of about 1.5 units compared to the simple conditional interval, while the upper limit of the unconditional interval is shifted up by only about 0.3 units at most. This is consistent with the unconditional interval's undercoverage in areas of greatest curvature, which the theoretical development leads us to expect. Because the bias-corrected interval addresses the bias problem directly, we would expect its pointwise coverage to be closer to 95% for all values of x and particularly at values of x with substantial curvature, where the unconditional interval is known to have low coverage. Hence, we should not need extraordinary expedients like averaging over x to get 95% coverage, i.e., using a global calibration for an interval that is interpreted pointwise.

Sun & Loader (1994), while presenting methods for simultaneous (not pointwise) confidence intervals, argued against this simple bias correction based on simulation results in their Section 5. Intuitively, their rationale was that bias is estimated so poorly that subtracting it from the fit just adds noise to the center of the confidence interval without improving the interval's coverage. Instead, they and researchers following them propose intervals centered on the spline fit, like the unconditional interval but with different methods for increasing the conditional covariance. Unfortunately, Section 5 of Sun & Loader (1994) is so terse I cannot determine how they generated their simulated data. If they simulated data by drawing a new value of the random effect for each simulated dataset — which is appropriate for an old-style random effect but not for the penalized spline's new-style $\mathbf{u}$ (Chapter 13 argues this point in detail) — then their case against the simple bias correction is in doubt. A more appropriate simulation (and perhaps Sun & Loader did this, though it seems unlikely) would fix a true $f(x)$ that has corners or peaks or valleys, where bias is worst, and draw simulated datasets by adding iid error to that $f(x)$. Such a simulation fits the nature of the penalized spline's random effect $\mathbf{u}$. (A simulation experiment of this sort is given as an exercise.)

If we simply view $\mathbf{u}$ as unknown — overlooking, for the moment, that the mixed linear model specifies $\mathbf{u} \sim N(\mathbf{0}, \mathbf{G})$ — then it is consistent with common statistical practice to treat $\mathbf{u}$ as fixed but unknown in deriving a confidence interval, i.e., to use the conditional variance (3.19). Standard asymptotics for maximum likelihood estimates (MLEs) do exactly that: When the asymptotic covariance matrix for the MLE of an unknown parameter θ is a function of θ, the MLE $\hat{\theta}$ is substituted into the approximate covariance and I've never seen anyone object to doing so. In my view, it is controversial to do exactly the same thing for the penalized spline's unknown $\mathbf{u}$ because of confusion arising from failure to distinguish old-style and new-style random effects.

3.3.2 *Brief Comments on Other Aspects of the Mixed-Linear-Model Analysis*

Ruppert et al. (2003, Chapter 6) give a well-rounded discussion of statistical inference for penalized splines based on the conventional (non-Bayesian) approach to analyzing mixed linear models. In particular, they discuss simultaneous confidence bands around a fit (Section 6.5) as distinct from the pointwise bands considered above in Section 3.3.1; testing whether the spline fits better than a parametric model (i.e., "Is $\sigma_s^2 = 0$?" in Section 6.6); and testing whether an effect fit with a spline is, in fact, present at all (Section 6.7). That material is outside the present book's scope and their presentation is excellent, so I refer you to them for more details.

A Bayesian analysis of penalized splines is a straightforward application of Bayesian methods to mixed linear models. For example, a simultaneous 95% posterior band around the fitted curve can be construced from MCMC output in many ways, e.g., by constructing pointwise intervals with probability mass α and increasing α until 95% of the MCMC draws give a curve that is entirely contained in the bands specified by α. Testing the spline versus a parametric fit, or testing whether the effect fit with the spline is absent, can be done using DIC or by using a prior distribution with a point mass of probability on $\sigma_s^2 = 0$.

Regarding the puzzle discussed at length in Section 3.3.1, Bayesian methods have no conceptual or practical advantage over conventional methods. By the familiar formula, the marginal posterior variance of $f(x)$ is $E\{\text{var}(f(x)|\theta)\} + \text{var}\{E(f(x)|\theta)\}$, where θ includes all the unknowns in the mixed-linear-model formulation. In most cases, this is larger than (3.19) because it accounts for uncertainty about $\beta, \mathbf{u}, \sigma_s^2$, and σ_e^2, though it is not necessarily larger than (3.19) or (3.21). Like the unconditional confidence interval, a Bayesian interval will have high coverage in areas of low curvature, where the fitted spline has low bias, and low coverage in areas of high curvature, where the fitted spline has high bias. This does not concern someone holding the traditional Bayesian view that bias and interval coverage over repeated sampling are irrelevant, but it should concern those who like Bayesian intervals because they account for uncertainty about σ_s^2 and σ_e^2.

Exercises

Regular Exercises

1. (Section 3.3.1) Reproduce the calculations for the penalized spline fit to the global-mean surface temperature data.
2. (Section 3.3.1) Derive the bias of the fitted spline $\hat{f}(x)$ conditional on $\mathbf{u}$ in equation (3.20). Assume that $f(x) = \mathbf{X}\beta + \mathbf{Z}\mathbf{u}$ for some true values of β and $\mathbf{u}$.
3. (Section 3.3.1) Show that the conditional covariance of $\hat{f}(x)$, equation (3.19), is less than or equal to the unconditional covariance of $\hat{f}(x)$, equation (3.21).
4. (Section 3.3.1) Draw $B = 20$ bootstrap samples from the GMST data, fit the penalized spline to each bootstrap sample, and plot all 20 bootstrap fits on the same plot as the fit to the actual data. Compare the informal uncertainty band produced by these bootstrap fits to the point-wise 95% confidence intervals derived from mixed-linear-model theory. Note that your bootstrap must preserve the data's time-series stucture. The obvious way to do this is a parametric bootstrap. An alternative, semi-parametric bootstrap fits the spline, computes the residuals, draws bootstrap samples from the residuals, and adds those to the spline fit.

Open Questions

1. (Section 3.2.2) Examine penalizing splines using the lasso penalty or some other penalty (e.g., the l_0 penalty) instead of a quadratic-form penalty. The first problem is choosing the penalty constant λ. An informal approach could use lasso-style trace plots of the fitted coefficients as a function of λ. A more formal approach could formulate the spline as a hierarchical model as in Section 3.3 and choose λ as part of the fit. Now, however, $\mathbf{u}$'s distribution is Laplacian (double exponential) not Gaussian, and the restricted likelihood is not so simple. A Bayesian analysis

of this hierarchical model is not hard to code in BUGS or JAGS but it sacrifices the lasso's property of forcing some β_{2j} to zero, which may be acceptable. However, you could frame the problem of automatic knot selection — for which there is a substantial literature — as a lasso problem by using many knots and having the lasso machinery set $u_k = 0$ for most of them.

2. (Section 3.3.1) Do a simulation experiment to test whether the simple bias-corrected confidence interval, using the conditional variance of $\hat{f}(x)$ in equation (3.19), has better coverage than the confidence interval using the unconditional covariance in equation (3.21). In this simulation, the true $f(x)$ should be held fixed, with simulated datasets created by adding errors ε to that true $f(x)$, and coverage should be considered at several different x, especially at corners, peaks, or valleys in $f(x)$ where the spline fit has greatest bias. Consider several true $f(x)$ with different shapes. DO NOT generate such f as draws of $\mathbf{X}\beta + \mathbf{Zu}$, where $\mathbf{X}$ and $\mathbf{Z}$ are the penalized spline's design matrices and u_j is iid, because that gives f without corners, peaks, or valleys and such fs are inappropriate for the present purpose. (Simulate some $\mathbf{X}\beta + \mathbf{Zu}$ and see for yourself. Chapter 13 says a lot more about this.)

Chapter 4

Additive Models and Models with Interactions

Section 4.1 presents additive models represented as mixed linear models. This is another class of models with an extensive literature and as with penalized splines, additive models can be fit into the mixed-linear-model representation by representing them as sums of penalized splines at the price of sacrificing some techniques in that extensive literature, although this pruning appears to sacrifice little if any modeling power.

The sequence of additive models presented here naturally leads to models that may appear additive but which in fact include interactions. Section 4.2 begins with some such models, completing the progression of additive models based on penalized splines. These can be understood as models with an interaction of a categorical (grouping) variable with a continuous variable, with penalized splines used to fit smooth functions of the continuous variable. This is followed by analyses with interactions of two or more categorical variables, with the fits shrunk or smoothed by formulating them as analyses of mixed linear models. Such models, called "smoothed ANOVA" here, need not simply shrink categories toward a common center. As the example in Section 5.2.3 shows, when the levels of a categorical variable can be represented as a graph's nodes and edges, e.g., a spatial model where the outcomes y_i are aggregations over spatial regions, the smoothing can take advantage of the graph's structure.

The variety of models that can be represented with these additive and interaction models illustrates what Ruppert et al. (2003, Section 12.3.1) call the "modularity of spline models":

> [C]oncepts like main effects, interaction effects, generalized regression, and the mixed model formulation with smoothing parameter selection by REML can be viewed as modules and put together into an almost endless variety of statistical models. ... [O]ne can easily tailor a model to a specific application. (pp. 226–227)

Even this could be considered an understatement, because "by REML" could be expanded to "by REML or Bayes" and, as the following sections and chapters will illustrate, the modules can include spatial, time-series, or other models as well as penalized splines.

4.1 Additive Models as Mixed Linear Models

I present this material using a dental-research dataset that I originally analyzed using simpler methods. First the example is introduced, then additive models are defined, then several additive models are introduced and fit to the example's dataset. This sequence of models is not intended to demonstrate an efficient or even sensible data analysis, but merely to show the modularity of additive models expressed as mixed linear models, to show how these models look when fit to a dataset for which each is at least plausible, and to show some things that can happen in such fits. Many other models could be fit to these data; an exercise invites you to do so. As noted, the literature on additive models is quite large and this section is not intended to summarize it. Thorough, friendly treatments include Hastie & Tibshirani (1990) and Ruppert et al. (2003, especially Chapters 7, 8, 9, and 12).

4.1.1 The Pig Jawbone Example

Example 6. Mechanical properties of pig jawbone. Dental implants are placed in human jaw bones and used to support "permanent" prosthetic teeth or dentures. At the time this research was done, the standard of care was to drill holes in the jaw bone, place the implants, wait several months (usually 4) for the bone to heal so the implants were osseointegrated into the jaw, and only then to attach prosthetic teeth to the implants and allow the patient to load the implants by chewing. However, if bone is not stressed, it tends to lose calcium or volume or both; a good deal of bone volume or density can be lost in 4 months, particularly by the middle-aged or older people who are most likely to receive dental implants. The laboratory of Ching-Chang Ko, then at the University of Minnesota's Dental School, studied an alternative approach: Begin loading the implants much earlier, perhaps as early as a month after surgery, to speed the bone remodeling that produces osseointegration while also reducing bone loss from inactivity.

The dataset used here was partly reported in Ko et al. (2003) and Chang et al. (2003). Seven mini-pigs had specific teeth extracted and replaced by an implant. Each implant was covered by a device that protected it from loading until a specified healing time had elapsed, and then loaded it in a manner that simulated chewing. Two pigs' implants were loaded after 1 month of healing, two others were loaded after 2 months of healing, and the other three waited 4 months before loading. After a specified regimen of loading, each pig was sacrificed and samples of jawbone near the implants were subjected to nano-indentation testing to measure two mechanical properties, elastic modulus and hardness. The analyses presented below consider elastic modulus measurements from one mini-pig (Whitey) whose implant rested for two months before loading.

Figure 4.1 is a schematic of the data collection for one bone sample from Whitey, based on a photograph in Chang et al. (2003). Nanoindentations were made along *transects* that extended out from the implant's surface, roughly perpendicular to the implant. There were 9 such transects, 3 each in the coronal, middle, and apical thirds of the sample; the coronal 3 transects were mostly in cortical bone, the dense outer shell of the jawbone, while the other 6 transects were mostly in trabecular bone, the

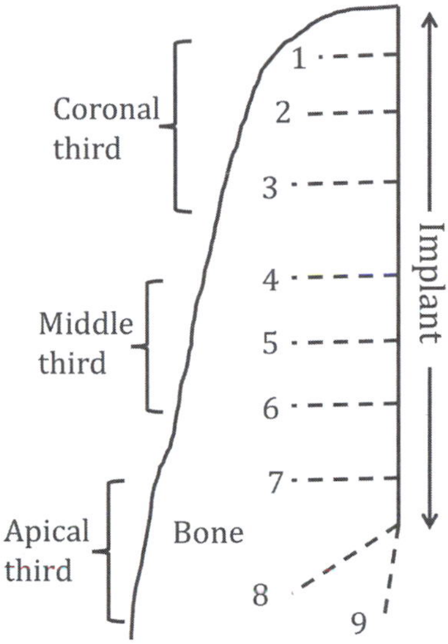

Figure 4.1: Schematic of data collection for the pig jawbone data. This is a cross section of the pig's lower jawbone: The top is the top of the bone, the right is the implant, and the rounded left margin is the outer margin of bone. The numbered dashed lines are the transects along which measurements were taken.

sparser structure inside the shell of cortical bone. Along each transect, a measurement was taken every 15 microns (0.015 mm) measuring out from the implant, when bone tissue was present; nano-indentation measurements could not be taken where no bone tissue was present to be nano-indented. Each transect extended to 1.5 mm, if the sample permitted. The scientific question was: How does elastic modulus depend on distance from the implant and does that relationship differ for the coronal, middle, or apical thirds or for the different transects within those sections? (And, considering all the pigs, how does elastic modulus depend on healing time?) The transects were not evenly spaced on the sample and each measurement location was only identified according to its distance from the implant along its transect, so a quantitative two-dimensional description of measurement locations was not available. Preliminary analyses shown below suggest that straight lines do not always describe how elastic modulus changes with distance from the implant, so a smooth curve might be more appropriate. (In the analyses in Chang et al. 2003, I divided distances into four groups and treated distance as a categorical factor.)

4.1.2 Additive Models Defined

In the preceding chapter, penalized splines were used to fit models of the form $y = f(x) + [\text{error}]$, where x is scalar and f is a smooth function of x. Additive models are useful when the predictor or regressor is a vector, in which case the model would begin as $y = f(\mathbf{x}) + [\text{error}]$, where $\mathbf{x} = (x_1, \ldots, x_p)'$ is a p-vector of known regressors. Additive models are the simplest version of this model that still allows a flexible shape for $f(\mathbf{x})$. For the case where the x_k are all measured on continuous scales, an additive model can be defined as

$$y = \alpha + \sum_{k=1}^{p} f_k(x_k) + \varepsilon, \tag{4.1}$$

where each $f_k(x_k)$ is a smooth function of its argument and the errors ε are independent with mean zero and constant variance (e.g., Hastie & Tibshirani 1990, p. 86). If each $f_k(x_k) = x_k\beta_k$, this is just a linear model without interactions, so an additive model can be viewed as a smooth model in which the predictors do not interact.

If x_1 (say) is categorical, e.g., transect in the pig jawbone data, then the sum in (4.1) runs from $k = 2$ to p and α is replaced by $\alpha + f_1(x_1)$, where $f_1(x_1)$ is allowed take a different value for each possible value of x_1.

Usually some further condition is required to ensure identifiability. For the case of categorical x_1, this might be $\alpha = 0$ or $f_1(x_1) = 0$ for one of x_1's categories.

4.1.3 Additive Models Fit to the Pig Jawbone Data

For the pig jawbone data, we have elastic modulus measurements y_{ij} taken at distance x_{ij} from the implant along the ith transect. In the additive models considered here, the first regressor is categorical, indicating transect, while the second regressor, distance along transect, is treated as continuous. The models considered in this section involve plugging in different forms for f_1 and f_2 in the definition of an additive model. All fits with random effects were computed using maximum restricted-likelihood estimates of ϕ_R and ϕ_G.

Figure 4.2 is a trellis plot of the elastic modulus versus distance from the implant, with separate panels for the 9 transects. In each panel, the dashes are a smooth of the transect's data using the R function loess.smooth with defaults; the straight solid line in each panel will be explained later. Elastic modulus as a function of distance does not seem to be the same for all 9 transects, and for some the relationship is arguably non-linear, particularly transects 5 and 6 and less obviously transect 7. Other aspects of these data that are visible in the plots will be considered later.

The simplest additive model gives each transect its own intercept but fits a common slope:

$$y_{ij} = \sum_{l=1}^{9} \beta_l I(l = i) + \beta_d x_{ij} + \varepsilon_{ij}, \tag{4.2}$$

where β_d is the common slope in distance and the β_l are the transect-specific intercepts, selected by the indicator function $I(l = i)$. The ε_{ij} are iid normal with mean zero and common variance σ_e^2. This is an additive model: Transect number and distance do not interact. Everything in this fit is a fixed effect, so this is not yet a mixed linear model. Figure 4.2 shows the fit, which has 10 DF (nine intercepts and one

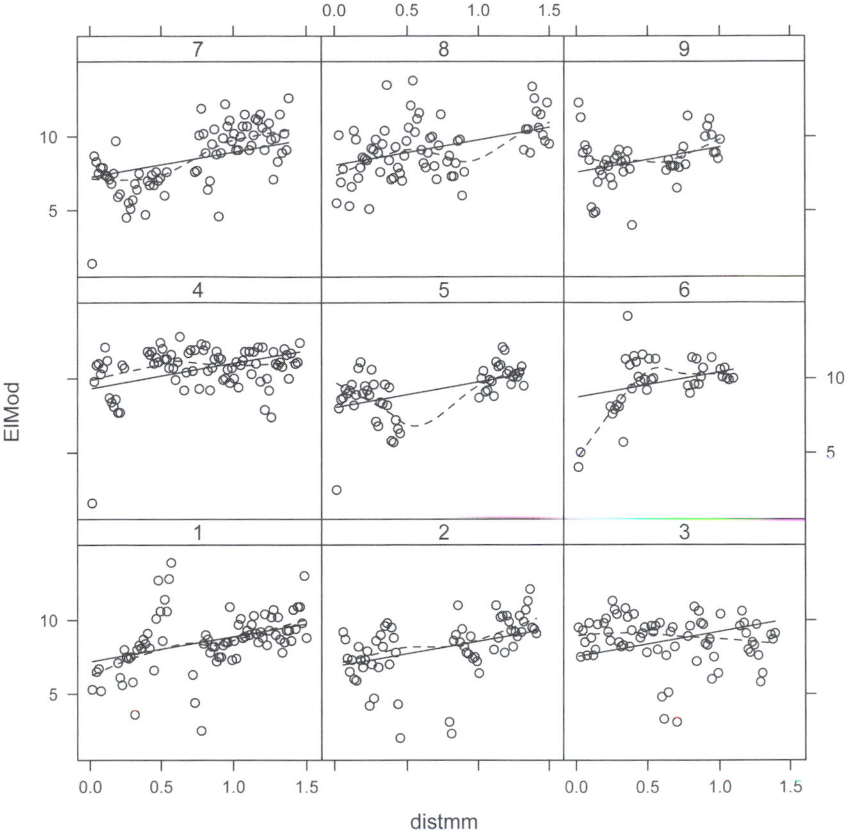

Figure 4.2: Pig jawbone data: Elastic modulus for each of the 9 transects. The dashed curve is a loess fit. The solid line is the fitted additive model with common slope in distance and a different intercept for each transect.

slope). The common slope seems to fit well for many transects, except possibly transects 3, 5, and 6.

The next model demonstrates the modularity of splines by unplugging the linear term for distance in (4.2) and in its place, plugging in a penalized spline. For a truncated-quadratic spline with knots κ_k, the equation for this new model is

$$y_{ij} = \sum_{l=1}^{9} \beta_l I(l = i) + \beta_{11} x_{ij} + \beta_{12} x_{ij}^2 + \sum_{k=1}^{K} u_k (x_{ij} - \kappa_k)_+^2 + \varepsilon_{ij}, \tag{4.3}$$

where the u_k are iid $N(0, \sigma_s^2)$. I follow the default in Ruppert et al. (2003) by using 25 equally spaced knots, there being 100 distinct distance (x_{ij}) values. This is an additive model — there is no interaction between transect and distance — and is represented as a mixed linear model as follows. If n_i is the number of observations on transect i and $\mathbf{x}_i$ is the column vector of x_{ij} for transect i, the fixed effects design matrix and parameters are

$$\mathbf{X} = \left[\begin{array}{cccc|cc} \mathbf{1}_{n_1} & \mathbf{0}_{n_1} & \dots & \mathbf{0}_{n_1} & \mathbf{x}_1 & \mathbf{x}_1^2 \\ \mathbf{0}_{n_2} & \mathbf{1}_{n_2} & \dots & \mathbf{0}_{n_2} & \mathbf{x}_2 & \mathbf{x}_2^2 \\ & & \vdots & & \vdots & \\ \mathbf{0}_{n_9} & \mathbf{0}_{n_9} & \dots & \mathbf{1}_{n_9} & \mathbf{x}_9 & \mathbf{x}_9^2 \end{array} \right], \quad \beta = \begin{bmatrix} \beta_1 \\ \vdots \\ \beta_9 \\ \beta_{11} \\ \beta_{12} \end{bmatrix} \tag{4.4}$$

and the random effects design matrix is

$$\mathbf{Z} = \begin{bmatrix} \mathbf{B}_1 \\ \vdots \\ \mathbf{B}_9 \end{bmatrix} \quad \text{for } \mathbf{B}_i = \begin{bmatrix} (x_{i1} - \kappa_1)_+^2 & \dots & (x_{i1} - \kappa_K)_+^2 \\ & \vdots & \\ (x_{in_i} - \kappa_1)_+^2 & \dots & (x_{in_i} - \kappa_K)_+^2 \end{bmatrix}, \tag{4.5}$$

with random effect $\mathbf{u} = (u_1, \dots, u_K)'$. The covariance matrices are $\mathbf{G} = \sigma_s^2 \mathbf{I}_K$ and $\mathbf{R} = \sigma_e^2 \mathbf{I}_n$. Using a different basis and knots would imply different fixed and random effects.

Figure 4.3 shows a fit of this model to Whitey's data. This fit has 11.9 degrees of freedom total; since there are 9 DF for intercepts, this means the spline smooth in distance has 2.9 DF, 2 DF for the linear and quadratic fixed effects and 0.9 DF for the spline's random effects. The fitted values in Figure 4.3 show a modest waver in the fitted smooth in distance consistent with the small DF in the random-effect part of the spline fit. This modest waver is a compromise between the pronounced effect for transect 6 and the considerably gentler effects in the other transects.

Plugging in a spline fit for distance cost 1.9 extra degrees of freedom compared to the common-slope line in Figure 4.2. The spline fit could be compared to the linear fit using a test of the null hypothesis that $\sigma_s^2 = 0$; the test is an exercise, but it's hard to imagine how this model could be preferable to the simpler one.

The obvious alternative at this point is to fit distinct smooth functions for each transect. I'll do that later, but first I want to show how easily other modules can be plugged into the additive model.

So far the intercepts for the 9 transects have been treated as fixed effects, each with its own full DF. But in Figures 4.2 and 4.3, most transects seem to have similar fitted intercepts. Perhaps we can save some degrees of freedom by treating the transect intercepts as a random effect, shrinking them toward a common value. If we take the previous model (4.3), unplug the 9 fixed effect intercepts and plug in a single fixed-effect intercept (β_0) with a random effect for intercepts, the new model is

$$y_{ij} = \beta_0 + \beta_{11} x_{ij} + \beta_{12} x_{ij}^2 + u_{0i} + \sum_{k=1}^{K} u_{1k} (x_{ij} - \kappa_k)_+^2 + \varepsilon_{ij}, \tag{4.6}$$

where the u_{0i} are iid $N(0, \sigma_{s0}^2)$; the random effects in the penalized spline are now

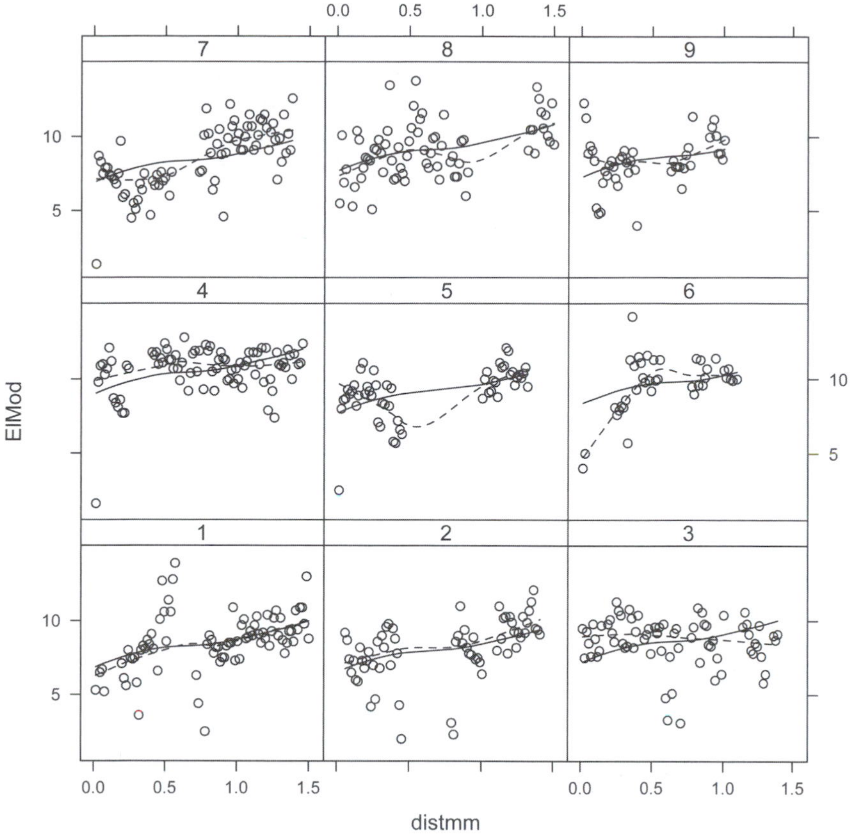

Figure 4.3: Pig jawbone data: The solid curve is the fitted additive model with a common spline fit in distance and a different intercept for each transect. The dashed curve is a loess fit.

called u_{1k} and are iid $N(0, \sigma_{s1}^2)$. The mixed-linear-model representation now has

$$\mathbf{X} = \begin{bmatrix} \mathbf{1}_{n_1} & \mathbf{x}_1 & \mathbf{x}_1^2 \\ \mathbf{1}_{n_2} & \mathbf{x}_2 & \mathbf{x}_2^2 \\ & \vdots & \\ \mathbf{1}_{n_9} & \mathbf{x}_9 & \mathbf{x}_9^2 \end{bmatrix} \qquad \beta = \begin{bmatrix} \beta_0 \\ \beta_{11} \\ \beta_{12} \end{bmatrix} \tag{4.7}$$

and the random effects design matrix is

$$\mathbf{Z} = \left[\begin{array}{cccc|c} \mathbf{1}_{n_1} & \mathbf{0}_{n_1} & \cdots & \mathbf{0}_{n_1} & \mathbf{B}_1 \\ \mathbf{0}_{n_2} & \mathbf{1}_{n_2} & \cdots & \mathbf{0}_{n_2} & \mathbf{B}_2 \\ & & \vdots & & \vdots \\ \mathbf{0}_{n_9} & \mathbf{0}_{n_9} & \cdots & \mathbf{1}_{n_9} & \mathbf{B}_9 \end{array} \right] \tag{4.8}$$

with $\mathbf{B}_j$ as above. The random effect is $\mathbf{u} = (u_{01}, \ldots, u_{09}, u_{11}, \ldots, u_{1K})'$, and the covariance matrices are $\mathbf{G} = \text{blockdiag}(\sigma_{s0}^2 \mathbf{I}_9, \sigma_{s1}^2 \mathbf{I}_K)$ and $\mathbf{R} = \sigma_e^2 \mathbf{I}_n$.

Somewhat surprisingly, the new fit differs little from the fit in Figure 4.3: Pearson's correlation between the fitted values from these two models is 0.998, so I haven't shown a plot of the new fitted values. The DF of the new fit is 11.3, hardly reduced at all from the previous fit even though the intercepts appear similar in the fit of (4.3). This 11.3 DF is composed of 3 DF for the fixed effects (overall intercept and linear and quadratic terms in distance), 7.4 DF for the intercept random effect, and 0.9 DF for the random-effect part of the spline, as in the preceding fit. So the total DF for intercepts has only come down from 9 to 8.4, even though the intercepts look quite similar in Figures 4.2 and 4.3 and might have been expected to shrink close to a common value.

Perhaps the intercepts will shrink more with a random walk down the transects from the most coronal (transect 1) to the most apical (transect 9). That is, unplug the module placing an unstructured random effect on the intercepts and plug in a module that imposes some structure. This model is

$$y_{ij} = \beta_0 + \beta_{11} x_{ij} + \beta_{12} x_{ij}^2 + \xi_i + \sum_{k=1}^{K} u_{1k}(x_{ij} - \kappa_k)_+^2 + \varepsilon_{ij}, \tag{4.9}$$

where the ξ_i are smoothed using a random walk: $\xi_i = \xi_{i-1} + \delta_i, i = 2, \ldots, 9$, and the δ_i are iid $N(0, \sigma_{s0}^2)$ (note that δ_i's index i runs from 2 to 9). To fit this into the mixed-linear-model formulation, rewrite the model for ξ_i as $\xi_i = \xi_1 + \sum_{l=2}^{i} \delta_l, i = 2, \ldots, 9$. Now we have two intercepts, β_0 and ξ_1, so set $\xi_1 = 0$, making $\xi_i = \sum_{l=2}^{i} \delta_l, i = 2, \ldots, 9$. The fixed effects design matrix $\mathbf{X}$ and parameters $(\beta_0, \beta_{11}, \beta_{12})'$ are unchanged, as is the error covariance $\mathbf{R}$. The random effects design matrix is now

$$\mathbf{Z} = \left[\begin{array}{cccc|c} \mathbf{0}_{n_1} & \mathbf{0}_{n_1} & \ldots & \mathbf{0}_{n_1} & \mathbf{B}_1 \\ \mathbf{1}_{n_2} & \mathbf{0}_{n_2} & \ldots & \mathbf{0}_{n_2} & \mathbf{B}_2 \\ \mathbf{1}_{n_3} & \mathbf{1}_{n_3} & \ldots & \mathbf{0}_{n_2} & \mathbf{B}_2 \\ & & \vdots & & \vdots \\ \mathbf{1}_{n_9} & \mathbf{1}_{n_9} & \ldots & \mathbf{1}_{n_9} & \mathbf{B}_9 \end{array} \right] \tag{4.10}$$

with $\mathbf{B}_j$ as before. The random effect is $\mathbf{u} = (\delta_2, \ldots, \delta_9, u_{11}, \ldots, u_{1K})'$, and the random-effect covariance matrix is $\mathbf{G} = \text{blockdiag}(\sigma_{s0}^2 \mathbf{I}_8, \sigma_{s1}^2 \mathbf{I}_K)$.

This latest fit, however, looks hardly different from the last two: Its fitted values have Pearson's correlation 0.995 or higher with each previous set of fitted values, so again I haven't plotted it. This fit is a bit more economical in terms of DF, with 10.8 DF total: 3 DF for the fixed effects again, 0.9 DF again for the random-effect part of the spline, and 6.9 DF for the random-walk part of the intercepts. With a total of 7.9 DF for intercepts, we have shaved a total of 1.1 DF off the fitted intercepts without any perceptible effect on the fitted curves.

In a further attempt to simplify the intercepts, we can unplug the intercept β_0 and plug in a fixed effect for the coronal, middle, and apical sections of this sample from Whitey's jaw, consisting of transects 1–3, 4–6, and 7–9 respectively. The only change to (4.9) is to replace β_0 and the corresponding column of $\mathbf{X}$ with one intercept column for each of the sections and the corresponding parameters $(\beta_0, \beta_1, \beta_2)'$. The resulting fitted values have Pearson's correlation 0.998 with the preceding model's fitted values so again, no plot. Examining the DF and the estimates for the ξ_i shows that this fit has done little besides shift 2 DF from the random-walk random effect into the fixed effect for sections.

Before leaving additive models, consider one further module change: Unplugging the iid error covariance $\mathbf{R} = \sigma_e^2 \mathbf{I}$ and plugging in an $\mathbf{R}$ with errors serially correlation within transects. This can be done in several ways; for illustration, consider a block diagonal $\mathbf{R}$ with blocks $\mathbf{R}_i$ having $(j,l)^{\text{th}}$ element $\sigma_e^2 \exp(-\theta d_{jl})$, where $d_{jl} = |x_{ij} - x_{il}|$ is the distance in mm between measurements j and l on transect i. Given the deficiencies in the additive fit for some transects, visible in Figure 4.3, we might expect that errors will appear highly autocorrelated in this fit, and they do. The maximum restricted-likelihood estimate of θ is 42.6, so two adjacent measurements, 0.015 mm apart, are estimated to have errors with correlation 0.53. Figure 4.4 shows the new fit. Comparing Figure 4.4 to Figure 4.3, the fit is detectably different, although the fitted values from this latest model have correlation 0.97 with the fitted values from (4.3), the simplest additive model with a spline. This model, however, has achieved rather more economy than the preceding models. Total DF in the fit is 8.2, with 5 DF for fixed effects, 2.4 DF for the random walk in intercepts (so 5.4 DF for intercepts total, with 3 in the section effect), and 0.8 for the random-effect part of the spline. This economy in intercepts is reflected in rather more similar intercepts within each of the three sections (transects 1–3, transects 4–6, and transects 7–9). By what mechanism do autocorrelated errors have this effect? No one knows, but perhaps you can discover that mechanism. An exercise at the end of this chapter suggests a hypothesis.

The next step, long delayed, is to fit a distinct curve for each of the 9 transects. Such a model is no longer additive: The elastic modulus is a different function of distance for each transect, which is the definition of an interaction. Models with interactions are the subject of the next section.

4.2 Models with Interactions

An interaction can involve a categorical predictor and a continuous predictor, as in the pig-jawbone example. Section 4.2.1 fits one interaction model of this type and briefly discusses other possible models. An interaction can also involve two categorical predictors; Section 4.2.2 presents an approach to shrinking or smoothing such interactions. Finally, an interaction can involve two continuous predictors. This is equivalent to two-dimensional smoothing, a special case of spatial smoothing, which is discussed in Chapter 5. (Ruppert et al. 2003, Section 12.4, show how to represent varying-coefficient regression, a special case of a continuous-by-continuous interaction, as a penalized spline and thus as a mixed linear model.)

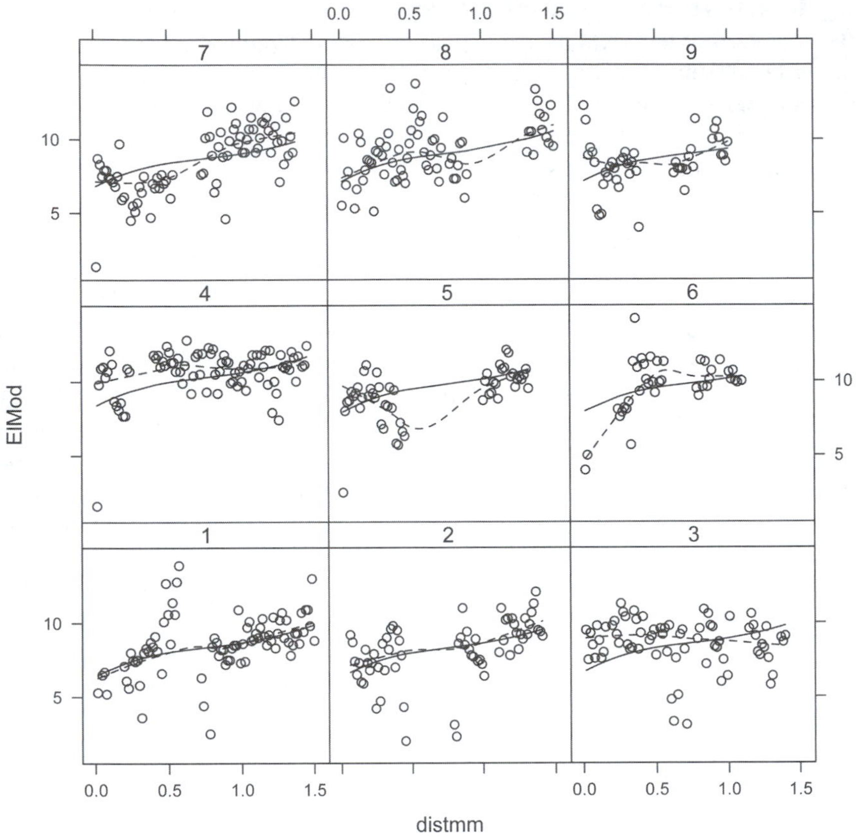

Figure 4.4: Pig jawbone data: The solid curve is the fitted additive model with a common spline fit in distance, a random walk for intercepts, and serially correlated errors within transects. The dashed curve is a loess fit.

4.2.1 Categorical-by-Continuous Interactions

Let's start from model (4.3), the additive model with a fixed-effect intercept for each transect and a single smooth curve for distance. The transect-specific loess fits in Figure 4.3 suggest fitting different curves in distance for the different transects. The following model fits a different spline smooth for each transect. In place of (4.3), write

$$\begin{aligned} y_{ij} &= \sum_{l=1}^{9} I(l=i)\left\{\beta_{0i} + \beta_{1i}x_{ij} + \beta_{2i}x_{ij}^2\right\} \\ &\quad + \sum_{l=1}^{9} I(l=i)\left\{\sum_{k=1}^{K} v_k^l (x_{ij} - \kappa_k)_+^2\right\} + \varepsilon_{ij}, \end{aligned} \tag{4.11}$$

where the $\mathbf{v}^l = (v_1^l, \ldots, v_K^l)', l = 1, \ldots, 9$ are distributed independently as $N(\mathbf{u}, \Sigma_{vl})$ for $\mathbf{u} = (u_1, \ldots, u_K)'$. ($\Sigma_{vl}$ will be simplified below.) The mean vector $\mathbf{u}$ amounts to the main effect of distance and the covariance matrices Σ_{vl} describe how the transect-specific fits vary around this main effect. Equation (4.11) can be rewritten with explicit main effects and interactions, writing $\mathbf{w}^l = \mathbf{v}^l - \mathbf{u}$ to give

$$\begin{aligned} y_{ij} &= \sum_{l=1}^{9} I(l=i)\left\{\beta_{0i} + \beta_{1i}x_{ij} + \beta_{2i}x_{ij}^2\right\} && (4.12)\\ &+ \sum_{k=1}^{K} u_k (x_{ij} - \kappa_k)_+^2 && (4.13)\\ &+ \sum_{l=1}^{9} I(l=i)\left\{\sum_{k=1}^{K} w_k^l (x_{ij} - \kappa_k)_+^2\right\} + \varepsilon_{ij}, && (4.14) \end{aligned}$$

where (4.13) is the main effect of distance, (4.14) is the interaction of distance with transect along with the error term, and the $\mathbf{w}^l$ are iid $N(\mathbf{0}, \Sigma_{vl})$. Now $\mathbf{u}$ needs some penalty or constraint or shrinkage to avoid overfitting. This can be done by adding a model (or prior, if you prefer) $\mathbf{u} \sim N(\mathbf{0}, \Sigma_u)$. Estimating Σ_u or the Σ_{vl} from a single draw is obviously hopeless, so some further modeling of these matrices is necessary. Ruppert et al. (2003, Section 12.3) assume $\Sigma_u = \sigma_u^2 \mathbf{I}_K$ and $\Sigma_{vl} = \sigma_{vl}^2 \mathbf{I}_K$. This is consistent with their preferred penalties for splines and allows each transect's fitted curve to differ from the main effect to a degree specific to the transect. Obviously other models are possible. (My parameterization differs slightly from Ruppert et al. 2003, Chapter 12. I am not an expert here, so this may not be innocuous.)

Expressed as a mixed linear model, the fixed-effect terms in (4.12) imply 3 columns in $\mathbf{X}$ for each transect, intercept, linear, and quadratic terms, with zeroes in those columns for observations in all other transects. The spline for the main effect of distance in (4.13) has 25 design columns in $\mathbf{Z}$, one per knot, identical to those in the simple additive model (4.3). The design matrix for the interaction of transect and distance in (4.14) is block diagonal, with one block of 25 columns for each transect. Each transect's block consists of that transect's rows in the main effect's design columns.

This model is naïve in important respects. First, it uses the same 25 knots as in the fit with a single spline in distance, model (4.3). For that fit, the number 25 was the default of Ruppert et al. (2003), {number of distinct distance values}/4. However, applying the same default to the 9 transects individually gives knot counts ranging from 10 for transect 6 up to 21 for transect 4. The individual transects have less than 100 distinct values of distance because of empty spaces in trabecular bone and because some transects are less than 1500 microns long. Also, I look at model (4.12, 4.13, 4.14) and see potential identification problems because the first 25 columns of $\mathbf{Z}$, for the main effect for distance (4.13), are exactly collinear with a simple linear combination of the last 225 columns of $\mathbf{Z}$, for the interaction (4.13). The model (4.12, 4.13, 4.14) is identified only because of the model (or prior, if you prefer) on $\mathbf{u}$ and the $\mathbf{w}^l$, specifically the covariance matrix $\mathbf{G}$: if the unknown variances in $\mathbf{G}$ are allowed to be large, the model is no longer identified. A Bayesian analysis can create "good" behavior in the fit by constraining the variances σ_u^2 and σ_{vl}^2 with prior distributions that nudge posterior probability away from large values and weak identification. However, with the current state of knowledge and software, setting such a prior for the Bayesian analysis would require non-trivial tinkering to ensure that it constrained the fit enough but not too much.

Although this model is naïve, I fit it anyway using restricted likelihood, and the results are somewhat instructive. Not surprisingly, maximizing the restricted likelihood was a bit fussy: I got different results with two fairly similar sets of starting values, one ending in true convergence while the other ended in false convergence (according to the R function nlminb) at a somewhat smaller log restricted likelihood. A preliminary fit had showed that maximizing the restricted likelihood would produce zero variance estimates, so I used a lower bound of 0.0001 for all the variance estimates. (The R package nlme uses a similar constraint to avoid numerical problems arising from zero variance estimates.) Figure 4.5 shows the fitted values from the starting values that gave true convergence.

Maximizing the restricted likelihood gave an estimate of 2.2 for the error variance, compared to about 2.9 for all the additive models. The variance controlling the main effect for distance, σ_u^2, was estimated at the lower bound 0.0001, while the interactions for transects 1 to 9 had variances σ_{vl}^2 estimated at 5793, 0.0001, 0.0001, 15.0, 1754, 40.3, 22.8, 44.4, and 8022 respectively. Apart from the zeroes, the effect of these variances depends on the scales of the respective design-matrix columns, so the variances are impossible to interpret. Thus, consider DF instead. This fit had 43.2 total DF — compared to 8 to 12 DF for the additive fits — with 27 DF for fixed effects and 16.2 DF for random effects. The spline for the function of distance in each of the 9 transects had 3 fixed-effect DF, one each for intercept, linear, and quadratic terms. DF beyond these fixed effects were 0.00006 for the main effect, effectively zero, and for transects 1 to 9, 5.6, 0.00002, 0.00002, 1.2, 2.7, 0.7, 1.2, 1.5, and 3.2 respectively. The fits for transects 1, 5, and 9 over-fit apparently chance irregularities at the beginning or end of a gap in distance.

The other set of starting values, which produced a false convergence, stopped at a rather different set of variances. However, these variances gave very similar total DF in the fit, 43.4 DF, allocated somewhat differently than in the truly converged fit, mostly notably with 2.5 DF for the random-effect part of the main effect in distance, compared to zero for the truly converged fit. The fitted values for the falsely converged fit (Figure 4.6) are similar to those from the truly converged fit, the main difference being that transect 5 is less grossly overfit than in the truly converged fit.

Several less naïve models come to mind readily, for example with fewer knots. The derivation above and the model in Ruppert et al. (2003, Chapter 12) use the same number of knots for each transect. Presumably this is not necessary, though its simplicity is attractive. We could also assume $\sigma_{vl}^2 \equiv \sigma_v^2$ for all l, although this would force the same degree of smoothing for all transects; the loess and spline fits in Figure 4.5 suggest this might overfit some transects and underfit others. These possibilities are left for exercises. A further possibility is to fit the model with different v_{kl} but to treat the v_{kl} as draws from, say, an inverse gamma distribution with unknown parameters. This would shrink the v_{kl} toward a common value and is thus intermediate between

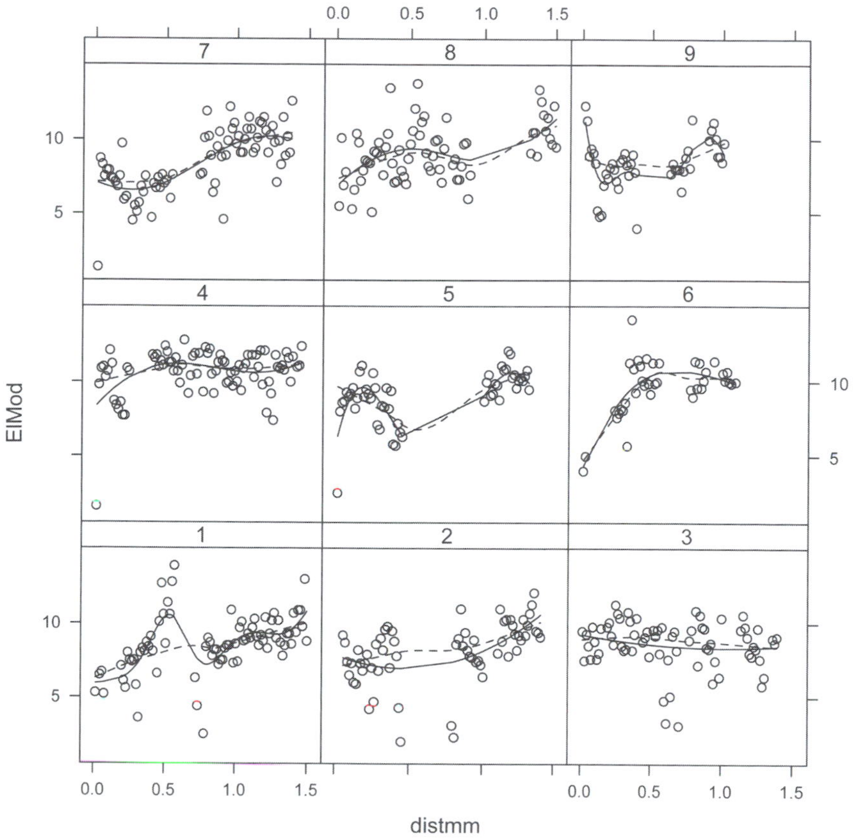

Figure 4.5: Pig jawbone data: The solid curve is the fitted model with fixed effects for each transect's intercepts, a separate truncated-quadratic spline for each transect's function of distance, and uncorrelated errors. The dashed curve is the loess fit.

the model fit above and a model with $\sigma_{vl}^2 \equiv \sigma_v^2$. This is no longer a mixed linear model, although a Bayesian analysis using BUGS or JAGS would be straightforward to code. The MCMC could be ill-behaved depending on sample size and possibly other aspects of the dataset or design. Lee et al. (2006, Chapter 11) discuss fitting models like these using their approach to analysis.

In describing fits of models like (4.12, 4.13, 4.14), Ruppert et al. (2003, Chapter 12) suggest there are no particular problems as long as the $(\beta_{0i}, \beta_{1i}, \beta_{2i})$ are treated as a fixed effect and, presumably, you avoid naïve behavior like using too many knots. However, for the pig-jawbone data or other datasets it could make sense to treat the $(\beta_{0i}, \beta_{1i}, \beta_{2i})$ as a random effect, so that each transect in Whitey's data is a randomly drawn "subject" with a subject-specific curve in distance. Ruppert et al. (2003, Section 9.3) give one way to specify and fit the resulting model as a mixed linear model, but note that "[a]lthough this approach to fitting subject-specific curves

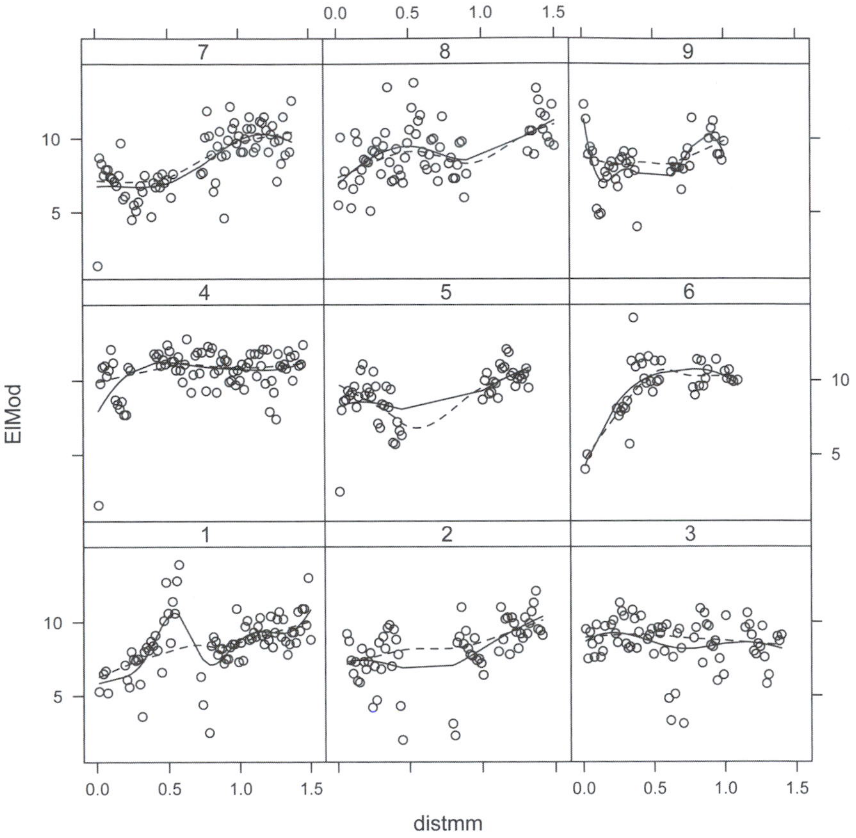

Figure 4.6: Pig jawbone data: The solid curve is the fit arising from the false-convergence solution for the model with fixed effects for each transect's intercepts, a separate truncated-quadratic spline for each transect's function of distance, and uncorrelated errors. The dashed curve is the loess fit.

seems straightforward, more work needs to be done on implementation" (p. 192), for example by centering the continuous predictor (distance, in our example) to avoid poor identification. I infer that these experts find interaction models of this sort to be somewhat fragile, at least in the current state of understanding. An exercise asks you to fit this model to the pig jawbone data.

4.2.2 Categorical-by-Categorical Interactions (Smoothed ANOVA)

At this point, it seems a natural progression to shrink or smooth interactions of categorical variables. Statistical folklore and experience suggest that interactions are often absent or small, but it's unwise to assume that any specific interaction is absent or small. Interactions are often modeled in a stepwise way, with significance tests used to delete or retain effects, but stepwise methods are outperformed by model-averaging and smoothing methods (Leamer 1978; Freedman 1983; Derksen & Keselman 1992; Raftery, Madigan, & Hoeting 1993, Hodges et al. 2007). This suggests shrinking or smoothing interactions as an alternative to a binary include/exclude choice.

Apart from that general motivation, the earliest specific motivation I know is Dixon & Simon's (1991) approach to subgroup analysis in clinical trials, in which treatment effects in subgroups of patients, say males and females, are considered after first shrinking the treatment-by-sex interaction. This has the effect of filtering out some error and shrinking, to some extent, spurious apparent differences between subgroups. More recently, Nobile & Green (2000) proposed a version of smoothed two-way analysis of variance (ANOVA) in this spirit, aiming to gather the ANOVA's levels into clusters and thus simplify reporting and interpretation of results. Their model cannot be represented as a mixed linear model. Gelman (2005a) proposed a version of smoothed ANOVA much like the one presented here, treating all effects as random effects, replacing F-tests with variance-component estimation, and making error-term selection implicit and automatic. (One might argue that mixed-linear-model theory already does the latter, but Gelman had bigger fish to fry.) Gelman intended to re-interpret ANOVA as "a tool for data exploration . . . used to construct useful models," and the version of smoothed ANOVA presented below might be seen as providing some useful models.

More recently, ANOVA-like models have been introduced into the literature growing out of the lasso on simultaneous variable selection and shrinkage; one example is Bondell & Reich (2009). Obviously the idea of analyzing variance into components attributed to factors is hugely fertile and again, the present section is not intended to survey this growing body of work. Instead, this section's goal is to present a version of smoothed ANOVA that emphasizes modeling the effects being shrunk or smoothed while staying within the mixed-linear-model representation. Thus, while the example below just shrinks effects toward zero, Section 5.2.3 shows a two-way ANOVA with a spatial factor in which the corresponding main effect and interaction are smoothed spatially.

Although my collaborators and I were motivated by subgroup analysis to develop our version of smoothed ANOVA, it turns out to serve unexpected purposes specific to particular applications. These are discussed in the context of the examples used below.

4.2.3 Smoothed ANOVA, Balanced Design with a Single Error Term

This section considers only balanced designs with a single error term. A later section briefly considers more general designs.

Example 7. Denture-liner material. Soft denture liners are fabricated on a hard denture base, then polished and finished. Polishing and finishing can create or widen a gap between the liner and base. Such gaps harbor *Candida* and other oral pathogens, which are especially hazardous for older people with fragile oral mucosa, the usual indication for a soft denture liner. This study (Pesun et al. 2002) compared gaps, measured in microns (μm), for a new and a standard soft-liner material (factor M) under all 32 combinations of four polishing methods (factor P) and eight finishing methods (factor F), with no replication within design cells. The primary interest is how much the materials differ in gap size. Based on standard diagnostics, we analyzed the common logarithm of gap size, $\log_{10}$gap.

Three issues arise in this analysis. The standard analysis of unreplicated designs uses the highest-order interaction as the error term (e.g., Scheffé 1959, Section 4.2; this is the analysis published in Pesun et al. 2002). Table 4.1 shows the first 35 rows of the dataset (the other columns in this table are explained below). This dataset has an egregious outlier on the raw scale which also fails outlier tests on the log scale, though not by much. It would thus be desirable to keep in the fitted model part of the three-way interaction — the outlier — while also deriving an error measure from this and perhaps other interactions. Using the three-way interaction as the error term, including the outlier gives P-values 0.12, 0.097, and 0.15 for the M×P, M×F, and P×F interactions respectively; removing the outlier changes these P-values to 0.096, 0.004, and 0.16 respectively, so the outlier differentially affects the tests. Second, the P×F and M×P×F interactions have 21 degrees of freedom (DF) each. At most a few of these contrasts are truly present, and they could be masked by the many null contrasts. Finally, the interactions of special interest are M×P and M×F, with 3 and 7 DF respectively. Shrinking (smoothing) the material comparison across the four polishing methods and eight finishing methods would reduce clutter. As we'll see, smoothing ANOVA addresses all these issues, unreplicated designs, masking in effects with many DF, and subgroup analysis, where treatment-by-subgroup interactions capture subgroup treatment effects as in Dixon & Simon (1991).

The subsections to follow lay out the machinery for smoothed ANOVA (SANOVA), and then apply that machinery to the denture-liner material example.

4.2.3.1 Notation and Other Machinery

This section illustrates the notation using a 2^3 design with 6 replicates per cell. The details were presented in Hodges et al. (2007) so we omit many of them.[1] In this section we only smooth/shrink the interactions, but we could use the same machinery to smooth/shrink the main effects. In Section 5.2.3's spatial example, one of the main effects is smoothed. The analysis is Bayesian; prior distributions have an important role here, as we will see.

Suppose a balanced design has c cells and $m \geq 1$ observations per cell, cm in total. In the 2^3 example with 6 replicates per cell, $c = 2^3 = 8, m = 6$ and $cm = 48$. Write the ANOVA as a linear model with each effect having design-matrix columns orthogonal to each other and to columns for other effects, i.e., an orthogonal parameterization. Effects with two or more DF have infinitely many such design matrices related by orthogonal transformation. Group the p columns for the grand mean and main effects into a $cm \times p$ matrix $\mathbf{X}$, and the q columns for interactions into a $cm \times q$ matrix $\mathbf{Z}$.

[1] Much of this section's text comes directly from Hodges et al. (2007).

Table 4.1: The Denture-Liner Data and Design Matrix Used in the Smoothed ANOVA

gap (μm)	Material	Polishing			Finishing						
5.02	1	3	0	0	7	0	0	0	0	0	0
8.84	1	3	0	0	−1	6	0	0	0	0	0
3.61	1	3	0	0	−1	−1	5	0	0	0	0
10.55	1	3	0	0	−1	−1	−1	4	0	0	0
3.90	1	3	0	0	−1	−1	−1	−1	3	0	0
5.64	1	3	0	0	−1	−1	−1	−1	−1	2	0
98.95	1	3	0	0	−1	−1	−1	−1	−1	−1	1
10.75	1	3	0	0	−1	−1	−1	−1	−1	−1	−1
2.91	1	−1	2	0	7	0	0	0	0	0	0
3.00	1	−1	2	0	−1	6	0	0	0	0	0
5.94	1	−1	2	0	−1	−1	5	0	0	0	0
8.64	1	−1	2	0	−1	−1	−1	4	0	0	0
16.33	1	−1	2	0	−1	−1	−1	−1	3	0	0
7.44	1	−1	2	0	−1	−1	−1	−1	−1	2	0
11.26	1	−1	2	0	−1	−1	−1	−1	−1	−1	1
16.35	1	−1	2	0	−1	−1	−1	−1	−1	−1	−1
4.75	1	−1	−1	1	7	0	0	0	0	0	0
3.93	1	−1	−1	1	−1	6	0	0	0	0	0
4.90	1	−1	−1	1	−1	−1	5	0	0	0	0
13.44	1	−1	−1	1	−1	−1	−1	4	0	0	0
2.82	1	−1	−1	1	−1	−1	−1	−1	3	0	0
6.44	1	−1	−1	1	−1	−1	−1	−1	−1	2	0
20.88	1	−1	−1	1	−1	−1	−1	−1	−1	−1	1
9.30	1	−1	−1	1	−1	−1	−1	−1	−1	−1	−1
178.22	1	−1	−1	−1	7	0	0	0	0	0	0
1.95	1	−1	−1	−1	−1	6	0	0	0	0	0
3.70	1	−1	−1	−1	−1	−1	5	0	0	0	0
18.11	1	−1	−1	−1	−1	−1	−1	4	0	0	0
16.40	1	−1	−1	−1	−1	−1	−1	−1	3	0	0
9.61	1	−1	−1	−1	−1	−1	−1	−1	−1	2	0
36.52	1	−1	−1	−1	−1	−1	−1	−1	−1	−1	1
14.88	1	−1	−1	−1	−1	−1	−1	−1	−1	−1	−1
18.68	−1	3	0	0	7	0	0	0	0	0	0
49.02	−1	3	0	0	−1	6	0	0	0	0	0
4.55	−1	3	0	0	−1	−1	5	0	0	0	0
⋮	⋮		⋮					⋮			

Note: The first 35 observations and corresponding design columns are shown. The left-most column is the gaps on the original scale. The second column is the design-matrix column for the material main effect, the next 3 columns are for the polishing main effect, and the 7 right-most columns are for the finishing main effect. The latter 10 columns repeat for the last 32 observations in the dataset. Columns for an interaction are obtained by multiplying the columns for the two main effects. Design-matrix columns have *not* been scaled to have the same Euclidean length.

Scale the columns of $\mathbf{X}$ and $\mathbf{Z}$ so $\mathbf{X}'\mathbf{X} = cm\mathbf{I}_p$ and $\mathbf{Z}'\mathbf{Z} = cm\mathbf{I}_q$. (For the models shown in this section, where contrast coefficients θ_k sharing a smoothing precision η_j are exchangeable, any choice of contrasts with the same span will give the same posterior distribution for the vector of precisions and thus the same fitted values and the same posterior distribution for DF in the fit. This is not necessarily true for more general smoothing schemes.)

Hodges et al. (2007) used the constraint-case formulation (Section 2.1) to present SANOVA; here we'll use the mixed-linear-model formulation. The usual ANOVA in linear model form is simply $\mathbf{y} = [\mathbf{X}|\mathbf{Z}]\Theta + \varepsilon$, where $\mathbf{y}$ is the cm-vector of data, $\Theta = (\theta_1, \ldots, \theta_{p+q})$ is the vector of unknown mean-structure parameters, and ε is cm-dimensional normal with mean zero and covariance $\mathbf{R} = \frac{1}{\eta_0}\mathbf{I}_{cm}$, the error precision η_0 being unknown. In the 2^3 example, one choice of $[\mathbf{X}|\mathbf{Z}]$ is $H \otimes \mathbf{1}_6$, where $\otimes$ is the Kronecker product and

$$H = \left[\begin{array}{cccc|cccc} +1 & +1 & +1 & +1 & +1 & +1 & +1 & +1 \\ +1 & +1 & +1 & -1 & +1 & -1 & -1 & -1 \\ +1 & +1 & -1 & +1 & -1 & +1 & -1 & -1 \\ +1 & +1 & -1 & -1 & -1 & -1 & +1 & +1 \\ +1 & -1 & +1 & +1 & -1 & -1 & +1 & -1 \\ +1 & -1 & +1 & -1 & -1 & +1 & -1 & +1 \\ +1 & -1 & -1 & +1 & +1 & -1 & -1 & +1 \\ +1 & -1 & -1 & -1 & +1 & +1 & +1 & -1 \end{array}\right]. \tag{4.15}$$

H's first four columns, Kronecker-multiplied by $\mathbf{1}_6$, make up $\mathbf{X}$, for the grand mean and main effects; H's last four columns, Kronecker-multiplied by $\mathbf{1}_6$, make up $\mathbf{Z}$, for the interactions. Partition Θ as $(\Theta_1', \Theta_2')'$ conforming to $\mathbf{X}$ (main effects) and $\mathbf{Z}$ (interactions). The coefficients for the interactions $\theta_k, k = p+1, \ldots, p+q$ are shrunk or smoothed by means of the random-effect covariance matrix $\mathbf{G}$. Specifically, Θ_2 is modeled as $\theta_k|\phi_k \sim N(0, 1/\phi_k), k = p+1, \ldots, p+q$, independent of the other θ_j given the ϕ_k. (Obviously other $\mathbf{G}$ are possible, but at the price of some of the tidy results to follow.) The unknown precisions ϕ_k control shrinkage of their respective θ_k. Choosing a model for the ϕ_k is the key step in smoothing ANOVA.

The precisions ϕ_k need not all be distinct. The user specifies a set of distinct random-effect precisions $\{\eta_1, \ldots, \eta_s\}, s \leq q$, and a deterministic assignment function $j(k)$ such that $\phi_k = \eta_{j(k)}$, so $\mathbf{G}^{-1} = \text{diag}(\eta_{j(1)}, \ldots, \eta_{j(q)})$. This groups the θ_k and their associated columns in $\mathbf{Z}$, with each group's θ_k smoothed using its own η_j. Define $\eta = (\eta_0, \eta_1, \ldots, \eta_s)$, the vector consisting of the error precision η_0 and the distinct smoothing precisions. Let q_j be the number of ϕ_k mapping to η_j; $\sum_{j=1}^{s} q_j = q$. In the 2^3 design, $\mathbf{Z}$ has $q = 4$ columns, one three-way and three two-way interactions. If each interaction is smoothed separately, each ϕ_k is distinct, so $s = 4$ and $q_j = 1, j = 1,2,3,4$. Alternatively, the two-way interactions can be smoothed using a single η_j, while the three-way interaction keeps its own η_j. Then $s = 2$ and, referring to (4.15), $j(1) = j(2) = j(3) = 1$, $j(4) = 2$, so $q_1 = 3$ and $q_2 = 1$.

This can be made more general by setting $\phi_k = d_k \eta_{j(k)}$ for known constants d_k and proceeding as above. The spatial example in Section 5.2.3 uses this generalization.

The random effects here are new-style random effects. There is no population from which these random effects could have been drawn and these specific levels are of central interest in the analysis. These random effects are just devices to smooth/shrink the interactions. This has implications for the design of the simulation experiment discussed below, which was intended to compare priors for the ϕ_k according to the resulting bias, mean-squared error, and interval coverage. Chapter 13 discusses simulation experiments for new-style random effects in greater detail.

For the rest of this section, I assume a flat prior on the θ_k for the grand mean and main effects. This is not required. The main effects can be smoothed/shrunk in the same manner as the interactions, or informative (non-hierarchical) priors can be added.

With this specification, it is straightforward to write down the conditional and marginal posterior distributions for the unknowns Θ and η for use in MCMC algorithms, including Rao-Blackwellized algorithms. These are given in Hodges et al. (2007, Section 2.2, Appendix A.1). For the balanced, single-error-term case described here, these can be simplified to avoid matrix inversions.

4.2.3.2 DF in Effects, Prior Distributions, SANOVA Table

Degrees of freedom. Standard ANOVA uses DF in F-tests. Smoothed ANOVA emphasizes estimation and a thoroughly Bayesian approach eschews F-tests. Still, DF as defined in Chapter 2 can be used to specify priors on the unknown precisions η and to describe the extent of smoothing.

For given covariance matrices $\mathbf{R}$ and $\mathbf{G}$, straightforward algebra gives the DF in the whole fit of the SANOVA model as

$$DF = p + \sum_{k=1}^{q} \frac{cm\eta_0}{cm\eta_0 + \phi_k} = p + \sum_{j=1}^{s} \frac{q_j cm\eta_0}{cm\eta_0 + \eta_j} = p + \sum_{j=1}^{s} \frac{q_j cm}{cm + r_j} = p + \sum_{j=1}^{s} \rho_j, \tag{4.16}$$

where $r_j = \eta_j/\eta_0$ (this ratio of precisions equals error variance over smoothing variance); $\rho_j = (q_j cm)/(cm + r_j) \in [0, q_j]$ is the DF controlled by the ratio r_j. In this highly structured problem with orthogonal design columns, the DF in the whole fit partitions naturally. For each grouping of columns in $\mathbf{Z}$ with coefficients controlled by the same η_j, each column in the grouping gets an equal share of the DF. The proof is an exercise. As always, ρ_j depends on the precisions η only through the ratio $r_j = \eta_j/\eta_0$, so a prior on r_j induces a prior on ρ_j and vice versa.

Prior distributions. A prior distribution on η completes a Bayesian specification of the SANOVA model. In non-Bayesian terms, the model in Section 4.2.3.1 plus a prior on η defines a procedure; different priors define different procedures, which can be assessed in a frequentist way, e.g., by mean squared error of point estimates. As discussed in Chapter 2, we consider two kinds of priors: Unconditional priors and conditional priors that fix the fit's smoothness, or the smoothness of some part of the fit, at or below a certain number of DF. Several examples of unconditional priors are given just below. Examples of conditional priors are given when we return to the denture-liner example below. Other types of priors are possible and may be advantageous, for example, lasso-like procedures use a double-exponential prior and the posterior mode as a point estimate to shrink and select θ_k.

Regarding performance of different priors, Hodges et al. (2007, Section 3 and Appendix A.4) presented a simulation experiment comparing four unconditional priors to each other and to two common procedures that do not smooth/shrink. The simulated data had the form of the 2^3 design with 6 replications per cell that was used above as an example. This simulation compared 4 prior distributions to an analysis without smoothing (called "no-smoothing" below), and a two-step procedure ("drop-non-sig") that first fit the ANOVA with all interactions included, then re-fit excluding interactions that had $P > 0.05$ in the first step. For all prior distributions, each of the 4 interactions was smoothed by its own precision η_j, so in a sense this simulation put smoothed ANOVA at a disadvantage by emphasizing the contribution of the priors. The priors considered were Gamma(0.001,0.001) on each precision η_j ("gamma"); ρ_j distributed as uniform on $(0,1), j = 1,2,3,4$, with η_0 having an improper flat prior ("flat"); ρ_j distributed as beta(0.5,0.5), $j = 1,2,3,4$, with η_0 having an improper flat prior ("beta"); and a two-point prior with $\rho_j = 0.001$ or 0.999 each with probability 0.5, and η_0 having an improper flat prior ("two-point"). The beta prior on DF favors complete inclusion or shrinkage of an effect more than the flat prior on DF, which is indifferent among degrees of shrinkage. The two-point prior was included because it is a step in a Bayesian direction from the familiar "drop-non-sig" procedure.

The simulation experiment was itself a factorial design, with factors being the true 2^3 mean structure (specifically, the number of truly present interactions), the true error precision η_0, and the analysis procedure (the four priors and two other procedures listed above). As befits a simulation experiment for a new-style random effect (see Chapter 13), when an interaction was truly present, its coefficient was fixed at 1 in simulating the data; the coefficient was *not* a simulated draw from the new-style random effect used to shrink the interaction.

Briefly, the results were as follows. First, bias, mean-squared error, and interval coverage for smooth priors — gamma, flat, and beta — differed surprisingly little. Their performance degraded gracefully as error precision decreased, while procedures that guess the right interactions (two-point and drop-non-sig) degraded sharply as error precision decreased (i.e., error variance increased). Second, when 0, 1, or 2 interactions were truly present, the smooth priors provided notable gains in performance over unsmoothed ANOVA, while giving up little or nothing when more interactions were truly present. Some procedures showed performance gains over no-smoothing even when 3 interactions were present. Given the common presumption — dare we say prior belief? — that real data usually have few truly present interactions, the gamma, flat, and beta priors seem reasonable candidates for general use. The two-point and drop-non-sig procedures, however, are hazardous when error precision is small (error variance is large).

SANOVA table. The usual ANOVA table analyzes — in that word's literal sense — the dependent variable's sum of squares into pieces for each effect. In smoothing single-error-term ANOVAs, the smoothed θ_k are shrunk toward zero, so part of each smoothed effect is removed from the fitted model and counted as error. A SANOVA table records this division of each smoothed/shrunk effect's DF and sum of squares into a part retained in the fit and a part regarded as error. Smoothed ANOVA emphasizes estimation over testing, but a SANOVA table is still a useful bookkeeping device, in particular showing how information about error variation is derived from replication and from variation smoothed out of shrunken/smoothed effects. I show SANOVA tables for three analyses below, when we return to the denture-liner example.

Above we showed the portion of an effect's DF retained in the fit for a given **R** and **G**; the rest of the effect's DF is deemed error. Hodges et al. (2007, Appendix A.2) derived an analogous partition of an effect's sum of squares allotted to the fit and to error. These elements suffice to construct a SANOVA table accounting for a dataset's DF and sum of squares. The partition of each effect's SS and DF is a function of the precisions η, but single number summaries are convenient. Hodges et al. (2007, Appendix A.2) argued for using the posterior expected DF and sum of squares because this preserves the bookkeeping function of the ANOVA table. Further discussion is deferred to the following section, which analyzes the denture-liner data.

4.2.3.3 SANOVA Applied to the Denture-Liner Example

As noted, deleting the outlier had a substantial effect on the results, but the investigators had no reason to consider the outlying measurement defective, so deleting it is hard to justify. Smoothed ANOVA is a smooth alternative to a binary (include/exclude) choice. This section presents three smoothed analyses; Table 4.1 gives the design matrix we used for the main effects, **X**, from which the design matrix for the interactions **Z** was computed. Note that Table 4.1's design-matrix columns have *not* been scaled to have squared Euclidean length *cm* as assumed in preceding sections. Also note that each design column except the intercept is a contrast in the data because it sums to 0. Thus, the following discussion refers to columns in the design matrix, the components of each effect, as "contrasts." Computing methods for this example are in Hodges et al. (2007, Appendix A.3).

In all three smoothed analyses presented here, the M×P and M×F interactions were smoothed by giving each θ_k its own η_j, i.e., smoothing each contrast separately. The P×F interaction is of secondary interest because it does not involve the materials, so its 21 contrasts were smoothed using a single η_j.

The three smoothed analyses differ in their handling of the M×P×F interaction, which is most directly affected by the outlier. Analysis 1 gave each of M×P×F's 21 contrasts its own η_j, i.e., smoothed them separately. Analysis 2 smoothed all 21 contrasts using a single η_j, giving posterior expected DF of 6.75. To distinguish the effect of using a single η_j from the effect of having few total DF, Analysis 3 fixed M×P×F's total DF at the same 6.75 by conditioning the prior on this effect to have DF = 6.75, but allowed each contrast to be smoothed by its own η_j subject to this constraint. (We also did an analysis like Analysis 3 but fixing M×P×F's total DF at 9.83, Analysis 1's posterior mean. The results were nearly identical to Analysis 1.) Finally, we put a flat prior on η_0 and flat priors on each interaction's ρ_j, subject to conditioning as described.

Figure 4.7 summarizes the results for M×P×F. In both panels, M×P×F's contrasts are sorted from left to right in increasing order of their unsmoothed estimates' absolute values. Figure 4.7a shows, for each contrast, the posterior mean DF in the fit. In Analysis 1, where each contrast has its own η_j, posterior mean DF increases as the unsmoothed contrast increases, for a posterior mean of 9.83 total DF. The right-most contrast retained 0.8 DF in the fit, reflecting the outlier's effect. The left-most contrast, with unshrunk estimate near zero, still had posterior mean DF about 0.4 compared to a prior mean of 0.5 — the prior matters a good deal, when each θ_k is shrunk using its own η_j. In Analysis 2, all 21 contrasts were smoothed with the same η_j and have the same posterior mean DF, 0.32. All 21 contrasts were smoothed more than in Analysis 1, some much more, leaving 6.75 total DF. Analysis 3's prior fixed M×P×F's DF at the same 6.75 but allowed each contrast to be smoothed by its own η_j subject to this condition. It smoothed the four largest contrasts less than Analysis 2, while the other 17 contrasts were smoothed a bit more. The right-most contrast, reflecting the outlier, changed the most, from 0.32 DF in Analysis 2 to 0.68 in Analysis 3. Thus, although Analyses 2 and 3 both gave M×P×F 6.75 DF, Analysis 3 unmasked the outlier by allowing different contrasts to be smoothed differently and forcing them to compete for those 6.75 DF.

Figure 4.7b shows absolute values of the unsmoothed contrast estimates and posterior means from Analyses 1, 2, and 3. These largely reflect Figure 4.7a's posterior mean DF. However, although small contrasts have fewer DF in Analysis 3 compared to Analysis 2, their smoothed estimates hardly change, while the largest contrast's estimate increases substantially.

Table 4.2 gives the SANOVA tables for Analysis 1, 2, and 3. The Unsmoothed section, for the grand mean and main effects, is the same as in the usual ANOVA table. The Smoothed section shows the partition of each effect's DF and sum of squares (SS) between the model fit (columns headed "Model") and error (columns headed "Error"). As noted, these are posterior expectations of the respective SS and DF. The Model and Error halves of the table include a column for mean squares, which as usual are SS divided by DF. A smoothed effect's error mean square describes the effect's contribution of information about error variation. Pure Error's SS is zero because the design is unreplicated; a replicated design would have DF and SS for error from replication as well as from smoothing, with the total error SS and DF being the sums of the two sources. Unlike the standard analysis, in which only the three-way interaction is deemed error, in Analysis 1 a bit more than half of the 23.08 DF for error come from variation smoothed out of two-way interactions and about half of the three-way interaction is retained in the fit.

The SANOVA tables for Analyses 2 and 3 have the same Unsmoothed section as Analysis 1, so Table 4.2b gives only their Smoothed sections. Comparing Analyses 1 and 2, a single η_j for all 21 M×P×F contrasts (Analysis 2) inflates the error mean square from 0.11 to 0.13 and forces M×P×F to shrink more (posterior mean DF 6.75 vs. 9.83). This happens because the contrast reflecting the outlier is smoothed as much as the other 20 contrasts (Figure 4.7a). Pushing this variation into error inflates the estimate of error variance and induces further smoothing, resulting in about 3 more DF smoothed out of M×P×F. Smoothing all M×P×F contrasts with

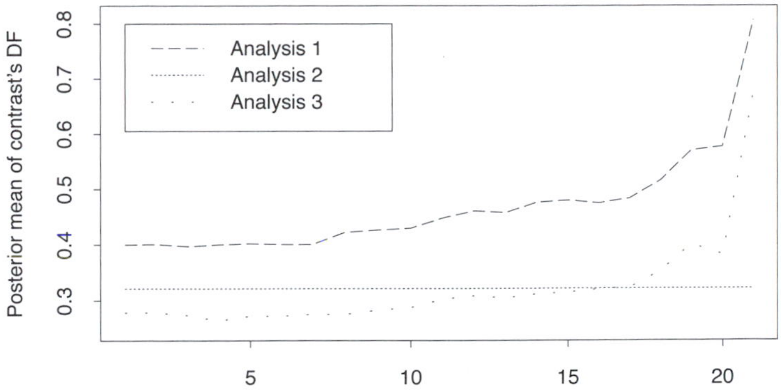

(a) Contrasts sorted by absolute value of unsmoothed estimate

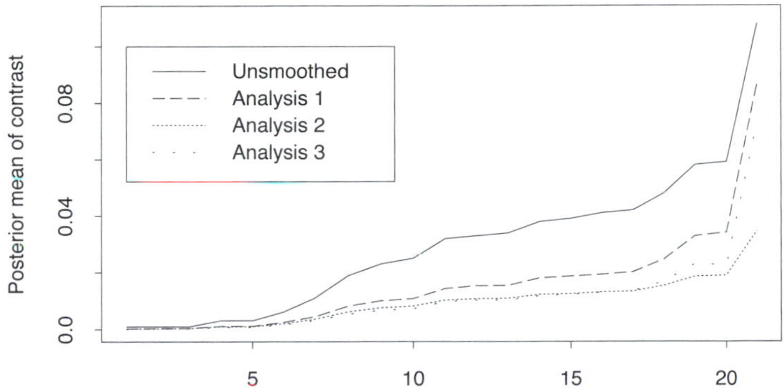

(b) Contrasts sorted by absolute value of unsmoothed estimate

Figure 4.7: Denture-liner data: Posterior summaries for the 21 contrasts in the M×P×F interaction. Panel (a): E($DF|\mathbf{y}$) for the three smoothed analyses. Panel (b): Absolute values of contrast estimates for the unsmoothed and smoothed analyses, i.e., E($\theta_k|\mathbf{y}$).

a single η_j indirectly forces P×F to shrink more, with posterior mean DF in the fit only 10.50 in Analysis 2 vs. 13.65 in Analysis 1. (Recall that P×F was smoothed using a single η_j.)

A subgroup analysis is illustrated using the M×F interaction, which addresses whether the material difference, standard minus new, varies between levels of F (finishing methods). Figure 4.8 shows the unsmoothed estimates and 95% confidence intervals for standard minus new (using M×P×F as the error term), and Analysis 2's smoothed estimates and intervals. The three smoothed analyses give nearly identical posterior means for standard minus new despite their differences for M×P×F.

Table 4.2: Denture-Liner Data: SANOVA Tables for Analyses 1, 2, and 3

	Model			Error		
	SS	DF	MS	SS	DF	MS
Analysis 1						
Unsmoothed						
Grand mean	75.54	1.00	75.54			
M	1.12	1.00	1.12			
P	0.38	3.00	0.13			
F	1.92	7.00	0.27			
Smoothed						
M×P	0.48	1.59	0.30	0.17	1.41	0.12
M×F	0.88	3.85	0.23	0.52	3.15	0.17
P×F	2.13	13.65	0.16	1.15	7.35	0.16
M×P×F	1.26	9.83	0.13	0.79	11.17	0.07
Error						
Pure				0.00	0.00	
Smoothing				2.63	23.08	0.11
Total				2.63	23.08	0.11
Analysis 2						
Smoothed						
M×P	0.43	1.52	0.28	0.22	1.48	0.15
M×F	0.78	3.53	0.22	0.62	3.47	0.18
P×F	1.64	10.50	0.16	1.64	10.50	0.16
M×P×F	0.66	6.75	0.10	1.39	14.25	0.10
Error						
Pure				0.00	0.00	
Smoothing				3.87	29.70	0.13
Total				3.87	29.70	0.13
Analysis 3						
Smoothed						
M×P	0.45	1.54	0.29	0.20	1.46	0.13
M×F	0.82	3.63	0.22	0.58	3.37	0.17
P×F	1.81	11.64	0.16	1.46	9.36	0.16
M×P×F	0.95	6.75	0.14	1.10	14.25	0.08
Error						
Pure				0.00	0.00	
Smoothing				3.34	28.44	0.12
Total				3.34	28.44	0.12

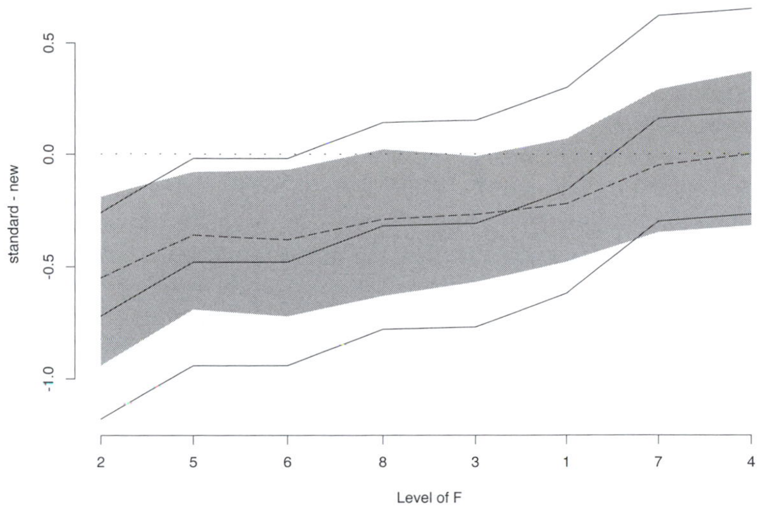

Figure 4.8: Denture-liner data: Subgroup analysis, standard material minus new material for the levels of factor F. Unsmoothed analysis: Solid lines, point estimates and 95% confidence intervals. Smoothed Analysis 2: Dashed line, posterior mean; gray region, 95% posterior intervals. The dotted line indicates no difference between standard and new.

The SANOVA tables reflect this: M×F's row in the Smoothed section is similar in all three analyses. In Figure 4.8, the smoothed subgroup-specific differences are shrunk toward $-0.26 \log_{10} \mu$m, the material main effect. M×F's seven contrasts were smoothed using different η_j, so F's levels 1 and 3 were shrunk toward the material main effect by larger fractions than was level 2. The intervals from the unsmoothed analysis are wider, at 0.92 $\log_{10} \mu$m, than Analysis 2's intervals, which range in width from 0.55 to 0.75 (median 0.65; Analyses 1's and 3's are narrower by about 0.1). Based on the simulation experiment in Hodges et al. (2007), I conjecture that these intervals have nearly nominal coverage despite being much narrower. Finally, smoothing simplifies interpretation: while the unsmoothed analysis has a scattered group of estimates, Analysis 2 suggests that F's levels 1, 4, and 7 have no treatment difference, that level 2 has a standard-to-new ratio of about 1/4 ($\log_{10} 0.25 = -0.6$), and that F's other levels cluster around a ratio of about 1/2 ($\log_{10} 0.5 = -0.3$).

4.2.4 Smoothed ANOVA for More General Designs

Smoothed ANOVA was developed in the preceding section for balanced, single-error-term ANOVAs. Cui (2008, Chapter 4; available from this book's web site as Cui & Hodges 2009) extended smoothed ANOVA to more general designs, including unbalanced designs. For general balanced designs, including arbitrary random and fixed effects, Cui (2008) generalized the smoothed ANOVA table to analyze variation in the outcome according to the effect in which it originates and the effects into which variation is shifted after being shrunk/smoothed out of the effect in which it originates. Cui (2008) illustrated this using a design with a random effect for subjects and an ordinary error term, and two factors, one varying between subjects and one varying within subjects so the interaction of the two factors was a within-subject effect. In the resulting smoothed ANOVA table, variation smoothed/shrunk out of the between-subject fixed effect was shifted to the subject effect and the error term in proportions determined by the model's between-subject and error variances. By contrast, variation smoothed/shrunk out of the within-subject factor and the interaction was shifted entirely to the error term. As in Section 4.2.3, this tells us which effects provide information about the variance components. Apart from its interest in specific applications, such a decomposition may be helpful in understanding the as yet ill-understood problem of how the data provide information about ϕ_G and ϕ_R, the unknowns in the mixed linear model's two covariance matrices.

For general unbalanced designs, smoothed ANOVA as derived in Cui (2008) has no particular structure because of the imbalance and amounts to a special case of mixed linear models. Some have argued (e.g., Speed 1987, Section 9) that in the absence of balance ANOVA is not well-defined, so it is arguably reasonable to restrict the term "ANOVA" to balanced designs. However, approaching unbalanced designs as a smoothed ANOVA suggests some alternative priors on the DF in the various effects, arising out of different ways to embed the unbalanced design in a larger balanced design, which gives a simpler expression for DF and thus allows a simpler DF-based prior. Interested readers are referred to Cui (2008) for further details.

Exercises

Regular Exercises

1. (Section 4.1) Fit the models in this section to the pig jawbone data, including tests of whether each model is preferable to the preceding model, e.g., whether (4.3) is preferable to (4.2). If you find this boring and pointless, analyze these data as you see fit, but fit at least one additive model with a penalized spline in it.
2. (Section 4.2.1) Fit interaction models to Whitey's data with fewer knots. You might try (say) 2 different reduced numbers of knots to see how they differ from each other and from the 25-knot fit shown in this section. Also, fit an interaction model with $v_{kl} \equiv v_k$ for all l, using 25 knots and also using the same choices for numbers of knots you just used to fit the model with different v_{kl}. This will let you see the effects of the two aspects of the fitting machinery that you have tuned: number of knots and common versus line-specific v_{kl}.

3. (Section 4.2.1) Fit to the pig jawbone data a model with a transect-by-distance interaction in which the $(\beta_{0i}, \beta_{1i}, \beta_{2i})$ are modeled as a random effect with a 3-variate normal distribution. As the text suggests, you may have some trouble doing this analysis.
4. (Section 4.2.3.2) Derive the DF in the whole fit of the smoothed ANOVA model and show that the DF for the columns in $\mathbf{Z}$ smoothed by η_j is ρ_j.
5. (Section 4.2.3.3) Fit smoothed ANOVA models to the denture-liner material data, either the ones fit here or other ones you think are interesting.

Open Questions

1. (Section 4.1.3) By what mechanism does adding autocorrelated errors cause the intercepts for the 9 transects to shrink closer to their section main effects? One hypothesis is that replacing independent errors ε_{ij} with autocorrelated errors reduces each transect's effective sample size, so there is less information regarding each transect's intercept, which implies the intercepts will be shrunk more. Autocorrelated errors also imply somewhat less information for the spline fit on distance, though the effect should be smaller than the effect on transect-specific intercepts because data from all 9 transects are used to estimate the spline's smoothing parameter. This arm-waving explanation is reflected in rather more shrinkage for intercepts compared to the shrinkage in the uncorrelated-errors model, but only a bit more smoothing for the spline. One simple test of this hypothesis is to re-fit this model but without the fixed effect for sections; if reduced effective sample size is part of the mechanism here, the total DF in the intercepts should be reduced even further without the fixed effect for sections. It might be possible to determine whether this is the entire mechanism by returning to independent errors but forcing the error variance to be larger than the restricted-likelihood estimate from model (4.3). However, for this purpose it is not obvious exactly how large the error variance should be.

Chapter 5

Spatial Models as Mixed Linear Models

Spatial statistics is a large area with two main branches. In one branch, analyses are focused on *locations at which certain events occur*, for example, locations of trees of a given species. In the other branch, analyses are focused on *quantities measured at specific locations* such as concentrations of a soil pollutant measured at many point locations in a given area, or counts of stomach cancers in each of the regions that partition a state.

This chapter is about the second of these two branches. The distinctive feature of analyses of this sort is a concern for spatial correlation of the quantities measured at the specific locations. Models used in these analyses deploy particular mathematical models to represent vague intuition about spatial association, most commonly that measurements taken at locations near each other tend to be more similar than measurements taken at locations far from each other.

Spatially referenced data are usually divided into two types, geostatistical (point-referenced) and areal. For geostatistical data, each measurement of a quantity is taken at a specific point usually in two-dimensional space, though sometimes locations are in one- or three-dimensional space and sometimes time is treated as another dimension. Thus, each measurement's location is identified by a location in some coordinate system, interpreted literally as distances from an origin along axes. Interpolation between observed points is meaningful and is often the main goal. Such interpolation amounts to two-dimensional smoothing and penalized splines can be used, as well as smooths arising from models of spatial covariance. Both approaches are discussed below.

For areal data, a measurement's location is usually an area or region and the measurement is a total or average over the region. Distance is usually not used explicitly; instead, spatial nearness is specified by defining pairs of neighboring regions. Once neighbor pairs are defined, the map is reduced to a lattice or graph, with regions as vertices or nodes, a measurement attached to each node, and edges between nodes representing neighbor pairs. Abstracted this way, areal models are now being applied to non-spatial problems such as social networks (e.g., O'Malley & Marsden 2008, Section 4) and gene-gene interactions (e.g., Wei & Pan 2012). Without an explicit distance measure, interpolation is no longer meaningful; the goal of analyses with areal models is often smoothing a map or "borrowing strength" from neighbors to improve estimates.

Both geostatistical and areal models can used as part of regression models with explicit predictors or regressors. Another common goal of representing spatial correlation is to account for spatial correlation in computing standard errors or posterior standard deviations for regression coefficients, in effect to discount the sample size to account for correlation among observations. The specific purpose of an analysis matters here; this point is discussed at length in Section 10.1.

Here is another class of models that can be represented as mixed linear models at the price of pruning away parts of an extensive literature. Again, this chapter's purpose is not to present a comprehensive survey but to show how to represent important spatial models as mixed linear models. The first widely circulated overview of spatial methods appears to be Cressie (1991), but I find this quite hard to use. I find Banerjee et al. (2004) friendlier though less comprehensive. Ripley (2004), Waller & Gotway (2004), and Schabenberger & Gotway (2004) are other popular overviews.

Section 5.1 considers geostatistical models, which are most readily represented as mixed linear models. Section 5.2 considers areal models and is mostly about the intrinsic or improper conditionally autoregressive (ICAR) model, which fits especially neatly into the mixed-linear-model framework. Finally, Section 5.3 considers two-dimensional penalized splines.

5.1 Geostatistical Models

The small amount I know about geostatistical models I have learned from colleagues who are experts. These colleagues are entranced by the mathematics of geostatistical models, especially Michael Stein's work on Gaussian processes (e.g., Stein 1999). If you've ever thought math can be intriguing or beautiful, it's easy to understand the appeal of Stein's work. In this approach, you begin by specifying a "process," a particular kind of probabilistic model; a realization of or draw from a process is a function from the two-dimensional plane (usually) to the real numbers. Under certain assumptions, a process is characterized by a covariance function describing the covariance of process realizations at pairs of locations. When n locations have been chosen, a Gaussian process model implies a particular multivariate normal distribution for realizations of the process at those n locations, so Gaussian processes can be represented, for analysis purposes, as mixed linear models.

Philistine that I am, I prefer starting with the multivariate normal distribution and specifying its mean and covariance; this brief section takes that approach. This preference highlights the difference between the focus of process-model enthusiasts — on properties of draws from the process's probability model — and the focus of someone who uses these models to analyze data. This difference is clearest for a Bayesian analysis, in which the data are treated as fixed — so the draw from the process is fixed — and the process model becomes the likelihood, a function of its unknown parameters. I would suggest that until someone shows that properties of draws from Gaussian processes tell us something useful about likelihoods arising from Gaussian process models, then focusing on draws is a distraction, perhaps a hazardous distraction. Chapter 9 and Section 10.1 give an example of one potential hazard. When I asked Michael Stein about the distinction between process realizations and data analysis, he readily admitted that little is known about how Gaussian process models behave as data-analysis engines, to his considerable credit. However, in my perhaps too-limited experience, few process-model enthusiasts seem to appreciate this distinction.

A very common form for geostatistical models is

$$y(\mathbf{s}) = \mathbf{x}(\mathbf{s})\beta + w(\mathbf{s}) + \varepsilon(\mathbf{s}), \tag{5.1}$$

where $\mathbf{s}$ is a vector of coordinates identifying a location in space, $y(\mathbf{s})$ is a scalar outcome or dependent variable measured at $\mathbf{s}$, $\mathbf{x}(\mathbf{s})$ is a row vector of regressors or predictors describing $\mathbf{s}$, including an intercept, $w(\mathbf{s})$ is the scalar realization of a Gaussian process at $\mathbf{s}$, and $\varepsilon(\mathbf{s})$ is an error process. The error process $\varepsilon(\mathbf{s})$ is most often so-called white noise — in practice, iid $N(0, \sigma_e^2)$ — but it can also be the realization of a Gaussian process, although a model with Gaussian processes for both w and ε is identified only under certain conditions.

If data have been observed at n locations $\mathbf{s}_1, \ldots, \mathbf{s}_n$, then (5.1) immediately gives a mixed linear model. This model has one row in each of $\mathbf{y}$, $\mathbf{X}$, $\mathbf{Z}$, and ε for each of the n locations $\mathbf{s}_i$. The row vectors $\mathbf{x}$ provide the rows of the fixed effect design matrix $\mathbf{X}$ and β in (5.1) is the mixed linear model's fixed effects β. The random effect design matrix $\mathbf{Z}$ is $\mathbf{I}_n$, $\mathbf{u}$ is $(w(\mathbf{s}_1), \ldots, w(\mathbf{s}_n))'$, and $\mathbf{G}$ is the covariance matrix implied by the Gaussian process for w (some examples are given below). $\mathbf{R}$ is the covariance matrix of the error process $\varepsilon(\mathbf{s})$. For the most common iid case, this is $\sigma_e^2 \mathbf{I}_n$; otherwise it is the covariance matrix implied by the Gaussian process for $\varepsilon(\mathbf{s})$.

So far, it would seem that any legal $\mathbf{G}$ or $\mathbf{R}$ could be used in model (5.1) and indeed many are available in standard software. For example, SAS's MIXED procedure (version 9.2) offers 12 distinct spatial covariance matrices for both $\mathbf{G}$ and $\mathbf{R}$. However, many spatial specialists are persuaded by Stein's advocacy of the Matérn family of covariance functions for Gaussian processes. A Matérn covariance function for w implies $\mathbf{G}$ with $(i,j)^{\text{th}}$ element

$$\sigma_s^2 \frac{1}{2^{\nu-1}\Gamma(\nu)} \left(\frac{\delta_{ij}}{\rho}\right)^{\nu} \mathsf{K}_\nu \left(\frac{\delta_{ij}}{\rho}\right), \tag{5.2}$$

where Γ is the gamma function, δ_{ij} is the distance between $\mathbf{s}_i$ and $\mathbf{s}_j$, and K_ν is the modified Bessel function of order $\nu > 0$. The constant ν controls the smoothness of correlation as a function of distance, while ρ scales distance. K_ν has a closed form if $\nu = m + 1/2$ for m a non-negative integer: For $m = 0$, $\nu = 0.5$ and $G_{ij} = \sigma_s^2 \exp(-\delta_{ij}/\rho)$, while for $m > 0$, this G_{ij} is multiplied by a polynomial of order m in δ_{ij}/ρ.

In practice, ν is often chosen on *a priori* grounds, although it is also often estimated as part of the analysis. Stein (1999) recommended estimating the unknowns in this model by maximizing the restricted likelihood, in contrast to the more *ad hoc* estimation methods historically used in spatial analyses. Unfortunately the likelihood alone and thus the restricted likelihood do not identify the three unknowns (σ_s^2, ν, ρ) in the covariance (5.2); (σ_s^2, ρ) is not identified even if ν is known (Zhang 2004). As Zhang (2004) points out, this does not create a problem for predictions (interpolation), but it can make for awkward computations. These problems are glossed over in the Bayesian approach by placing informative priors on these troublesome parameters.

I am puzzled by the strong preference my spatial-maven colleagues have for Matérn covariances. As far as I can tell, this preference is not based on their performance as data-analytic devices. Ruppert et al. (2003, p. 247) say "well-behaved likelihood surfaces" are an argument in favor of Matérn, though it is not clear how this squares with the identification problem just mentioned. One reader suggested that Matérn *per se* is not preferred, it is just better than popular choices like $G_{ij} = \sigma_s^2 \exp(-\delta_{ij}^2/\rho)$, which is the limit of (5.2) as $\nu \to \infty$.

The model (5.1) captures spatial correlation in $w(\mathbf{s})$ while the error term $\varepsilon(\mathbf{s})$ allows an extra "nugget" of variation that is rationalized in various ways, for example as measurement error. The same marginal covariance for $\mathbf{y}$ is obtained by writing $\mathbf{y} = \mathbf{X}\beta + \varepsilon$ and setting $\text{cov}(\varepsilon) = \mathbf{G} + \mathbf{R}$, where $\mathbf{G}$ and $\mathbf{R}$ are the random-effect and error covariances from the mixed-linear-model formulation. However, standard software for mixed linear models may not fit such models, e.g., apparently SAS's MIXED procedure (version 9.2) does not.

In models like (5.1), is $w(\mathbf{s})$ an old-or new-style random effect? Sometimes it is old-style and sometimes new-style, as Chapter 13 discusses at length. Briefly, if (5.1) is used to smooth one set of measurements, say for interpolation, or to capture local bias in a model with regressors $\mathbf{x}\beta$, then $w(\mathbf{s})$ is part of the model's mean not part of its variance, and $w(\mathbf{s})$ is a new-style random effect. However, suppose the dataset consists of many days of air-pollution data measured in a given area and the interest is in the average spatial trend and not in the individual days. If $w_j(\mathbf{s})$ captures variation between days j, then $w(\mathbf{s})$ may be an old-style random effect.

5.2 Models for Areal Data

We begin by illustrating two key ideas with an atypical areal-data example that recurs later in this book. Figure 5.1 shows attachment-loss measurements for one dental patient. In Figure 5.1, the rectangles are teeth, 14 in the upper jaw (maxillary) and 14 in the lower jaw (mandibular). Attachment loss — the distance down the tooth's root that was originally attached to the surrounding bone but has become detached — is measured at 6 sites on each tooth. In Figure 5.1, measurement sites are indicated by circles and each site's attachment-loss measurement is depicted using a grey scale.

Two key ideas in analyses of areal data are neighbor pairs and islands. In Figure 5.1, two measurement sites are a neighbor pair or simply neighbors if they are physically adjacent to each other, either on the same side of the tooth (the cheek side or the tongue side) or in the same gap between teeth (interproximal region). In Figure 5.1, two sites are neighbors if their circles touch or if they are joined by a line. In graph-theory terms, measurement sites are nodes and neighbor pairs are joined by edges; the sites and neighbor pairs form a lattice. Figure 5.1's periodontal lattice has two islands, one each for maxilla and mandible. They are islands because no maxillary site is joined to any mandibular site by a neighbor pairing. In graph theory terms, each island is a connected graph, but islands are not connected to each other. In the spatial analysis literature, graphs are almost always connected (have one island) but

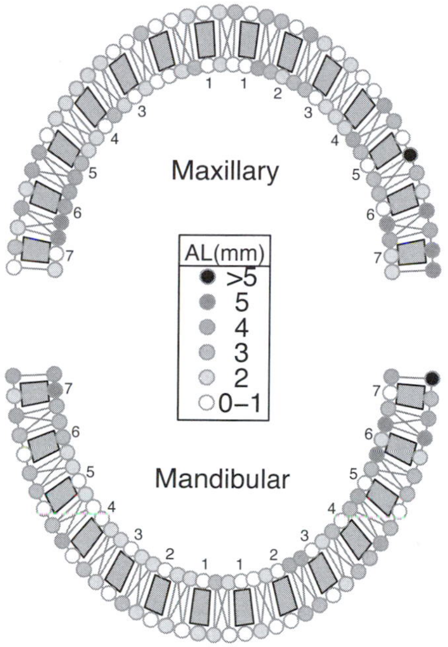

Figure 5.1: Schematic of attachment-loss data. Circles are measurement sites, with the attachment-loss measurement depicted using a grey scale. Neighboring pairs of sites are circles that either touch each other or are joined by a line.

in more recent applications such as gene-gene interactions, there may be hundreds of islands.

5.2.1 Common Areal Models: SAR, CAR, and ICAR

The most common models for areal data are the simultaneously and conditionally autoregressive models, SAR and CAR respectively, and the intrinsic or improper CAR (ICAR). I first introduce the CAR and SAR models following Wall (2004), then show how they can be included in a mixed linear model. I introduce the ICAR here but develop it in Section 5.2.2.

SAR model. A random variable $\mathbf{u} = (u_1, \ldots, u_N)$ on a lattice with N measurement sites follows a SAR model if

$$u_i = \mu_i + \sum_{j=1}^{N} b_{ij}(u_j - \mu_j) + \varepsilon_i \tag{5.3}$$

where $\varepsilon = (\varepsilon_1, \ldots, \varepsilon_N)' \sim N(\mathbf{0}, \Lambda)$ with diagonal covariance matrix Λ and the b_{ij} are either known or unknown constants (more below) with $b_{ii} = 0$. The expected value of $\mathbf{u}$ is $\mu = (\mu_1, \ldots, \mu_N)$, which often has a regression structure with unknown coefficients. Define $\mathbf{B} = (b_{ij})$ as an $N \times N$ matrix of the b_{ij}. The joint distribution of $\mathbf{u}$ is then

$$\mathbf{u} \sim N(\mu, (\mathbf{I}_N - \mathbf{B})^{-1}\Lambda(\mathbf{I}_N - \mathbf{B})^{-1\prime}) \tag{5.4}$$

where $\mu = (\mu_1, \ldots, \mu_N)'$. The b_{ij} and thus $\mathbf{B}$ are determined by the spatial lattice's neighbor pairs. Commonly $\mathbf{B} = \rho_s \mathbf{W}$, where $w_{ij} = 1/m_i$ if i and j are neighbors, for $m_i = \{\text{number of area } i\text{'s neighbors}\}$, and 0 otherwise. The scalar ρ_s is often called a spatial correlation or spatial dependence parameter and is commonly treated as unknown while the neighbor pairings are treated as known.

CAR model. A different model for $\mathbf{u}$ uses this conditionally autoregressive (CAR) specification:

$$u_i|\mathbf{u}_{(-i)} \sim N(\mu_i + \sum_{j\sim i} \rho_c(u_j - \mu_j)/m_i, \sigma_s^2/m_i) \tag{5.5}$$

where $\mathbf{u}_{(-i)}$ is $\mathbf{u}$ without area i, $j \sim i$ means areas i and j are neighbors, and again m_i is the number of area i's neighbors. (More general versions of this model exist.) The scalar ρ_c is also often called a spatial correlation or spatial dependence parameter and treated as unknown. Besag (1974) showed that this conditional specification gives a proper joint specification, normal with mean μ and covariance $\mathbf{G}$, where the precision matrix $\mathbf{G}^{-1}$ has i^{th} diagonal element m_i and $(i, j)^{\text{th}}$ off-diagonal element $-\rho_c$ if i and j are neighbors and 0 otherwise.

Wall (2004) examined the strange way in which correlations between neighboring areas vary with ρ_s and ρ_c in the SAR and CAR models. Assunçao & Krainski (2009) gave an intriguing re-interpretation of Wall's results for the CAR model, as a random walk around the lattice.

These distributions were originally developed as distributions for outcomes $\mathbf{y}$ but now are more often used (by statisticians, at least) as distributions for unobserved latent variables or random effects. That's why I defined them using $\mathbf{u}$ and not $\mathbf{y}$. Either model for the vector $\mathbf{u}$ can be introduced directly into a mixed linear model in the same manner as the Gaussian process:

$$\mathbf{y} = \mathbf{X}\beta + \mathbf{I}_N\mathbf{u} + \varepsilon, \tag{5.6}$$

where the random-effects design matrix $\mathbf{Z}$ has been set to $\mathbf{I}_N$ and $\mathbf{G}$ is the covariance matrix appropriate to either the SAR or CAR model. In specifying a CAR or SAR as part of a mixed linear model, customarily $\mu = \mathbf{0}$, i.e., $\mathbf{u}$ is centered around zero and any explicit mean structure is placed in $\mathbf{X}\beta$. (But as we'll see in Section 10.1, this appearance is deceiving!) For the SAR, the unknowns ϕ_G in $\mathbf{G}$ are ρ_s and any unknowns in Λ; for the CAR model, the unknowns ϕ_G are ρ_c and σ_s^2.

At this level of generality, for the present purpose little more can be said about these areal models. They can be expressed as mixed linear models as above and fit using the usual tools. SAS's MIXED procedure currently does not include either of these models. The GeoBUGS add-on module for BUGS includes various versions of the CAR model but not the SAR model, although the SAR model can always be coded directly in BUGS. The SAR model can be fit using user-contributed software in R and the STATA system and probably others; SAR is more in favor among economists and I know nothing about their software.

Is **u** in (5.6) an old-style or new-style random effect? Depending on the problem, it could be either; the argument is similar to the one for Gaussian processes. This is discussed in Chapter 13.

Much more can be done in the mixed-linear-model framework with a popular model that is closely related to the CAR, the intrinsic or improper CAR (ICAR). The ICAR can be derived in many ways (Besag et al. 1991); I'll use δ for an ICAR-distributed random variable instead of **u**, for reasons that will become clear. The ICAR's relationship to the CAR is seen most clearly by beginning with equation (5.5) and letting $\rho_c \to 1$, giving

$$\delta_i|\delta_{(-i)} \sim N(\sum_{j\sim i} \delta_j/m_i, \sigma_s^2/m_i), \tag{5.7}$$

where μ has been set to $\mathbf{0}$. Here, the conditional mean of δ_i is just the average of its neighbors. This conditional specification gives a partially specified probability density for δ,

$$f(\delta|\sigma_s^2) \propto \exp\left(-\frac{1}{2\sigma_s^2}\delta'\mathbf{Q}\delta\right), \tag{5.8}$$

where the matrix $\mathbf{Q}$ has diagonal elements $Q_{ii} = m_i$ and off-diagonals $Q_{ij} = -1$ if i and j are neighbors and 0 otherwise. This joint (partially specified) density can also be written profitably in a pairwise-difference form:

$$f(\delta|\sigma_s^2) \propto \exp\left(-\frac{1}{2\sigma_s^2}\sum_{i\sim j}(\delta_i - \delta_j)^2\right). \tag{5.9}$$

Derivation of these two forms is an exercise. The obvious intuition of this specification is that σ_s^2 controls the similarity of δ_i for neighboring areas: Small σ_s^2 forces neighbors to be similar, while large σ_s^2 allows neighbors to be different.

The density (5.8) or (5.9) is improper. For any map, $\mathbf{1}_N$ is an eigenvector of $\mathbf{Q}$ with eigenvalue 0, so $\mathbf{Q}\mathbf{1}_N = 0$ and (5.8) and (5.9) are flat (improper) along the line $\delta = \alpha\mathbf{1}_N$. If the lattice of neighbor pairs has I islands, $\mathbf{Q}$ has I zero eigenvalues; one basis for $\mathbf{Q}$'s null space has k^{th} basis vector equal to $\mathbf{1}_{N_k}$ for the N_k areas in island k and zero for other areas. (Proofs are exercises.)

The densities (5.8) and (5.9) are missing something that is usually part of a normal density, namely a function of σ_s^2 that multiplies the exponential. The reason for this omission is that the conditional specification of the ICAR is incomplete in the sense that it is consistent with uncountably many such multiplicative functions, each of which defines a legal probability distribution (Lavine & Hodges 2012). Although this was apparently not unknown to everyone, it is or was unknown to most authors, several of whom (including me) have used fallacious arguments to "prove" that the multiplicative function is either $(\sigma_s^2)^{N/2}$ or $(\sigma_s^2)^{(N-I)/2}$. Lavine & Hodges (2012) cataloged these "proofs" and showed their fallacies. For example, $(\sigma_s^2)^{N/2}$ can be obtained by letting $\rho_c \to 1$ in the proper density derived from (5.5), while $(\sigma_s^2)^{(N-I)/2}$ can be rationalized as the dimension of the subspace on which the ICAR specifies a proper normal distribution (e.g., Hodges et al. 2003). Lavine & Hodges (2012) show that either of these functions and many others can be obtained by completing the ICAR's specification in different ways. I prefer $(\sigma_s^2)^{(N-I)/2}$ because as Lavine & Hodges (2012) show, it implies that the smoothing parameter σ_s^2 constrains or shrinks *only* contrasts in δ, while $(\sigma_s^2)^{m/2}$ for $m \in \{N-I+1,\ldots,N\}$ implies that σ_s^2 also constrains or shrinks the average of δ on $I-(N-m)$ of the I islands. Accordingly, this book always uses the following density for the ICAR:

$$f(\delta|\sigma_s^2) = K(\sigma_s^2)^{(N-I)/2} \exp\left(-\frac{1}{2\sigma_s^2}\delta'\mathbf{Q}\delta\right), \tag{5.10}$$

where K is a constant.

The ICAR's precision matrix $\mathbf{Q}/\sigma_s^2$ is simple in a way that permits much more to be done with it in the mixed-linear-model framework than could be done with the SAR or proper CAR. The following subsection shows how and gives examples, one of them a smoothed ANOVA in which terms are smoothed spatially.

Parenthetically, another interpretation of the proper CAR and ICAR models is that they are specified by conditional independence and dependence relations. If $\mathbf{P}$ is the precision matrix of δ (the inverse of δ's covariance), then

$$\begin{aligned} \text{correlation}(\delta_i, \delta_j|\delta_{(-ij)}) &= -P_{ij}/\sqrt{P_{ii}P_{jj}} \text{ and} \\ \text{precision}(\delta_i|\delta_{(-i)}) &= P_{ii}, \end{aligned} \tag{5.11}$$

where the subscript "$(-ij)$" means "without i and j." Proofs are in Rue & Held (2005), Section 2.2. In particular, if $P_{ij}=0$, then δ_i and δ_j are independent conditional on all the other entries in δ. In the CAR and ICAR models, areas i and j that are not neighbors have $P_{ij}=0$ and are conditionally independent. Conditional dependence structures are readily represented as graphs and models with simple conditional dependence structures have sparse $\mathbf{P}$, which allows rapid computing. This is the beauty of Gaussian Markov random fields (GMRFs), of which the CAR and ICAR are examples. GMRFs are the basis of an alternative syntax for richly parameterized models developed by Rue & Held (2005), discussed in Chapter 7.

5.2.2 More on the ICAR Model/Prior

A SAR or proper CAR distribution for $\mathbf{u}$ fits easily into this mixed linear model

$$\mathbf{y} = \mathbf{X}\beta + \mathbf{I}_n\mathbf{u} + \varepsilon \tag{5.12}$$

because the SAR and proper CAR specify normal distributions for $\mathbf{u}$ with finite covariance matrices. An ICAR model for δ, however, does not have a finite covariance matrix. Instead, the multivariate normal density for δ, equation (5.8), is specified in terms of its singular precision matrix $\mathbf{Q}/\sigma_s^2$. Thus, a tiny bit of cleverness is needed to fit the ICAR into a mixed linear model.

The key is re-expressing the improper density (5.8) or (5.10). Let $\mathbf{Q}$ have spectral decomposition $\mathbf{Q}=\mathbf{VDV}'$, where $\mathbf{V}$ is an orthogonal matrix with columns containing $\mathbf{Q}$'s eigenvectors and $\mathbf{D}$ is diagonal with non-negative diagonal entries. If the spatial map has I islands, then $\mathbf{D}$ has I zero diagonal entries, one of which corresponds to the eigenvector $\frac{1}{\sqrt{N}}\mathbf{1}_N$, by convention the Nth (right-most) column in $\mathbf{V}$. Define a new parameter $\theta = \mathbf{V}'\delta$, so θ has dimension N and precision matrix $\mathbf{D}/\sigma_s^2$. Giving an N-vector θ a normal model or prior with mean zero and precision $\mathbf{D}/\sigma_s^2$ is equivalent to giving $\delta = \mathbf{V}\theta$ an ICAR model or prior with precision $\tau\mathbf{Q}$. (I should qualify this with "if you use the multiplicative function of σ_s^2 that I prefer," but from here forward I omit such provisos.)

By convention, the eigenvalues on $\mathbf{D}$'s diagonal are sorted in decreasing order, so the first $N-I$ entries in $\mathbf{D}$'s diagonal are positive and the last I are zero. Partition $\mathbf{D}$ so its upper left $(N-I)\times(N-I)$ submatrix has all positive values on the diagonal and its lower right submatrix is all zeroes, and partition $\mathbf{V}$ and θ conformably:

$$\mathbf{D} = \left[\begin{array}{c|c} \mathbf{D}_1 & \mathbf{0} \\ \hline \mathbf{0} & \mathbf{0} \end{array}\right]; \quad \mathbf{V} = [\mathbf{V}_1|\mathbf{V}_2]; \quad \theta = \left[\begin{array}{c} \theta_1 \\ \hline \theta_2 \end{array}\right] \tag{5.13}$$

where $\mathbf{D}_1$ is $(N-I)\times(N-I)$, $\mathbf{V}_1$ is $N\times(N-I)$, $\mathbf{V}_2$ is $N\times I$, θ_1 is $(N-I)\times 1$, and θ_2 is $I\times 1$. Then $\theta_1 = \mathbf{V}_1'\delta \sim N_{N-I}(0, \sigma_s^2\mathbf{D}_1^{-1})$ and $\theta_2 = \mathbf{V}_2'\delta$ has a normal distribution with zero precision, i.e., a flat prior that makes θ_2 a fixed effect implicit in the ICAR model.

Thus in the simplest case without regressors,

$$\begin{aligned} \mathbf{y} &= \delta + \varepsilon \\ &= \mathbf{V}\mathbf{V}'\delta + \varepsilon \\ &= [\mathbf{V}_1|\mathbf{V}_2]\left[\begin{array}{c} \theta_1 \\ \hline \theta_2 \end{array}\right] + \varepsilon \\ &= \mathbf{V}_2\theta_2 + \mathbf{V}_1\theta_1 + \varepsilon. \end{aligned} \tag{5.14}$$

This is a mixed linear model with $\mathbf{X} = \mathbf{V}_2$, $\beta = \theta_2$, $\mathbf{Z} = \mathbf{V}_1$, and $\mathbf{u} = \theta_1 \sim N_{N-I}(0, \sigma_s^2\mathbf{D}_1^{-1})$. In this model with no other regressors, the ICAR implicitly specifies a fixed effect for the I island means, or for the intercept if there is only $I = 1$ island, i.e., if the spatial map is connected.

This model is easily extended to include fixed effects other than the island means. If $\mathbf{X}^*$ is a design matrix for other fixed effects (*not* including an intercept) and β^* is the corresponding vector of coefficients, then equation (5.12), the full mixed linear model with an ICAR model for δ, becomes

$$\mathbf{y} = \mathbf{V}_2\theta_2 + \mathbf{X}^*\beta^* + \mathbf{V}_1\theta_1 + \varepsilon, \tag{5.15}$$

where the fixed effects are specified by $\mathbf{X} = [\mathbf{V}_2|\mathbf{X}^*]$ and $\beta = (\theta_2', \beta^{*'})'$.

This re-expression draws attention to the random effects implicit in the ICAR model, which have design matrix $\mathbf{V}_1$ and random-effect precision matrix $\mathbf{D}_1/\sigma_s^2$. Because $\mathbf{1}_N$ is always an eigenvector of $\mathbf{Q}$ with a zero eigenvalue, for any number of islands $I \geq 1$ the columns of $\mathbf{V}_1$ are contrasts, i.e., they sum to zero. The positive diagonal elements of $\mathbf{D}$ and thus of $\mathbf{D}_1$, d_j, $j = 1, \ldots, N-I$, differentially shrink the random effects $\mathbf{u} = \theta_1$ toward zero. For any σ_s^2, eigenvectors of $\mathbf{Q}$ (contrasts, or columns of $\mathbf{V}_1$) with large eigenvalues d_j have coefficients $u_j = \theta_{1j}$ that are shrunk more toward zero than are coefficients of eigenvectors of $\mathbf{Q}$ with small eigenvalues

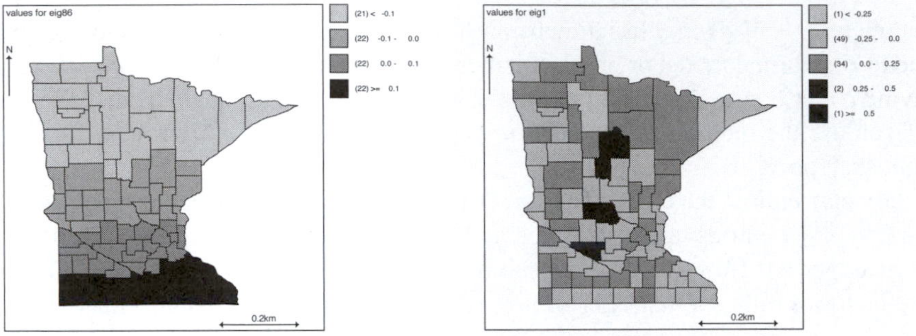

(a) Least-shrunk coefficient

(b) Most-shrunk coefficient

Figure 5.2: Minnesota's counties: Two columns of $\mathbf{V}_1$, canonical design-matrix columns arising as eigenvectors of $\mathbf{Q}$. Panel (a): $\mathbf{V}_{1,86}$, the design-matrix column for which the coefficient $\theta_{1,86}$ is shrunk least toward zero; Panel (b): $\mathbf{V}_{1,1}$, the design-matrix column for which the coefficient $\theta_{1,1}$ is shrunk most toward zero.

d_j. Variation in $\mathbf{y}$ in the space spanned by contrasts with large d_j is mostly shrunk into error, while variation in $\mathbf{y}$ in the space spanned by contrasts with small d_j is mostly retained in the fit. The exact degree of "mostly" is determined by the smoothing parameter σ_s^2.

An ICAR model is thus equivalent to a collection of canonical design-matrix columns, and each column's coefficient has a canonical degree of relative shrinkage d_j (actual shrinkage depends on $r = \sigma_e^2/\sigma_s^2$, which affects all coefficients). The canonical design-matrix columns and degrees of relative shrinkage are determined entirely by the neighbor pairings in the spatial map.

Are these canonical design-matrix columns interpretable?[1] I am not aware of any general theory for interpreting the ICAR's implicit design-matrix columns — though one might be in the graph-theory literature, waiting for someone to translate it into our language — but various examples give very similar intuitive interpretations, and we turn to these now.

Figure 5.2 shows two maps of the $N = 87$ counties of Minnesota, which form a connected graph ($I = 1$) when neighbor pairs are defined as counties sharing part of a boundary. Panel a's map depicts the 86th column of $\mathbf{V}_1$, which has the smallest positive eigenvalue d_{86} and thus precision d_{86}/σ_s^2; its coefficient $u_{86} = \theta_{1,86}$ is shrunk toward zero *least* of all the random effects. Panel b's map depicts the 1st column of $\mathbf{V}_1$, which has the largest positive eigenvalue d_1 and thus precision d_1/σ_s^2; its coefficient $u_1 = \theta_{1,1}$ is shrunk toward zero *most* of all the random effects.

Panel a's eigenvector or contrast is very close to a linear gradient in the north-south long axis of the map defined by Minnesota's counties. Panel b's eigenvector or contrast is the difference between three counties and the other 84 counties; the

[1] Some of the following material is taken from Reich & Hodges (2008a).

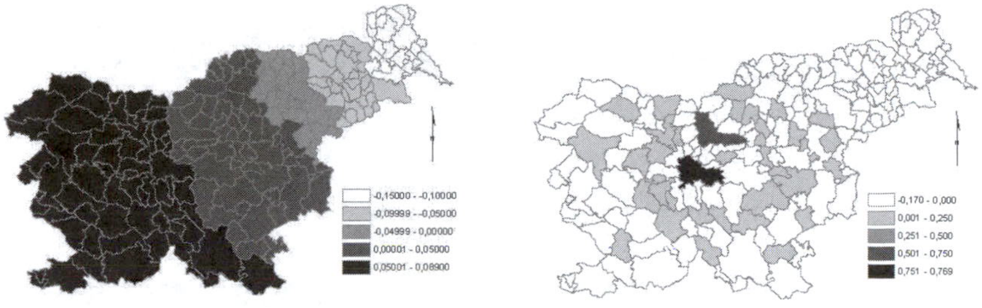

(a) Least-shrunk coefficient (b) Most-shrunk coefficient

Figure 5.3: Slovenia's municipalities: Two columns of $\mathbf{V}_1$, canonical design-matrix columns arising as eigenvectors of $\mathbf{Q}$. Panel (a): $\mathbf{V}_{1,193}$, the design-matrix column for which the coefficient $\theta_{1,193}$ is shrunk least toward zero; Panel (b): $\mathbf{V}_{1,1}$, the design-matrix column for which the coefficient $\theta_{1,1}$ is shrunk most toward zero.

distinguishing feature of these three counties is that they have the largest numbers m_i of neighboring counties.

Figure 5.3, like Figure 5.2, shows two maps of the 194 municipalities that partitioned Slovenia in 2005, which form a connected graph ($I = 1$) when neighbor pairs are defined as municipalities sharing part of a boundary. Again, panel a's map shows the column of $\mathbf{V}_1$ with the least-shrunk random-effect coefficient, while panel b's map shows the column of $\mathbf{V}_1$ with the most-shrunk random effect coefficient. The contrast (design-matrix column) depicted in panel a is, again, very close to linear in the long axis of Slovenia's map, while panel b's contrast is roughly the difference between municipalities having many neighbors and municipalities having few neighbors; the two municipalities with the largest entries in this design-matrix column (the two darkest municipalities) are distinguished by having the most neighbors.

Figure 5.4 shows four design-matrix columns for the spatial map arising from a simplified version of the attachment-loss lattice shown in Figure 5.1. Figure 5.4 shows only one arch of 14 teeth, which forms a connected graph. For each design-matrix column in panels a through d, the 14 shaded boxes represent the teeth and the 6 triangles around each tooth are the measurement sites. The gray-scale represents the absolute value of the design-matrix column at that site, with darker shading indicating larger absolute value, while white indicates zero. A triangle pointing up indicates a positive entry and a triangle pointing down indicates a negative entry. Panel a shows the eigenvector (design matrix column) with the largest eigenvalue (prior relative precision); its coefficient is shrunk most of all the coefficients. All entries in this design-matrix column are zero except those for the second tooth from the left. This contrast is the difference between the two sides of the tooth, of a particular linear combination of the three sites on a side. Eleven other contrasts, one for each of the 11 other interior teeth, have analogous contrasts with the same eigenvalue, while the two exterior teeth have analogous contrasts with slightly smaller eigen-

Figure 5.4: Attachment-loss data, simplified lattice of 14 teeth. Shaded squares represent teeth, triangles represent measurement sites. Each triangle's shade indicates the absolute value of the design-matrix entry for that site, darker indicating larger absolute value and white indicating zero. Triangles pointing upward are positive entries, triangles pointing downward are negative entries.

values. Panel d shows the design-matrix column with the least-shrunk coefficient; it appears to be linear down the two sides of the row of teeth. For a periodontal lattice with two arches having 14 teeth each, there are two such design-matrix columns, one for each arch, with identical d_j. Panel c shows the design-matrix column with the coefficient that is shrunk second-least; it appears to be quadratic down the two sides of the row of teeth. As you progress through the list of eigenvalues starting from the smallest positive eigenvalue, the corresponding eigenvectors (design-matrix columns) resemble polynomials of increasing degree or sinusoidal curves of increasing frequency, although this progression breaks down after a while. Panel b shows a design-matrix column with a moderate-to-large eigenvalue, which turns out to be important in Chapters 14 and 16. It is the difference between the average of the direct sites (sites in the middle of a tooth) and the average of the interproximal sites (sites on the gap between two teeth).

Based on these examples and others that give similar results, I venture some tentative generalizations. Columns of $\mathbf{V}_1$ (eigenvectors of $\mathbf{Q}$) describing low-frequency or large-scale trends have coefficients that are shrunk least among all the coefficients, so variation in $\mathbf{y}$ in the space of these columns is mostly retained in the fit. Columns of $\mathbf{V}_1$ describing high-frequency or small-scale features have coefficients that are shrunk most among all the coefficients, so variation in $\mathbf{y}$ in the space of these columns is mostly pushed into error. Again, the precise extent of "mostly" is determined by the variances σ_s^2 and σ_e^2. I used the term "low-frequency"; the contrasts I described as almost exactly linear or quadratic could equally well be interpreted as almost exactly a half or full cycle of a sine wave. I hazard the guess that as the lattice becomes larger and more regular, these contrasts become more like sine curves. One of the exercises explores a theoretically based approach to interpreting $\mathbf{Q}$'s eigenvectors in a modest degree of generality.

One implication of interpreting the ICAR as we have here is that the large-scale contrasts with least-shrunk coefficients correspond to trends such as gradients along the long axis of the map, which can be collinear with explicit fixed effects in $\mathbf{X}^*$ in equation (5.15). As with confounding in ordinary linear models, such spatial confounding can obliterate obvious associations between the outcome $\mathbf{y}$ and a fixed effect in $\mathbf{X}^*$. Chapter 9 and Section 10.1 give an example of spatial confounding that arose in a real problem, explains the mechanics of it, discusses how interpretation of and proper response to spatial confounding depends on the interpretation of the different parts of the mixed linear model — in particular, whether the ICAR random effect is an old- or new-style random effect — and shows one way to avoid it.

In models with Poisson observables, the ICAR model is usually specified along with an iid heterogeneity component:

$$\begin{aligned} Y_i &\sim \text{Poisson with mean } \mu_i \\ \log(\mu_i) &= \log(E_i) + \mathbf{x}\beta + \delta_i + \phi_i \end{aligned} \tag{5.16}$$

where $\log(E_i)$ is an offset (e.g., log population of area i), $\delta = (\delta_1, \ldots, \delta_N)'$ has an ICAR model, and the ϕ_i are iid $N(0, \sigma_h^2)$. Customarily, δ is described as capturing spatial clustering, while the ϕ_i capture unstructured heterogeneity. This combination is necessary in non-normal models such as Poissons, in which the variance is a function of the mean, because without ϕ_i, if the ICAR's parameter σ_s^2 is made small, the model has no way to account for extra-Poisson variability.

In models with normal errors and a single measurement for each area, such as (5.12), this heterogeneity term is not necessary because the error term ε captures heterogeneity. However, in some uses of the ICAR model with normal errors, such as periodontal measurements or gene expression, it is possible to have multiple measurements at a given "location," so it may be desirable or necessary to replace (5.12) with a clustering-plus-heterogeneity formulation like

$$\mathbf{y} = \mathbf{X}\beta + \mathbf{I}_n\delta + \mathbf{I}_n\phi + \varepsilon. \tag{5.17}$$

Section 12.1.1 considers this as a model for attachment-loss data with repeat measurements at each measurement site; see also Cui et al. (2010).

5.2.3 Smoothed ANOVA with Spatial Smoothing

Chapter 4 celebrated the modularity of mixed linear models by replacing linear modules with penalized-spline modules. This section continues the celebration, demonstrating the modularity of spatial models by smoothing an ANOVA using one factor's spatial structure. In the example used in Zhang et al. (2010) and developed below, we had data on 3 types of cancer in each of Minnesota's 87 counties. In Zhang et al. (2010), the data were counts of the 3 cancers in each county in a certain period and were modeled as Poisson observables, while smoothed ANOVA was defined in

Section 4.2.2 for normal observables. The spatially smoothed ANOVA model that follows can be used for either kind of outcome with a suitable link function.

This example uses the re-expression of the ICAR model developed in Section 5.2.2. In that re-expression, one of the zero eigenvalues of $\mathbf{Q}$ corresponds to the eigenvector $\frac{1}{\sqrt{N}}\mathbf{1}_N$, where N is the number of areas in the map. By convention, this is the Nth (right-most) column in $\mathbf{V}$. In defining the new parameter $\theta = \mathbf{V}'\delta$, $\theta_N = \frac{1}{\sqrt{N}}\mathbf{1}_N'\delta = \sqrt{N}\,\bar{\delta}$, the scaled average of the δ_i, while the other $N-1$ elements of θ are $N-1$ contrasts in δ, which are orthogonal to $\frac{1}{\sqrt{N}}\mathbf{1}_N$ by construction. Thus the ICAR model or prior is informative (has positive precision) only for contrasts in δ while putting zero precision on $\theta_{GM} \equiv \theta_N = \frac{1}{\sqrt{N}}\mathbf{1}_N'\delta$.

This reparameterization allows the ICAR model to fit into the ANOVA framework. θ_{GM} corresponds to the ANOVA's grand mean and the rest of θ, θ_{Reg}, corresponds to $\mathbf{V}^{(-)'}\delta$, where $\mathbf{V}^{(-)}$ is $\mathbf{V}$ excluding the column $\frac{1}{\sqrt{N}}\mathbf{1}_N$, consisting of $N-1$ orthogonal contrasts among the N areas and giving the $N-1$ degrees of freedom in the usual ANOVA accounting:

$$\begin{aligned} \delta &= (\delta_1, \delta_2, \ldots, \delta_N)' \\ &= \mathbf{V}\theta \\ &= \begin{bmatrix} \mathbf{V}^{(-)} & \frac{1}{\sqrt{N}}\mathbf{1}_N \end{bmatrix} \begin{bmatrix} \theta_{Reg} \\ \theta_{GM} \end{bmatrix}. \end{aligned} \tag{5.18}$$

Giving δ an ICAR model or prior is equivalent to giving θ a $N(\mathbf{0}, \mathbf{D}/\sigma_s^2)$ model or prior, where $\mathbf{D}/\sigma_s^2$ is a precision matrix; in mixed linear model terms, the θ_{Reg} are random effects and θ_{GM} is a fixed effect. The precision $\mathbf{D}_{NN} = 0$ for the overall level is equivalent to a flat prior on θ_{GM}, though θ_{GM} could be given a normal prior with mean zero and specified finite variance.

Here is a two-factor smoothed ANOVA model using the example of 3 cancers in Minnesota's counties. Zhang et al. (2010) constructed this model as a competitor for a more complex spatial model, the multivariate ICAR (MCAR) model. This smoothed ANOVA is equivalent to a particular point prior distribution (i.e., degenerate prior) on a part of the MCAR model's covariance structure. However, this smoothed ANOVA model competed well with the MCAR model in a simulation experiment even when that point prior was incorrect as a description of the true covariance structure. Interested readers are referred to Zhang et al. (2010) for further details.

Consider the Minnesota map with $N = 87$ counties and suppose each county has data y_{ij} for $m = 3$ cancers indexed as $j = 1, 2, 3$. County i has an m-vector of parameters describing the m cancers, $\delta_i = (\delta_{i1}, \ldots, \delta_{im})'$; define the Nm vector δ as $\delta = (\delta_1', \ldots, \delta_N')'$. Assume the $N \times N$ matrix $\mathbf{Q}$ encodes neighbor pairs among counties as before. The smoothed ANOVA model for this problem is a 2-way ANOVA with factors cancer ("*CA*", m levels) and county ("*CO*", N levels) and no replication within design cells. As in Section 4.2.3's smoothed ANOVA, model δ with a saturated linear model and put the grand mean and the main effects in their traditional positions as in ANOVA (matrix dimensions and definitions appear below the equation):

$$\delta = (\delta_1', \delta_2', \ldots, \delta_N')' = [\mathbf{X}|\mathbf{Z}]\,\theta =$$

$$\left[\underbrace{\frac{1}{\sqrt{Nm}}\mathbf{1}_{Nm}}_{\substack{\text{Grand mean}\\ Nm\times 1}} \quad \underbrace{\frac{1}{\sqrt{N}}\mathbf{1}_N \otimes H_{CA}}_{\substack{\text{Cancer}\\ \text{main effect}\\ Nm\times(m-1)}} \middle| \quad \underbrace{\mathbf{V}^{(-)} \otimes \frac{1}{\sqrt{m}}\mathbf{1}_m}_{\substack{\text{County}\\ \text{main effect}\\ Nm\times(N-1)}} \quad \underbrace{\mathbf{V}^{(-)} \otimes H_{CA}{}^{(1)} \ldots \mathbf{V}^{(-)} \otimes H_{CA}{}^{(m-1)}}_{\substack{\text{Cancer}\times\text{County}\\ \text{interaction}\\ Nm\times(N-1)(m-1)}} \right] \begin{bmatrix} \Theta_{GM} \\ \theta_{CA} \\ \theta_{CO} \\ \theta_{CO\times CA} \end{bmatrix}$$

where H_{CA} is an $m \times (m-1)$ matrix whose columns are contrasts among cancers, so $\mathbf{1}_m' H_{CA} = \mathbf{0}_{m-1}'$, and $H_{CA}' H_{CA} = \mathbf{I}_{m-1}$; $H_{CA}^{(j)}$ is the jth column of H_{CA}; and $\mathbf{V}^{(-)}$ is $\mathbf{V}$ without its Nth column $\frac{1}{\sqrt{N}}1_N$, so it has $N-1$ columns, each a contrast among counties, i.e., $\mathbf{1}_N'\mathbf{V}^{(-)} = \mathbf{0}_{N-1}'$, and $\mathbf{V}^{(-)'}\mathbf{V}^{(-)} = \mathbf{I}_{N-1}$. The user must select the contrasts among cancers in H_{CA}'s columns. The column labeled "Grand mean" corresponds to the ANOVA's grand mean and has parameter θ_{GM}; the other blocks of columns labeled as main effects and interactions correspond to the analogous ANOVA effects and to their respective parameters $\theta_{CA}, \theta_{CO}, \theta_{CO\times CA}$.

Defining random-effect and prior distributions on θ completes the smoothed ANOVA specification. Zhang et al. (2010) put independent flat priors on θ_{GM} and θ_{CA}, which are therefore treated as fixed effects, i.e., not smoothed or shrunk. This is equivalent to putting a flat prior on each of the m cancer-specific (main effect) means. To specify the random effect variance matrix $\mathbf{G}$ in the mixed-linear-model formulation, let the county main effect parameter θ_{CO} have distribution $\theta_{CO} \sim N_{N-1}(\mathbf{0}, \tau_0\mathbf{D}^{(-)})$, where $\tau_0\mathbf{D}^{(-)}$ is a precision matrix, $\mathbf{D}^{(-)}$ is $\mathbf{D}$ without its Nth row and column, and $\tau_0 > 0$ is unknown. Similarly, for the jth group of columns in the cancer-by-county interaction, $\mathbf{V}^{(-)} \otimes H_{CA}^{(j)}$, let the coefficient have distribution $\theta_{CO\times CA}^{(j)} \sim N_{N-1}(\mathbf{0}, \tau_j\mathbf{D}^{(-)})$, where again $\tau_j\mathbf{D}^{(-)}$ is a precision matrix and $\tau_j > 0$ is unknown. Each of the distributions on θ_{CO} and the $\theta_{CO\times CA}^{(j)}$ specifies an ICAR model; the overall level of each ICAR, with precision zero, has been included in the grand mean and cancer main effects.

Thus the main effect for counties is smoothed using the spatial map and an ICAR model with precision-like parameter τ_0, and the cancer-by-county interaction is smoothed by separately smoothing each contrast in cancers, $H_{CA}^{(j)}$, using its own ICAR model and precision-like parameter τ_j. This smoothed ANOVA can now be fit using methods for mixed linear models, with $\phi_G = (\tau_0, \tau_1, \ldots, \tau_{m-1})'$ as the unknowns in the random-effect covariance matrix $\mathbf{G}$.

Other spatial models could be used here in place of the ICAR model, e.g., a proper CAR model. However, the re-expression used here for the ICAR model does not necessarily apply because in general $\mathbf{V}$ and $\mathbf{D}$ in the spectral decomposition depend on the unknowns in the spatial model's covariance structure, so that the smoothed ANOVA's design matrix will change in every MCMC or RL-maximization iteration. Such models can be specified and fit using standard software (e.g., SAS's MIXED procedure or BUGS), though not as tidily as the ICAR model allows.

5.3 Two-Dimensional Penalized Splines

At first glance, two-dimensional penalized splines appear to be simply a way to draw a smooth surface representing a function of two variables, and thus quite different from geostatistical models. In fact, two-dimensional splines are quite closely related to geostatistical models arising from certain process models. Once again, this area has an enormous literature that I will not try to summarize. Instead, as before the goal here to show how two-dimensional penalized splines can be represented as mixed linear models, and as always, this representation has a price.

Like the one-dimensional penalized splines discussed in Chapter 3, a two-dimensional penalized spline is specified by specifying a basis, knots, and a penalty. The presentation here emphasizes the basis and penalty, for reasons that will become clear. This section draws heavily on Chapter 13 of Ruppert et al. (2003).

When working in more than one dimension, there are at least two qualitatively different types of basis. Ruppert et al. (2003) make this point using two examples. In the first example, the dataset consists of counts of scallops caught in Long Island Sound, where each count is made at a specific location identified by x-y coordinates, as in geostatistical data. In the second example, the dataset consists of incidence of AIDS in Italian men who have sex with men, and the two predictors are calendar year and age at diagnosis, each treated as a continuous variable. In a certain technical sense, these two problems are identical: for each outcome, the two predictors are on continuous scales, and the object is to select a smooth surface describing a function of the two predictors. However, these problems differ in an important way. In the AIDS example, the two predictors are on meaningful scales and it would change the interpretation of the results to change their scales. Also, although in this example both scales have years as their units, in general the two predictors are on incommensurable scales. In the scallops example, by contrast, the coordinate axes are purely conventional and could be changed by translation or rotation with no loss of meaning.

This difference affects the choice of basis. The tensor-product basis, presented first below, is defined in terms of the original predictors and thus seems more natural for problems like the AIDS example. However, the tensor-product basis depends on the original coordinate axes and the fitted smooth will change with a different coordinate system. This is undesirable for problems like the scallops example, where the original coordinates are entirely conventional. This difficulty is avoided with a radial basis, which is presented after the tensor-product basis. This section concludes with a brief comment on the two kinds of bases.

5.3.1 Tensor-Product Basis

The tensor-product basis is the obvious generalization of the penalized spline models for category-by-continuous interactions considered in Section 4.2.1. As in that section, consider fitting the model $y_i = f(s_i, t_i) + \varepsilon_i$ where s and t are continuous predictors, but now without requiring additivity. For simplicity, I'll use a truncated-line basis for each of s and t, but this is easily replaced by other bases. The spline model is

$$\begin{aligned} y_i = \quad & \beta_0 && +\beta_s s_i + \textstyle\sum_{k=1}^{K^s} u_k^s (s_i - \kappa_k^s)_+ \\ & && +\beta_t t_i + \textstyle\sum_{k=1}^{K^t} u_k^t (t_i - \kappa_k^t)_+ \\ & +\beta_{\text{st}} s_i t_i && +\textstyle\sum_{k=1}^{K^s} v_k^s s_i (t_i - \kappa_k^t)_+ \\ & && +\textstyle\sum_{k=1}^{K^t} v_k^t t_i (s_i - \kappa_k^s)_+ \\ & && +\textstyle\sum_{k=1}^{K^s} \sum_{k'=1}^{K^t} v_{kk'}^{\text{st}} (s_i - \kappa_k^s)_+ (t_i - \kappa_k^t)_+ \\ & +\varepsilon_i, \end{aligned} \tag{5.19}$$

where $\kappa_k^s, k = 1, \ldots, K^s$ are knots for s and $\kappa_k^t, k = 1, \ldots, K^t$ are knots for t. The first two rows in (5.19) are for main effects, while the next three rows are for interactions. As the notation suggests, the β's are treated as fixed effects, while the u's and v's are treated as random effects. The basis vectors for the main effects are the same as for one-dimensional splines, while the basis vectors for the interactions, $s_i(t_i - \kappa_k^t)_+$, $t_i(s_i - \kappa_k^s)_+$, and $(s_i - \kappa_k^s)_+(t_i - \kappa_k^t)_+$ involve both predictors and their respective knots. Figure 13.4 in Ruppert et al. (2003) shows these basis vectors for a simple case of two knots for each of s and t. The obvious covariance matrix $\mathbf{G}$ for the random effects would be diagonal with different variances for $\mathbf{u}^s$, $\mathbf{u}^t$, $\mathbf{v}^s$, $\mathbf{v}^t$, and $\mathbf{v}^{\text{st}}$, though other choices are possible.

5.3.2 *Radial Basis*

This section develops a more general version of splines with one-dimensional radial bases, introduced in Section 3.2.1, which then generalize trivially to higher dimensions. I follow Ruppert et al. (2003), Section 13.4, which seems to involve no small amount of *ad hoc* tinkering. They start with a familiar spline formulation and make changes as needed to achieve properties that seem desirable. I see nothing wrong with this approach, unlike my more purist colleagues, but it does seem to suggest there's room for exploration and improvement.

The bases in this approach are built using a function $C(d)$ where d is a distance measure taking values in the positive real numbers and $C(\bullet)$ is real-valued. Because $C(d)$ depends only on distance and not orientation, a basis and penalty constructed using C are invariant to translations and rotations of the coordinate system.

For simplicity, start with a full-rank truncated-line basis for a one-dimensional spline fit of $y_i = f(x_i) + \text{error}$. The design matrix of the unpenalized effects, $\mathbf{X}$, has rows $[1\ x_i]$. For now, consider a full-rank basis in which each unique value of x is a knot, so the design matrix for the penalized effects, $\mathbf{Z}$, has entry $(x_i - x_j)_+$ in row i and column j. The two end-most knots are redundant (the right-most gives a column of zeros in $\mathbf{Z}$) but because the fit is penalized, this creates no numerical problems and it's convenient to retain those columns. The columns of $\mathbf{X}$ and $\mathbf{Z}$ are shown in Figure 5.5 for a particular set of 20 distinct x_i, iid draws from a $U[0,1]$ distribution. If the smoothing constant λ^2 is known, the penalized spline fit is $\hat{\mathbf{y}} = \mathbf{X}\hat{\beta} + \mathbf{Z}\hat{\mathbf{u}}$, where $(\hat{\beta}, \hat{\mathbf{u}})$ solves

$$\text{argmin}_{\beta,\mathbf{u}} \left\{ (\mathbf{y} - \mathbf{X}\beta + \mathbf{Z}\mathbf{u})'(\mathbf{y} - \mathbf{X}\beta + \mathbf{Z}\mathbf{u}) + \lambda^2 \mathbf{u}'\mathbf{u} \right\}. \tag{5.20}$$

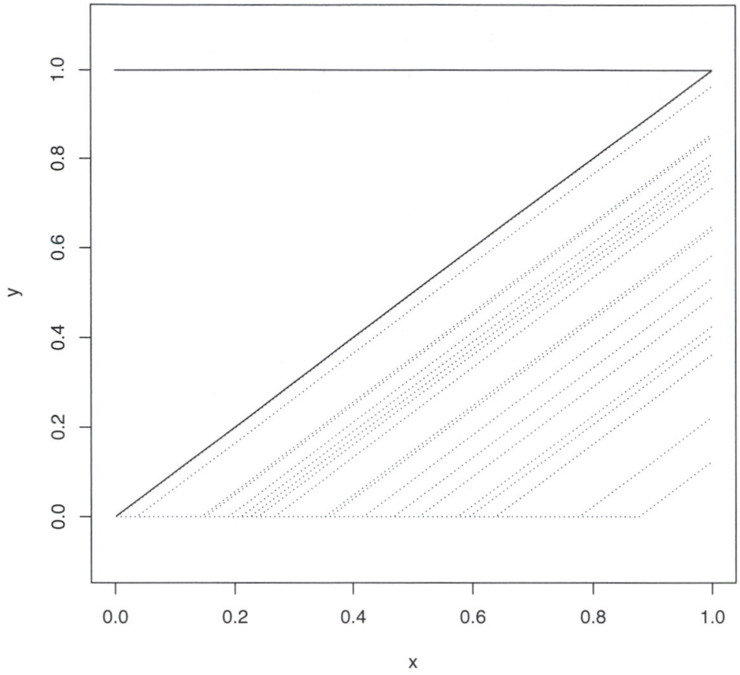

Figure 5.5: Full-rank truncated-line basis for a one-dimensional penalized spline. The solid lines are the basis vectors corresponding to the intercept and the linear term in **X**; the dashed lines are the basis vectors for the smoothed part of the curve in **Z**, one for each knot. These basis vectors turn up from zero at their respective knots, which are the x_i.

We'd like to have a radial basis for the penalized effects in **Z** instead of this truncated-line basis, while leaving the unpenalized effects **X** unchanged. If the sample size is n and each x_i is unique, then the joint design matrix used in (5.20), $[\mathbf{X}|\mathbf{Z}]$, is $n \times (n+2)$ and there exists an $(n+2) \times (n+2)$ transformation matrix $\mathbf{L}$ such that $[\mathbf{X}|\mathbf{Z}_R] = [\mathbf{X}|\mathbf{Z}]\,\mathbf{L}$, where the columns of $\mathbf{Z}_R$ are a radial basis and $\mathbf{Z}_R$ itself is symmetric. (Derivation of $\mathbf{L}$ is an exercise.) Figure 5.6 shows the resulting radial basis.

In terms of this new basis, the fitted values are $\hat{\mathbf{y}} = \mathbf{X}\hat{\beta}_R + \mathbf{Z}_R\hat{\mathbf{u}}_R$, where $(\hat{\beta}_R, \hat{\mathbf{u}}_R)$ solves

$$\text{argmin}_{\beta_R, \mathbf{u}_R} \left\{ (\mathbf{y} - \mathbf{X}\beta_R + \mathbf{Z}_R\mathbf{u}_R)'(\mathbf{y} - \mathbf{X}\beta_R + \mathbf{Z}_R\mathbf{u}_R) + \lambda^2 \mathbf{u}_R'\mathbf{L}'\mathbf{D}\mathbf{L}\mathbf{u}_R \right\}, \qquad (5.21)$$

where the matrix $\mathbf{D}$ is diagonal with diagonal elements $(0, 0, \mathbf{1}_n')'$.

This change is partly successful because $\mathbf{Z}_R$ is now radially symmetric: Its columns depend only on the distance of the corresponding x_i from the knots, not on their orientations relative to the knots. However, the penalty $\lambda^2 \mathbf{u}_R'\mathbf{L}'\mathbf{D}\mathbf{L}\mathbf{u}_R$ is not radially symmetric, so the fit will still change with translation or rotation of the coordinate system. Also, the penalty does not generalize readily to higher dimensions.

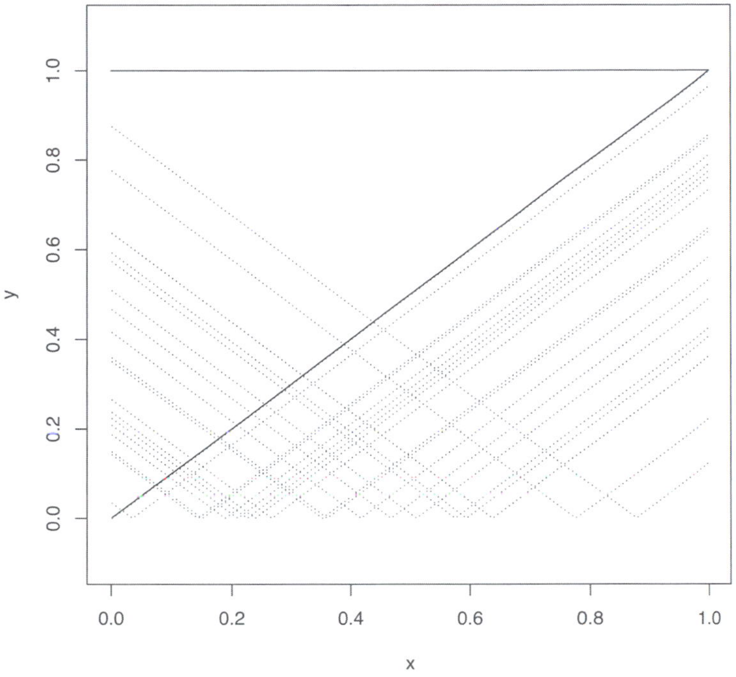

Figure 5.6: Full-rank radial basis for a one-dimensional penalized spline, resulting from transformation of the truncated-line basis. The solid lines are the basis vectors corresponding to the intercept and the linear term in $\mathbf{X}$; the dashed lines are the basis vectors in $\mathbf{Z}_R$ for the smoothed part of the fit. These basis vectors reach their minima at their respective knots.

This motivates the first *ad hoc* adjustment: Change the penalty so it *is* radially symmetric. Specifically, change the penalty to $\lambda^2 \mathbf{u}\mathbf{Z}_R\mathbf{u}$; now both the basis and penalty are invariant. With this new penalty, the spline fit is now the solution of

$$\text{argmin}_{\beta,\mathbf{u}} \left\{ (\mathbf{y} - \mathbf{X}\beta + \mathbf{Z}_R\mathbf{u})'(\mathbf{y} - \mathbf{X}\beta + \mathbf{Z}_R\mathbf{u}) + \lambda^2 \mathbf{u}\mathbf{Z}_R\mathbf{u} \right\}. \tag{5.22}$$

This is a member of the so-called thin-plate spline family of smoothers, which has a large literature (e.g., Green & Silverman 1994). Unfortunately, this penalized spline cannot be written as a mixed linear model because that would require $\mathbf{G} = \text{cov}(\mathbf{u}) = \sigma_s^2 \mathbf{Z}_R^{-1}$, and $\mathbf{Z}_R$ is not necessarily positive definite. (Counterexample: Let x_i take values $1, 2, \ldots, 100$; $\mathbf{Z}_R$ has 99 negative eigenvalues.)

This motivates a second *ad hoc* adjustment. Ruppert et al. (2003) offer three equally *ad hoc* options for changing this full-rank spline so it can be expressed as a legal mixed linear model. I'll skip that step and go to their next step, which combines one of these three choices with reduction in the rank of the spline, which is desirable for various reasons (e.g., computing speed).

Define $\mathbf{Z}_C = [C(|x_i - \kappa_k|)]$, where C is a function from the positive real numbers to the real numbers; $\mathbf{Z}_C$ has rows indexed by $i = 1, \ldots, n$ and columns indexed by $k = 1, \ldots, K$, for knots κ_k. C may depend on unknown parameters. Setting $C(r) = r$ and $K = n$ gives the radial basis we have used until now. The penalized spine fit is obtained by fitting this mixed linear model:

$$\begin{aligned} \mathbf{y} &= \mathbf{X}\beta + \mathbf{Z}_C\mathbf{u} + \varepsilon, \qquad \mathbf{R} = \sigma_e^2\mathbf{I}_n \\ \mathbf{G} &= \sigma_s^2(\Omega_K^{-0.5})(\Omega_K^{-0.5})' \\ \text{where } \Omega_K &= [C(|\kappa_k - \kappa_{k'}|)] \text{ for } k, k' = 1, \ldots, K. \end{aligned} \tag{5.23}$$

To obtain $\Omega_K^{-0.5}$, let Ω_K have singular value decomposition $\Omega_K = \mathbf{UDV}'$, where $\mathbf{D} = \text{diag}(\mathbf{d})$ and the diagonal entries $\mathbf{d}$ are positive. $\mathbf{U}$ and $\mathbf{V}$ have orthonormal columns and dimensions that conform to $\mathbf{D}$. Then $\Omega_K^{-0.5} = \mathbf{U}\text{diag}(\mathbf{d}^{-0.5})\mathbf{V}'$, where $\mathbf{d}^{-0.5}$ contains the reciprocals of the square roots of the entries in $\mathbf{d}$. This is a legal mixed linear model.

To make the model in (5.23) look like the penalized splines in Chapters 3 and 4, we take one last step and re-parameterize the random effect. Replace $\mathbf{Z}_C$ with $\mathbf{Z} = \mathbf{Z}_C\Omega_K^{-0.5}$, giving

$$\mathbf{y} = \mathbf{X}\beta + \mathbf{Z}\mathbf{u} + \varepsilon, \qquad \mathbf{R} = \sigma_e^2\mathbf{I}_n, \qquad \mathbf{G} = \sigma_s^2\mathbf{I}_K, \tag{5.24}$$

recalling that $\mathbf{Z}_C = [C(|x_i - \kappa_k|)]$ is $n \times K$ and $\Omega_K = [C(|\kappa_k - \kappa_{k'}|)]$ is $K \times K$.

Up to this point, the development has been for one-dimensional splines. The extension to $p > 1$ dimensions is now trivial: In the definition of $\mathbf{Z}_C$, replace $|x_i - \kappa_k|$ with $\|x_i - \kappa_k\|$, and in the definition of Ω_K, replace $|\kappa_k - \kappa_{k'}|$ with $\|\kappa_k - \kappa_{k'}\|$, where $\|\bullet\|$ is a p-dimensional distance measure and x_i and κ_k are now p-dimensional. The result is a penalized spline with a basis and penalty that are invariant to translations or rotations in the coordinate system.

The user must choose the function $C(\bullet)$, which determines $\mathbf{Z}$. You get an approximate thin-plate spline by using a polynomial of order s to specify the rows of $\mathbf{X}$ and using $C(\|\mathbf{r}\|) = \|r\|^{2m-p}$ when the dimension p is odd, and $C(\|\mathbf{r}\|) = \|\mathbf{r}\|^{2m-p}\log\|\mathbf{r}\|$ when the dimension p is even, where $m > s$ is a constant you choose (Ruppert et al. 2003, p. 254). Ruppert et al. (2003) also describe C arising from covariance functions in the Matérn family, which was prominent in Section 5.1 above. In this family, the two simplest C are $C(\|\mathbf{r}\|) = \exp(-\|\mathbf{r}\|/\rho)$, corresponding to $\nu = 1/2$, and $C(\|\mathbf{r}\|) = \exp(-\|\mathbf{r}\|/\rho)(1 + \|\mathbf{r}\|/\rho)$, corresponding to $\nu = 3/2$.

Finally, the user must choose knot locations. A rectangular lattice is wasteful if the data are not on a rectangular grid. Ruppert et al. (2003, Section 13.4.6) suggest mimicking the shape of the predictor space by using a space-filling algorithm to select knot locations.

By now, readers encountering radial-basis splines for the first time may be thirsty for some intuition. Unfortunately, although I have performed and seen a few fits of two-dimensional radial-basis splines, I have almost no intuition about them and even my heroes Ruppert et al. (2003) offer very little. The development above provides no sense what the basis $\mathbf{Z} = \mathbf{Z}_C\Omega_K^{-0.5}$ looks like, in stark contrast to truncated-polynomial bases or the random-effect design matrix $\mathbf{Z}$ arising from an ICAR model.

One of this chapter's exercises involves constructing $\mathbf{Z}$ for some simple cases and looking at its columns. I have done this and the results were interesting but still gave little intuition. It is not clear, for example, how the basis in $\mathbf{Z}$ changes as C changes, or as the tuning constants m or ν change in the approximate thin-plate and Matérn families respectively. Note that the C functions for the approximate thin-plate and Matérn families differ substantially: The former are increasing in $||\mathbf{r}||$ while the latter are decreasing in $||\mathbf{r}||$. However, if you do the exercise for the one-dimensional case and look at the columns of $\mathbf{Z}$ resulting from $C(|r|) = |r|$ and $C(|r|) = \exp(-|r|/25)$ for 10 knots, you will find that the two bases do not seem to differ much. From my position of fairly thorough ignorance, if these two splines behaved differently in practice, I would be very concerned.

5.3.3 A Comment on Tensor-Product vs. Radial Basis

Early in this section, it was argued that a tensor-product basis is more natural for problems like the AIDS example, in which the predictors are not spatial coordinates on an arbitrary scale but rather measures with meaningful scales, especially if those scales are incommensurable. The obvious counter-argument is that you can standardize the original predictors — for each predictor, subtract its average and divide by its standard deviation, so each standardized predictor has average 0, standard deviation 1, and no units — then do a radial-basis fit with the standardized predictors and transform the fit back to the original scale.

However, after an admittedly modest search, I have found no mention of this standardization trick. For example, Ruppert et al. (2003) discuss radial bases at length and the accompanying R package SemiPar allows them, but neither the book nor SemiPar's manual mentions standardization. Similarly, SAS's GLIMMIX procedure does radial basis fits but the documentation (version 9.2) says nothing about standardizing the predictors either before calling GLIMMIX or in GLIMMIX itself. Perhaps the presumption is that users will use a radial basis only for a spatial coordinate system. However, I have seen analyses that used a radial basis for non-spatial predictors like those in the AIDS example.

Exercises

Regular Exercises

1. (Section 5.2.1) Derive the two forms of the joint partially specified density for the ICAR model.
2. (Section 5.2.1) Show that if the ICAR's spatial lattice has I islands, then the matrix $\mathbf{Q}$ has I zero eigenvalues and that one basis for $\mathbf{Q}$'s null space has basis vectors that are $\mathbf{1}_{N_k}$ for the N_k areas in island k and zero for other areas. (Hint: Use the pairwise form of the density, equation 5.9.)
3. (Section 5.3.2) Derive the transformation matrix $\mathbf{L}$ that transforms the truncated-line basis into a radial basis. Hint: Start with the singular-value decomposition

of $[\mathbf{X}|\mathbf{Z}]$. Compute this for some special cases and see if you can develop some intuition; I haven't been able to.

4. (Section 5.3.2) Derive the basis matrix $\mathbf{Z} = \mathbf{Z}_C\Omega_K^{-0.5}$ for some particular cases and see if you can develop some intuition about this basis. For example, consider a one-dimensional case with 100 observations and 10 equally spaced knots, consider two very different C functions, $C(|r|) = |r|$ and $C(|r|) = \exp(-|r|/25)$, and plot the columns of $\mathbf{Z}$. For the Matérn C, compare bases arising from $\nu = m + 1/2$ for $m = 0, 1, 2$; for the approximate thin-plate C, compare bases for a fixed polynomial dimension s and different m. Then consider a two-dimensional case with observations on a grid, and knots spaced equally on the grid.

Open Questions

1. (Section 5.2.2) Interpret the eigenstructure of $\mathbf{Q}$ on theoretical grounds, i.e., not just using examples like I did. One way to make this tractable is to take a special case like a rectangular or square lattice with a regular structure of neighbor pairings. In such a case, it might be possible to approximate the eigenstructure of $\mathbf{Q}$ by appealing to a spectral approximation like the one for Gaussian processes (e.g., Paciorek 2007; see also Chapter 17). You might also look in the graph-theory literature. Some of my students have found useful results there; Reich & Hodges (2008a), Section 3.3 and the Appendix, are suggestive.

Chapter 6

Time-Series Models as Mixed Linear Models

Methods for analyzing time-series data have developed independently in different areas, producing disparate collections of methods that appear to have little relationship. These include analyses in the frequency domain; autoregressive-moving-average (ARMA) models, sometimes called Box-Jenkins models; and state-space models, also known as dynamic linear models.

This chapter focuses on dynamic linear models (DLMs), which are a close relative of the Kalman filter. DLMs can be used two different ways, as filters or as smoothers. An inertial navigation system for an aircraft is an example of filtering. The system uses a state-space model to update the most recent estimates of location and velocity, then combines those updated estimates with new measurements (with error) from the aircraft's sensors, giving an estimate of current location and velocity that filters out at least some of the measurement error. For a DLM used as a filter, the object is to produce improved real-time estimates of a system's state. By contrast, when a DLM is used as a smoother, the complete series of observations is in hand and the model is used to produce a smooth estimate of the underlying state of the system at each time.

This chapter is about using DLMs as smoothers. First we revisit a simple example used in an earlier chapter, then re-express DLMs as mixed linear models in some generality, then explore an example given to me by Michael Lavine. Again, this chapter is not intended to be a complete survey. West & Harrison (1997) is an encyclopedic treatment of dynamic models used for filtering and forecasting including many models that cannot be expressed as mixed or generalized mixed linear models, e.g., multi-state models. ARMA models can be written as DLMs — see, e.g., West & Harrison (1997, Section 9.4); functions in the R package dlm (Petris 2010) automatically express ARMA models in DLM form — but this chapter does not discuss them explicitly.

6.1 Example: Linear Growth Model

Section 2.2.2.4 used the linear growth model to smooth the global-mean surface temperature series and exemplify degrees of freedom. We now revisit that example.

The series of global average temperature deviations from 1881 to 2005, y_t, $t = 0,\ldots,124$, was shown in Figure 2.1. There, we smoothed y_t using the linear growth model. In DLM jargon, the linear growth model has two parts, the *observation equation* and the *state equation(s)*. The observation equation is a model for the observed data y_t as a function of the unobserved state μ_t:

$$y_t = \mu_t + n_t, \quad t = 0,\cdots,T, \tag{6.1}$$

where n_t is the observation error at time t, modeled as $n_t \sim N(0,\sigma_n^2)$. The state equation is a model for the state's evolution over time, with systematic and random components:

$$\begin{aligned} \mu_t &= \mu_{t-1} + \theta_{t-1} + w_{1,t}, \quad t = 1,\cdots,T, & (6.2)\\ \theta_t &= \theta_{t-1} + w_{2,t}, \quad t = 1,\cdots,T-1, & (6.3) \end{aligned}$$

where μ_t is now interpretable as the current or local level and θ_t is the current or local slope or time trend. The local level μ_t is updated by adding the local time trend θ_t and a random term $w_{1,t}$ modeled as $w_{1,t} \sim N(0,\sigma_1^2)$; the local time trend evolves in a random walk with random term $w_{2,t}$ modeled as $w_{2,t} \sim N(0,\sigma_2^2)$. (Note that (6.3) specifies a one-dimensional ICAR model, where each year's neighbors are the preceding and following years.) Let $\mathbf{n}$, $\mathbf{w}_1$, and $\mathbf{w}_2$ be the vectors of n_t, $w_{1,t}$, and $w_{2,t}$ respectively. If $\mathbf{w}_1 = \mathbf{0}_T$ and $\mathbf{w}_2 = \mathbf{0}_{T-1}$, this model is simply a straight line with intercept μ_0 and slope θ_0. The two random components, controlled by the variances σ_1^2 and σ_2^2, allow the local level and local slope to vary over time.

In the DLM literature, it is customary to add a fully specified distribution for (μ_0,θ_0), which is sometimes called a prior distribution even by those who do a maximum-likelihood analysis. A prior for (μ_0,θ_0) is necessary for filtering, specifically to allow Bayesian updating of the posterior for the state (μ_t,θ_t) at each time t. (In filtering, the variances are usually treated as known, so this updating is very simple.) In smoothing applications, μ_0 and θ_0 are often given normal distributions with mean zero and large variances. Letting that variance go to infinity is equivalent to using a flat prior for (μ_0,θ_0) or simply omitting the prior for (μ_0,θ_0). We return to this point below.

It is considerably easier to express this model in the constraint-case formulation of Section 2.1 than as a mixed linear model. To do so, take equation (6.1) as it is, and rewrite equations (6.2) and (6.3) in the manner of Section 2.1, to give

$$\begin{array}{rclllll} y_t &=& \mu_t & & +n_t, & t=0,\cdots,T \\ 0 &=& \mu_{t-1}-\mu_t & +\theta_{t-1} & +w_{1,t} & t=1,\cdots,T \\ 0 &=& & +\theta_{t-1}-\theta_t & +w_{2,t} & t=1,\cdots,T-1. \end{array} \tag{6.4}$$

(Note that the ranges of the index t are different for the three equations.) Assembled in the form of a linear model as in Section 2.1, this becomes

$$
\left[\begin{array}{c} \mathbf{y} \\ \hline \mathbf{0}_T \\ \hline \mathbf{0}_{T-1} \end{array}\right]
=
\left[\begin{array}{cccccc|cccccc}
\multicolumn{6}{c|}{\mathbf{I}_{T+1}\mathbf{0}_{(T+1)\times T}} & & & & & & \\ \hline
1 & -1 & 0 & \dots & 0 & 0 & & & & & & \\
0 & 1 & -1 & \dots & 0 & 0 & & & & & & \\
 & \vdots & & \ddots & & \vdots & \multicolumn{6}{c}{\mathbf{I}_T} \\
0 & 0 & 0 & \dots & 1 & -1 & & & & & & \\ \hline
 & & & & & & 1 & -1 & 0 & \dots & 0 & 0 \\
 & & & & & & 0 & 1 & -1 & \dots & 0 & 0 \\
\multicolumn{6}{c|}{\mathbf{0}_{(T-1)\times(T+1)}} & & \vdots & & \ddots & & \vdots \\
 & & & & & & 0 & 0 & 0 & \dots & 1 & -1
\end{array}\right]
\times
\left[\begin{array}{c} \mu_0 \\ \mu_1 \\ \vdots \\ \mu_T \\ \hline \theta_0 \\ \theta_1 \\ \vdots \\ \theta_{T-1} \end{array}\right]
+
\left[\begin{array}{c} \mathbf{n} \\ \hline \mathbf{w}_1 \\ \hline \mathbf{w}_2 \end{array}\right].
\tag{6.5}
$$

With a re-parameterization, this can be rewritten as a mixed linear model as in Section 2.2.2.4 and Cui et al. (2010). First, re-parameterize the θ_t as follows:

$$
\begin{aligned}
\theta_1 &= \theta_0 + w_{2,1} \\
\theta_2 &= \theta_1 + w_{2,2} = \theta_0 + \sum_{i=1}^{2} w_{2,i} \\
\theta_3 &= \theta_2 + w_{2,3} = \theta_0 + \sum_{i=1}^{3} w_{2,i} \\
&\vdots \\
\theta_t &= \theta_0 + \sum_{i=1}^{t} w_{2,i}.
\end{aligned}
\tag{6.6}
$$

Now do a similar re-parameterization for μ_t using (6.6):

$$
\begin{aligned}
\mu_1 &= \mu_0 + w_{1,1} + \theta_0 \\
\mu_2 &= \mu_1 + w_{1,2} + \theta_1 = \mu_0 + \sum_{i=1}^{2} w_{1,i} + 2\theta_0 + w_{2,1} \\
\mu_3 &= \mu_2 + w_{1,3} + \theta_2 = \mu_0 + \sum_{i=1}^{3} w_{1,i} + 3\theta_0 + 2w_{2,1} + w_{2,2} \\
&\vdots \\
\mu_t &= \mu_0 + \sum_{i=1}^{t} w_{1,i} + t\theta_0 + \sum_{i=1}^{t-1} (t-i) w_{2,i}.
\end{aligned}
\tag{6.7}
$$

Interpreting equations (6.6) and (6.7) in mixed-linear-model terms, the fixed effects are $\beta = (\mu_0, \theta_0)'$ and the random effects are $\mathbf{w}_1$ and $\mathbf{w}_2$; this is how the linear growth model was written in equations (2.38) and (2.39) in Section 2.1. The prior distribution on (μ_0, θ_0) in the DLM literature is just an informative prior on the fixed effects implicit in the DLM. Thus for smoothing purposes, the informative prior is not necessary, though it may improve the behavior of numerical algorithms by conditioning the likelihood's tails.

Section 2.1 used the mixed-linear-model formulation to fit the linear growth model to the global-mean surface temperature data. There we noted that the combined design matrix $[\mathbf{X}|\mathbf{Z}]$ is exactly collinear: If $\mathbf{Z}_1$ and $\mathbf{Z}_2$ are the random-effect design matrices corresponding to $\mathbf{w}_1$ and $\mathbf{w}_2$ respectively in equations (2.38) and (2.39), then $\mathbf{Z}_2 = \mathbf{Z}_1\mathbf{H}$ for a particular matrix $\mathbf{H}$. (Derivation of $\mathbf{H}$ is an exercise.) Another way to interpret this is that if σ_1^2 becomes large, the μ_t are a saturated model, while if σ_2^2 becomes large, the θ_t are a saturated model, so in that sense the entire model is doubly saturated. Thus, users often set $\sigma_1^2 = 0$, which makes this model equivalent to a penalized spline with a truncated-line basis and a knot at each time point. (The proof is an exercise.)

This radical collinearity or double saturation appears — note the careful wording — to have a variety of important consequences. For example, Section 2.1's Bayesian analyses used flat priors on the DF in the local level and local trend but constrained them to have no more than K DF, for small values of K. The reason was that a flat prior on DF with no constraint gives a fairly flat posterior on DF, so that the posterior expectation of the fit is quite wiggly, with about 45 DF out of a maximum of 125. In other words, the posterior is not much different from the prior, i.e., the data provide little information about the ratios $(\sigma_e^2/\sigma_1^2, \sigma_e^2/\sigma_2^2)$ that determine smoothing. This is also an example in which the approximate DF, given in Section 2.2.2.3, performs poorly.

The example in Section 6.3 below suggests DLMs have this much collinearity in considerable generality. West & Harrison (1997) avoid this difficulty in many if not all cases by specifying values for variance ratios like σ_e^2/σ_1^2, which amounts to a point (degenerate) prior on the variance ratio or DF. This may be the only feasible choice when DLMs are used to forecast from short series; priors on the DF in each component of a DLM can be viewed as specifying these same variance ratios but allowing the data to have some influence. Section 6.3 below and Chapters 9 and 12 show some strange things that DLM fits can do. Chapter 12 argues that these arise at least in part from the radical collinearity of these models, but at present this is poorly understood.

6.2 Dynamic Linear Models in Some Generality

A DLM for an r-dimensional outcome $\mathbf{y}_t$ can be written with observation equation

$$\mathbf{y}_t = \mathbf{F}_t\theta_t + \mathbf{n}_t, \quad \mathbf{n}_t \sim N_r(0, \Sigma_t^n), \tag{6.8}$$

where $\mathbf{F}_t$ is $r \times p$, θ is $p \times 1$, $\mathbf{n}_t$ is $r \times 1$, and Σ_t^n is $r \times r$. $\mathbf{F}_t$ is known, θ and $\mathbf{n}_t$ are treated as unknown, and Σ_t^n can be treated as known or unknown or modeled with a few unknowns. The state equation describing the evolution of the state θ_t is

$$\theta_t = \mathbf{H}_t \theta_{t-1} + \mathbf{w}_t, \quad \mathbf{w}_t \sim N_p(0, \Sigma_t^w), \tag{6.9}$$

where $\mathbf{H}_t$ and Σ_t^w are $p \times p$; $\mathbf{H}_t$ is known and again, Σ_t^w can be treated as known or unknown or modeled with a few unknowns. The DLM literature usually adds a fully specified p-variate normal prior distribution for the initial state θ_0.

This framework defines a huge class of models with covariate effects, intervention effects, flexible cyclic and quasi-cyclic effects, and so on. The linear growth model for the global mean surface temperature data had $r = 1$, $p = 2$, $\theta_t = (\mu_t, \theta_t)'$, $\mathbf{F}_t = [1\ 0]$,

$$\mathbf{H}_t = \begin{bmatrix} 1 & 1 \\ 0 & 1 \end{bmatrix}, \Sigma_t^n = \sigma_n^2, \text{ and } \Sigma_t^w = \begin{bmatrix} \sigma_1^2 & 0 \\ 0 & \sigma_2^2 \end{bmatrix}. \tag{6.10}$$

As for the linear growth model, the more general DLM is expressed easily in the alternative formulation. Duncan & Horn (1972) used such an expression to prove Gauss-Markov type theorems about DLMs. To write the DLM in this form, re-write (6.9) as

$$\mathbf{0}_p = -\theta_t + \mathbf{H}_t \theta_{t-1} + \mathbf{w}_t. \tag{6.11}$$

Then the DLM can be rewritten in constraint-case form as

$$\left[\begin{array}{c} \mathbf{y}_1 \\ \mathbf{y}_2 \\ \vdots \\ \mathbf{y}_T \\ \hline \mathbf{0}_p \\ \mathbf{0}_p \\ \vdots \\ \mathbf{0}_p \end{array}\right] = \left[\begin{array}{cccccc} \mathbf{0}_{r\times p} & \mathbf{F}_1 & \mathbf{0}_{r\times p} & \dots & \mathbf{0}_{r\times p} & \mathbf{0}_{r\times p} \\ \mathbf{0}_{r\times p} & \mathbf{0}_{r\times p} & \mathbf{F}_2 & \dots & \mathbf{0}_{r\times p} & \mathbf{0}_{r\times p} \\ & \vdots & & \ddots & & \vdots \\ \mathbf{0}_{r\times p} & \mathbf{0}_{r\times p} & \mathbf{0}_{r\times p} & \dots & \mathbf{0}_{r\times p} & \mathbf{F}_T \\ \hline \mathbf{H}_1 & -\mathbf{I}_p & \mathbf{0}_{r\times p} & \dots & \mathbf{0}_{r\times p} & \mathbf{0}_{r\times p} \\ \mathbf{0}_{r\times p} & \mathbf{H}_2 & -\mathbf{I}_p & \dots & \mathbf{0}_{r\times p} & \mathbf{0}_{r\times p} \\ & \vdots & & \ddots & & \vdots \\ \mathbf{0}_{r\times p} & \mathbf{0}_{r\times p} & \mathbf{0}_{r\times p} & \dots & \mathbf{H}_T & -\mathbf{I}_p \end{array}\right] \begin{bmatrix} \theta_0 \\ \theta_1 \\ \vdots \\ \theta_T \end{bmatrix} + \left[\begin{array}{c} \mathbf{n}_1 \\ \vdots \\ \mathbf{n}_T \\ \hline \mathbf{w}_1 \\ \vdots \\ \mathbf{w}_T \end{array}\right], \tag{6.12}$$

with $\text{cov}\left([\mathbf{n}_1, \dots, \mathbf{n}_T, \mathbf{w}_1, \dots, \mathbf{w}_T]'\right) = \text{blockdiag}\,[\Sigma_1^n, \dots, \Sigma_T^n, \Sigma_1^w, \dots, \Sigma_T^w]$.

This DLM can be re-written as a mixed linear model with the same trick used for the linear growth model. First, re-parameterize the θ_t as follows, using the state equation (6.9):

$$\begin{aligned} \theta_1 &= \mathbf{H}_1 \theta_0 + \mathbf{w}_1 \\ \theta_2 &= \mathbf{H}_2 \theta_1 + \mathbf{w}_2 \\ &= \mathbf{H}_2 \mathbf{H}_1 \theta_0 + \mathbf{H}_2 \mathbf{w}_1 + \mathbf{w}_2 \\ &\vdots \\ \theta_t &= \mathbf{H}_t \dots \mathbf{H}_2 \mathbf{H}_1 \theta_0 + \sum_{i=1}^{t-1} (\mathbf{H}_t \dots \mathbf{H}_{i+1}) \mathbf{w}_i + \mathbf{w}_t. \end{aligned} \tag{6.13}$$

Substituting this into the observation equation (6.8) gives

$$
\begin{aligned}
\mathbf{y}_t &= \mathbf{F}_t \theta_t + \mathbf{n}_t \\
&= \mathbf{F}_t \mathbf{H}_t \dots \mathbf{H}_2 \mathbf{H}_1 \theta_0 && (6.14) \\
&\quad + \sum_{i=1}^{t-1} \mathbf{F}_t \mathbf{H}_t \dots \mathbf{H}_{i+1} \mathbf{w}_i && (6.15) \\
&\quad + \mathbf{F}_t \mathbf{w}_t + \mathbf{n}_t. && (6.16)
\end{aligned}
$$

Thus in mixed-linear-model terms, the fixed effect is θ_0 and the random effects are the $\mathbf{w}_t$. The fixed-effect design matrix **X** would be constructed using the row above labeled (6.14), while the random-effects design matrix **Z** would be constructed using the rows labeled (6.15) and (6.16). I won't do so because the result would put you to sleep and provide no intuition. However, once again the initial state θ_0 is seen to be the fixed effect implicit in the DLM, so an informative prior on the initial state is an informative prior on the mixed linear model's fixed effects. As noted earlier, this informative prior is not necessary, though it may be helpful as a conditioning device.

Are the random effects in DLMs old-style or new-style? For a DLM used to smooth a single series, it is hard to see how its random effects differ materially from those in the mixed-linear-model representation of a penalized spline, which are plainly new-style. However, if a DLM is used to smooth several series simultaneously, the random effects for the individual series could be either old- or new-style depending on whether the smoothed fits for the individual series are of interest. Chapter 13 discusses this in more detail.

6.3 Example of a Multi-component DLM

Example 8. Localizing epileptic activity. Surgical treatments for intractable epilepsy require identifying or localizing the part of the brain responsible for the epilepsy and then removing it. Current methods for localizing the affected part of the brain are unpleasant for the patient, cumbersome, costly, and error prone. An experimental alternative uses the fact that when neurons become active, blood flow increases in nearby vessels, which affects the amount of light absorbed and scattered at certain wavelengths. Thus, if the exposed brain is optically imaged while a suitable stimulus is applied, it may be possible to use the resulting optical data to identify the affected area. Lavine et al. (2011) describe a method that uses a DLM to process optical imaging data collected during direct electrical stimulation of the brain, to extract a particular kind of signal. Data were collected from the exposed cerebral cortexes of human subjects undergoing surgical treatment for medically intractable epilepsy. Here and in Chapters 9 and 12 we describe analyses reported in an unpublished preliminary version of Lavine et al. (2011). The data series used here and 11 others are available on the book's web site, along with the preliminary version of Lavine et al. (2011).

For a specific region of one patient's brain, 650 measurements ($T = 649$) were made at intervals (time steps) of 0.28 seconds. The scalar outcome y_t is the percent change in average pixel value in that region relative to the average pixel value at time

0, for light of wavelength 535 nm. The observations y_t are plotted in Figure 6.1a. The stimulus was applied during time steps $t = 75$ to 94 and a response was expected after time step 75. Extraction of the response from the data was complicated by quasi-cyclic artifacts arising from heartbeat and respiration, with periods of about 2–4 and 15–25 time steps respectively, and by unsystematic error.

To filter out the cyclic artifacts and unsystematic error and give a smoothed estimate of the response, Lavine and his colleagues used a DLM with observation equation

$$y_t = s_t + h_t + r_t + v_t, \tag{6.17}$$

where s_t was the smoothed response, the object of this analysis, while h_t and r_t were components for heartbeat and respiration respectively and v_t was unsystematic error, distributed as iid $N(0, W_v)$.

The state equation of the DLM had distinct parts for each of the three components s_t, h_t, and r_t. The response s_t had state equation

$$\begin{pmatrix} s_t \\ \text{slope}_t \end{pmatrix} = \begin{bmatrix} 1 & 1 \\ 0 & 1 \end{bmatrix} \begin{pmatrix} s_{t-1} \\ \text{slope}_{t-1} \end{pmatrix} + \mathbf{w}_{s,t}, \tag{6.18}$$

where $\mathbf{w}'_{s,t} = (0, w_{slope,t})$ and $w_{slope,t} \sim$ iid $N(0, W_s)$. This is the linear growth model with the evolution variance of the local level s_t set to zero, allowing only the local slope to vary.

The state equations for the two quasi-cyclic components are structured identically, differing only in their period (wavelength), so I'll develop the state equation for the heartbeat artifact. The general form of a quasi-cyclic term's state equation, in this case heartbeat, is

$$\begin{pmatrix} b_t \cos \alpha_t \\ b_t \sin \alpha_t \end{pmatrix} = \begin{bmatrix} \cos \delta & \sin \delta \\ -\sin \delta & \cos \delta \end{bmatrix} \begin{pmatrix} b_{t-1} \cos \alpha_{t-1} \\ b_{t-1} \sin \alpha_{t-1} \end{pmatrix} + \mathbf{w}_{h,t}, \tag{6.19}$$

where $\mathbf{w}'_{h,t} = (w_{h1,t}, w_{h2,t}) \sim$ iid $N_2(0, \mathbf{W}_h)$. In general, $\mathbf{W}_h$ can be a 2×2 covariance matrix; Lavine et al. used $\mathbf{W}_h = W_h \mathbf{I}_2$. The systematic part of this state equation, the matrix $\begin{bmatrix} \cos \delta & \sin \delta \\ -\sin \delta & \cos \delta \end{bmatrix}$ rotates a 2-vector through an angle of δ radians, where $\delta = 2\pi/\text{period}$, "period" being the period of the cyclic term. The state that enters the observation equation, h_t, is $h_t = b_t^h \cos \alpha_t^h$. For heartbeat, Lavine et al. (2011) used a wavelength of 2.78 time steps chosen by inspection, giving $\delta_h = 2\pi/2.78$. If the evolution variance W_h is zero, this equation describes movement around a circle of radius $b_t \equiv b$ with steps of angle δ. With a positive evolution variance W_h, $b_t \neq b_{t-1}$ and $\alpha_t \neq \alpha_{t-1} + \delta$, so this state equation can accomodate functions that are not sinusoidal with constant amplitude and period, while shrinking/smoothing the fit toward such a function. (As Lavine et al. 2011 noted, because quasi-cyclic DLM components can adjust for inexact specification of the period, it is not critical to treat the period as unknown and specifying it directly gives a much simpler analysis.)

The state equation for respiration has the same form as above but with period 18.75 time steps, again chosen by inspection. Thus $\delta_r = 2\pi/18.75$; the state that enters the observation equation is $r_t = b_t^r \cos\alpha_t^r$. Again, Lavine et al. modeled the evolution variation $\mathbf{w}'_{r,t} = (w_{r1,t}, w_{r2,t}) \sim$ iid $N_2(0, W_r\mathbf{I}_2)$.

In the notation of Section 6.2, the state θ is $\theta' = [s_t, \text{slope}_t, b_t^h \cos\alpha_t^h, b_t^h \sin\alpha_t^h, b_t^r \cos\alpha_t^r, b_t^r \sin\alpha_t^r]$, with pairs of elements for s_t, h_t, and r_t respectively. $\mathbf{F}_t = \begin{bmatrix} 1 & 0 & 1 & 0 & 1 & 0 \end{bmatrix}$,

$$\mathbf{H}_t = \left[\begin{array}{cc|cc|cc} 1 & 1 & \multicolumn{2}{c|}{\mathbf{0}_{2\times 2}} & \multicolumn{2}{c}{\mathbf{0}_{2\times 2}} \\ 0 & 1 & & & & \\ \hline \multicolumn{2}{c|}{\mathbf{0}_{2\times 2}} & \cos\delta_h & \sin\delta_h & \multicolumn{2}{c}{\mathbf{0}_{2\times 2}} \\ & & -\sin\delta_h & \cos\delta_h & & \\ \hline \multicolumn{2}{c|}{\mathbf{0}_{2\times 2}} & \multicolumn{2}{c|}{\mathbf{0}_{2\times 2}} & \cos\delta_r & \sin\delta_r \\ & & & & -\sin\delta_r & \cos\delta_r \end{array}\right], \tag{6.20}$$

and $\text{cov}(\mathbf{w}_t) = \text{diag}(0, W_s, W_h, W_h, W_r, W_r)$. (Re-expressing a quasi-cyclic component in a mixed-linear-model formulation is given as an exercise.)

For exploratory analyses, the four variances were estimated by maximum likelihood using the R package dlm (Petris 2010). (Note: *Not* the restricted likelihood. For DLMs, maximum likelihood is the conventional analysis, presumably because it is easier to compute.) The model was parameterized using the logs of the variances, starting values for the maximization were the defaults (zero), and the prior for θ_0 was the default normal with mean 0 and variance 10^7. The resulting estimated variances were $\hat{W}_v = 9.56 \times 10^{-8}$, $\hat{W}_s = 0.00158$, $\hat{W}_h = 0.0176$, and $\hat{W}_r = 0.234$. These estimates were then treated as if they were known to be true, allowing posterior distributions to be computed easily for s_t, h_t, and r_t. Figure 6.1 shows the data and the fit. Figure 6.1a shows the fitted $s_t + h_t + r_t$, the jagged line — which fits the data so closely that the data are not shown separately — and the fitted response s_t, the smooth line. The smoothed response does seem to be a smoothed version of the data, and dips at about the time the stimulus was applied, time steps 75 to 94. Figure 6.1b shows the fitted local slope of the model's response component, slope_t, with a 90% pointwise posterior interval. The slope is clearly negative around the time the stimulus was applied, but it makes similar-sized changes elsewhere in rough cycle with a period of about 117 time steps (about 30 seconds). Fitted heartbeat and respiration are, as expected, cyclic but with much variation in the amplitude, period, and shape of the cycle. The DF in the fit were 32.7, 180.9, and 436.4 DF for response, heartbeat, and respiration respectively, with 2 DF in each component for fixed effects and the rest for random effects. The large number of DF in the fitted respiration component reflects its substantial deviation from a sinusoidal shape, visible in Figure 6.1d. (In later models, Lavine et al. added a second harmonic to this component.) The three-component model, as estimated by maximum likelihood, fit the data almost exactly, with 0.0003 DF for error.

The long-period quasi-cycle visible in Figure 6.1a,b had no ready explanation and motivated Lavine et al. to add another quasi-cyclic component to model (6.17) with a period of 117. When they did so, they got a big surprise, which is explored in Chapters 9 and 12.

Model (6.17) has a striking feature that may prove problematic. This model, like the linear growth model, is highly overparameterized. In the linear growth model,

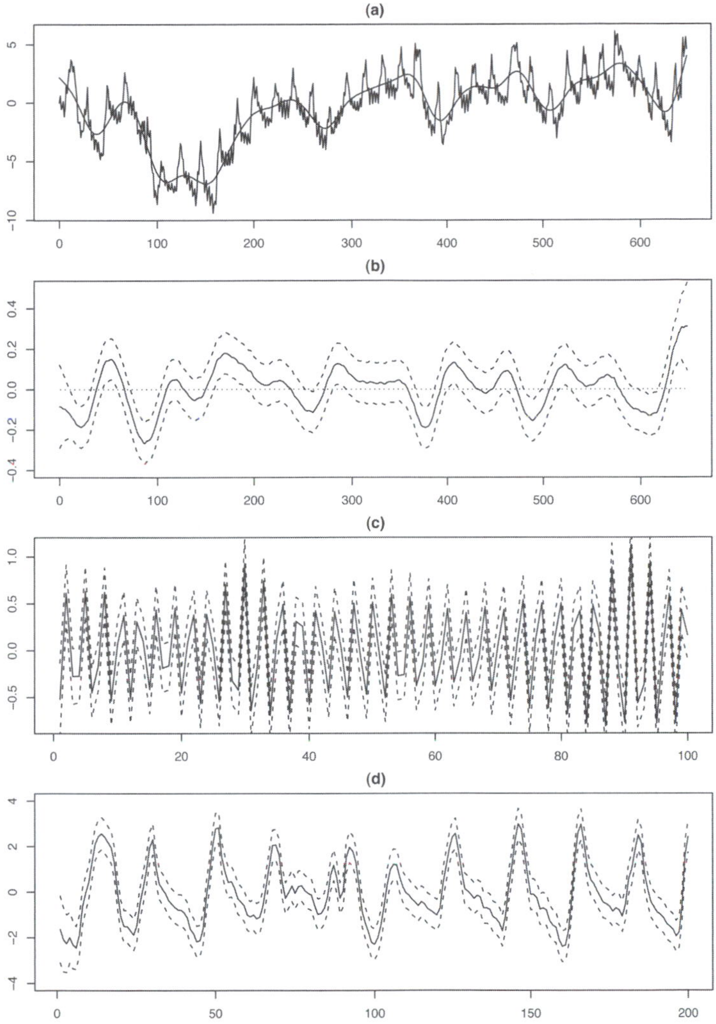

Figure 6.1: DLM fit to the optical-imaging data. Panel (a): Data and full fit (jagged line) and the smoothed response s_t (smooth line). Panel (b): Fitted local slope, slope$_t$. Panel (c): Fitted heartbeat component, first 100 time steps. Panel (d): Fitted respiration component, first 200 time steps. Panels b, c, and d show pointwise 90% posterior regions, treating the variance estimates as known.

both the local level and local slope had design matrices that defined saturated models, so the model was doubly saturated in that sense and the two effects were identified only because of substantial smoothing, i.e., small evolution variances $w_{1,t}$ and $w_{2,t}$. For model (6.17), the response component s_t is only singly saturated because the local level's evolution variance was fixed at zero. However, each of the heartbeat and respiration components is doubly saturated so that as a whole, model (6.17) is saturated five times over and the three components compete with each other for the 650 DF available in the data. If any of the three evolution variances W_s, W_h, or W_r is allowed to grow large enough — whatever "enough" might mean — the other two components are necessarily either squeezed out of the fit or not identifiable. For the fit in Figure 6.1, which gave 436 DF out of 650 to the heartbeat component, there is reason to be concerned, although the overall fit and its components seem reasonable enough.

Further exploration by Lavine et al. found two strange things that appear to arise from the extensive collinearity. When Lavine et al. added a quasi-cyclic term of period 117 to capture the mysterious long-cycle variation visible in Figure 6.1a,b, the model's fit changed radically. It is not clear why this happened; Chapters 9 and 12 explore some possible explanations. Lavine also re-fit the original model (6.17) using 500 different starting values and got 500 different "maxima" for the likelihood. It is hard to visualize a likelihood surface for the four unknown log variances; an array of small plots, each showing a bivariate contour plot, suggests this likelihood surface has a weird shape with flat stretches and multiple maxima. A log variance less than (say) -8 may be effectively zero, so the likelihood is essentially flat for small log variances; this accounts for some of the variation between Lavine's 500 different "maxima." However, Lavine found that the 500 maxima fell into three clusters such that fits within a cluster were similar, but different clusters had quite different fits. One cluster fit the data almost exactly, as in Figure 6.1a, while the other two clusters had far smoother fits. Chapter 12 shows similar behavior for these models and artificial data. Based on the modest evidence in Chapter 12, it appears that the fit really did change a lot when the new component was added, i.e., it is not the case that the information about the original four variances W_s, W_h, W_r, and W_v was similar in the likelihoods from the two models but the numerical maximization routine simply found different maxima in fitting the two models.

Exercises

Regular Exercises

1. (Section 6.1) For the linear growth model, if $\mathbf{Z}_1$ and $\mathbf{Z}_2$ are the random-effect design matrices corresponding to $\mathbf{w}_1$ and $\mathbf{w}_2$ respectively, then derive the matrix $\mathbf{H}$ such that $\mathbf{Z}_2 = \mathbf{Z}_1\mathbf{H}$.
2. (Section 6.1) For the linear growth model, show that if you set $\sigma_1^2 = 0$, this model is equivalent to a penalized spline with the truncated-line basis and a knot at each time point.
3. (Section 6.3) Re-express one of the quasi-cyclic components of this model in a mixed-linear-model formulation, identifying the fixed and random effects corresponding to this component. The results turn out to be highly structured.

Open Questions

1. (All sections) Are DLMs in general subject to the exact collinearity that was present in the examples in Sections 6.1 and 6.3? Or can you specify conditions identifying DLMs that have this collinearity and DLMs that do not?

Chapter 7

Two Other Syntaxes for Richly Parameterized Models

This book uses mixed linear models as a syntax for richly parameterized linear models. At least two other systems have been proposed, one based on Gaussian Markov random fields, proposed by Rue & Held (2005), and another proposed by Lee et al. (2006), which might be viewed as extending mixed linear models although the analytic approach is somewhat different. This chapter gives a brief overview of these two syntaxes and their associated analytic approaches, to contrast them with mixed linear models and to give interested readers a place to begin. Each alternative approach is treated briefly, so it may seem that I've given undue emphasis to their controversial aspects. However, this is inevitable in a brief overview and disgruntled readers might recall that I've been more critical of mixed-linear-model theory and methods than any other treatment I've ever seen. Section 7.1 gives a schematic comparison of the three competing syntaxes; this is followed by a section for each alternative syntax.

7.1 Schematic Comparison of the Syntaxes

Main syntax: Linear models with random effects (mixed linear models).

- *Key idea*: Express a large class of models as mixed linear models through the choice of fixed effects $\mathbf{X}\beta$, random effects $\mathbf{Zu}$, and covariance matrices $\mathbf{G}$ and $\mathbf{R}$. This extends fairly readily to error distributions in the exponential family (e.g., Ruppert et al. 2003, Chapters 11, 14).
- *Key tools*:
  - Mixed linear model theory, methods, and computing, and ideas adapted from ordinary linear models (i.e., with one error term and no random effects).
  - The conventional analysis is unabashedly *ad hoc*, using the restricted likelihood and relying on large-sample approximations and bootstrapping. Generalized linear model extensions rely on further approximations.
  - Bayesian methods rely on Markov chain Monte Carlo (MCMC) and are exact to that extent.

Alternative #1: Gaussian Markov random fields (Rue & Held 2005).

- *Key idea*: Represent components of models and priors as Gaussian Markov random fields (GMRFs), taking advantage of conditional dependence and independence.
- *Key tools*:
  - Model the mean structure in a modular fashion, with components being GMRFs or simple effects without further structure (e.g., regressors or fixed effects). Rue & Held (2005) covers Gaussian distributions but later papers extend their computational approach to exponential family error distributions, among other things.
  - Rue, Held, and their collaborators propose Bayesian analyses only, while acknowledging that non-Bayesian analyses are possible for at least many models expressible in their syntax.
  - Their "exact" analyses use MCMC that is relatively fast because it exploits a sparse-matrix representation of precision matrices.
  - Their approximate analyses use the integrated nested Laplace approximation (INLA), which is much faster than their off-the-shelf MCMC analysis.
  - Most models discussed in Part II of the present book, along with many others, can be represented in this syntax. For the exceptions, e.g., penalized splines, at worst closely analogous models exist that *can* be represented in this syntax.

Alternative #2: Likelihood inference for models with unobservables (Lee et al. 2006).

- *Key ideas*: Extend generalized linear models in several directions, with a non-Bayesian analytic approach using likelihood-like functions.
- *Key tools*:
  - Modular modeling of the observation error distribution (within the exponential family), the linear predictor (i.e., observation mean structure), the dispersion of the error distribution, and the dispersion of the random effects.
  - Besides random effects, this class of models includes other types of random variables that are not observable, such as missing data or predictions of future observations.
  - Analysis is by maximization of functions derived or extended from the traditional likelihood, using theory extended from traditional likelihood theory. Point estimates are maxima of these functions, and uncertainty is described using curvatures at these maxima.

All three syntaxes have modest (at best) toolkits for understanding fits. In that sense, each approach represents about half of a theory of richly parameterized models.

7.2 Gaussian Markov Random Fields (Rue & Held 2005)

The key idea is that in the Gaussian (normal) density for a vector $\mathbf{x} = (x_1, \ldots, x_n)'$, the precision matrix (inverse of the covariance matrix) directly expresses conditional dependence and independence relations among the x_i. Specifically, if $\mathbf{x}$ is distributed as Normal with mean μ and precision matrix $\mathbf{P}$ — which is not necessarily invertible — then

$$\text{covariance}(x_i, x_j | \mathbf{x}_{(-ij)}) = 0 \text{ if and only if } P_{ij} = 0, \tag{7.1}$$

where $\mathbf{x}_{(-ij)}$ is the vector $\mathbf{x}$ without its ith and j^{th} elements and P_{ij} is the $(i,j)^{\text{th}}$ element of $\mathbf{P}$. More generally, as noted in Chapter 5,

$$\begin{aligned} \text{correlation}(x_i, x_j | x_{(-ij)}) &= -P_{ij}/\sqrt{P_{ii}P_{jj}} \text{ and} \\ \text{precision}(x_i | \mathbf{x}_{(-i)}) &= P_{ii}. \end{aligned} \tag{7.2}$$

Proofs are in Rue & Held (2005), Section 2.2. If many entries in $\mathbf{x}$'s precision matrix $\mathbf{P}$ are 0 then $\mathbf{P}$ is called "sparse" and computing for models with Gaussian (normal) likelihoods or latent variables can be sped up considerably by exploiting this sparsity.

Many familiar models have sparse precision matrices $\mathbf{P}$, especially when the vector $\mathbf{x}$ includes the outcome $\mathbf{y}$ as well as unknown parameters. Here are some examples.

One-way random effects model. Write this model as $y_{ij} = \theta_i + \varepsilon_{ij}, \theta_i = \mu + \delta_i$, with ε_{ij} and δ_i having normal distributions with finite variances, all mutually independent. Then

$$\begin{aligned} \text{cov}(y_{ij}, \theta_{i'} | \theta_i) &= 0 \text{ for } i \neq i' \\ \text{cov}(y_{ij}, \mu | \theta_i) &= 0 \\ \text{cov}(\theta_{i'}, \theta_i | \mu) &= 0. \end{aligned} \tag{7.3}$$

If $\mathbf{x}$ includes $\mathbf{y}$, θ, and μ, $\mathbf{x}$'s precision matrix $\mathbf{P}$ has many zero elements and is thus sparse.

Intrinsic (or improper) CAR model (ICAR). In the simplest version of this model with no regressors, $y_i = \delta_i + \varepsilon_i$ for areas $i = 1, \ldots, n$ with the ε_i distributed as normal with mean zero and finite variance independently of each other and of $\delta = (\delta_1, \ldots, \delta_n)'$, which has an ICAR distribution with precision matrix $\mathbf{Q}/\sigma_s^2$. Then

$$\begin{aligned} \text{cov}(y_i, \delta_j | \delta_i) &= 0 \text{ for } i \neq j \\ \text{cov}(\delta_j, \delta_i | \delta_{(-ij)}) &= 0 \text{ if } i \text{ and } j \text{ are not neighbors} \\ &= -1/\sigma_s^2 \text{ if } i \text{ and } j \text{ are neighbors,} \end{aligned}$$

recalling that $\mathbf{Q}/\sigma_s^2$ has diagonal elements m_i/σ_s^2, where m_i is the number of area i's neighbors, and off-diagonals $Q_{ij} = -1/\sigma_s^2$ if i and j are neighbors and 0 otherwise. Actual neighbor pairs are generally a small minority of the potential neighbor pairs, so $\mathbf{Q}$ is sparse. Thus if $\mathbf{x}$ includes $\mathbf{y}$ and δ, this model has a sparse precision matrix $\mathbf{P}$.

Autoregressive model of order 1, AR(1). Suppose $x_t = \phi x_{t-1} + \varepsilon_t$ with $|\phi| < 1$ and the ε_t having independent standard normal distributions. Now $x_t | x_1, \ldots, x_{t-1} \sim N(\phi x_{t-1}, 1)$, so $x_t | x_{t-1}$ is conditionally independent of $x_1, \ldots, x_{t-2}$ and it is easy to show that $x_t | x_{t-1}, x_{t+1}$ is independent of $x_{t'}$ for $t' \notin \{t-1, t, t+1\}$. If the marginal distribution of x_1 is normal with mean zero and variance $1/(1-\phi)$, then $\mathbf{x}$ is a GMRF with precision matrix

$$\begin{bmatrix} 1 & -\phi & 0 & 0 & & 0 & 0 & 0 \\ -\phi & 1+\phi^2 & -\phi & 0 & \dots & 0 & 0 & 0 \\ 0 & -\phi & 1+\phi^2 & -\phi & & 0 & 0 & 0 \\ & \vdots & & & \ddots & & \vdots & \\ 0 & 0 & 0 & 0 & & -\phi & 1+\phi^2 & -\phi \\ 0 & 0 & 0 & 0 & \dots & 0 & -\phi & 1 \end{bmatrix}. \tag{7.4}$$

This precision matrix is again mostly zeroes. The proof is an exercise.

Dynamic linear models. For dynamic linear models as defined in Section 6.2, the observation equation is $y_t = \mathbf{F}_t\theta_t + \mathbf{n}_t$ where $\mathbf{n}_t \sim N_r(0, \Sigma_t^n)$ independently of each other. The state equation is $\theta_t = \mathbf{H}_t\theta_{t-1} + \mathbf{w}_t$ where $\mathbf{w}_t \sim N_p(0, \Sigma_t^w)$ independently of each other and of the $\mathbf{n}_t$. This too gives a sparse precision matrix for the data y_t and the unknown states θ_t because

$$\begin{aligned} \text{cov}(y_t, \theta_{t'}|\theta_t) &= 0 \text{ if } t \neq t', \\ \text{cov}(\theta_t, \theta_{t'}|\theta_{t-1}) &= 0 \text{ if } t' < t-1, \text{ and} \\ \text{cov}(\theta_t, \theta_{t'}|\theta_{t-1}, \theta_{t+1}) &= 0 \text{ if } t' \notin \{t-1, t, t+1\}. \end{aligned} \tag{7.5}$$

In this approach, modeling involves adding together components for different features of the data, as in Section 6.3's optical-imaging example. In the precision matrix for the combined vector of outcomes $\mathbf{y}$ and unknowns, including fixed effects and each component's unknown state vector, the components are unconditionally independent of each other and of the fixed effects. This creates still more sparsity in the joint precision matrix of the outcomes and unknowns.

A quite large class of models can be represented in this syntax. A simple random effect, as noted, is just a GMRF with a diagonal precision matrix. Among time series models, state-space models are GMRFs; ARMA models can be expressed as state-space models and thus as GMRFs. Longitudinal analyses typically involve either random effects or a time-series model for the errors and can thus be expressed as GMRFs. Models written in graphical form, where edges between nodes represent conditional dependence relations, are easily represented as MRFs. Rue et al. (2009, Section 5) describe other less familiar models that can be represented as MRFs, such as stochastic volatility models. From my reading of Chapter 3 of Rue & Held (2005) it does not appear that penalized splines *per se* can be written as GMRFs. Instead, Rue & Held (2005) develop GMRF models for observations on regular and irregular lattices using differences and the Weiner process respectively. These models have various connections to splines, though not necessarily penalized splines that can be written as mixed linear models (see the bibliographic notes in their Section 3.6).

Geostatistical models are accommodated in this system but perhaps not as seamlessly as most of the preceding models. The most popular geostatistical models — including the Ornstein-Uhlenbeck process on the real line and Gaussian processes with

Matérn covariance matrices for which $\nu + d/2$ is an integer, where ν is the Matérn's smoothness parameter and d is the spatial dimension — do have a Markovian property. However, in general the Markov property is lost for $d \geq 2$ when considering the joint distribution of the process at n locations. As a result, the joint distribution of observations or latent variables at n specific locations has a dense precision matrix; see Simpson et al. (2012) for a thorough discussion. Lindgren et al. (2011) show the Markov property can be recovered by viewing the continuously indexed process as the solution of a stochastic partial differential equation (SPDE), and using the SPDE to construct a "projection" to a finite dimensional representation, where the joint distribution for the n weights has a sparse precision matrix. This representation can then be incorporated into Rue et al.'s system for the purpose of computing a posterior or posterior predictive distribution. Gaussian process mavens that I have consulted[1] accept the foregoing but note that a Gaussian process model as represented in Rue et al.'s system is not identical to the original Gaussian process model, which has realizations in real d-dimensional space while Rue et al.'s representation is an interpolated surface represented by n weights in d-dimensional space. In particular, I am advised that it is not clear this approach retains the smoothness properties of the original process, and Gaussian process mavens consider these important for prediction and for inferences about gradients of the realized latent surface, among other things. Rue has said "It is safe to say that there are issues here that are not yet fully understood" (personal communication), though this does not diminish the interest of this ingenious scheme.

Thus I believe it's true that any model that can be represented as a mixed linear model has at least an analogous model that can be represented in the syntax of Rue & Held, though in some cases these are not identical to the mixed-linear-model version.

As noted in the schematic overview, Rue & Held (2005) propose only Bayesian analyses and make no claims about bias, mean-squared error, or confidence interval coverage. Computing for "exact" analyses uses MCMC. Because the necessary conditional distributions have sparse precision matrices, the vector of unknowns can be re-ordered so the resulting precision matrix $\mathbf{P}$ is banded, i.e., all the non-zero elements are close to the diagonal. Special computational routines take advantage of this banding to speed up MCMC calculations considerably.

These MCMC routines can still be quite slow — as fast as computers become, problems always grow to make our algorithms too slow — so Rue et al. (2009) developed an off-the-shelf approximate method, integrated nested Laplace approximations (INLA), for models with an exponential-family error distribution and a Gaussian random field for latent variables. INLA involves some iteration to find modes but not repeated sampling as in MCMC, and in many problems at least it is orders of magnitude faster than off-the-shelf MCMC routines. The discussion of Rue et al. (2009) points out possible weaknesses of this paper's argument for INLA: Whether MCMC routines are necessarily as slow as the authors claim; assessment of INLA's accuracy when long MCMC runs are not feasible; performance for multi-modal posterior

[1] This work is far outside my area of expertise, so I am relying on the judgment of expert colleagues who will remain unnamed so they cannot be blamed for my errors.

distributions; and MCMC's ability to provide draws from the posterior of an arbitrary function of a model's unknowns. Rue and co-authors acknowledged most of these difficulties in their response to the discussion. Still, this is an encouraging example of what can be accomplished by focusing on a large class of models defined by a few assumptions — in this case, exponential-family errors, a GMRF for latent variables, and a few unknowns controlling the latent variables' distribution — instead of addressing very general classes of models, as in BUGS and much of the MCMC literature. The point is not that BUGS is wrong-headed but rather that settling for less generality can sometimes produce very useful tools. It seems likely that INLA will be useful for exploratory modeling and, if the accuracy claims of Rue et al. (2009) can be justified in some generality, in Bayesian analyses much more broadly.

Rue, Held, and their collaborators have developed computing methods for some standard Bayesian model diagnostics like DIC and predictive measures (Rue et al. 2009, Section 6.3) but as far as I know they have not used their framework for exploring puzzles of the sort considered in the present book's Parts III and IV. I am not prepared to conjecture whether their framework can be useful for this purpose, but an optimistic outlook is reasonable.

7.3 Likelihood Inference for Models with Unobservables (Lee et al. 2006)

Lee et al. (2006) describe their syntax and analysis approach as growing out of generalized linear models (McCullagh & Nelder 1989). The key elements of the latter are

- The error distribution, from the one-parameter exponential family;
- The linear predictor, connected to the mean of the error distribution by the link function; and
- Analysis by maximum likelihood, using standard large-sample approximate theory and iterated reweighted least squares as the all-purpose computing method.

The system of Lee et al. (2006) builds on this with several extensions.

- Adding random effects to the linear predictor. When the random effects have a normal distribution, the result is a so-called generalized linear mixed model; when the random effects have a distribution that is conjugate to the error distribution, the result is a so-called hierarchical generalized linear model.
- Modeling the dispersion parameter of the error distribution. The dispersion parameter can have its own generalized linear model, estimated simultaneously with the model for the linear predictor; both the linear predictor and the dispersion parameter's model can include random effects.
- Models in the included class may have other types of "unobservables," by which Lee & Nelder (2009) mean random variables (in the non-Bayesian sense) that are unobservable. These include missing data and future observables that are being predicted.

- Analysis uses the so-called h-likelihood, on which various computations are performed so that a model with all the pieces described in the preceding items is analyzed as a series of linked generalized linear models.

Regarding model criticism, Lee et al. (2006) emphasize residuals and goodness-of-fit measures based on deviance. Their approach to residuals is discussed briefly in Chapter 8.

This modeling syntax may seem like a straightforward extension of generalized linear mixed models as described in, for example, Ruppert et al. (2003, Chapters 11 & 14). However, various commentators (e.g., discussants of Lee & Nelder 2009) agree that Lee, Nelder and their collaborators have introduced new models and some unification of existing models, as well as a unified analysis approach. Commentators mainly disagree with Lee and Nelder over the properties and value of that analytic approach.

The main claim to novelty, e.g., in Lee & Nelder (2009), is that this analytic approach is unified in much the same way as the conventional analysis of generalized linear models, in its model syntax, its theory of analysis, and its computing method. The syntax is uncontroversial and the computing method is relatively fast, involving maximization and computing curvatures at maxima without repeated sampling as in MCMC. The controversy arises from the theory of analysis.

Lee & Nelder assert that their theory of analysis is principled, based on probabilistic models and likelihood theory (the Likelihood Principle); that it "avoids prior probabilities" and is thus superior to a Bayesian approach; and that it solves all problems in analysis of "models with unobservables" apart from minor problems that, they claim, will be solved by higher-order approximations. (It is hard to write this without seeming unfair to Lee and Nelder, but the previous sentence's last segment is a fair summary of passages in Lee & Nelder 2009; see also Lee 2007.) Readers interested in this approach should consider subtle, erudite critiques such as Meng's (2009) discussion of Lee & Nelder (2009). Here, I will just make a few simple comments.

First, Lee & Nelder (2009) describe their approach as firmly founded in likelihood theory. They note that Fisher's likelihood theory only handles situations in which all random variables are observable and are produced by probabilistic mechanisms determined by unknown but fixed parameters. This class of situations does not include situations in which random variables are unobservable, for example random effects. This awkward feature of likelihood-based theories is well-known and was noted in the present book's Chapter 1. Thus an analysis based on the likelihood but avoiding prior distributions requires some extension of Fisher's likelihood and these authors claim their scheme is a principled extension of this kind.

I find this claim unconvincing. Instead, the relevant sections of Lee et al. (2006) or Lee & Nelder (2009) seem like a series of *ad hoc* patches. Consider this quote:

> Lee and Nelder [citations deleted] propose maximizing the h-likelihood h for the estimation of v [the unobservable random variables], the marginal likelihood l for the ML [maximum likelihood] estimators for β [ordinary fixed effects] and the restricted likelihood $p_\beta(l)$ for the dispersion parameters σ^2 [unknown variance-structure parameters]. Thus our position is consistent with

> the likelihood principle by using the marginal likelihood for inferences about θ [apparently the conventional unknown parameters]. However, when l is numerically hard to obtain, we propose to use adjusted profile h-likelihoods (APHLs) $p_v(h)$ and $p_{\beta,v}(h)$ as approximations to l and $p_\beta(l)$; $p_{\beta,v}(h)$ approximates the restricted log-likelihood. (Lee & Nelder 2009, p. 260, right column)

The obvious question is why different unknown quantities should be treated differently and the plain answer is that if you restrict yourself by not using prior distributions and by maximizing likelihood-like functions, then treating all unknowns the same way gives results with well-known deficiencies. To make things work reasonably well under these restrictions, some unknowns must be integrated out against implicit flat priors while others are profiled out (replaced by their maximizing values given other unknowns) and the only criterion that seems to distinguish these groups of unknowns is the properties of the resulting procedures. This epitomizes *ad hoc*kery. Ruppert et al. (2003), by contrast, do not try to rationalize the restricted likelihood as a principled extension of Fisher's likelihood; it is simply a fix that solves some well-known problems.

The *ad hoc* nature of this theory of analysis is also suggested by the assortment of modifiers these authors apply to the word *likelihood*. In the first few chapters of Lee et al. (2006), I noted 10 such modifiers before I stopped counting. This begins to look like epicycles within epicycles and suggests that a more basic overhaul is needed, for example using integrals instead of maximization and conditioning unruly likelihoods with prior distributions. But this would imply a Bayesian form of analysis and these authors reject Bayesian methods, which is puzzling given the variety of *ad hoc* patches with which they are comfortable.

Now, I don't consider an *ad hoc* approach defective as long as it has demonstrably good properties. But Lee & Nelder's analysis scheme cannot have the all-purpose good properties they claim. First, this approach relies on maximizing functions that can have multiple maxima. Second, in analyses using the restricted likelihood, maxima commonly occur at a boundary value like a zero variance or a correlation of ± 1. Neither Lee et al. (2006) nor Lee & Nelder (2009) mentions either possibility. Third, in this approach intervals and measures of uncertainty (analogous to standard errors) are based on the curvature of some function at its maximum. Lee & Nelder (2009) defend their *point estimates* based on simulation experiments — "We have indeed carried out extensive simulations with a wide variety of data comparing our estimates with those of other methods, and so far h-likelihood estimates have been often uniformly better in terms of mean-square error" (p. 295) — but I have not found a similar defense of using curvature to describe uncertainty. The reason is simple: This cannot work well in general for realistic sample sizes because the relevant functions are often highly skewed and often maximized at a boundary value. Curvature measures are ultimately rationalized as large-sample approximations and in contrast to Ruppert et al. (2003), there is (to my knowledge) no mention in Lee et al.'s work of the extensive gaps in understanding of large-sample properties of likelihood-based methods for this class of models. For one thing, curvatures at estimates on a boundary do not have desirable properties even in large samples.

Thus, while this approach does provide analyses of a wide range of models and may prove valuable, the claims made for it are not convincing and as far as I can tell, it does not address the difficulties that I find most troublesome in practice.

Exercises

Regular Exercises

1. (Section 7.2) Derive the precision matrix (7.4) for the AR(1) model.

Part III

From Linear Models to Richly Parameterized Models: Mean Structure

Part III Introduction

A full theory of richly parameterized models would have two components, a syntax expressing a large class of models, with associated statistical and computing methods, and tools for exploring and explaining analyses that use the syntax and associated methods. This book's Parts I and II discussed one such syntax, mixed linear models; Parts III and IV use that syntax to suggest how to start building the rest of a theory of richly parameterized models.

What might we want in "tools for exploring and explaining" fitted models? I would say everything the powerful, beautiful theory of linear models does and more, so we can explain and perhaps avoid strange or undesirable results that arise using mixed linear models. The theory of linear models is mostly about the linear model's mean, that is, the regression coefficients. (Even when it focuses on the errors, as with case influence or outliers, linear model theory mainly concerns coefficient estimates.) Part III takes a step from linear model theory toward a theory of richly parameterized models by focusing on the mixed linear model's mean structure, $\mathbf{X}\beta + \mathbf{Zu}$. Part IV takes a further step by focusing on the novel aspect of mixed linear models, the variance structure.

Part III's Chapter 8 describes an attempt (Hodges 1998) to adapt linear model theory wholesale to provide diagnostics for mixed linear models. Some methods for ordinary linear models adapt readily, but adding a second variance to an ordinary linear model has some unexpected consequences; some linear model tools can be adapted safely only if we are aware of these consequences. One idea from ordinary linear models, collinearity, received little attention in Hodges (1998) but turns out to be very helpful in explaining puzzles that arise in using richly parameterized models. Chapter 9 introduces four such puzzles; Chapters 10 through 12 explore collinearity as explanations for them. These puzzles are problems of collinearity in that they arise from correlation among columns in the mixed linear model's two design matrices $\mathbf{X}$ and $\mathbf{Z}$.

The views just described appear to be somewhat unusual in that most statistical writers treat the random effects $\mathbf{Zu}$ as qualitatively distinct from the fixed effects $\mathbf{X}\beta$. Thus, with one exception, I have never heard anyone else discuss collinearity or confounding of $\mathbf{X}\beta$ with $\mathbf{Zu}$, at least not in so many words. (The exception is so-called spatial confounding, discussed in Sections 9.1.1 and 10.1.) Many people who are adept in random-effect models seem to find this idea difficult: They are used to thinking of $\mathbf{u}$ as an old-style random effect which belongs not to the model's mean but rather to an error term consisting of $\mathbf{Zu} + \varepsilon$. This view is not incorrect for old-style random effects, although it may lead people to overlook collinearity as an explanation for some puzzles. (Sections 9.1.2 and 10.2 discuss an example with clustered observations.) However, when a random effect is used as a device to fit a penalized spline, this view of $\mathbf{Zu}$ seems irrelevant and misleading. Instead, for penalized splines and other models with new-style random effects, $\mathbf{X}\beta + \mathbf{Zu}$ describes a linear model with some coefficients ($\mathbf{u}$) shrunk toward zero and some (β) left unshrunk, so that everything we know about collinearity in ordinary linear models still applies. The only novelty here is that because $\mathbf{G}$ is a function of unknowns, the degree to which

u is shrunk toward zero is determined as part of the fit. Shrinkage and confounding are thus determined simultaneously and affect each other. Chapter 10 explores the effects of confounding for a fixed value of $r = \sigma_e^2/\sigma_s^2$, which controls smoothing/shrinkage, because that suffices to explain Chapter 10's two puzzles. Chapters 11 and 12 consider how the data determine r as well.

Recall the two premises discussed in the Preface, which were largely implicit in Parts I and II but are especially salient in Part III. Using a model to analyze a dataset is equivalent to specifying a function from the data to inferential or predictive summaries. However that model is rationalized or interpreted, it is essential to understand how, mechanically, that function turns the data into summaries. Also, we must distinguish between the model that *we choose* to analyze a given dataset and the process that *we imagine* produced the data. Choosing a model with random effects does not imply that those random effects correspond to any random mechanism out there in the world. This distinction is related to the recurring theme of old-versus new-style random effects; Chapter 13 sums up the book's material about old- and new-style random effects and describes some practical consequences of the difference between them.

Chapter 8

Adapting Diagnostics from Linear Models

This chapter considers an attempt (Hodges 1998) to adapt linear model theory to mixed linear models using the constraint-case formulation described in Section 2.1. Here, as in linear model theory, the purpose is to do the following:

- Seek discrepant features of the data; linear model theory does this using residuals.
- Seek deviations from model assumptions; linear model theory uses residuals to check for non-linearity in the mean structure and non-constant error variance, and to consider transformations of the outcome $\mathbf{y}$ that are better suited to a homoscedastic linear model.
- Seek observations with a large influence on estimates; linear model theory uses case influence.
- Assess evidence for adding predictors; linear model theory uses added variable plots.
- Understand ill-determined estimates, i.e., competition among predictors (regressors, right-hand-side variables, effects) to capture variation in the outcome $\mathbf{y}$; in linear model theory, the relevant ideas are collinearity, confounding, and variance inflation.

As we will see, some linear model methods adapt quite well to mixed linear models, while others adapt less gracefully, highlighting differences between ordinary and mixed linear models. Until recently,[1] Hodges (1998) was (as far as I know) the only attempt at "a general approach (*process*) to diagnostics [for richly parameterized models], rather than offering yet another *procedure* that would need fixes ('further research') for non-standard settings" (Longford 1998). For that reason, this material serves as a first step from the theory of ordinary linear models toward a theory of richly parameterized models.

This chapter uses the constraint-case formulation. Many, perhaps all, of these ideas could be re-expressed using the standard formulation of mixed linear models.

[1] Yuan & Johnson (2012) gave a very different approach to diagnostics for hierarchical models. I have a few qualms, e.g., it does not distinguish old- and new-style random effects, but it really uses the structure of hierarchical models and is thus of great interest. It is, however, so different from this book's approach that considering it would take me too far afield, so I don't discuss it further.

James Hilden-Minton's doctoral dissertation (Hilden-Minton 1995) used the standard formulation to develop ideas parallel to several presented here but as far as I can determine no papers were published from that dissertation. This chapter describes some of its ideas; Snijders & Bosker (2012, Chapter 10) describes some in more depth. Hodges (1998, Section 3.1 and the discussion) gives a catalog as of 1998 of what Longford called "yet another *procedure*," i.e., non-systematic diagnostic methods. Snijders & Bosker (2012, Chapter 10) give an approach to checking the various assumptions of hierarchical (multi-level) models with normal errors like those described in this chapter, though not a catalog of published methods. Finally, for consistency with the rest of this book, the notation used here differs somewhat from Hodges (1998).

8.1 Preliminaries

The following assumes the main analysis is Bayesian and that a model has been fit using Markov chain Monte Carlo (MCMC), so we have a sequence of MCMC draws on all unknowns.

In linear model theory as I studied it in the early 1980s, the emphasis was on *diagnostics*, procedures applied to a fitted model to diagnose problems and examine possible fixes for them. Hodges (1998) therefore emphasized specifying diagnostics in the spirit of Weisberg (1983), who argued that these four considerations should guide construction of diagnostics:

- A diagnostic should aim to detect a specific problem.
- Diagnostics should compute quickly.
- A diagnostic should have a corresponding plot, so you can assess the effect of individual observations (or "cases").
- Graphics should allow users to look at the data as directly as possible.

Thus although the main analysis is Bayesian, the diagnostics are not Bayesian because that would violate Weisberg's fourth principle: Every Bayesian diagnostic I've ever seen is either non-graphic or puts an extra layer of math between the analyst and the data. Combining the Bayesian formalism with non-Bayesian diagnostics is not as paradoxical as it may seem. Prominent advocates of the Bayesian approach (e.g., Smith 1986; Berger 1992, 1993) have argued that although the Bayesian formalism is the only acceptable way to draw inferences, exploratory analysis and diagnosis are less formal and non-Bayesian methods are acceptable. Among non-Bayesians, it is now commonplace to view the Bayesian formalism as a means for generating procedures whose frequentist properties can then be assessed, usually by simulation. From this view, the present chapter takes a particular approach to fitting hierarchical models and derives diagnostics for that approach, most of which can be used with other approaches because they take estimates of variance parameters as fixed. The conventional theory for mixed linear models relies on large-sample approximations; a frequentist view of a Bayesian analysis also considers that an approximation.

This chapter uses a particular example throughout.

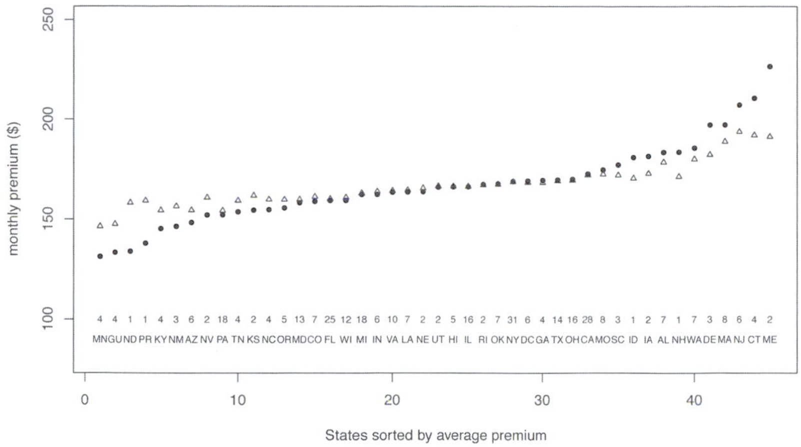

Figure 8.1: The HMO premium data with summaries and fitted values: •, state average premium; △, posterior mean from the simple random-effects model. Each state's data are labeled at the bottom of the figure by the state's two-letter abbreviation and the number of plans in the state.

Example 9. HMO premiums. As part of estimating the cost of moving military retirees and dependents from a Defense Department health plan to private plans serving U.S. government employees, a dataset was assembled in the mid-1990s describing 341 health maintenance organizations (HMOs) serving U.S. Government employees. These plans operated in 42 states, the District of Columbia, Guam, and Puerto Rico; for simplicity, I call these all "states." The dataset has between 1 and 31 plans per state with a median of 5. A quantity of particular interest is each HMO's monthly premium for individual subscribers, in U.S. dollars. Figure 8.1 summarizes the HMO premium data, with each dot representing a state's average monthly premium. The dataset also includes variables describing plans and states. The plan-level variables are a plan's counts of individual and family enrollees among U.S. government employees; the state-level variables are a state's average expenses per hospital admission, population, and region (one of 7 regions partitioning the U.S.).

The development begins with a simple unbalanced one-way random-effects model for these data expressed in the constraint-case form, as in Section 2.1. This model is usually written as

$$y_{ij} = \theta_i + \varepsilon_{ij} \tag{8.1}$$

$$\theta_i = \mu + \delta_i \tag{8.2}$$

where $i = 1, \ldots, 45$ indexes states, $j = 1, \ldots, n_i$ indexes plans within states, $\varepsilon_{ij} \sim N(0, \sigma_e^2)$ and $\delta_i \sim N(0, \sigma_s^2)$, independently conditional on their variances. A Bayesian analysis would add prior distributions for σ_e^2, σ_s^2, and μ, as noted earlier; the development here uses a flat prior for μ and inverse-gamma priors for σ_e^2

and σ_s^2. Rewrite equation (8.2) as

$$0 = -\theta_i + \mu + \delta_i \tag{8.3}$$

Equations (8.1) and (8.3) have the form of a linear model. The left-hand side of each equation is known and the right-hand side is a linear function of unknown parameters with additive errors:

$$\left[\begin{array}{c}\mathbf{y}\\ \hline \mathbf{0}_{45\times 1}\end{array}\right] = \left[\begin{array}{ccc|c}\mathbf{1}_{n_1} & \cdots & \mathbf{0}_{n_1} & \\ \vdots & \ddots & \vdots & \mathbf{0}_{341\times 1}\\ \mathbf{0}_{n_{45}} & \cdots & \mathbf{1}_{n_{45}} & \\ \hline & -\mathbf{I}_{45} & & \mathbf{1}_{45}\end{array}\right]\left[\begin{array}{c}\theta_1\\ \vdots\\ \theta_{45}\\ \hline \mu\end{array}\right] + \left[\begin{array}{c}\varepsilon\\ \hline \delta\end{array}\right] \tag{8.4}$$

where $\mathbf{y}$ and ε are column 341-vectors of y_{ij} and ε_{ij} respectively, ordered with j changing most quickly, and δ is the column 45-vector of δ_i.

The Bayesian analyses of the HMO premium data use these prior distributions: for μ, a flat (improper) prior; for $1/\sigma_e^2$, a flat (improper) prior; and for $1/\sigma_s^2$, a gamma prior with mean 11 and variance 110. (Analyses of an artificial dataset below use a different prior for σ_e^2.) To obtain mean 11 and variance 110, the gamma density parameterized as $\pi(x|\alpha,\beta) \propto x^{\alpha-1}e^{-\beta x}$ has $\alpha = 1.1$, $\beta = 0.1$. I can't remember why I thought this was a good choice, but I use it here for consistency. Also, it is generally not advisable to use a flat prior for $1/\sigma_e^2$, although in the present case it is probably innocuous because within-state variation provides strong information about σ_e^2. As we will see (Chapter 19), the prior on $1/\sigma_s^2$ is not so innocuous.

When fit using these prior distributions, the simple random-effects model gave posterior means for each state's θ_i shown as triangles in Figure 8.1. The overall mean μ had posterior mean 167 and posterior standard deviation 2.5. The between-state variance σ_s^2 had posterior mean 179 and 95% equal-tailed interval (61,339); the analogous estimate and interval for σ_e^2 were 502 and (424,592) respectively. For Hodges (1998) I computed these using a Rao-Blackwellized Gibbs sampler run for so few iterations that I'm embarrassed to report it; nonetheless, I report these results here. (For this simple model, I got nearly identical results in a Rao-Blackwellized Gibbs run with far more iterations.) The posterior means differ from the state averages in the expected way: Each state's posterior mean is shrunk toward μ with less shrinkage for states that have more plans.

To derive and present results in more generality, I will use the following form for models in the constraint-case formulation (here, including prior cases), which was used in Hodges (1998):

$$\left[\begin{array}{c}\mathbf{y}\\ \hline \mathbf{0}\\ \hline \mathbf{M}\end{array}\right] = \left[\begin{array}{c|c}\mathbf{K} & \mathbf{0}\\ \hline \mathbf{S}_1 & \mathbf{S}_2\\ \hline \mathbf{W}_1 & \mathbf{W}_2\end{array}\right]\left[\begin{array}{c}\Theta_1\\ \hline \Theta_2\end{array}\right] + \left[\begin{array}{c}\varepsilon\\ \hline \delta\\ \hline \xi\end{array}\right] \tag{8.5}$$

where $\mathbf{S}_1$, $\mathbf{S}_2$, $\mathbf{W}_1$, and $\mathbf{W}_2$ have suitable dimensions. Recall that parts of this form can be null, e.g., Θ_2 and the corresponding parts of the design matrix are absent in Section 2.1's Example 4 (Experimental plots in a long row). A somewhat different general form was given in equation (2.13), which directly re-expresses a mixed linear model. This general form could be used in place of (8.5), giving expressions equivalent to those shown below.

In the usual notation, (8.5) can be summarized as

$$\mathbf{Y} = \mathbf{H}\Theta + \mathbf{E} \tag{8.6}$$

where $\mathbf{Y}$ and $\mathbf{H}$ are known, Θ is unknown, and $\mathbf{E}$ is an unobserved error term. For the HMO premium dataset, $\mathbf{E}$ has mean zero and a diagonal covariance matrix Γ, the first 341 diagonal elements of Γ being σ_e^2 and the last 45 σ_s^2; the prior cases are omitted in these analyses.

This model now looks like a heteroscedastic linear model, and recall from Section 2.1 that any mixed linear model can be written in this form. This linear model form suggests applying diagnostics for ordinary linear models, with modifications as needed, to this constraint-case formulation. The sections that follow do that, presenting first diagnostics that seem to extend to the present case without difficulty and progressing to diagnostics that extend less gracefully.

8.2 Added-Variable Plots

Added-variable plots allow a visual check of whether a variable should be added to a linear model. A common added-variable plot for an ordinary linear model with homoscedastic errors is derived as follows (Cook and Weisberg 1982, p. 44; Atkinson 1985, section 5.2). Suppose the explanatory variables currently in the model fill the columns of $\mathbf{A}$ and that $\mathbf{B}$ is a candidate variable. In the usual matrix notation, the model is

$$\mathbf{y} = \mathbf{A}\beta + \mathbf{B}\phi + \varepsilon. \tag{8.7}$$

Premultiplying both sides of (8.7) by $\mathbf{I} - \mathbf{A}(\mathbf{A}'\mathbf{A})^{-1}\mathbf{A}'$ yields

$$\hat{\mathbf{e}} = (\mathbf{I} - \mathbf{A}(\mathbf{A}'\mathbf{A})^{-1}\mathbf{A}')\mathbf{B}\phi + (\mathbf{I} - \mathbf{A}(\mathbf{A}'\mathbf{A})^{-1}\mathbf{A}')\varepsilon \tag{8.8}$$

where $\hat{\mathbf{e}}$ is the residuals from the least squares fit of $\mathbf{y}$ on $\mathbf{A}$. If $E(\varepsilon) = 0$, as usual, then

$$E(\hat{\mathbf{e}}) = (\mathbf{I} - \mathbf{A}(\mathbf{A}'\mathbf{A})^{-1}\mathbf{A}')\mathbf{B}\phi. \tag{8.9}$$

Thus if $\mathbf{B}$ should enter the model linearly, a plot with $\hat{\mathbf{e}}$ on the vertical axis and $(\mathbf{I} - \mathbf{A}(\mathbf{A}'\mathbf{A})^{-1}\mathbf{A}')\mathbf{B}$ on the horizontal axis will show points clustered around a line through the origin with slope ϕ.

In a richly parameterized model written in constraint-case form, variables can be added for either data cases like (8.1) or constraint cases like (8.2, 8.3). The added-variable plot for ordinary linear models can be applied directly to either kind of variable using four steps (with explanations of steps 1, 3, and 4 to follow): (1) Reformulate the candidate variable and call it $\mathbf{B}$, (2) write the model with the candidate variable as in (8.10) below, with $\mathbf{H}$ taking the place of $\mathbf{A}$ in (8.7), (3) premultiply the resulting equation by $\Gamma^{-1/2}$, where Γ takes a suitable value, to obtain a homoscedastic-errors problem, and (4) draw the usual added-variable plot.

Step 1. **B**'s format depends on the level to which the candidate variable belongs. Consider adding plan enrollment to the data-case model for the HMO premium data. If $\mathbf{B}_1$ is the column 341-vector of plan enrollments, $\mathbf{B}$ is $[\mathbf{B}_1', \mathbf{0}_{1\times 45}]'$, conforming to (8.4). To add state average expenses per hospital admission to the constraint-case model, let $\mathbf{B}_2$ be the column 45-vector of average expenses per admission; then $\mathbf{B}$ is $[\mathbf{0}_{1\times 341}, \mathbf{B}_2']'$.

Step 3. The added-variable plot does not appear to be sensitive to the choice of Γ. To see this, write the hierarchical model and candidate variable as

$$\mathbf{Y} = \mathbf{H}\Theta + \mathbf{B}\phi + \mathbf{E} \tag{8.10}$$

and pre-multiply by $\Gamma^{-1/2}$ to give

$$\mathsf{Y} = \mathsf{H}\Theta + \mathsf{B}\phi + \mathsf{E} \tag{8.11}$$

where the sans-serif font indicates pre-multiplication by $\Gamma^{-1/2}$. Now pre-multiply (8.11) by $\mathbf{I}-\mathbf{V}$, for $\mathbf{V} = \mathsf{H}(\mathsf{H}'\mathsf{H})^{-1}\mathsf{H}'$, to give

$$\hat{\mathsf{E}} = (\mathbf{I}-\mathbf{V})\mathsf{Y} = (\mathbf{I}-\mathbf{V})\mathsf{B}\phi + (\mathbf{I}-\mathbf{V})\mathsf{E}. \tag{8.12}$$

Thus, for any Γ, if $\mathbf{B}$ should enter the model as in (8.10), the added-variable plot will show the correct slope ϕ. Changing Γ only affects the plot by changing the scale of (8.12) and by changing the shrinkage through $\mathbf{I}-\mathbf{V}$, that is, by changing the relative position of the data- and constraint-case points along the plot's regression line. When I wrote Hodges (1998), it seemed reasonable to use Γ's posterior mean as a point estimate for this purpose, so I did (and none of the commentators disagreed in print). Later work, cited in Section 1.3.2.1, strongly suggests that posterior means of variances (or precisions or standard deviations) are generally not good estimators; posterior medians are preferable. Because use of posterior means appears innocuous in analyzing the HMO premium dataset, I follow Hodges (1998) in doing so for this chapter.

Step 4. The residuals in this plot are scaled: The vertical axis is $\hat{\mathsf{E}} = \Gamma^{-1/2}\hat{\mathbf{E}} = \Gamma^{-1/2}(\mathbf{Y}-\mathbf{H}\hat{\Theta})$. This puts data- and constraint-case residuals (and prior-case residuals, if included) on the same scale.

It may appear odd to use all cases to judge a variable appearing in just one level of the model, but the data and constraint cases convey distinct information, as is easily seen by fitting lines separately to the data and constraint cases in an added variable plot. Nonetheless, the data- and constraint-case residuals are linearly related. For any choice of Γ, write $\hat{\mathsf{E}}$ as $(\hat{\mathsf{E}}_d, \hat{\mathsf{E}}_c)'$, where the subscripts indicate data cases and constraint cases, respectively, and note that $\mathsf{H}'\hat{\mathsf{E}} = 0$. Therefore

$$\left[\frac{\mathbf{K}'}{\mathbf{0}}\right]\Gamma_1^{-1/2}\hat{E}_d = -\left[\frac{\mathbf{S}_1'}{\mathbf{S}_2'}\right]\Gamma_2^{-1/2}\hat{E}_c. \tag{8.13}$$

Equation (8.13) also holds if $\hat{\mathsf{E}}$ is replaced by $(\mathbf{I}-\mathbf{V})\mathsf{B}$. The data- and constraint-case residuals vary subject to this constraint.

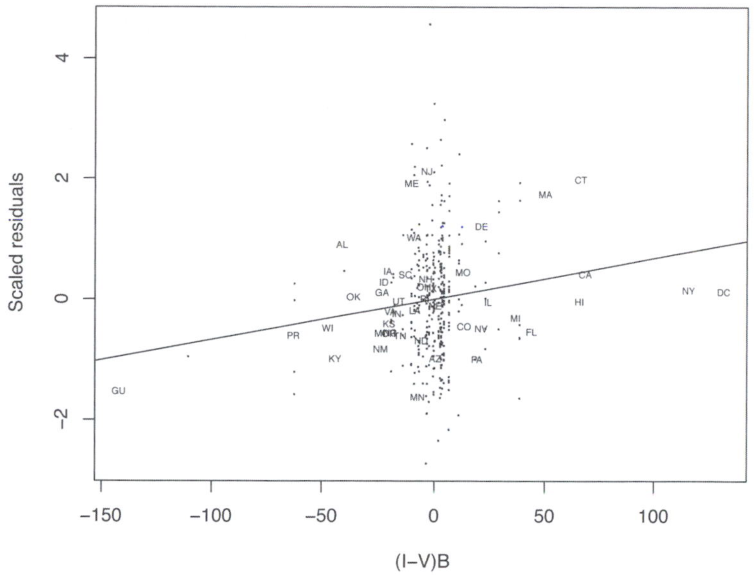

Figure 8.2: The HMO premium data: Added-variable plot for the state-level variable average expenses per hospital admission. Data cases are represented by dots and constraint cases by their state's two-letter abbreviation. The line is the ordinary least-squares fit to this plot.

Hilden-Minton (1995, section 3.1.1) gave an added-variable plot for mixed-effect models expressed in the mixed-linear-model form.

Figure 8.2 shows the added-variable plot checking whether average expenses per hospital admission should be added to the simple random-effects model. It shows an OLS slope of 0.007 (standard error 0.002), suggesting that expenses per admission should enter the model but it also suggests the slope may be largely determined by Guam's (GU) constraint and data cases at the lower left (the dot near "GU" is the data-case plotting point from one of Guam's plans).

8.3 Transforming Variables

Transforming the outcome or explanatory variables can remove curvature or interactions and can make residuals look normally distributed. For the usual linear model in equation (8.7), it is common to consider Box-Cox transformations for the outcome variable:

$$y^{(\lambda)} = \begin{cases} (y^{\lambda} - 1)/\lambda & \lambda \neq 0, \\ \log(y) & \lambda = 0. \end{cases} \tag{8.14}$$

A graphical method to determine likely values of λ is as follows: Expand $y^{(\lambda)}$ as a linear Taylor series approximation around $\lambda = 1$ to create a new explanatory variable **B** whose coefficient is linear in λ, then draw an added variable plot for **B**.

One possibility for **B** has l^{th} coordinate $\mathbf{B}_l = \hat{y}_l \log(\hat{y}_l) - \hat{y}_l + 1$, where $\hat{y}_l$ is the fitted value for the lth case from the least squares fit of the untransformed y. (Cook and Weisberg 1982, section 2.4.3, describes this **B**, due to Andrews.) Another possibility, due to Atkinson (1985, section 6.4) has $\mathbf{B}_l = \hat{y}_l(\log(\hat{y}_l/\dot{\hat{y}}) - 1)$, where $\dot{x} = (\prod_{k=1}^{n} x_k)^{1/n}$ is the geometric mean. If either of these **B** is added to the linear model, it has coefficient $1 - \lambda$.

A similar constructed variable and plot are easily derived to examine transformations of explanatory variables; this will not be discussed here.

This method is readily applied to richly parameterized models in the form (8.6). To draw the graphic, construct **B** as above for the data cases, then treat it like any other candidate variable for the data cases and draw an added variable plot as in Section 8.2.

Figure 8.3 shows the added-variable plot checking for a transformation of the outcome for the simple random-effect model fit to the HMO premiums. The ordinary least-squares fit to this plot has intercept -0.002 (standard error 0.04) and slope 0.33 (SE 0.04), which suggests a transformation with $\lambda = 0.67$ (95% confidence interval 0.59 to 0.75). Support for this λ is spread throughout the dataset and seems somewhat stronger in the constraint cases than in the data cases. As we will see, this is consistent with the residual plots, which show some skewing to the right; Wakefield's (1998) re-analysis found evidence of a long upper tail in the distribution of random effects for states.

8.4 Case Influence

The two diagnostics discussed so far, added variable plots for a new variable and for transforming the outcome y, could be applied without any noteworthy modification to the constraint-case formulation of a richly parameterized model. For the next two diagnostics, case influence and residuals, some changes are needed.

Case influence diagnostics show how inference or prediction summaries change when cases (observations) are deleted. In the usual linear model $\mathbf{y} = \mathbf{A}\beta + \varepsilon$, where the errors ε_i are iid normal with mean zero and variance σ^2, case-influence diagnostics take advantage of a handy updating formula (Weisberg 1980, section 5A.1; Atkinson 1985, section 2.2). If the subscript (I) indicates deleting one or more cases indexed by I, then

$$(\mathbf{A}'_{(I)}\mathbf{A}_{(I)})^{-1} = (\mathbf{A}'\mathbf{A})^{-1} + (\mathbf{A}'\mathbf{A})^{-1}\mathbf{A}'_I(\mathbf{I} - \mathbf{V}_I)^{-1}\mathbf{A}_I(\mathbf{A}'\mathbf{A})^{-1}, \tag{8.15}$$

where $\mathbf{A}_I$ is a matrix composed of the deleted rows of **A** and $\mathbf{V}_I = \mathbf{A}_I(\mathbf{A}'\mathbf{A})^{-1}\mathbf{A}'_I$. If a single case is deleted, this formula simplifies even more. Thus, if case l is deleted and the subscript (l) now indicates deletion of case l,

$$\hat{\beta}_{(l)} - \hat{\beta} = -(\mathbf{A}'\mathbf{A})^{-1}a_l r_l/(1 - h_l) \tag{8.16}$$

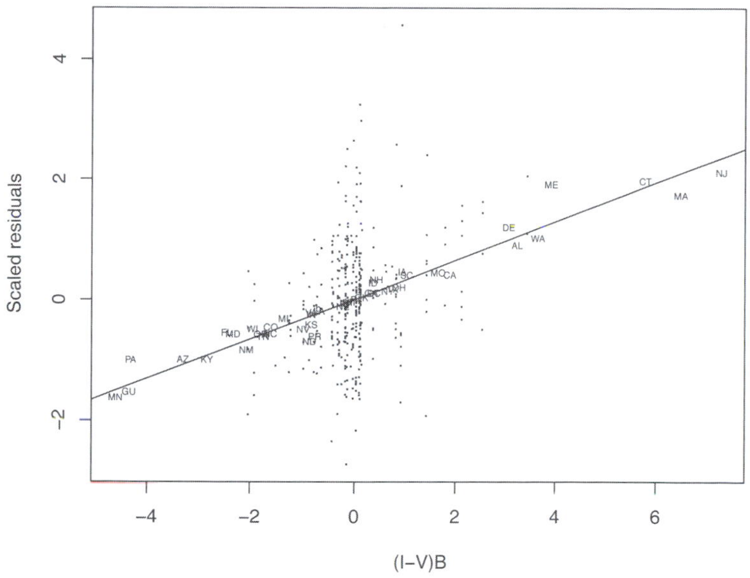

Figure 8.3: The HMO premium data: Added-variable plot to check for a transformation of the outcome y in the simple random-effects model. Data cases are represented by dots and constraint cases by their state's two-letter abbreviation. The line is the ordinary least-squares fit to this plot.

where a_l is the lth row of $\mathbf{A}$, r_l is the residual from the lth case in the full-data fit, and $h_l \in (0,1)$ is the lth diagonal element of the so-called hat matrix for the full dataset, $\mathbf{A}(\mathbf{A}'\mathbf{A})^{-1}\mathbf{A}'$. These formulas lead to Cook's distance, which measures the effect on $\hat{\beta}$ of deleting case l,

$$(\hat{\beta}_{(l)} - \hat{\beta})'\mathbf{A}'\mathbf{A}(\hat{\beta}_{(l)} - \hat{\beta})/p\hat{\sigma}^2 \tag{8.17}$$

$$= r_l^2 \frac{h_l}{(1-h_l)^2} \frac{1}{p\hat{\sigma}^2} \tag{8.18}$$

where p is the number of columns of $\mathbf{A}$, $\hat{\sigma}^2 = \mathbf{y}'(\mathbf{I} - \mathbf{A}(\mathbf{A}'\mathbf{A})^{-1}\mathbf{A}')\mathbf{y}/(n-p)$ is the full-data estimate of the error variance and n is the number of cases in the full dataset. Cook's distance is large if both the squared residual r_l^2 and the leverage h_l are large.

The case-deletion formula also gives a simple formula for the change in the error variance estimate from deleting case l:

$$\hat{\sigma}^2_{(l)} - \hat{\sigma}^2 = \frac{1}{n-p-1}(\hat{\sigma}^2 - \frac{r_l^2}{1-h_l}). \tag{8.19}$$

Under the assumed model, $E(r_l^2) = (1-h_l)\sigma^2$ and $E(r_l) = 0$, so deleting case l reduces the estimate of error variance substantially if r_l^2 is large relative to its variance under the assumed model.

Applying these ideas to richly parameterized models in the constraint-case formulation raises several issues. For hierarchical models, constraint- and prior-case influence are of interest. Deleting a prior case amounts to setting the corresponding parameter's prior variance to infinity, i.e., replacing an informative prior with an improper flat prior. Deleting constraint cases is somewhat more complicated. For the random-effects model applied to the HMO dataset, deleting Minnesota's constraint case means that Minnesota's posterior mean is no longer shrunk toward μ. If this has a large effect on the estimate of Minnesota's θ_i, one might revisit the choice of a random-effects model, but otherwise this is not an especially useful fact. Deleting Minnesota's constraint case may also affect the posterior distributions of θ_i from other states, because their shrinkage may change if they no longer borrow strength (or weakness!) from Minnesota. When the state-level model is more complicated, deleting Minnesota's constraint case is equivalent to using a mean-shift outlier model (Cook and Weisberg 1982, section 2.2.2; Atkinson 1985, section 5.5) in the constraint cases. The influence of Minnesota's constraint case on Minnesota's θ_i is now somewhat more interesting, as it reflects on the entire state-level model and not simply on the decision to use random effects.

One might object that data cases are part of the data while the constraint cases are part of the model, so deleting one is not like deleting the other. But any modeling choice inserts information into the analysis, just as an observation y_i inserts information. Thus, although the data cases on the one hand and constraint or prior cases on the other hand may be qualitatively distinct kinds of information, the constraint-case formulation allows us to use the same case influence diagnostic to determine the effect of inserting these distinct kinds of information into the analysis.

Common influence measures may have problems if they are applied directly to richly parameterized models using the constraint-case formulation. First, the local influence methods of Cook (1986) and Kass et al. (1989) use the mode of the likelihood or posterior respectively, which can be misleading or problematic if a fit has more than one mode. (Adding some state-level predictors to the random effect-model produced a bi-modal posterior; see Section 5.2 of Hodges 1998 and the re-analysis in Wakefield 1998). Williams's (1987) partial influence measure (extended by Davison and Tsai 1992) has the same difficulty. Second, Cook's distance and similar measures (Atkinson 1985, section 2.2; Cook and Weisberg 1982, section 3.5) describe a case's influence on the entire parameter vector. However, as "richly parameterized" suggests, the parameter vector has many elements and the influence of a given case is often felt strongly by one parameter and weakly by the others. Thus, Cook's distance can miss important effects (Bradlow and Zaslavsky 1997). Finally, hierarchical models have at least two unknown error variances; deleting a case changes the information about Γ and thus affects the posterior mean of Θ non-linearly. This non-linear effect is illustrated below. But if Cook's distance is used with Γ fixed, as in the preceding two diagnostics considered here, case deletion can only affect the parameter estimates linearly.

These considerations might suggest re-running the MCMC to re-fit the model for each deleted case, but the computing would generally be prohibitive and thus violate Weisberg's principles for diagnostics. Thus I consider two less computer-intensive methods, the first a single-parameter analog to Cook's distance (henceforth "the linear-approximation method") and the second an importance sampling method that re-uses the original MCMC draws.

To derive the linear approximation method, fix Γ at an estimate — I would now use the posterior median, but Hodges (1998) used the posterior mean and I do so here — and apply the updating formula for ordinary linear models. The approximate change in Θ from deleting case l is

$$\hat{\Theta}_{(l)} - \hat{\Theta} \approx -(\mathsf{H}'\mathsf{H})^{-1} a_l' \hat{\mathsf{E}}_l/(1-h_l) \tag{8.20}$$

where $\hat{\Theta}$ is the posterior mean of Θ using the full dataset, $\hat{\Theta}_{(l)}$ is the analogous quantity with the lth case deleted, $\hat{\mathsf{E}}_l$ is the residual for the lth case, i.e., the lth row of $\hat{\mathsf{E}} = \mathsf{Y} - \mathsf{H}\hat{\Theta}$, a_l is the lth row of H, and h_l is the lth diagonal element of $\mathbf{V} = \mathsf{H}(\mathsf{H}'\mathsf{H})^{-1}\mathsf{H}'$, recalling that the sans-serif font indicates pre-multiplication by $\Gamma^{-1/2}$. The change in the kth element of $\hat{\Theta}$ from deleting the lth case is the kth element of (8.20). Equation (8.20) is approximate because deleting the lth case changes the information about Γ, but (8.20) keeps Γ fixed.

The second approach, suggested by Gelfand et al. (1992) (see also Bradlow and Zaslavsky 1997, Peruggia 1997, MacEachern and Peruggia 2000) uses the MCMC sequence as an importance sample. Label the MCMC sequence by $k = 1, \ldots, m$. Let $\mathbf{Y}_{(l)}$ denote $\mathbf{Y}$ with the lth case deleted, let $\Psi = (\Theta, \Gamma)$, and let $g(\Psi)$ be a function of Ψ. Then

$$\begin{aligned} E(g(\Psi)|\mathbf{Y}_{(l)}) &= \int g(\Psi) f(\Psi|\mathbf{Y}_{(l)}) d\Psi \\ &= \int g(\Psi) \frac{f(\Psi|\mathbf{Y}_{(l)})}{f(\Psi|\mathbf{Y})} f(\Psi|\mathbf{Y}) d\Psi \\ &\approx \frac{1}{m} \sum_{k=1}^{m} g(\Psi_k) \frac{f(\Psi_k|\mathbf{Y}_{(l)})}{f(\Psi_k|\mathbf{Y})}, \end{aligned} \tag{8.21}$$

where $\Psi_k, k = 1, \ldots, m$, is the MCMC sequence. The proportionality constants of $f(\Psi|\mathbf{Y})$ and $f(\Psi|\mathbf{Y}_{(l)})$ are not necessary if the weights $f(\Psi_k|\mathbf{Y}_{(l)})/mf(\Psi_k|\mathbf{Y})$ are re-scaled to sum to one. For data, constraint, and prior cases, $f(\Psi_k|\mathbf{Y}_{(l)})/f(\Psi_k|\mathbf{Y})$ is simple, corresponding to a single row of (8.5), and will not be given.

For either the linear-approximation method or the importance-sampling method, the change in $g(\Psi)$ induced by a case deletion can be calibrated using the relative change:

$$RC(g(\Psi); l) = [E(g(\Psi)|\mathbf{Y}_{(l)}) - E(g(\Psi)|\mathbf{Y})]/psd(g(\Psi)|\mathbf{Y}), \tag{8.22}$$

where *psd* means "posterior standard deviation" and, for the linear approximation, $g(\Psi)$ is an element of Θ. By the usual folk wisdom, $|RC| > 2$ suggests an influential case.

The linear-approximation method should work well as long as the posterior distribution of Γ isn't changed much by deleting the lth case. Similarly, importance sampling should work well as long as no unknown quantity moves to an extreme part of the range of its MCMC samples. Unfortunately, the entire purpose of case-influence methods is to find cases that induce a large change when deleted. Thus the question is: How large must a change be before these approximate methods fail? Can they find cases that induce large changes even if the approximations do fail?

To illustrate what can happen when a case is deleted — and also one way in which richly parameterized models differ from linear models — consider an artificial example based on the HMO data. This example is intended to bend both methods until they break; I discuss its relevance to practice below. Suppose each state has 8 plans and that the state averages are the same as in the actual dataset except for Maine, which has the highest average in the actual dataset. Suppose also that for the 44 states other than Maine, the sum over states of the sum of squared deviations around the state averages is $44 \times 7 \times 501$, mimicking the actual data. Set the prior variance for μ to infinity and put gamma priors with mean 11 and variance 110 on $1/\sigma_e^2$ and $1/\sigma_s^2$. (Any continuous proper priors for σ_e^2 and σ_s^2 will give qualitatively the same result.) Label Maine as $i = 1$ and assume one of its HMOs, labeled $j = 1$, has premium y_{11} equal to Maine's average and that for this value of y_{11}, the sum of squared deviations of Maine's 8 plans around its state average is also 7×501. Now increase y_{11} to show the effect of an outlying plan. Figure 8.4 shows the posterior means of the θ_i as y_{11} increases; the horizontal axis is the number of within-state standard deviations ($\sigma_e = 22.4$ from the actual HMO premium data) by which y_{11} is increased above the average of Maine's other plans. The upper solid line is Maine's average, the dotted line is Maine's posterior mean, and the cluster of lines describes the posterior means of the other 44 states. As y_{11} increases, Maine's posterior mean increases at a damped rate while the other posterior means shrink slowly toward μ. When y_{11} reaches about $45\sigma_e$ from its starting point, the state posterior means approach μ rapidly. If y_{11} were originally $60\sigma_e$ greater than the average of Maine's other plans and then it were deleted, Maine's posterior mean would *increase* to the value at the extreme left of the plot. (Note: In re-doing this computation for the present chapter, I found a small error in the computation in Hodges 1998, so the results here differ slightly.)

Now consider the importance-sampling method, and suppose y_{11} is about $65\sigma_e$ from the average of Maine's other HMOs. The other 44 θ_i all have posterior mean about 170, while the posterior mean of Maine's θ_1 is about 175. The MCMC draws for θ_1 will thus be within a few posterior standard deviations of 175, and so will the importance-sampling computation of $E(\theta_1|\mathbf{Y}_{(I)})$, which reweights the MCMC draws. If we delete the constraint case for Maine, i.e., for θ_1, the true value of $E(\theta_1|\mathbf{Y}_{(I)})$, given by the diagonal solid line in Figure 8.4, is about 400, far from $E(\theta_1|\mathbf{Y})$, but the importance-sampling method cannot detect this because the MCMC draws are all near $E(\theta_1|\mathbf{Y})$. If we remove y_{11}, the posterior mean of θ_1 changes to about 215 but again, the MCMC draws are all too small to allow the importance sampler to be close to this.

Therefore, we can already conclude that the importance sampler will understate case-deletion effects when it fails. Although this example is artificial, it is easy to make this happen in real data when a θ_i has a small posterior standard deviation and you delete its constraint case.

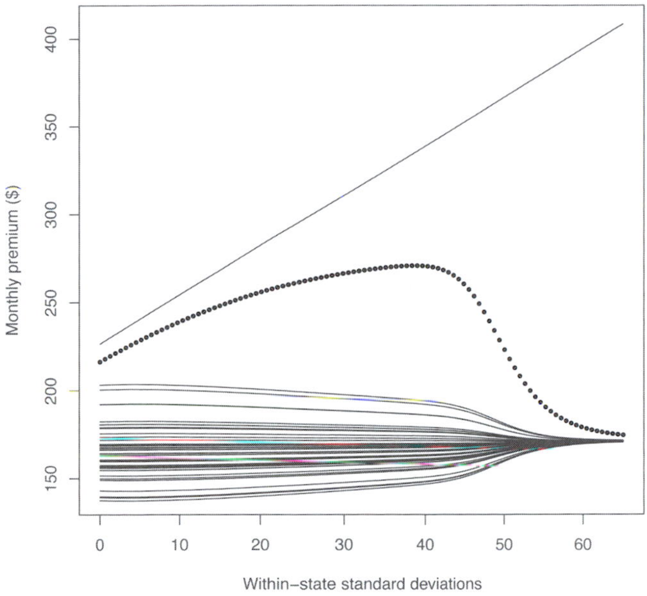

Figure 8.4: The effect of increasing y_{11} in the artificial dataset. The horizontal axis is the number of within-state standard deviations $\sigma_e = 22.4$ by which y_{11} is greater than the average of the other y_{1j}. The diagonal solid line is the average of all the y_{1j}; the dotted curve is the posterior mean of θ_1; the curves in the lower half of the figure are the posterior means of the other states' θ_i.

Figure 8.5 describes in more detail the performance of the linear approximation in this artificial example. The diagonal solid line is, as in Figure 8.4, the average of Maine's HMOs including y_{11}. It is also the posterior mean of θ_1 if the constraint case for θ_1 is deleted. The adjacent dashed line is the linear approximation to the posterior mean of θ_1 under the same deletion. The linear approximation is quite good until y_{11} reaches about $30\sigma_e$ above the average of the other y_{1j}, after which it grossly overstates the effect of deleting Maine's constraint case.

The horizontal solid line in Figure 8.5 is the posterior mean of θ_1 when y_{11} is deleted. The line of large dots is $E(\theta_1|\mathbf{Y})$, as in Figure 8.4. Twined around these lines is a line of small dots indicating the linear approximation to the posterior mean of θ_1 when y_{11} is deleted. This approximation is also quite good until y_{11} reaches about $30\sigma_e$, after which it erroneously indicates that deleting y_{11} has no effect on the posterior mean of θ_1.

(On re-doing these calculations, I noticed that the marginal posterior of (σ_e^2, σ_s^2) is bi-modal when y_{11} is much larger than the average of Maine's other plans. When y_{11} is about $50\sigma_e$ above Maine's other plans, the posterior of the shrinkage factor $n\sigma_e^2/(n\sigma_e^2 + \sigma_s^2)$, for $n = 8$ plans per state, has about equal mass in the two modes,

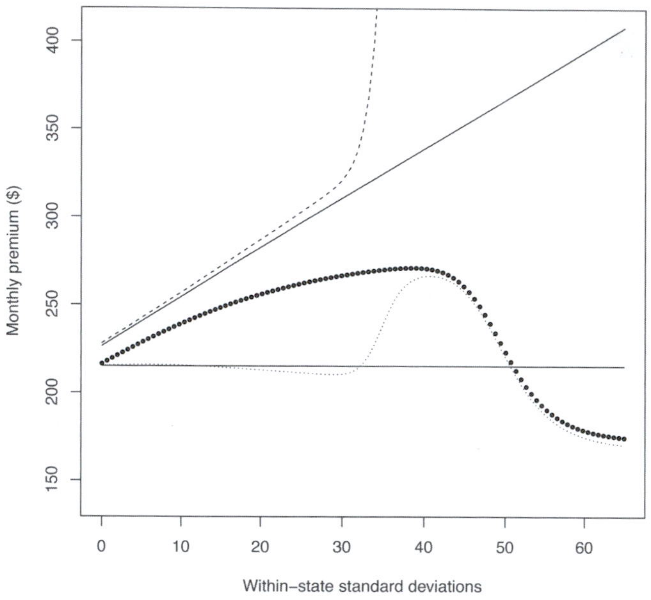

Figure 8.5: Performance of the linear approximation in the artificial dataset. The horizontal axis is the number of within-state standard deviations σ_e by which y_{11} is greater than the average of the other y_{1j}. The diagonal solid line is the average of all the y_{1j} and also the true posterior mean of θ_1 if the constraint case for θ_1 is deleted. The curving dashed line is the linear approximation to the posterior mean if the constraint case for θ_1 is deleted. The curve of bold dots is the posterior mean of θ_1; the horizontal line is the posterior mean of θ_1 if y_{11} is deleted; the curve of small dots is the linear approximation to the posterior mean of θ_1 if y_{11} is deleted.

one of which is very close to zero and sharply peaked, while the other is centered on 0.5 and spread out with a peak about 7 logs lower than the mode close to zero. When y_{11} is about $65\sigma_e$ above Maine's other plans, the mode centered away from zero has only about 5% of the posterior mass.)

For deleting θ_1's constraint case, the linear approximation is, from (8.20)

$$E(\theta_1|\mathbf{Y}_{(l)}) \approx (1+v)E(\theta_1|\mathbf{Y}) - vE(\mu|\mathbf{Y}), \tag{8.23}$$

where $v = \sigma_e^2/n\sigma_s^2 \geq 0$, for $n = 8$ plans per state. For the computation shown here, I replaced $1/\sigma_e^2$ and $1/\sigma_s^2$ by their full-data posterior means. If an outlying plan in Maine causes $E(\sigma_s^2|\mathbf{Y})$ to be small and $E(\sigma_e^2|\mathbf{Y})$ to be large, then deleting θ_1's constraint case increases the former and decreases the latter, so (8.20) overshoots $E(\theta_1|\mathbf{Y}_{(i)})$. The degree of inaccuracy depends on the amount that the deleted constraint case changes the posterior distributions of σ_e^2 and σ_s^2.

For deleting y_{11} itself, the linear approximation for the posterior mean of θ_1 is

$$E(\theta_1|\mathbf{Y}_{(l)}) \approx (1+\nu)E(\theta_1|\mathbf{Y}) - \nu y_{11}. \tag{8.24}$$

where now $\nu = (Nn+s)/(Nn(n-1+s)-s)$, for $N = 45$ states and $n = 8$ plans per state, and $s = \sigma_e^2/\sigma_s^2$, and again to compute this I replaced $1/\sigma_e^2$ and $1/\sigma_s^2$ by their full-data posterior means. Because $\nu \geq 0$, the linear approximation moves $E(\theta_1|\mathbf{Y}_{(i)})$ away from $E(\theta_1|\mathbf{Y})$ in the direction opposite y_{11}. This works well until y_{11} becomes very extreme.

This artificial example used an extreme outlier to exercise the two approximate influence methods. Constraint-case deletions readily produce similar effects when the posterior variance of the affected θ_i is small. Thus both the linear approximation and importance sampling approximations have weaknesses, but the linear-approximation method seems less flawed because it overstates influence when it fails while the importance-sampling method understates influence when it fails.

For data cases, the effects shown above only occur when an outlier is extreme relative to the data-case error variance. As a referee to Hodges (1998) pushed me to note, it is hard to imagine an outlier that was extreme enough to make the linear approximation fail that would also escape detection on a casual examination of the data. For smaller datasets, the same effect can be obtained with outliers that are less extreme but which would still quite easy to detect in a residual plot. Thus, the linear approximation seems adequate for data cases.

MacEachern and Peruggia (2000) gave a modified importance sampler intended to avoid the difficulty noted here. However, their method uses posterior modes and thus would be problematic for hierarchical models.

I know of no way to completely avoid these problems short of re-running the MCMC for each case deletion or, for a non-Bayesian analysis, re-fitting the model for each case deletion. An exercise to this chapter suggests a variant on the above approach that uses the case-deletion formula for the error-variance estimate, equation (8.19), to select cases that might induce a big change when deleted, so that the model could be re-fit without these selected cases instead of relying on the linear approximation.

For the *actual* HMO premium dataset (Hodges 1998, Section 5.1), it was practical for research purposes to re-fit the simple random-effect model deleting each case in turn and thus to compute exact (to within Monte Carlo error) relative changes RC for each case. Four cases had true $|RC| > 2$, each of which was the effect of deleting a constraint case on its own θ_i: Maine ($RC = 3.4$), Connecticut ($RC = 2.1$), Puerto Rico ($RC = -2.1$), and North Dakota ($RC = -2.4$). Deleting Maine's constraint case, for example, increased Maine's posterior mean from \$191 to \$231. Each of these states had few plans and an extreme average premium and thus was shrunk considerably, hence the large constraint-case influence. For each of these cases, the importance weights were dominated by a single Gibbs draw. For cases with true $|RC| < 1$, the importance-sampling approximation was accurate but for larger true $|RC|$ this approximation understated the true value. By contrast, the linear approximation was effectively exact for all cases. Section 5.2 of Hodges (1998) computed these case-influence measures for a larger model with two predictors at the state level. In that analysis, the linear-approximation $|RC|$ values were all too large by a factor of about 3.4, while the importance-sampling approximate $|RC|$ were too low by factors of 5–10. Unfortunately, this analysis was flawed because this posterior distribution is bi-modal and my far-too-short Gibbs sampler run only found one mode. That mode happened to have a very small σ_s^2, which accounts for the linear approximation's inaccuracy. (Wakefield 1998 found both modes.)

The relative changes RC describe the effect of deleting a case on the estimate of an individual parameter in Θ. Cook's distance, by contrast, describes the effect of deleting a case on all of Θ. In analyzing the fit of the simple random effect model to the HMO premium data, Cook's distance was dominated by elements of Θ that were affected little by case deletions, as expected. Thus, for example, deleting Maine's constraint case changed Maine's θ_i with $RC = 3.4$, and deleting North Dakota's constraint case changed North Dakota's θ_i with $RC = -2.4$, but Cook's distance for these two deletions were 0.38 and 0.15 respectively, neither of which suggests the magnitude of the change in the most-affected θ_i.

8.5 Residuals

In the ordinary linear model with $\mathbf{y} = \mathbf{A}\beta + \varepsilon$, the residuals are $\hat{\varepsilon} = (\mathbf{I} - \mathbf{A}(\mathbf{A}'\mathbf{A})^{-1}\mathbf{A}')\mathbf{y}$. In the familiar plot with residuals on the vertical axis and fitted values on the horizontal axis (e.g., Cook and Weisberg 1982, section 2.3.1), the residuals are internally studentized — each raw residual is divided by its estimated standard deviation under the assumed model, $\hat{\sigma}(1-h_i)^{0.5}$ — so that the studentized residuals all have the same variance under the assumed model.

In ordinary linear models, residuals are used for the aforementioned plot of residuals versus fitted values, which shows outliers and evidence of non-constant variance. Residuals are also plotted against candidate predictors to check whether to add the predictor (an added variable plot), to check for non-linearity, and to check for non-constant variance that is a function of the predictor. Quantile plots of studentized residuals are a check of the assumed normal distribution for the errors. In all these uses, we ignore correlations among the residuals $\hat{\varepsilon}$, which are usually small.

If we try to directly adapt the plot of studentized residuals versus fitted values to richly parameterized models in the constraint-case formulation, we immediately see a severe defect. Fix Γ and proceed as for the three previous diagnostics. If data and constraint cases are plotted together, the points for the constraint cases are necessarily uninformative because the residuals for the constraint cases are $0 - (\mathbf{S}_1\hat{\Theta}_1 + \mathbf{S}_2\hat{\Theta}_2)$, while the fitted values are $\mathbf{S}_1\hat{\Theta}_1 + \mathbf{S}_2\hat{\Theta}_2$. These *are* the correct residuals for this plot but the correct fitted values are $\mathbf{S}_2\hat{\Theta}_2$. Thus, unlike the previous three diagnostics, the data and constraint cases need separate residual plots.

Now consider the usual residual plot for the data cases only. Another problem surfaces, this one intrinsic to hierarchical and other richly parameterized models. In ordinary linear models, the vector of residuals $\hat{\varepsilon}$ is orthogonal to the vector of fitted values $\mathbf{A}\hat{\beta}$ by construction. In richly parameterized models, data-case residuals are generally not orthogonal to the fitted values because the data-case fitted values $\hat{\mathbf{y}}$ are not an orthogonal projection of the data $\mathbf{y}$. In the mixed-linear-model formulation, this is perhaps clearest in equation (1.11). In the constraint-case formulation, beginning at equation (8.13) and recalling that sans-serif font indicates pre-multiplication by $\Gamma^{-1/2}$, note that

$$\begin{aligned}
\hat{\mathsf{y}}'\hat{\mathsf{E}}_d &= \hat{\Theta}_1'\mathsf{K}'\hat{\mathsf{E}}_d \\
&= -\hat{\Theta}_1'\mathsf{S}_1'\hat{\mathsf{E}}_c \qquad \text{from (8.13)} \\
&= \hat{\Theta}_1'\mathsf{S}_1'\mathsf{S}_1\hat{\Theta}_1 + \hat{\Theta}_1'\mathsf{S}_1'\mathsf{S}_2\hat{\Theta}_2. \qquad (8.25)
\end{aligned}$$

When S_2 is null, as in the spatial model of Example 4 (Section 2.1), $\hat{\mathsf{y}}'\hat{\mathsf{E}}_d \geq 0$, that is, the data-case residuals and fitted values are positively correlated except in trivial cases where the smoothing variance σ_s^2 is infinite. For the one-way random-effects model, $\sigma_s^2\hat{\mathsf{y}}'\hat{\mathsf{E}}_d = \hat{\Theta}_1'\hat{\Theta}_1 - N\hat{\mu}\bar{\Theta}_1$ where $\bar{\Theta}_1$ is the simple average of the $\hat{\theta}_i$. Usually $\hat{\mu} \approx \bar{\Theta}_1$, in which case $\sigma_s^2\hat{\mathsf{y}}'\hat{\mathsf{E}}_d \approx (\hat{\Theta}_1 - \bar{\Theta}_1)'(\hat{\Theta}_1 - \bar{\Theta}_1) \geq 0$. For both models, $\hat{\mathsf{y}}'\hat{\mathsf{E}}_d \geq 0$ for each value of Γ, so it is non-negative marginally as well. For these models, then, a plot of studentized residuals versus fitted values will show a positive slope even for artificial data generated from the model being fit.

To make this more concrete, consider Minnesota's 4 plans in the HMO premium data, plotted in Figure 8.6. Minnesota has the lowest average premium of the 45 states. In the fit with no shrinkage, where Minnesota's fitted value is the average of its 4 plans' premiums ("MN average"), the fitted value is centered among the 4 observations and the residuals sum to zero by construction. In the random-effects fit, Minnesota's fitted value, the posterior mean ("MN post mean"), is shrunk upward toward the center of all the states ("E(mu|Y)"), so the average of Minnesota's residuals is negative. In this case with a simple random effect, residuals tend to be smallest (most negative) for states with the smallest fitted values and to be largest (most positive) for states with the largest fitted values. In other words, a plot of residual versus fitted will show a trend upward to the right.

More complicated models will have more complicated patterns in their residuals. For Example 4 (Experimental plots in a long row, Section 2.1.1), if the treatment effect is substantial, so that the fitted values for the treatment groups are separated, the plot of residuals versus fitted will show the same pattern just described for the simple random-effects model, but will show it separately for each of the treatment and control groups. (Drawing this plot for simulated data is an exercise.)

So residuals deviate from the orthogonality that is present in simple linear models. In fact this deviation can be broken into two distinct pieces, one arising from a form of regression to the mean and one arising from bias in estimating random effects. To isolate the regression-to-the-mean effect, consider the simple random-effects model for the HMO premium data, let $n_i = n$, and suppose the constraint-case errors δ_i — the random effects — are in fact all zero, so there is no bias in estimating them. This implies each state has true mean premium $\theta_i = \mu$. Assume, as above, that μ has a flat prior and fix the two variances at finite positive values. If $\bar{y}_{i.}$ is the average of y_{ij} over j, $\bar{y}_{..}$ is the average of y_{ij} over i and j, and $\bar{\varepsilon}_{i.}$ and $\bar{\varepsilon}_{..}$ are defined analogously, then

$$\hat{y}_{ij} = \frac{n\sigma_s^2}{n\sigma_s^2 + \sigma_e^2}\bar{y}_{i.} + \frac{\sigma_e^2}{n\sigma_s^2 + \sigma_e^2}\bar{y}_{..} = \mu + \bar{\varepsilon}_{..} + \alpha(\bar{\varepsilon}_{i.} - \bar{\varepsilon}_{..}) \qquad (8.26)$$

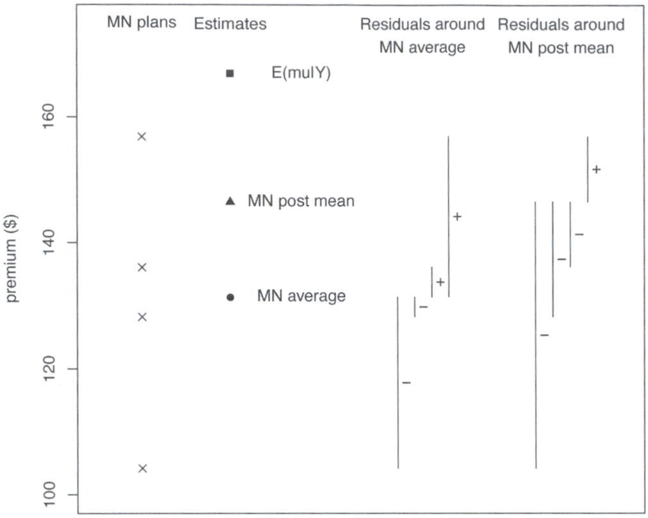

Figure 8.6: HMO premium dataset: Minnesota's 4 plans and the residuals around the state average premium ("MN average," an unshrunk fit) and around the posterior mean from the random-effect fit ("MN post mean," a fit with shrinkage).

for $\alpha = n\sigma_s^2/(n\sigma_s^2 + \sigma_e^2)$, so that

$$\hat{\varepsilon}_{ij} = y_{ij} - \hat{y}_{ij} = \varepsilon_{ij} - \bar{\varepsilon}_{i.} + (1-\alpha)(\bar{\varepsilon}_{i.} - \bar{\varepsilon}_{..}). \tag{8.27}$$

The term $\varepsilon_{ij} - \bar{\varepsilon}_{i.}$ is the usual residual variation around the state average; the term $(1-\alpha)(\bar{\varepsilon}_{i.} - \bar{\varepsilon}_{..})$ is peculiar to hierarchical models. Over repeated realizations of ε_{ij}, $\bar{\varepsilon}_{i.} - \bar{\varepsilon}_{..}$ will have mean zero because $E(\varepsilon_{ij}) = 0$. However, for any given realization of ε_{ij}, the $\bar{\varepsilon}_{i.}$ are spread around $\bar{\varepsilon}_{..}$ and because shrinkage is proportional, the $\hat{y}_{ij}$ that are largest in absolute value will be shrunk the most in absolute terms. Thus, the plot of studentized residuals versus fitted values shows a positive slope.

In practice, this effect depends on the estimates for the two variances σ_s^2 and σ_e^2. If $\hat{\sigma}_s^2 = 0$ — which is common with restricted-likelihood maximization — then $\alpha = 0$, the state means are all estimated to be $\hat{\mu}$, and the plot of residual versus fitted has no spread on its horizontal axis. For $\hat{\sigma}_s^2 > 0$, however, this regression effect is present.

Now suppose at least one δ_i is not zero, i.e., the random effect is not null, and again fix σ_s^2 and σ_e^2 at finite positive values. The mean of the data-case residuals, across repeated realizations of ε_{ij} holding δ_i fixed, is

$$E(\hat{\mathrm{E}}_d|\delta,\xi) = (\mathbf{I} - \mathbf{V})\Gamma^{-1/2} \left[\begin{array}{c} \mathbf{0} \\ \hline \delta \\ \hline \xi \end{array} \right] \tag{8.28}$$

which is non-zero in general. That is, the residuals are biased. (Hilden-Minton 1995, section 4.1 gives an alternative derivation.) This bias is the same as the bias in the fitted values described in Chapter 3's equation (3.20) for the mixed-linear-model formulation. Thus, the data-case residuals display three effects: the usual residual variation, a regression effect, and bias.

These patterns in the residuals are not an artifact of the reformulation, but are intrinsic to hierarchical and other richly parameterized models. Empirical Bayes residuals (e.g., Hilden-Minton 1995) display the same patterns because they are computed as above with a particular Γ. The joint posterior distribution of the ε_{ij} (Chaloner 1994) will also show the regression and bias effects: $E(\varepsilon_{ij}|\mathbf{Y})$ is simply $E(\varepsilon_{ij}|\mathbf{Y},\Gamma)$ integrated against the posterior distribution for Γ, but $E(\varepsilon_{ij}|\mathbf{Y},\Gamma)$ is identical to $\hat{\varepsilon}_{ij}$ computed with the same Γ.

Hilden-Minton (1995, sections 4.1, 4.2) offers two ways around this problem. The first, in terms of the HMO premium data, is to fit a separate model for each state — in effect, fit an unshrunk model — and use the residuals from these fits. This avoids the bias and regression effects just described, producing residuals with the same properties as in the simple linear model, which is what this fit is. Unfortunately, this fix cannot be applied to all models that can be represented as mixed linear models, for example to ICAR models. Also, it would be rather strange to fit an unpenalized spline to assess how well the penalized spline fits. In the HMO premium data, the separate models in each state may have unstable fits, and states with 1 or 2 plans will have no residuals at all. Nonetheless, this approach can be useful for truly hierarchical models with adequate cluster sizes because it allows the lowest-level (e.g., plan-level) model to be evaluated free from the effects of deficiencies in the higher-level model.

Hilden-Minton's second solution is to construct linear transforms of the data-case residuals, $l'\hat{\mathsf{E}}_d$, that are independent of each other and have no bias. This construction is easily adapted to the reformulation given here and in the cases I have examined, the bias-free residuals are also free of the regression effect. Unfortunately, transformation to $l'\hat{\mathsf{E}}_d$ can obliterate indications such as multiple outliers which are manifest in a plot of the untransformed residuals. Finally, effects in the $l'\hat{\mathsf{E}}_d$ are not easy to attribute to individual observations, which is usually an important goal.

Lee et al. (2006), using the mixed-linear-model formulation, propose the following fix (p. 163): Plot the residuals from the full fit, $\hat{e}_l = y_l - X_l\hat{\beta} - Z_l\hat{\mathbf{u}}$, where X_l and Z_l are the lth rows of the design matrices $\mathbf{X}$ and $\mathbf{Z}$ respectively, versus the fitted *fixed effects* $X_l\hat{\beta}$. They claim "this successfully removes the unwanted trend" but do not give a proof; following their citations produces a dead end and this result appears not to hold in the generality they claim. In particular, their claim appears to require that $\mathbf{R}$, the mixed linear model's error covariance, be proportional to the identity $\mathbf{I}$ and that the fixed-effect design matrix $\mathbf{X}$ include an intercept, which is not the case for, e.g., the ICAR model. Even if their result were true in the generality they claim, for many richly parameterized models the fitted fixed effects $X_i\hat{\beta}$ are almost completely divorced from the fitted values of interest. For a penalized spline, the fitted fixed effects are the smooth fit to the data up to the first knot, extended over the entire range of the predictor x. Similarly, for the linear growth model in Chapter 2's equations (2.38, 2.39), the fixed effects are the initial level and slope extended over the time span of the dataset, while in the actual fit the level and slope change at every observation. Even if the local-level evolution variance is fixed at zero, as in Example 8 (Localizing epileptic activity, Section 6.3), the local trend changes at each time step. For either model, it is hard to imagine how the fixed-effect fitted values can have any relevance to the whole model's fit.

One possible fix, as yet unexplored, is to estimate the bias in the fitted values or equivalently the bias in the residuals, and compute bias-corrected residuals for the plot of residuals versus fitted values. As the preceding paragraph suggests, such an approach would seem to make sense for penalized splines or state-space models. An exercise explores this idea.

Given the flaws in proposed fixes for the artifacts in residuals, what should we do in practice? We routinely ignore structure in residuals from ordinary linear models, for example, that estimated residuals are correlated and can behave more like draws from a normal distribution than the true ε themselves (i.e., they can display "supernormality," Cook and Weisberg 1982, section 2.3.4). If we are to use the data-case plot of studentized residuals versus fitted values, we must learn to ignore still more structure than we ignore in residual plots for ordinary linear models. That is, we must be content to find grosser outliers and heteroscedasticity than we expect to find in ordinary linear regressions. Ruppert et al. (2003) implicitly follow this advice by making no mention (that I found) of these artifacts in residual plots. For the class of models and the purposes they discuss, this may be reasonable. They are interested in residuals mainly for two things: First, for detecting non-constant variance, which should be visible in an unmodified plot of data-case residuals versus fitted values despite the artifacts, and second, for assessing non-linearity in plots of residuals versus candidate predictors, which the artifacts described above do not appear to affect.

For constraint-case residuals, a plot of studentized residuals versus the correct fitted values $\mathbf{S}_2\hat{\Theta}_2$ does not have this non-orthogonality problem if the prior variances of the fixed effects are infinite (equivalently, prior cases are omitted from the constraint-case formulation). This follows from equation (8.13). However, if the fixed effects have a proper normal prior then the constraint-case residual plot has the same non-orthogonality problem as the data-case residual plot.

Variances for the data- and constraint-case residuals, taking as given an estimate of Γ, are easily derived; Hodges (1998, Section 4.5) gives formulas.

Figure 8.7 is the plot of data-case residuals versus fitted values for the HMO premium data; I studentized the residuals by setting Γ to its posterior mean. As expected, the residuals lie along the sloping solid line. Ignoring this slope, the plot shows an outlier with $\hat{y}_i$ about 180, a plan in Washington state. There is no strong suggestion of a funnel opening to the right, as would be expected if the error variance increased with the fitted value. However, Figure 8.7 shows somewhat more points in the range 2 to 4 than in the range -4 to -2, and a normal quantile plot of the studentized data-case residuals (not shown) also suggests a slightly long upper tail. Hodges (1998), Figure 5, shows studentized residuals from a model that simply fits a mean for each state, as suggested by Hilden-Minton (1995); it adds nothing to Figure 8.7. Figure 8.8 shows a normal quantile plot of the studentized constraint-case residuals, which also

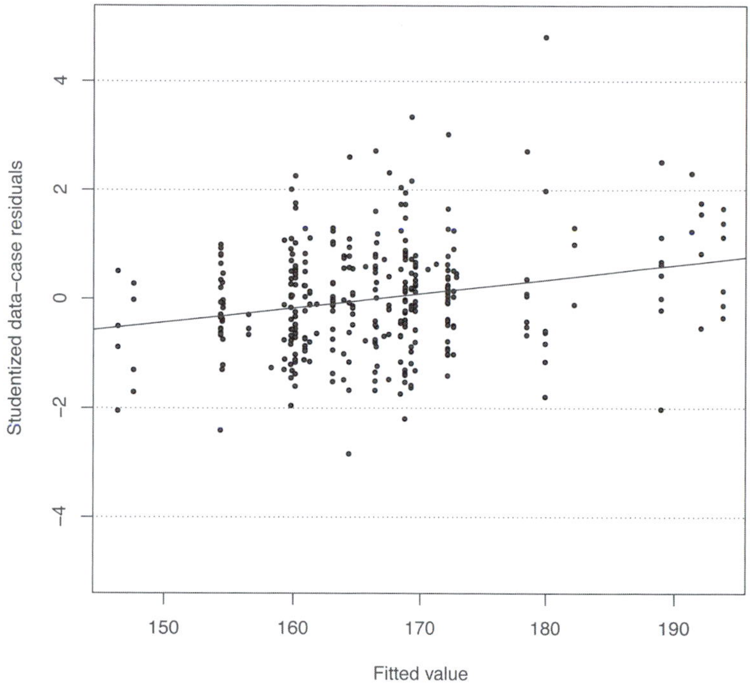

Figure 8.7: HMO premium dataset: Plot of studentized residuals versus fitted values for the simple random-effects fit. The solid line is the ordinary least-squares regression of studentized residuals on fitted values.

suggests a long upper tail. (Wakefield 1998 obtained a similar finding with a somewhat different plot.)

Exercises

Regular Exercises

1. Fit the simple random-effect model to the HMO premium dataset using your favorite MCMC software and compute the diagnostics shown here and any others that interest you.
2. Fit the simple random-effect model to the HMO premium dataset using restricted likelihood or another non-Bayesian method, if you prefer. Use your point estimates to compute the diagnostic methods and compare them to the results in this chapter.
3. (Section 8.4) Fit the model in Hodges (1998) Section 5.2, which adds two state-level predictors to the model. Wakefield (1998) fit this model and unlike me, he ran his MCMC long enough to find the two posterior modes. Check the accuracy

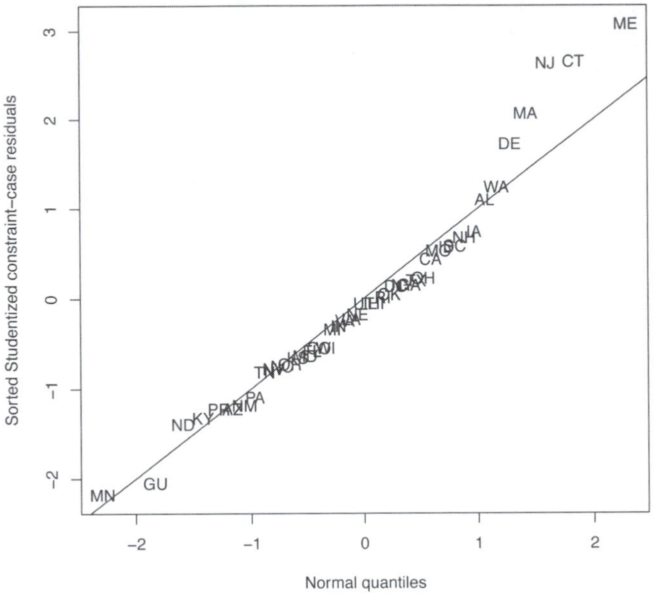

Figure 8.8: HMO premium dataset: Quantile plot of studentized constraint-case residuals from the simple random-effects fit.

of the linear-approximation and importance-sampling methods for case influence using the posterior mean and posterior median. (Using the posterior mode doesn't make sense here because this posterior is bimodal.)

4. (Section 8.5) For Example 4 (Experimental plots in a long row, Section 2.1.1), generate artificial data from the model. Use a fairly large treatment effect and a fairly large ratio of error variance to smoothing variance so there's a lot of shrinkage in the fit. Fit the model and plot the data-case residuals versus the fitted values. This plot should show the pattern described in Section 8.5.

Open Questions

1. (Section 8.4) For the linear-approximation method of case influence, the concern is that deleting a case might change the variance estimates substantially, so that the non-linear effects of deleting that case would cause the linear approximation to be inaccurate. Consider the following potential "patch" for models in which the errors ε have covariance $\sigma_e^2\mathbf{I}$. Write $\Gamma = \sigma_e^2\Sigma$ where

$$\Sigma = \begin{bmatrix} \mathbf{I} & \mathbf{0} \\ \mathbf{0} & \Sigma_2 \end{bmatrix}, \tag{8.29}$$

where $\Sigma_2 = \frac{1}{\sigma_e^2}\Gamma_2$ is re-parameterized as needed so it is no longer explicitly a function of σ_e^2. In the HMO premium example, $\Sigma_2 = s\mathbf{I}_{45}$, for $s = \sigma_s^2/\sigma_e^2$. Now re-derive the linear approximation method fixing Σ at its estimate instead of Γ, and apply the case-deletion formula for the error variance, equation (8.19), to $\hat{\sigma}_e^2$ in this new parameterization. If deleting a case induces a large change in $\hat{\sigma}_e^2$, then the linear approximation method may be inaccurate for that case. This could perhaps be used to screen for cases for which the model should be re-fit instead of relying on the linear approximation. The apparent advantage of this approach is that it reduces reliance on the linear approximation, though it too relies on fixing variance-structure quantities at their estimates, so it might not be any better than the linear approximation itself.

2. (Section 8.5) Residuals from fitting a richly parameterized model are biased because the fit is biased when the random effects are not null. Equation (8.28) gives an expression for the bias in the data-case residuals for models expressed in the constraint-case form; equation (3.20) gives an expression for the bias in the *fit* for models expressed in mixed-linear-model form. So consider bias-correcting these residuals. One way to start is to fit a model to the HMO premium data in either the constraint-case or mixed-linear-model form, compute the usual residuals, estimate the bias in the residuals (fixing the variances at some estimate), and compute bias-corrected residuals by subtracting the bias from the residuals. It should be straightforward to extend this to other models, e.g., Chapter 3's penalized splines. The key question is: Can the bias-corrected residuals allow you to find effects that you would miss in the usual residuals, with all their artifacts? One way to explore this is to consider models and simulated data with outliers in places where you'd expect the usual residuals to hide them. For example, for a penalized spline model, put an outlier at a peak or valley of the true model, where the bias is greatest. It's also of interest to see if the regression effect is still present in these bias-corrected residuals: Does bias correction also correct the regression effect?

Chapter 9

Puzzles from Analyzing Real Datasets

The present chapter gives four puzzles or oddities that arose in analyzing real datasets using models with random effects, and the following chapters use these puzzles to explore some aspects of collinearity in richly parameterized linear models. Some of these puzzles arose for models that do not have normally distributed errors but they are nonetheless relevant to the normal-normal models on which the present book is focused.

A reader commented that in a Bayesian analysis focusing on the posterior predictive distribution of $\mathbf{y}$, in which the parameters $(\beta, \mathbf{u}, \phi)$ merely serve the predictive purpose, collinearity is not a fundamental problem, though it may make computing harder. I am not so sure. I believe this is true in simpler problems but needs to be proved for mixed linear models, where fixed and random effects can affect the posterior predictive distribution differently, so that their collinearity — i.e., their competition to capture variation in the data — may not be as innocuous as in simpler problems.

9.1 Four Puzzles

9.1.1 Introducing Spatially Correlated Errors Makes a Fixed Effect Disappear

Example 10. Stomach cancer in Slovenia. Dr. Vesna Zadnik, a Slovenian epidemiologist, collected counts of stomach cancers in the 194 municipalities that partitioned Slovenia in 2004, for the years 1995 to 2001 inclusive. She was studying the possible association of stomach cancer with socioeconomic status as measured by a score calculated from 1999 data by Slovenia's Institute of Macroeconomic Analysis and Development. (Her findings were published in Zadnik & Reich 2006.) Figure 9.1a shows the standardized incidence ratio (SIR) of stomach cancer for the 194 municipalities; for municipality $i = 1, \ldots, 194$, $SIR_i = O_i/E_i$, where O_i is the observed count of stomach cancer cases and E_i is the expected count using indirect standardization, i.e., $E_i = P_i \sum_j O_j / \sum_j P_j$, P_j being municipality j's population. Figure 9.1b shows the socioeconomic scores for the municipalities, SEc_i, after centering and scaling so the SEc_i have average 0 and finite-sample variance 1. In both panels of Figure 9.1, dark colors indicate larger values. SIR and SEc plainly have a negative association: western municipalities generally have low SIR and high SEc while eastern municipalities generally have high SIR and low SEc.

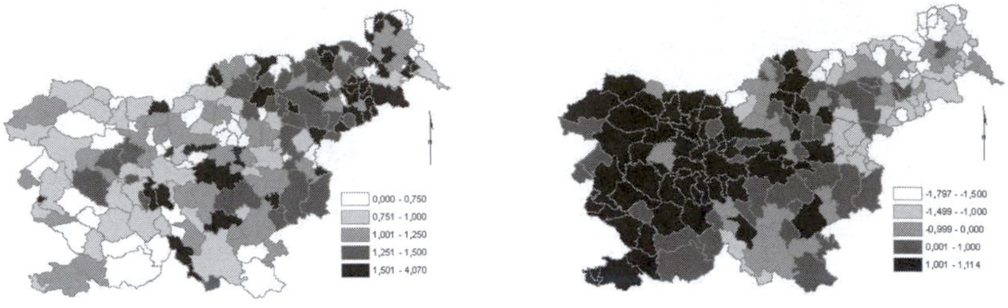

(a) Standardized incidence ratio, SIR (b) Socioeconomic status, SEc

Figure 9.1: For the Slovenian municipalities: Panel (a), observed standardized incidence ratio $SIR = O_i/E_i$; Panel (b), centered and scaled socioeconomic status SEc.

Following advice received in a spatial-statistics short course, Dr. Zadnik first did a non-spatial analysis assuming the O_i were independent Poisson observations with $\log\{E(O_i)\} = \log(E_i) + \alpha + \beta SEc_i$, with flat priors on α and β. This analysis gave the obvious result: β had posterior median -0.14 and 95% posterior interval $(-0.17, -0.10)$, capturing Figure 9.1's negative association.

Dr. Zadnik continued following the short course's guidance by doing a spatial analysis using the improper conditionally autoregressive (ICAR) model of Besag et al. (1991). Dr. Zadnik's understanding was that ignoring spatial correlation would make β's posterior standard deviation (standard error) too small, while the spatial analysis would in effect discount the sample size with little effect on the estimate of β, just as generalized estimating equations (GEE) adjusts standard errors for clustering but (in my experience) has little effect on point estimates unless the working correlations are very large. As we will see, other people have different reasons for introducing spatial correlation into this sort of analysis.

In this model with spatially correlated errors, the O_i are conditionally independent Poisson random variables with mean

$$\log\{E(O_i)\} = \log(E_i) + \beta SEc_i + S_i + H_i. \tag{9.1}$$

The intercept is now the sum of two random effects, $\mathbf{S} = (S_1, \ldots, S_{194})'$ capturing spatial clustering and $\mathbf{H} = (H_1, \ldots, H_{194})'$ capturing heterogeneity. The H_i are modeled as independent draws from a normal distribution with mean zero and precision (reciprocal of variance) τ_h. The S_i are modeled using an L_2-norm ICAR, discussed in Chapter 5. The ICAR represents the intuition that neighboring municipalities tend to be more similar to each other than municipalities that are far apart, where similarity of neighbors is controlled by an unknown parameter τ_s that is like a precision. (Chapter 5 parameterized the ICAR using $\sigma_s^2 = 1/\tau_s$.)

Zadnik's Bayesian spatial analysis used independent gamma priors for τ_h and τ_s with mean 1 and variance 100 and a flat prior for β, giving a posterior for β with mean -0.02 and central 95% interval $(-0.10, 0.06)$. Compared to the non-spatial analysis, the 95% interval was indeed wider and the deviance information criterion (DIC; Spiegelhalter et al. 2002) decreased from 1153 to 1082 even though the DIC's effective number of parameters (p_D) increased sharply, from 2.0 to 62.3. These changes were expected. The surprise was that the negative association, which is quite plain in the maps and the non-spatial analysis, had disappeared. What happened?

This spatial confounding effect, which was reported in Reich, Hodges, & Zadnik (2006), has been reported elsewhere but is not widely known. The earliest report is apparently Clayton et al. (1993), who used the term "confounding due to location" for a less dramatic but still striking effect in analyses of lung-cancer incidence in Sardinia. Those authors and Wakefield (2007) also reported a similar-sized effect in the long-suffering Scottish lip-cancer data.

This is the best-understood of the puzzles discussed in this book's Parts III and IV and a clear instance of confounding between the design matrices of the fixed and random effects. Section 10.1 describes what is known about it and what, if anything, should be done about it. Interpretation of this effect and appropriate choice of a remedy depend on whether the random effects **S** and **H** are old- or new-style.

9.1.2 Adding a Clustering Effect Changes One Fixed Effect but Not Another

The analysis in this example is a time-to-event analysis with clustering, but this puzzle is easily reproduced in data with normal errors, as we will see.

Example 11. Tooth crowns in children (kids and crowns). A badly decayed primary tooth ("baby tooth") is often capped with a stainless-steel crown so the child can retain the tooth until it is naturally exfoliated and replaced by a permanent tooth. Several types of crowns were on the market at the time of this study but no data were available to compare their performance after placement. Dr. Fadi Kass collected data on 167 children who had crowns placed under general anesthesia in one of 15 pediatric dental practices in the Minneapolis/Saint Paul area. The dataset includes three types of crowns, labeled Types I, III, and IV. (Few children had Type II crowns, so they are omitted.) Each child had between 1 and 4 crowns in the dataset and each child had crowns of only one type, so comparison of crown types is a between-child comparison. Dr. Kass collected several covariates but these do not affect the puzzle described here, so they are not considered. (Kass 2007 presents analyses done by my former student Yue Cui. The analyses presented here differ slightly.)

The analysis used a time-to-event outcome, where possible events included unrepairable damage to a crown or repairable events in which the crown fell off (so it could be re-attached) or suffered repairable damage. Thus a given child could have as many as 4 crowns in the dataset, and a given crown could have more than one "lifetime" and thus more than one event.

We fit models ignoring the clustering of teeth and lifetimes by child and other models that included clustering of lifetimes by child. I understand that analyses ignoring clustering by child are simply wrong, but soon I will give a very practical reason for considering such analyses. We also did analyses adding a tooth-specific random effect to capture clustering of lifetimes within teeth with repairable failures;

Table 9.1: Results of Fitting Cox Regressions to the Crown-Failure Data, Both without and with a Random Effect for Child

Crown Type	Random Effect?	Estimate	Standard Error	P-Value
III	Absent	0.48	0.20	0.015
	Present	0.22	0.41	0.59
IV	Absent	0.14	0.14	0.33
	Present	0.16	0.26	0.54

in all such analyses, the tooth-within-child random effect had estimated variance of zero, so the analyses presented here ignore clustering of lifetimes by tooth. The form of analysis was Cox regression, using Terry Therneau's R packages survival and coxme.

Table 9.1 shows the results from fitting this model with only the crown-type predictor, both without a random effect for child ("Absent") and with a random effect for child ("Present"). The parameterization used Type I crowns as the reference group, so the estimates are the log relative hazards of Types III and IV crowns versus Type I crowns. For both Type III and Type IV crowns, adding the random effect roughly doubled the standard error. This increase was expected because the crown-type regressor varies exclusively between children, not within children. The puzzle is in the estimates: For Type III crowns, the point estimate declined by half when the random effect was added, while for Type IV crowns it changed very little. Why did either estimate change much, and why didn't *both* change? This question is of real practical interest: My collaborators commonly present me with clustered data that they analyzed without accounting for clustering, which give rather different results in analyses that *do* account for clustering. I need to explain such changes, especially when, as in this case, a significant effect in the incorrect analysis becomes a lot smaller after accounting for clustering. ("It's the wrong analysis, just ignore it" tends not to be persuasive.)

When the child random effect was present, its standard deviation was estimated to be 1.17. This describes variation between children in log hazard, so the hazard ratio between the upper and lower 2.5% points of the random effect's distribution was about $e^{4.7} \approx 106$. This implies large variation between children in hazard or, alternatively, little shrinkage of children toward a common center. As in the example of stomach cancer in Slovenia, the random effect appears to be competing with a fixed effect, but in the present example the random effect is a simple clustering effect.

Section 10.2 considers this example and simplified versions of it with normal errors. It shows that, like the Slovenia example, the change from adding a random effect arises from a form of collinearity or confounding, specifically from a type of informative cluster size, though not the type usually considered in the informative-cluster-size literature. (Hoffman et al. 2001 was apparently the first in a series of papers about re-weighting generalized estimating equations [GEE] to adjust for informative cluster size. Dunson et al. 2003 gave a quite different latent-variable analysis.)

9.1.3 Differential Shrinkage of Effects with Roughly Equal Estimates and Standard Errors

Example 12. Colon cancer treatment effect. Sargent et al. (2001) and Gill et al. (2004) examined the effect of chemotherapy for treating colon cancer after resection surgery in elderly persons, using a dataset constructed by pooling datasets from seven randomized trials. Zhang (2009, Section 3.5 and Chapter 4) re-analyzed this dataset using smoothed ANOVA (as in the present book's Section 4.2.2) extended to Cox proportional-hazards regression. The seven trials had shown a clear treatment effect; the new question in Sargent et al. (2001) and Gill et al. (2004) was whether and how the treatment's effect depended on patient characteristics including age and the stage of the patient's cancer. The question in Zhang (2009) was whether prediction of patient outcomes could be improved by including all interactions and shrinking them using smoothed ANOVA.

The analyses described here were preliminary analyses that were not included in Zhang (2009). The dataset (which could not be made available for this book) consisted of about 2,900 persons after some exclusions for missing covariates. The outcome was time to the earliest of death or cancer recurrence (i.e., disease-free survival). The analyses were Bayesian but used Cox's partial likelihood instead of a likelihood from a parametric model. As in Section 4.2.2, the smoothed ANOVA included all main effects and all interactions of treatment group (chemotherapy versus no chemotherapy), age (4 categories), and cancer stage (II versus III). The design matrix in the Cox regression was parameterized as follows. The column for the treatment effect was coded as +1 for no chemotherapy and -1 for chemotherapy; the column for stage was coded analogously. The three columns for age took values $(1,1,1)$ for people aged less than 50, $(1,-1,-1)$ for people aged 50 to 60, $(-1,1,-1)$ for people aged 61 to 70, and $(-1,-1,1)$ for people aged over 70. Columns for interactions were constructed by multiplying the corresponding main effect columns. Finally, all columns in the design matrix were scaled to have the same Euclidean length.

This Cox regression was first fit without shrinking any interactions and then refit shrinking all interactions by treating them as random effects, as in Section 4.2.2. Table 9.2 shows the results. In the analysis with no shrinkage (second and third columns from left), the treatment-by-stage interaction and the three coefficients in the treatment-by-age interaction have similar estimates, about one-sixth of the stage main effect on the log-hazard scale, and have posterior 95% intervals of roughly equal width. In the analysis with shrinkage (smoothed ANOVA, right-most two columns), the four interaction coefficients all have estimates that are smaller in magnitude, as expected. The puzzle is that they have been shrunk toward zero to greatly differing extents — the treatment-by-age 1 coefficient is smaller by 37%, while the treatment-by-age 2 coefficient is smaller by 86% — and the posterior 95% intervals have analogous differences in width. The random effects that implemented shrinkage treated the coefficients as exchangeable (independent conditional on their variances, with identical priors on those variances) and all coefficients had roughly the same posterior standard deviation in the unshrunk analysis, so the obvious explanations for differential shrinkage do not appear to apply. In a balanced smoothed ANOVA

Table 9.2: Analyses of the Colon-Cancer Data, without and with Shrinkage of Interactions

	No shrinkage		Shrinkage	
Effect	Estimate	Posterior interval	Estimate	Posterior interval
treatment-by-stage	−4.2	(−8.3, 0.02)	−2.5	(−5.9, 0.3)
treatment-by-age 1	−4.6	(−8.8, −0.5)	−2.9	(−5.9, 0.3)
treatment-by-age 2	−4.2	(−8.3, 0.01)	−0.6	(−2.6, 0.6)
treatment-by-age 3	−4.8	(−9.0, −0.6)	−1.1	(−5.7, 1.8)
stage main effect	−25.9	(−30.1, −21.8)	−23.4	(−26.7, −20.0)

Note: "Estimate" is the posterior mean; "posterior interval" is an equal-tailed 95% interval.

with one error term and normal errors, this kind of differential shrinkage cannot happen. (The proof is an exercise.)

The obvious candidate explanation is that unlike the balanced smoothed ANOVA with normal errors, the columns in the present regression are not orthogonal, in two senses: The design is not balanced and the regression is non-normal so that each case's error variance depends on its design-cell mean and censoring status (i.e., the inner product defining "orthogonal" is not the same as in an iid error case). Thus I conjectured that this puzzle arose from collinearity between the effects in the analysis, possibly involving the large stage main effect. Chapter 11 explores this conjecture using a simplified version of this problem with normal errors.

9.1.4 *Two Random Effects Obliterated by Adding an Apparently Unrelated Random Effect*

Example 8 revisited. Localizing epileptic activity. Section 6.3 gave the background for this problem and described a dynamic linear model (DLM) fit by Michael Lavine and his colleagues, containing components for the signal they sought and for quasi-cyclic artifacts for heartbeat and respiration. Call that three-component model (signal, heartbeat, respiration) Model 1. As described in Section 6.3 and shown in Figure 6.1, the "signal" part of Model 1's fit showed an unexpected pattern that was roughly cyclic with period about 117 time steps. Lavine and his colleagues had no explanation for this mysterious quasi-cyclic behavior and decided to filter it out of the signal by adding a third quasi-cyclic component to their DLM to give Model 2:

$$y_t = s_t + h_t + r_t + m_t + v_t, \tag{9.2}$$

where m_t is the new mystery term and the other terms are as in Model 1, equation (6.17). The state equation for the mystery component m_t had the same form as the other quasi-cyclic components, h_t and r_t, with error $\mathbf{w}'_{m,t} = (w_{m1,t}, w_{m2,t}) \sim$ iid $N_2(0, W_m\mathbf{I}_2)$ and period 117 determined by inspection of the data, as before.

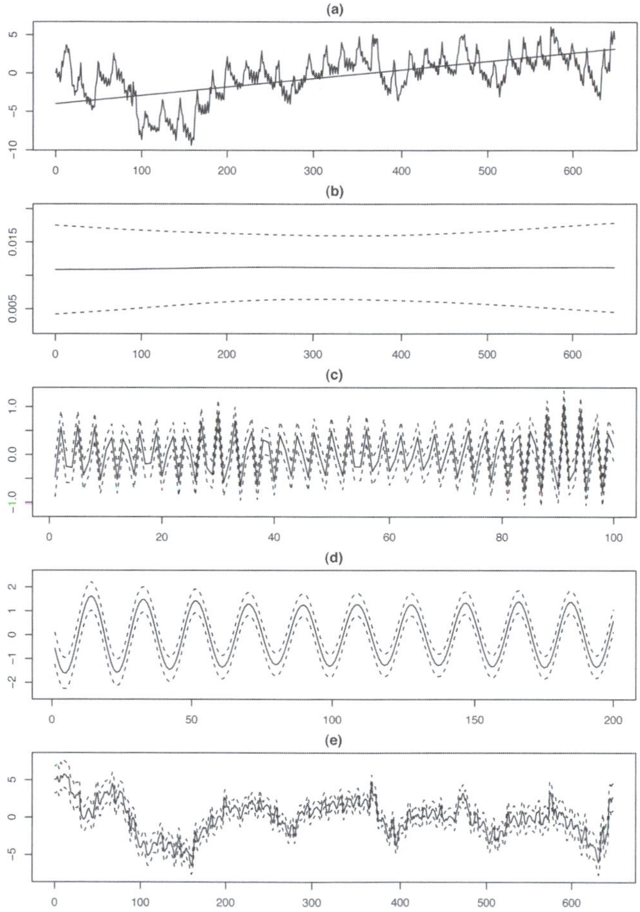

Figure 9.2: DLM fit to the optical-imaging data with mystery component (Model 2). Panel (a): Data and full fit (jagged line) and the smoothed response s_t (straight line). Panel (b): Fitted local slope, slope$_t$. Panel (c): Fitted heartbeat component, first 100 time steps. Panel (d): Fitted respiration component, first 200 time steps. Panel (e): Fitted mystery component. Panels b through e show pointwise 90% posterior regions, treating the variance estimates as known.

To everyone's great surprise, adding this mystery component changed the fit completely. Figure 9.2 shows the fit of Model 2 with the mystery component. The signal s_t has been radically smoothed so it is now almost identically a straight line (Figure 9.2a) with nearly constant local slope (Figure 9.2b). The respiration component, which had a complex fitted shape in Model 1, has been shrunk much closer to a regular, sinusoidal fit (Figure 9.2d). The heartbeat component, however, appears to have changed little from Model 1 (Figure 9.2c). Variation in the data that was previously captured by the signal and respiration components is now captured by the

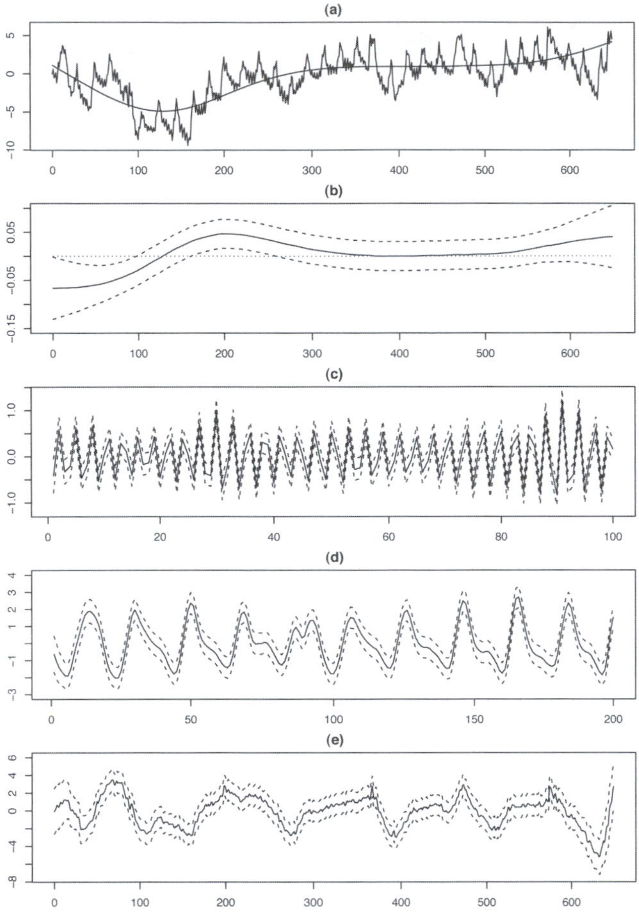

Figure 9.3: DLM fit to the optical-imaging data with two harmonics for respiration and mystery components (Model 3). Panel (a): Data and full fit (jagged line) and the smoothed response s_t (smooth line). Panel (b): Fitted local slope, slope$_t$. Panel (c): Fitted heartbeat component, first 100 time steps. Panel (d): Fitted respiration component, first 200 time steps. Panel (e): Fitted mystery component. Panels b through e show pointwise 90% posterior regions, treating the variance estimates as known.

mystery component (Figure 9.2e). The change in the fit is reflected in changes in the estimated variances, which are now $\hat{W}_v = 6.5 \times 10^{-9}$, $\hat{W}_s = 3.1 \times 10^{-8}$, $\hat{W}_h = 0.016$, $\hat{W}_r = 0.0039$, and $\hat{W}_m = 0.38$ and in changes in the DF in each component of the fit: 2.07 for signal (2 of which are from fixed effects, the level and slope at time zero), 147.6 for heartbeat (down from 180.9), 20.3 for respiration (down from 436.4), and 480.0 for the mystery component (up from zero). As with Model 1, the four components of Model 2 give a near-perfect fit to the data, so DF for error was zero to several decimal places.

Why did the fit change so much from adding the mystery component? Section 6.3 suggested two explanations. One is that the likelihood for the original four variances (Model 1) did not, in fact, change much from adding the mystery component (Model 2), but the maximizer happened to find a different peak when fitting Model 2. The other possible explanation is the spectacular collinearity among the random effects: Model 1 is saturated five times over, once for the signal and twice each for heartbeat and respiration, while Model 2 is saturated seven times over. Model 2's four components are obviously competing to capture aspects of the data; the puzzle is why the competition turned out the way it did. In particular, why did adding the mystery component cause the signal and respiration components to shrink radically but have little effect on the heartbeat component? It is probably important that for Model 1, the fitted respiration component is far from sinusoidal (Figure 6.1d); this lack of fit to the underlying cyclic model probably contributed to the change in fit from adding the mystery term. But how?

The non-sinusoidal shapes of Model 1's fitted respiration component and Model 2's fitted mystery term, among other things, motivated Lavine and his collaborators to change the respiration and mystery components by adding a second harmonic to each. In the new model, Model 3, the respiration component's part of the state equation (with the "r" superscript suppressed) is

$$\begin{pmatrix} b_{1,t}\cos\alpha_{1,t} \\ b_{1,t}\sin\alpha_{1,t} \\ b_{2,t}\cos\alpha_{2,t} \\ b_{2,t}\sin\alpha_{2,t} \end{pmatrix} = \begin{bmatrix} \cos\delta & \sin\delta & 0 & 0 \\ -\sin\delta & \cos\delta & 0 & 0 \\ 0 & 0 & \cos 2\delta & \sin 2\delta \\ 0 & 0 & -\sin 2\delta & \cos 2\delta \end{bmatrix} \begin{pmatrix} b_{1,t-1}\cos\alpha_{1,t-1} \\ b_{1,t-1}\sin\alpha_{1,t-1} \\ b_{2,t-1}\cos\alpha_{2,t-1} \\ b_{2,t-1}\sin\alpha_{2,t-1} \end{pmatrix} + \mathbf{w}_{r,t}, \tag{9.3}$$

where $\mathbf{w}_{r,t} \sim N_4(0, W_r\mathbf{I}_4)$. The mystery component's state equation changes analogously.

The resulting fit (Figure 9.3) is about what these investigators hoped they would get by adding the mystery component in the first place. The signal component is smooth (Figure 9.3a,b), perhaps a bit too smooth but it does capture the data's large-scale trend. The respiration component (Figure 9.3d) is like Model 1's fitted respiration component but somewhat smoother, while the heartbeat component has again changed little compared to either Model 1 or Model 2. DF in Model 3's fit are 6.5 for signal, 291.2 for heartbeat, 189.1 for respiration, 235.2 for mystery, and 0.015 for error. (DF for fixed effects are 2, 2, 4, and 4 for signal, heartbeat, respiration, and mystery component respectively.)

Again, the question is: Why did adding the second harmonics to the heartbeat and mystery components "fix" the problem of Model 2? The obvious answer is that some components of Model 2 fit poorly in the sense that they require relatively large state-evolution variances, while Model 3 fits better in the same sense. But why should this dataset's particular lack of fit produce these specific changes from Model 1 to Model 2 to Model 3? No one knows. Chapter 12 explores the collinearity hypothesis as a potential explanation.

9.2 Overview of the Next Three Chapters

I conjectured that each of these four puzzles could be explained at least in part as a consequence of collinearity involving the random effects design matrix **Z** or both design matrices **X** and **Z**. As it turns out, spatial confounding (Section 9.1.1) arises entirely and straightforwardly from collinearity of **X** and **Z** (Section 10.1). The change in fixed effect estimates from adding clustering (Section 9.1.2) arises from informative cluster size, which may be profitably understood as a form of confounding. Section 10.2 explores a simplified version of this problem with normal errors. Chapter 11 explores collinearity as an explanation for differential shrinkage (Section 9.1.3) using a simplified normal-errors model. It proves that collinearity cannot in fact produce differential shrinkage, at least not in this simplified normal-errors model. Other possible explanations for differential shrinkage are discussed at the end of Chapter 11 and approaches to exploring them are given as exercises. Finally, the conquering mystery component (Section 9.1.4) has two candidate explanations, collinearity or an ill-behaved likelihood function not caused by collinearity. Chapter 12 considers this example of competing model components of a model as well as simpler examples. Although the likelihood's odd behavior complicates interpretation of the results, Chapter 12 tentatively concludes that this puzzle arises from the extreme collinearity of the various components' random effect design matrices and a specific lack of fit in Model 1.

Exercises

Regular Exercises

1. (Section 9.1.3) For a balanced smoothed ANOVA with one error and normal errors, prove that if a group of coefficients θ_k are exchangeable *a priori* and have identical unshrunk estimates, they will have identical shrunken estimates.
2. Re-analyze the three available datasets from this chapter to see whether you can figure out why the puzzles happened, before you read the following chapters and see what my colleagues and I did.

Chapter 10

A Random Effect Competing with a Fixed Effect

This chapter first examines spatial confounding using the Slovenian stomach cancer dataset (Section 9.1.1) and then examines a type of informative cluster size effect that produced the puzzle in the kids-and-crowns dataset (Section 9.1.2). As we will see, both arise from collinearity of fixed and random effects.

10.1 Slovenia Data: Spatial Confounding

This section, which is adapted from Hodges & Reich (2010), first shows in mechanical terms how spatial confounding is in fact a problem of confounding and when it occurs. Section 10.1.2 then describes methods for removing spatial confounding. Section 10.1.3 argues that spatial confounding is not an artifact of the ICAR model but also occurs for geostatistical models and two-dimensional penalized splines. Derivations are for normal-errors models; Reich et al. (2006) extended most of these to models with error distributions in the exponential family. Section 10.1.4 then considers interpretations of spatial confounding that have been proposed, in the process debunking the common belief that a spatially correlated random effect adjusts fixed effect estimates for missing covariates with spatial structure (e.g., Clayton et al. 1993). This section also discusses situations in which it is appropriate or necessary to remove spatial confounding using the methods described in Section 10.1.2.

10.1.1 The Mechanics of Spatial Confounding

Although the approach here is Bayesian, these results are straightforwardly modified to apply to a conventional analysis.

10.1.1.1 The Model, Re-written as a Mixed Linear Model

For an n-dimensional observation $\mathbf{y}$, write the normal-errors analog to model (9.1) as

$$\mathbf{y} = \mathbf{F}\beta + \mathbf{I}_n\mathbf{S} + \varepsilon \tag{10.1}$$

where $\mathbf{y}$, $\mathbf{F}$, $\mathbf{S}$, and ε are $n \times 1$, β is scalar, $\mathbf{y}$ and $\mathbf{F}$ are known, and β, $\mathbf{S}$, and ε are unknown. In the Slovenian data, $n = 194$ and $\mathbf{F} = SEc$, which is centered and scaled to have average 0 and finite-sample variance 1. The derivation below generalizes easily to any full-rank **F** without an intercept column (Reich et al. 2006); the intercept is implicit in the ICAR model/prior for **S**, as noted in Chapter 5. The error term ε is n-dimensional normal with mean zero and precision matrix $\tau_e\mathbf{I}$, τ_e being the reciprocal of the error variance σ_e^2. The Bayesian analysis shown below puts a flat prior on β, but this is not necessary. Readers may wish to refresh their memory of the ICAR model by reviewing Section 5.2.2 before proceeding.

The model (10.1) was written with design matrix $\mathbf{I}_n$ for the random effect **S** to emphasize that it is over-parameterized and is identified only because **S** is smoothed or constrained by the ICAR model. To clarify the collinearity and identification problems, re-parameterize the ICAR part of (10.1) as in Chapter 5. The neighbor-pair matrix **Q** has spectral decomposition $\mathbf{Q} = \mathbf{VDV}'$, where **V** is $n \times n$ orthogonal and **D** is diagonal with diagonal elements $d_1 \geq \ldots \geq d_{n-I} > 0$ and $d_{n-I+1} = \ldots = d_n = 0$, where I is the number of islands (disconnected groups of regions) in the spatial map. **V**'s columns $V_1, \ldots, V_n$ are **Q**'s eigenvectors; **D**'s diagonal elements are the corresponding eigenvalues. Re-parameterize (10.1) as

$$\mathbf{y} = \mathbf{F}\beta + \mathbf{Vb} + \varepsilon, \tag{10.2}$$

where $\mathbf{b} = \mathbf{V}'\mathbf{S}$ is an n-vector with a normal distribution having mean zero and diagonal precision matrix $\tau_s\mathbf{D} = \tau_s\text{diag}(d_1, \ldots, d_{n-I}, 0, \ldots, 0)$. The precision-like parameter τ_s is the reciprocal of Chapter 5's variance-like parameter σ_s^2.

The spatial random effect **S** thus corresponds to a saturated collection of canonical regressors, the n columns of **V**, of which the coefficients **b** are shrunk toward zero to an extent determined by $\tau_s\mathbf{D}$. The smoothing parameter τ_s controls shrinkage of all n components of **b** and the d_i control the relative degrees of shrinkage of the b_i for a given τ_s. Both the canonical regressors **V** and the d_i are determined solely by the spatial map through **Q**. The first canonical coefficient, b_1, has the largest d_i and is thus shrunk the most for any given τ_s; for the Slovenian map, d_1 = 14.46. If the spatial map has I islands, the last I b_i, $b_{n-I+1}, \ldots, b_n$ are not shrunk at all because their prior precisions $\tau_s d_i$ are zero, so they are fixed effects implicit in the spatial random effect. The b_i with the smallest positive d_i, d_{n-I}, is shrunk least of all the shrunken coefficients. For the Slovenian map with $I = 1$, this is b_{193}, with d_{193} = 0.03, so its prior precision is smaller than b_1's by a factor of about 500 for any τ_s.

To understand the differential shrinkage of the b_i, we need to understand the columns of **V**, which the spatial map determines. $V_{n-I+1}, \ldots, V_n$, whose coefficients $b_{n-I+1}, \ldots, b_n$ are not shrunk at all, span the space of the means of (or intercepts for) the I islands in the spatial map, as discussed in Section 5.2.2. $\mathbf{Q1}_n = 0$ for any map, so the overall intercept $\mathbf{1}_n$ always lies in the span of $V_{n-I+1}, \ldots, V_n$ and is thus implicit in the ICAR specification. Thus without loss of generality, we set $V_n = \frac{1}{\sqrt{n}}\mathbf{1}_n$ so all other V_i are contrasts, i.e., $\mathbf{1}_n'V_i = 0, i = 1, \ldots, n-1$.

Section 5.2.2 discussed interpretation of the canonical regressors $V_1, \ldots, V_{n-I+1}$, using the Slovenian map, shown in Figure 5.3, as an example. Based on this and other examples, V_{n-I}, whose coefficient b_{n-I} has the smallest positive prior precision $\tau_s d_{n-I}$, can be interpreted loosely as the lowest *frequency* contrast in **S** among the shrunken contrasts. In the map of Slovenia, V_{193} is "low-frequency" in the sense that it is a roughly linear trend along the long axis of Slovenia's map (Figure 5.3a). Figure 5.3b shows V_1 for the Slovenian data; it is roughly the difference between the two municipalities with the most neighbors (the dark municipalities) and the average of their neighbors.

10.1.1.2 Spatial Confounding Explained in Linear-Model Terms

For the normal-errors model (10.2), it is straightforward to show that the posterior mean of β conditional on the precisions τ_e and τ_s is

$$E(\beta|\tau_e, \tau_s, \mathbf{y}) = (\mathbf{F}'\mathbf{F})^{-1}\mathbf{F}'\mathbf{y} - (\mathbf{F}'\mathbf{F})^{-1}\mathbf{F}'\mathbf{V}E(\mathbf{b}|\tau_e, \tau_s, \mathbf{y}), \tag{10.3}$$

where $E(\mathbf{b}|\tau_e, \tau_s, \mathbf{y})$ is *not* conditional on β, taking the value

$$E(\mathbf{b}|\tau_e, \tau_s, \mathbf{y}) = (\mathbf{V}'\mathbf{P}^c\mathbf{V} + r\mathbf{D})^{-1}\mathbf{V}'\mathbf{P}^c\mathbf{y}, \tag{10.4}$$

for $r = \tau_s/\tau_e$ (or, written in terms of variances instead of precisions, $r = \sigma_e^2/\sigma_s^2$) and $\mathbf{P}^c = \mathbf{I} - \mathbf{F}(\mathbf{F}'\mathbf{F})^{-1}\mathbf{F}'$, the familiar residual projection matrix from linear models. These expressions are correct for full-rank $\mathbf{F}$ of any dimension. (Derivations are given as exercises.)

In (10.3), the term $(\mathbf{F}'\mathbf{F})^{-1}\mathbf{F}'\mathbf{y}$ is the ordinary least-squares estimate of β and also β's posterior mean in a fit without $\mathbf{S}$, using a flat prior on β. Thus the second term $-(\mathbf{F}'\mathbf{F})^{-1}\mathbf{F}'\mathbf{V}E(\mathbf{b}|\tau_e, \tau_s, \mathbf{y})$ is the change in β's posterior mean from adding $\mathbf{S}$, conditional on (τ_e, τ_s). Note that $\mathbf{V}E(\mathbf{b}|\tau_e, \tau_s, \mathbf{y})$ is the fitted value of $\mathbf{S}$ given (τ_e, τ_s). Thus when $\mathbf{S}$ is added to the model, the change in β's posterior mean given (τ_e, τ_s) is -1 times the regression of $\mathbf{S}$'s fitted values on $\mathbf{F}$.

Because $\mathbf{F}$ is centered and scaled (so $\mathbf{F}'\mathbf{F} = n - 1$) and $\mathbf{V}$ is orthogonal, the correlations ρ_i of $\mathbf{F}$ and V_i, the ith column of $\mathbf{V}$, can be written as $\rho = (\rho_1, \ldots, \rho_{n-I}, 0, \ldots, 0)' = (n-1)^{-0.5}\mathbf{V}'\mathbf{F}$. Equation (10.3) can then be written as

$$E(\beta|\tau_e, \tau_s, \mathbf{y}) = \hat{\beta}_{OLS} - (n-1)^{1/2}\rho' E(\mathbf{b}|\tau_e, \tau_s, \mathbf{y}). \tag{10.5}$$

(Section 10.1.4.1 below gives a more explicit expression.) From (10.5), if ρ_i and $E(b_i|\tau_e, \tau_s, \mathbf{y})$ are large for the same i, then adding $\mathbf{S}$ to the model can induce a large change in β's posterior mean. This happens if four conditions hold: $\mathbf{F}$ is highly correlated with V_i; $\mathbf{y}$ has a substantial correlation with both $\mathbf{F}$ and V_i; $r = \tau_s/\tau_e$ is small; and d_i is small. The first two conditions define confounding in ordinary linear models; the last two conditions ensure that b_i is not shrunk much toward zero. (If more than one i meets these conditions, their effects on $E(\beta|\tau_e, \tau_s, \mathbf{y})$ can cancel each other.) These necessary conditions are all present in the Slovenian data: Figure 9.1 shows the strong association of $\mathbf{y}$ and $\mathbf{F} = SEc$, $\rho_{193} = \text{correlation}(SEc, V_{193}) = 0.72$, $d_{193} = 0.03$, and $\mathbf{S}$ is not smoothed much (the effective number of parameters in the spatial fit was $p_D = 62.3$).

This effect on β's estimate is easily understood in linear model terms as the effect of adding a collinear regressor to a linear model. If $\mathbf{S}$ were not smoothed at all — if the coefficients $\mathbf{b}$ of the saturated design matrix $\mathbf{V}$ were not shrunk toward zero — then β would not be identified. (In linear model terms, it would be inestimable).

This corresponds to setting the smoothing precision τ_s to 0, so the smoothing ratio $r = \tau_s/\tau_e = 0$. If the smoothing ratio r is small, β is identified but the coefficients of V_i with small d_i are shrunk very little, so if these V_i are collinear with $\mathbf{F}$, the estimate of β is subject to the same collinearity effects as in linear models.

Reich et al. (2006) also shows how β's posterior variance, conditional on τ_s and τ_e, changes when the ICAR-distributed $\mathbf{S}$ is added to the model. Interested readers are referred to Reich et al. (2006) or to Hodges & Reich (2010, p. 328, equation 9). Variance inflation is large under the same conditions that induce a large change in β's estimate when the spatially correlated random effect is added. Again, this differs from the analogous collinearity result in linear model theory only in that the b_i are shrunk.

10.1.1.3 Spatial Confounding Explained in a More Spatial-Statistics Style

The linear models explanation seems odd to many spatial-statistics mavens, who think of the ICAR as a model for the error covariance. This section gives a derivation more in line with this view.

Begin with the re-parameterized normal-errors model (10.2) but now rewrite it in a more spatial-statistics style as $\mathbf{y} = \mathbf{F}\beta + \psi$, where the elements of $\psi = \mathbf{V}\mathbf{b} + \varepsilon$ are not independent. $\mathbf{S} = \mathbf{V}\mathbf{b}$ does not have a proper covariance matrix under an ICAR model, so we must derive $\text{cov}(\psi)$ indirectly. Partition $\mathbf{V} = (\mathbf{V}^{(1)}|\mathbf{V}^{(2)})$, where $\mathbf{V}^{(1)}$ has $n-I$ columns and $\mathbf{V}^{(2)}$ has I columns, and partition $\mathbf{b}$ conformably as $\mathbf{b} = (\mathbf{b}^{(1)\prime}|\mathbf{b}^{(2)\prime})'$, so $\mathbf{b}^{(1)}$ is $(n-I)\times 1$ and $\mathbf{b}^{(2)}$ is $I \times 1$. Pre-multiply (10.2) by $\mathbf{V}'$, so (10.2) becomes

$$\mathbf{V}^{(1)\prime}\mathbf{y} = \mathbf{V}^{(1)\prime}\mathbf{F}\beta + \mathbf{e}_1, \qquad \text{precision}(\mathbf{e}_{1i}) = \tau_e(rd_i)/(1+rd_i) < \tau_e \quad (10.6)$$

$$\mathbf{V}^{(2)\prime}\mathbf{y} = \mathbf{V}^{(2)\prime}\mathbf{F}\beta + \mathbf{b}^{(2)} + \mathbf{e}_2, \quad \text{precision}(\mathbf{e}_{2i}) = \tau_e \quad (10.7)$$

where $\mathbf{e}_1 = \mathbf{b}^{(1)} + \mathbf{V}^{(1)\prime}\varepsilon$ and $\mathbf{e}_2 = \mathbf{V}^{(2)\prime}\varepsilon$, so (10.6) and (10.7) are independent, and recall that τ_e is ε's error precision in (10.2).

Suppose $I = 1$ as in the Slovenian data; $I > 1$ is discussed below. Then $\mathbf{V}^{(2)} \propto \mathbf{1}_n$ and $\mathbf{V}^{(2)\prime}\mathbf{F} = 0$ because $\mathbf{F}$ is centered, so all the information about β comes from (10.6). Fix τ_s and τ_e. Without $\mathbf{S}$ in the model, β's posterior precision (the reciprocal of β's posterior variance) is $\mathbf{F}'\mathbf{F}\tau_e = (n-1)\tau_e$, which is also β's information matrix (a scalar, in this case). By adding $\mathbf{S}$ to the model, the posterior precision of β decreases by $(n-1)\tau_e\sum_{i=1}^{n-1}\rho_i^2/(1+rd_i)$, where as before $\rho_i = \text{correlation}(\mathbf{F}, V_i) = (n-1)^{-1/2}V_i'\mathbf{F}$ and $r = \tau_s/\tau_e$. The information loss is large if r is small (relatively little spatial smoothing) and ρ_i is large for i with small d_i, as in the Slovenian data. Because different rows of (10.6) contribute different information about β, if row i of (10.6) differs from the other rows and row i is effectively deleted by the combination of large ρ_i and small d_i, then β's estimate can change markedly when $\mathbf{S}$ is added to the model.

If the spatial map has $I > 1$ islands, the ICAR model includes an implicit $I-1$ degree of freedom unsmoothed fixed effect for the island means, $\mathbf{b}^{(2)}$ in (10.7), in addition to the overall intercept. This island-mean fixed effect may be collinear with

F in the usual manner of linear models even if **S** is smoothed maximally within islands (i.e., τ_s is very large).

10.1.2 Avoiding Spatial Confounding: Restricted Spatial Regression

The preceding section showed that spatial confounding arises from collinearity, which suggests removing spatial confounding by adapting a simple trick sometimes used in linear models.

Suppose we are fitting an ordinary linear model with $\mathbf{y}$ as the dependent variable and $\mathbf{X}_1$ and $\mathbf{X}_2$ as independent variables. If $\mathbf{X}_1$ and $\mathbf{X}_2$ are highly correlated with each other, that collinearity can produce a variety of inconvenient side effects. Many regression classes teach the following trick to get rid of this collinearity: Regress $\mathbf{y}$ on $\mathbf{X}_2$ and $\mathbf{X}_1^* = (\mathbf{I} - \mathbf{X}_2(\mathbf{X}_2'\mathbf{X}_2)^{-1}\mathbf{X}_2')\mathbf{X}_1$ instead of on $\mathbf{X}_2$ and $\mathbf{X}_1$. $\mathbf{X}_1^*$ is the residuals of $\mathbf{X}_1$ regressed on $\mathbf{X}_2$, so $\mathbf{X}_1^*$ and $\mathbf{X}_2$ are orthogonal by construction and their estimated coefficients are uncorrelated. This trick attributes to $\mathbf{X}_2$ all variation in $\mathbf{y}$ over which $\mathbf{X}_2$ and $\mathbf{X}_1$ are competing, which can be hard to justify in particular cases.

The analog in a model with an ICAR random effect **S** is to restrict the spatial random effect **S** to the subspace of n-dimensional space orthogonal to the fixed effect **F**. Reich et al. (2006) called this "restricted spatial regression." The next few paragraphs do this for a one-dimensional **F**, which generalizes easily to higher dimensions (Reich et al. 2006, Sec. 3; Sec. 4 extended the method to non-normal observables). This attributes to the fixed effect **F** all variation in $\mathbf{y}$ over which **F** and **S** are competing; I argue below that this is appropriate, indeed necessary, in a large class of situations.

The simplest way to specify a restricted spatial regression is to replace model (10.1) with $\mathbf{y} = \mathbf{F}\beta + \mathbf{P}^c\mathbf{S} + \varepsilon$. The design matrix in front of **S** has changed from $\mathbf{I}_n$ to $\mathbf{P}^c = \mathbf{I}_n - \mathbf{F}(\mathbf{F}'\mathbf{F})^{-1}\mathbf{F}'$, the residual projection for a regression on **F**, but otherwise the model is unchanged. Written this way, **S** has one superfluous dimension, $(\mathbf{I}_n - \mathbf{P}^c)\mathbf{S}$, which necessarily contributes nothing to the fitted values of $\mathbf{y}$ and about which the data provide no information.

For the spatial models considered here, it is easy to reformulate this restricted spatial regression so it has no superfluous dimensions. I show this for $p = 1$ but it generalizes readily (see Reich et al. 2006). Let $\mathbf{P}^c$ have spectral decomposition $\mathbf{P}^c = (\mathbf{L}|\mathbf{K})\Phi(\mathbf{L}|\mathbf{K})'$, where Φ is a diagonal matrix with $n-1$ eigenvalues of 1 and one 0 eigenvalue, **L** has n rows and $n-1$ columns, and **K** has n rows and 1 column, with **K** proportional to **F** and $\mathbf{K}'\mathbf{L} = 0$. Then fit the following model:

$$\mathbf{y} = \mathbf{F}\beta + \mathbf{L}\mathbf{S}^* + \varepsilon, \tag{10.8}$$

where $\mathbf{S}^*$ is $(n-1)$-dimensional normal with mean 0 and precision matrix $\tau_s\mathbf{L}'\mathbf{Q}\mathbf{L}$, **Q** is the neighbor matrix from **S**'s ICAR model, and ε is iid normal with mean 0 and precision τ_e.

Using this model in either form and conditioning on (τ_s, τ_e), β has the same conditional posterior mean as in the analysis without the ICAR-distributed **S**, but has larger conditional posterior variance (Reich et al. 2006, Sec. 3). Thus, restricted spatial regression discounts the sample size to account for spatial correlation without changing β's point estimate conditional on τ_s and τ_e.

The foregoing shows that restricted spatial regression can be done but equation (10.8) is not the only way to do it and is probably not an especially good way. Hughes & Haran (2013) noted that equation (10.8) neglects the spatial map in favor of brute-force orthogonalization (my words, not theirs). This implies rather strange canonical predictors **L** for the random effect in (10.8) and a covariance matrix for $\mathbf{LS}^*$ that can have negative covariances, which violates the intuition behind models like ICAR, that regions are positively correlated with the correlation diminishing as the distance between regions increases. Hughes & Haran (2013) proposed a model that also restricts spatial correlation to the orthogonal complement of **F**'s column space while accomplishing two other objectives: Capturing only residual patterns that could arise under positive spatial dependence and reducing the random effect's dimension to speed up computing. Their model is motivated by Moran's I statistic and replaces **L** in (10.8) with the eigenvectors of a generalized Moran operator $\mathbf{P}^c(\text{diag}(\mathbf{Q}) - \mathbf{Q})\mathbf{P}^c$ that have large positive eigenvalues. This selects predictors capturing most of the positive spatial dependence in the original model that is orthogonal to **F**, while radically reducing computing time with little sacrifice in fidelity to the original ICAR model. Indeed, I would argue that this is no loss at all because the ICAR itself is just a handy expedient, rarely if ever motivated by more than the vague intuition that near regions tend to be more similar than far. This intuition is preserved by the foregoing model and by other models constructed using similar methods, e.g., replacing $(\text{diag}(\mathbf{Q}) - \mathbf{Q})$ with **Q** in the generalized Moran operator (Hughes & Haran 2013, Section 4).

10.1.3 Spatial Confounding Is Not an Artifact of the ICAR Model

Spatial confounding also occurs if, in model (10.1), **S** is given a proper multivariate normal distribution with any of several covariance matrices capturing the intuition that near regions are more similar than far. Wakefield (2007), for example, found very similar spatial-confounding effects in the Scottish lip-cancer data using the ICAR model and a geostatistical model.

For the Slovenian data, Hodges & Reich (2010) considered geostatistical models in which each municipality was represented by a point with east-west coordinate the average of the farthest east and farthest west boundary points, and analogous north-south coordinate. An example of a proper covariance matrix is $\text{cov}(\mathbf{S})$ having $(i,j)^{\text{th}}$ element $\sigma_s^2 \exp(-\delta_{ij}/\theta)$, with δ_{ij} being Euclidean distance between the points representing municipalities i and j and θ controlling spatial correlation. Each covariance matrix considered had an unknown parameter like θ, which for now is treated as fixed and known. Applying Section 10.1.1.2's approach to such models for **S** requires only one change, arising because the precision matrix $\text{cov}(\mathbf{S})^{-1}$ now has no zero eigenvalues: (10.2) must now have an explicit intercept. Therefore, holding θ fixed, spatial confounding will occur by the same mechanism as with the ICAR model for **S**. (For the Slovenian data, the smallest eigenvalue of $\text{cov}(\mathbf{S})^{-1}$ has eigenvector V_{194} which, while not constant over the map as for the ICAR model, nonetheless varies little. I have seen this in other geostatistical models; it explains why the intercept sometimes has a huge posterior variance in such models, but the generality of this intercept confounding is unknown.) Hodges & Reich (2010) show that for several such models, the spatial-confounding effect is present for a large range of the adjustable parameter θ.

Penalized splines are a quite different approach to spatial smoothing but they too produce the same confounding effect in the Slovenian data. In a class project (Salkowski 2008), a student fit a two-dimensional penalized spline to the Slovenian data, attributing each municipality's counts to a point as described above. He used the R package SemiPar, which accompanied Ruppert et al. (2003), to fit a model in which municipality i's count of stomach cancers O_i was Poisson with log mean

$$\log\{E_i\} + \beta_0 + \beta_{SEc} SEc_i + \beta_A A_i + \beta_N N_i + \sum_k u_k basis_{ki}, \tag{10.9}$$

where A_i is municipality i's east-west coordinate, N_i is its north-south coordinate (each coordinate was centered and both were scaled by a single scaling constant to preserve the map's shape), $basis_{ki}$ is the default basis in SemiPar (based on the Matérn covariance function), the u_k were modeled as iid normal, and the knots were SemiPar's default knots, derived using a space-filling algorithm. SemiPar's default fitting method, penalized quasi-likelihood, shrank the random effect term $\sum_k u_k basis_{ki}$ to zero but the two fixed effects implicit in the spline, A_i and N_i, remained in the model and produced a collinearity effect as in an ordinary linear model. Without the spatial spline, a simple generalized linear model fit gave an estimate for β_{SEc} of -0.137 (standard error 0.020), essentially the same as in Zadnik's Bayesian analysis, while adding just the fixed effects A_i and N_i changed β_{SEc}'s estimate to -0.052 (SE 0.028). As the spline fit was forced to be progressively less smooth, β_{SEc}'s estimate increased monotonically and eventually became positive. (Steinberg & Bursztyn 2004, p. 415, noted in passing a similar confounding effect in a different spline.)

Thus, spatial confounding is not an artifact of the ICAR model, but arises from other, perhaps all, specifications of the intuition that measures taken at locations near to each other are more similar than measures taken at distant locations.

10.1.4 Five Interpretations, with Implications for Practice

To discuss interpretations of spatial confounding, I need to distinguish between old- and new-style random effects. As noted in Chapter 1, old-style random effects fit the definition of Scheffé (1959, p. 238): The levels of the random effect are draws from a population and the draws are not of interest in themselves but only as samples from the larger population, which *is* of interest. As noted, in recent years "random effect" has come to be used in a different sense, for effects that have the mathematical form of an old-style random effect but which are quite different. For some of these new-style random effects, the levels are the entire population and are themselves of interest; the 194 municipalities in the Slovenian data are an example. A full discussion of random effect interpretations is given in Chapter 13. For now, I note that discussions of spatial random effects are generally either unclear about their interpretation or seem to treat them as old-style random effects. Finally, it is both economical and accurate to describe all new-style random effects as formal devices to implement a smoother, interpreting shrinkage estimation as a kind of smoothing, so for the rest of this section, I will do so.

With that preface, the five interpretations of spatial confounding to be considered are as follows:

- **S is a new-style random effect, a formal device to implement a smoother.**
  - (i) Spatially correlated errors remove bias in estimating β and are generally conservative (Clayton et al. 1993).
  - (ii) Spatially correlated errors can introduce or remove bias in estimating β and are not necessarily conservative (Wakefield 2007; Hodges & Reich 2010).
- **S is an old-style random effect.**
  - (iii) The regressors $\mathbf{V}$ implicit in the spatial effect $\mathbf{S}$ are collinear with the fixed effect $\mathbf{F}$, but neither estimate of β is biased (David B. Nelson, personal communication).
  - (iv) Adding the spatial effect $\mathbf{S}$ creates information loss, but neither estimate of β is biased (David B. Nelson, personal communication).
  - (v) Because error is correlated with the regressor SEc in the sense commonly used in econometrics, *both* estimates of β are biased (Paciorek 2010).

Except for (v), these interpretations treat SEc as measured without error and not drawn from a probability distribution. The following two subsections discuss these two groups of interpretations.

10.1.4.1 S *Is a New-Style Random Effect, a Formal Device to Implement a Smoother*

It is commonly argued (e.g., Clayton et al. 1993) that introducing spatially correlated errors into a model, as with $\mathbf{S}+\varepsilon$, captures the effects of spatially structured missing covariates and thus adjusts the estimate of β for such missing covariates even if we have no idea what those covariates might be. Interpretation (i) reflects this view. My former student Brian Reich reports hearing a somewhat different statement of this view: "I *know* I am missing some confounders, in fact I have some specific confounders in mind that I was unable to collect, but from experience I know they have a spatial pattern. Therefore, I will add $\mathbf{S}$ to the model to try to recover them and let the data decide how much can be recovered." In some fields, he reports, it is nearly impossible to get a paper published unless a spatial random effect is included for this purpose.

This view can be evaluated using the results in Section 10.1.1, which are a modest elaboration of linear model theory. Indeed, the only aspect of the present problem not present in linear model theory is that most of the canonical coefficients b_i are shrunk toward 0, although the b_i that produce spatial confounding are shrunk the least and thus deviate least from linear model theory.

To make the discussion concrete, consider estimating β using the model $\mathbf{y} = \mathbf{1}_n\alpha + \mathbf{F}\beta + \varepsilon$, where ε is iid normal error, then estimating β using a larger model, either $\mathbf{y} = \mathbf{1}_n\alpha + \mathbf{F}\beta + \mathbf{H}\gamma + \varepsilon$, where $\mathbf{H}$ is a supposed missing covariate, or model (10.1), which adds the ICAR-distributed spatial random effect $\mathbf{S}$. It is possible to derive explicit expressions, given τ_e and τ_s, for the adjustment in the estimate of β under either of these larger models and for the *expected* adjustment assuming the data were generated by the model

$$\mathbf{y} = \mathbf{1}_n\alpha + \mathbf{F}\beta + \mathbf{H}\gamma + \varepsilon. \tag{10.10}$$

(Derivations are an exercise.)

Specifically, consider these three models:

- **Model 0**: $\mathbf{y} = \mathbf{1}_n\alpha + \mathbf{F}\beta + \varepsilon$
- **Model H**: $\mathbf{y} = \mathbf{1}_n\alpha + \mathbf{F}\beta + \mathbf{H}\gamma + \varepsilon$
- **Model S**: $\mathbf{y} = \mathbf{F}\beta + \mathbf{V}\mathbf{b} + \varepsilon$,

where Model S, the spatial model, is the same as (10.1) and (10.2). Assume $\mathbf{H}$ is centered and scaled in the same manner as $\mathbf{F}$, so $\mathbf{1}_n'\mathbf{H} = 0$ and $\mathbf{H}'\mathbf{H} = n-1$, and assume $\mathbf{y}$ is centered but not scaled. Bayesian results below assume flat (improper) priors on α, β, and γ.

Under Model 0, the estimate of β — the posterior mean or least-squares estimate — is

$$\hat{\beta}^{(0)} = (\mathbf{F}'\mathbf{F})^{-1}\mathbf{F}'y = \rho_{FY}(\mathbf{y}'\mathbf{y})^{0.5} \tag{10.11}$$

where ρ_{AB} is Pearson's correlation of the vectors $\mathbf{A}$ and $\mathbf{B}$. Under Models H and S, the estimates of β given τ_s and τ_e — the conditional posterior mean for both models; for Model H this is also the least-squares estimate, while for Model S it is the conventional (non-Bayesian) estimate after maximizing the restricted likelihood — are

$$\hat{\beta}^{(H)} = \hat{\beta}^{(0)} - (\mathbf{F}'\mathbf{F})^{-1}\mathbf{F}'\mathbf{H}(\mathbf{H}'\mathbf{P}^c\mathbf{H})^{-1}\mathbf{H}'\mathbf{P}^c\mathbf{y} \tag{10.12}$$

$$\hat{\beta}^{(S)} = \hat{\beta}^{(0)} - (\mathbf{F}'\mathbf{F})^{-1}\mathbf{F}'\mathbf{V}(\mathbf{V}'\mathbf{P}^c\mathbf{V} + r\mathbf{D})^{-1}\mathbf{V}'\mathbf{P}^c\mathbf{y}, \tag{10.13}$$

where, as before, $r = \tau_s/\tau_e$, $\mathbf{P}^c = \mathbf{I} - \mathbf{F}(\mathbf{F}'\mathbf{F})^{-1}\mathbf{F}'$, and $\mathbf{D} = \mathrm{diag}(d_1, \ldots, d_{n-I}, 0, \ldots, 0)$ is the diagonal matrix containing the eigenvalues d_i of the spatial-neighbor matrix $\mathbf{Q}$. The estimate $\hat{\beta}^{(S)}$ was given in Section 10.1.1.2 and $\hat{\beta}^{(H)}$ is derived by a similar argument.

Define $\mathbf{B}_F = \mathbf{V}'\mathbf{F}/(n-1)^{0.5}$; the entries in $\mathbf{B}_F$ are the correlations between $\mathbf{F}$ and the columns of $\mathbf{V}$ (we called this ρ in Section 10.1.1.2). Define $\mathbf{B}_H$ analogously as $\mathbf{B}_H = \mathbf{V}'\mathbf{H}/(n-1)^{0.5}$. Finally, define $\mathbf{B}_y = \mathbf{V}'\mathbf{y}/[(n-1)(\mathbf{y}'\mathbf{y})]^{0.5}$; $\mathbf{B}_y$'s entries are the correlations between $\mathbf{y}$ and the columns of $\mathbf{V}$. Then the estimates of β under Models H and S, given τ_s and τ_e, can be shown to be

$$\hat{\beta}^{(H)} = \hat{\beta}^{(0)}\left[1 - \frac{\frac{\rho_{FH}\rho_{HY}}{\rho_{FY}} - \rho_{FH}^2}{1 - \rho_{FH}^2}\right] \tag{10.14}$$

$$\hat{\beta}^{(S)} = \hat{\beta}^{(0)}\left[1 - \frac{\frac{\rho_{FY}'}{\rho_{FY}} - q}{1 - q}\right],$$

where $\rho_{FY}' = \mathbf{B}_F'(\mathbf{I} + r\mathbf{D})^{-1}\mathbf{B}_y = \mathbf{F}'(I + r\mathbf{Q})^{-1}\mathbf{y}/((n-1)\mathbf{y}'\mathbf{y})^{0.5}$ and $q = \mathbf{B}_F'(\mathbf{I} + r\mathbf{D})^{-1}\mathbf{B}_F = \mathbf{F}'(I + r\mathbf{Q})^{-1}\mathbf{F}/(n-1)$. (Proofs are an exercise.)

In (10.14), the expressions in square brackets for $\hat{\beta}^{(H)}$ and for $\hat{\beta}^{(S)}$ have no necessary relation to each other. For example, if **F** and **H** are uncorrelated, so $\rho_{FH} = 0$, then $\hat{\beta}^{(H)} = \hat{\beta}^{(0)}$ but $\hat{\beta}^{(S)}$ can be larger or smaller than $\hat{\beta}^{(H)}$ in absolute value, depending on how $(\mathbf{I} + r\mathbf{D})^{-1}$ differentially downweights specific coordinates of $\mathbf{B}_F$ and $\mathbf{B}_y$. When $\rho'_{FY} \approx \rho_{FY}$, $\hat{\beta}^{(S)} \approx 0$. This happens if the ith coordinate of $\mathbf{B}_F$, the correlation of **F** and V_i, is large; the ith coordinate of $\mathbf{B}_y$, the correlation of **y** and V_i, is large; and d_i and r are small, as in the Slovenian data.

If the data are in fact generated by Model H, we can treat $\hat{\beta}^{(H)}$ and $\hat{\beta}^{(S)}$ as functions of **y**, holding τ_e and τ_s fixed, and compute the expected change in β's estimate from adding either **H** or **S** to Model 0, where the expectation is with respect to **y**. The expected changes are

$$E(\hat{\beta}^{(H)} - \hat{\beta}^{(0)} | \tau_e) = \rho_{FH}\gamma \qquad (10.15)$$

$$E(\hat{\beta}^{(S)} - \hat{\beta}^{(0)} | \tau_e, \tau_s) = \left[\frac{\frac{\rho'_{FH}}{\rho_{FH}} - q}{1-q} \right] \rho_{FH}\gamma \quad \text{if } \rho_{FH} \neq 0 \qquad (10.16)$$

$$= \frac{\rho'_{FH}}{1-q}\gamma \quad \text{if } \rho_{FH} = 0, \qquad (10.17)$$

where $\rho'_{FH} = \mathbf{B}'_F(\mathbf{I} + r\mathbf{D})^{-1}\mathbf{B}_H = \mathbf{F}'(I + r\mathbf{Q})^{-1}\mathbf{H}/(n-1)$. If $\gamma \neq 0$, the expected adjustment under Model S is biased, with the bias depending on how $(\mathbf{I} + r\mathbf{D})^{-1}$ differentially downweights specific coordinates of $\mathbf{B}_F$ and $\mathbf{B}_H$; the bias can be positive or negative. If $\mathbf{H} = V_j$, then $\rho'_{FH} = \rho_{FH}/(1 + rd_j)$, so (10.16) becomes

$$E(\hat{\beta}^{(S)} - \hat{\beta}^{(0)} | \tau_e, \tau_s) = \left[\frac{(1 + rd_j)^{-1} - q}{1-q} \right] \rho_{FH}\gamma. \qquad (10.18)$$

The expression in square brackets is less than 1 when $rd_j > 0$ and becomes negative if r or d_j is large enough, i.e., Model S adjusts $\hat{\beta}$ in the wrong direction from $\hat{\beta}^{(0)}$. If there is no missing covariate, $\gamma = 0$, then the expected adjustment is zero under both Model H and Model S.

To summarize, there is no necessary relationship between the adjustment to β's estimate arising from adding **S** to the model and the adjustment arising from adding the supposed missing covariate **H**. This is most striking if we suppose that **H** is uncorrelated with **F**, so that adding **H** to the model would not change the estimate of β. In this case, adding the spatial random effect **S** *does* adjust the estimate of β, and in a manner that depends not on **H** but on the correlation of **F** and **y** and on the spatial map. If the data are generated by (10.10), the expected adjustment in β's estimate from adding **S** to the model is not zero in general and can be biased in either direction. If there *is* no missing covariate ($\gamma = 0$), adding **S** nonetheless adjusts β's estimate in the manner just described although the expected adjustment is zero. It is fair to describe such adjustments as haphazard.

Now suppose **H** is correlated with **F**, so that adding **H** to the model changes the estimate of β. In this case, the adjustment to β's estimate under the spatial model is again biased relative to the correct adjustment from including **H**. The bias can be large and either positive or negative, with more smoothness generally implying larger bias, and depending haphazardly on **H** and on the spatial map. It can even happen that on average β's estimate becomes larger when it would become smaller if **H** were added to the model.

Therefore, adding spatially correlated errors is not conservative. A canonical regressor V_i that is collinear with **F** can cause β's estimate to increase in absolute value just as in ordinary linear models. Further, in cases in which β's estimate should not be adjusted, introducing spatially correlated errors will nonetheless adjust the estimate haphazardly.

From the perspective of linear model theory, it seems perverse to use an error term to adjust for possible missing confounders. The analog in ordinary linear models would be to move part of the fitted coefficients into error to allow for the possibility of as-yet-unconceived missing confounders. In using an ordinary linear model, we know that if missing confounders are correlated with included fixed effects and with **y**, variation in **y** that would be attributed to the missing confounders is instead attributed to the included fixed effects. We acknowledge that possibility in the standard disclaimer for observational studies: If we have omitted confounders, our coefficient estimates could be wrong. In spatial modeling, the analogy to this practice would be to make this acknowledgment and use restricted spatial regression, so that all variation in **y** in the column space of included fixed effects is attributed to those included effects instead of haphazardly re-allocated to the spatial random effect.

Therefore, interpretation (i) cannot be sustained and interpretation (ii) is correct, when the random effect **S** is a new-style random effect, a mere formal device to implement a smoother. Adding spatially correlated errors cannot be expected to capture the effect of a spatially structured missing covariate, but only to smooth fitted values and discount the sample size in computing standard errors or posterior standard deviations for fixed effects. Therefore, in such cases you should *always* use restricted spatial regression so the sample size can be discounted without distorting the fixed effect estimate. If you are concerned about specific unmeasured confounders, you should add to the model a suitable explicit fixed effect, not adjust haphazardly by means of a spatially correlated error. Finally, conclusions from such analyses should be qualified as in any other observational study, e.g., we have estimated the association of our outcome with our regressors accounting for measured confounders, and if we have omitted confounders, then our estimate could be wrong.

10.1.4.2 **S** *Is an Old-Style Random Effect*

For these situations, I have seen three interpretations, which are listed again for convenience.

- (iii) The regressors **V** implicit in the spatial effect **S** are collinear with the fixed effect **F**, but neither estimate of β is biased (David B. Nelson, personal communication).
- (iv) Adding the spatial effect **S** creates information loss, but neither estimate of β is biased (David B. Nelson, personal communication).

- (v) Because error is correlated with the regressor **F** in the sense commonly used in econometrics, *both* estimates of β are biased (Paciorek 2010).

Interpretations (iii) and (iv) treat the fixed effect **F** as measured without error and not otherwise drawn from a probability distribution ("fixed and known"), while interpretation (v) treats **F** as drawn from a probability distribution. Interpreting spatial confounding therefore depends on whether **F** is interpreted as fixed and known or as a random variable. This is a messy business, which seems to be determined in practice less by facts than by the department in which one was trained. My training in a statistics department inclines me to view **F** as fixed and known as a default, while econometricians, for example, seem inclined to the opposite default.

To see the difficulty, consider an example in which the random effect is hospitals selected as a random sample from a population of hospitals and the fixed effect indicates whether each hospital is a teaching hospital. My default is to treat teaching status as fixed and known. However, if we have drawn 20 hospitals at random, then hospital *i*'s teaching status is a random variable determined by the hospital-sampling mechanism, so **F** is drawn from a probability distribution. But what if, as often happens, sampling is stratified by teaching status to ensure that (say) 10 hospitals are teaching hospitals and 10 are not? Now teaching status is fixed and known. What if someone gives us the dataset and we don't know whether sampling was stratified by teaching status? One might argue that our ignorance disqualifies us from analyzing these data but that argument is not compelling to, for example, people who interpret the Likelihood Principle as meaning they can ignore the sampling mechanism or to many people who do not have a tenured, hard-money job.

A full discussion of this issue is beyond the present chapter's scope. It is also unnecessary because there are unarguable instances of each kind of fixed effect. An example of a fixed and known **F** could arise in analyzing air pollution measured at many fixed monitoring stations on each day in a year. The days could be interpreted as a old-style random effect, and the elevation of each monitoring station as a fixed and known **F**. For an example of **F** plainly drawn from a probability distribution, consider the hospitals example just above, where the morbidity score for each hospital's patients is a random variable for the period being studied.

So first assume **F** is fixed and known. It is then straightforward to show that both (iii) and (iv) are correct and indeed arguably identical, though I think they are worth distinguishing. Interpretation (iv) follows from Section 10.1.1.3 and the familiar fact that generalized least squares gives unbiased estimates even when the error covariance is specified incorrectly. For interpretation (iii), recall from equation (10.5) that the estimate of β in the spatial model (which, given τ_e and τ_s, is both the posterior mean and the usual estimate following maximization of the restricted likelihood) is

$$E(\beta|\tau_e, \tau_s, \mathbf{y}) = \hat{\beta}_{OLS} - (n-1)^{-1/2}\rho' E(\mathbf{b}|\tau_e, \tau_s, \mathbf{y}). \tag{10.19}$$

By (10.4), $(n-1)^{-1/2}\rho' E(\mathbf{b}|\tau_e, \tau_s, \mathbf{y})$ can be written as $\mathbf{KP}^c\mathbf{y}$ where $\mathbf{K}$ is a known square matrix. Recalling that $\mathbf{P}^c = \mathbf{I} - \mathbf{F}(\mathbf{F}'\mathbf{F})^{-1}\mathbf{F}'$, it follows that $\mathbf{P}^c\mathbf{y} = \mathbf{P}^c(\mathbf{Vb} + \varepsilon)$ has expectation 0 with respect to $\mathbf{b}$ and ε. Therefore, the spatial and OLS estimates of β have the same expectation and are unbiased based on the aforementioned familiar fact about generalized least squares.

Now assume **F** is a random variable. Paciorek (2010) interprets model (10.2) as $\mathbf{y} = \mathbf{F}\beta + \psi$ where the error term $\psi = \mathbf{Vb} + \varepsilon$ has a non-diagonal covariance matrix. Because $\mathbf{F}'\mathbf{V} \neq 0$, **F** is correlated with ψ, so by the standard result in econometrics, *both* the OLS estimate of β and the estimate of β using (10.2) are biased. Formulating the result this way is more precise than the common statement that bias arises when "the random effect is correlated with the fixed effect," because the distribution of "the random effect" depends on the parameterization: The random effect **b** in model (10.2) is independent of **F**, but the random effect **S** in model (10.1) is not.

For the present purpose, the main point of Paciorek (2010) is that "inclusion of a spatial residual term accounts for spatial correlation in the sense of reducing bias [in estimating β] from unmeasured spatial confounders [supposedly captured by **S**] only when there is unconfounded variability in the exposure [**F**] at a scale smaller than the scale of the confounding" (p. 122). I conjecture that this can be interpreted in Section 10.1.1.2's terms as meaning that bias is reduced if **F** is not too highly correlated with the low-frequency canonical regressors V_i, which have small d_i and hence little shrinkage in b_i.

Paciorek (2010) concludes that restricted spatial regression is either irrelevant to the issue of bias in estimating β or makes an overly strong assumption by attributing to **F** all of the disputed variation in **y**. The latter appears to presume that the spatial random effect **S** captures an unspecified missing covariate, which, I have argued, is difficult to sustain in general (though Paciorek 2010 shows an example where it seems to do just that). However, this area of research is just beginning and much remains to be developed.

10.1.5 Concluding Thoughts on Spatial Confounding

The preceding sections laid out the mechanics by which spatial confounding occurs, argued briefly that this is more general than the ICAR model, discussed analyses that remove spatial confounding, and considered different ways that spatial confounding has been interpreted, concluding that restricted spatial regression should be used routinely in the common situation when **S** is a formal device to implement a smoother.

The mechanics of spatial confounding are underdeveloped in certain respects. In debunking the belief that spatially correlated errors adjust for unspecified missing covariates, the derivation (Section 10.1.4.1) took the smoothing ratio $r = \tau_s/\tau_e$ as given. Although r's marginal posterior distribution is easily derived when τ_e has a gamma prior (τ_s need not have a gamma prior), it is harder to interpret than the posterior mean of β so its implications are as yet unclear. However, it should be possible to extract some generalizations that will allow us to identify and understand situations in which spatial confounding will and will not occur. (This problem is given as an exercise.) I hypothesize this will show that whenever both **y** and **F** are highly correlated with V_{n-I}, the canonical regressor with the least-smoothed coefficient, there will be little smoothing (r will be small) and β's estimate under the spatial model will be close to zero. In other words, the hypothesis is that in any map, when both **y** and **F** show a strong trend along the long axis of the map, adding a spatially correlated error will nullify the obvious association between **y** and **F** as it did in the

Slovenian data. The same is probably true when both **y** and **F** show a quadratic-like trend along the long axis of the map, as is the case in the Scottish lip-cancer data, where both lip-cancer incidence and AFF (the percent of employment in the outdoor occupations agriculture, forestry, and fishing) are high in the rural north and south of Scotland but low in the urban middle.

(The preceding paragraph was taken with little change from the concluding section of Hodges & Reich 2010. Since I wrote this chapter, I have used some tools introduced in Chapter 15 to examine the restricted likelihood and thus marginal posterior in this spatial confounding problem. These new results, described Section 15.2.3.2, support conclusions that differ somewhat from the hypotheses stated just above. They also elaborate the finding that adding an ICAR random effect cannot reliably adjust β for an omitted predictor **H**. Rather than rewrite the preceding paragraph to make myself look smarter than I was, I invite you instead to read Chapter 15.)

The theory is particularly underdeveloped for the situation in which **S** can be interpreted as an old-style random effect and **F**, or at least parts of it, can also be interpreted as random. Paciorek (2010) is a first step in what should be a rich area of research.

10.2 Kids and Crowns: Informative Cluster Size

I met this investigator shortly after my former student, Brian Reich, worked out the spatial-confounding mechanics in Reich et al. (2006), so I asked Brian why adding the child-specific random effect changed the estimate for Type III crowns but not Type IV crowns. He replied: "The only remnants of the [spatial-confounding] model are the within-kid sample size. Is there any relationship between the number of observations per kid and the treatment group?" There is, though it is not as simple or striking as in the Slovenian stomach-cancer data. In the following subsections, I first derive some theory for the normal-errors case, showing that collinearity is involved and laying out the mechanics of this informative cluster size effect, then apply the theory to some illustrative cases, then show some simple analyses that somewhat support Brian's conjecture. (But I'm convinced he's right because, as he noted, the problem has no other features that *could* explain the change in the estimate from adding the random effect).

10.2.1 Mechanics of Informative Cluster Size

This section uses a simplified version of the kids-and-crowns problem with two groups and normal errors. As in the spatial-confounding section, the analyses are Bayesian but the results are easily modified to apply to the conventional analysis.

Specifically, assume the following:

- There are two groups, Group 1 and Group 2.
- Each group has m clusters (this is not necessary).
- Cluster i has n_i units or observations in it.

- There are $n = \sum_{i=1}^{2m} n_i$ total observations, with $n^{(1)} = \sum_{i=1}^{m} n_i$ in Group 1 and $n^{(2)} = \sum_{i=m+1}^{2m} n_i$ in Group 2.

We need to distinguish the *model used for analysis* from the *model that generated the data*; this subsection considers only the former, while the following subsection considers both. In the model used for analysis, the group difference is parameterized with a design-matrix column using the $+1/-1$ parameterization, so in mixed-linear-model form the model is

$$\mathbf{y} = \mathbf{1}_n\beta_0 + \begin{bmatrix} \mathbf{1}_{n^{(1)}} \\ -\mathbf{1}_{n^{(2)}} \end{bmatrix} \beta_1 + \begin{bmatrix} \mathbf{Z}_1 & \mathbf{0} \\ \mathbf{0} & \mathbf{Z}_2 \end{bmatrix} \mathbf{u} + \varepsilon, \text{ where} \tag{10.20}$$

$$\mathbf{Z}_1 = \begin{bmatrix} \mathbf{1}_{n_1} & \mathbf{0} & \dots & \mathbf{0} \\ \mathbf{0} & \mathbf{1}_{n_2} & \dots & \mathbf{0} \\ \vdots & & \ddots & \vdots \\ \mathbf{0} & \mathbf{0} & \dots & \mathbf{1}_{n_m} \end{bmatrix} \text{ and } \mathbf{Z}_2 = \begin{bmatrix} \mathbf{1}_{n_{m+1}} & \mathbf{0} & \dots & \mathbf{0} \\ \mathbf{0} & \mathbf{1}_{n_{m+2}} & \dots & \mathbf{0} \\ \vdots & & \ddots & \vdots \\ \mathbf{0} & \mathbf{0} & \dots & \mathbf{1}_{n_{2m}} \end{bmatrix}. \tag{10.21}$$

To finish the specification, $\mathbf{G} = \text{cov}(\mathbf{u}) = \sigma_s^2 \mathbf{I}_{2m}$ and $\mathbf{R} = \sigma_e^2 \mathbf{I}_n$, where σ_s^2 and σ_e^2 are unknown. The difference between groups is $2\beta_1$. The results below use flat priors for β_0 and β_1 and condition on (σ_e^2, σ_s^2) (i.e., treat them as known), so a prior for (σ_e^2, σ_s^2) is not needed.

The random-effect design matrix $\mathbf{Z}$ is collinear with the fixed-effect design matrix $\mathbf{X}$ in two ways, an obvious way that turns out not to matter and a non-obvious way that does matter. To see this, re-parameterize $\mathbf{Zu}$ as follows. For an orthogonal matrix $\mathbf{C}$ to be defined just below, let $\mathbf{Zu} = (\mathbf{ZC})(\mathbf{C}'\mathbf{u}) = \mathbf{Hb}$ for $\mathbf{H} = \mathbf{ZC}$ and $\mathbf{b} = \mathbf{C}'\mathbf{u}$. Thus in the re-parameterized model, $\mathbf{G} = \text{cov}(\mathbf{b})$ is still $\sigma_s^2 \mathbf{I}_{2m}$. Define the columns of $\mathbf{C}$ as follows. $\mathbf{C}$'s first column is $q\mathbf{1}_{2m}$ for $q = 1/\sqrt{2m}$. $\mathbf{C}$'s second column is $q\left[\mathbf{1}'_m | -\mathbf{1}'_m\right]'$. Thus the first two columns of $\mathbf{H} = \mathbf{ZC}$ are proportional to the first two columns of the fixed-effect design matrix $\mathbf{X}$ in equation (10.20). $\mathbf{C}$ has 3rd through $2m^{\text{th}}$ columns

$$\begin{bmatrix} \mathbf{K} & \mathbf{0} \\ \mathbf{0} & \mathbf{K} \end{bmatrix}, \tag{10.22}$$

where $\mathbf{K}$ is $m \times (m-1)$ with columns that are orthonormal and sum to zero, so $\mathbf{K}'\mathbf{K} = \mathbf{I}_{m-1}$ and $\mathbf{1}'_m\mathbf{K} = \mathbf{0}$, i.e., $\mathbf{K}$'s columns are contrasts in each group's cluster averages.

The re-parameterized model for analysis is thus

$$\mathbf{y} = \mathbf{X}\beta + \mathbf{Hb} + \varepsilon, \text{ where} \tag{10.23}$$

$$\mathbf{X} = \begin{bmatrix} \mathbf{1}_{n^{(1)}} & \mathbf{1}_{n^{(1)}} \\ \mathbf{1}_{n^{(2)}} & -\mathbf{1}_{n^{(2)}} \end{bmatrix} \text{ and } \mathbf{H} = \mathbf{ZC} = \left[q\mathbf{X} \;\middle|\; \begin{matrix} \mathbf{Z}_1\mathbf{K} \\ \mathbf{0} \end{matrix} \;\middle|\; \begin{matrix} \mathbf{0} \\ \mathbf{Z}_2\mathbf{K} \end{matrix} \right]; \tag{10.24}$$

the b_j in $\mathbf{b}$ are iid normal with mean zero and variance σ_s^2 and the errors ε_i are iid normal with mean zero and variance σ_e^2. The potential for collinearity is clear, but how does this lead to the informative cluster size effect seen in the kids-and-crowns data?

The derivations made for spatial confounding in Section 10.1 can be applied to the model as formulated in (10.23). I'll give the results, which are admittedly somewhat opaque, and then show how the change in estimates from adding the random effect arises from collinearity between the fixed-effect design matrix $\mathbf{X}$ and the random-effect design matrix $\mathbf{Z}$.

If $\hat{\beta}_{OLS} = (\mathbf{X}'\mathbf{X})^{-1}\mathbf{X}'\mathbf{y}$ is the ordinary least squares estimate of the fixed effects ignoring the clustering — which is also the posterior mean of β if the random effect $\mathbf{Zu}$ or $\mathbf{Hb}$ is omitted — then by the same derivation that led to the results for spatial confounding, the posterior mean of β accounting for the clustering, conditional on σ_e^2 and σ_s^2, is

$$E(\beta|\sigma_e^2,\sigma_s^2,\mathbf{y}) = \hat{\beta}_{OLS} - (\mathbf{X}'\mathbf{X})^{-1}\mathbf{X}'\mathbf{H}E(\mathbf{b}|\sigma_e^2,\sigma_s^2,\mathbf{y}) \quad (10.25)$$

$$= \hat{\beta}_{OLS} - qE\left(\begin{bmatrix} b_1 \\ b_2 \end{bmatrix}|\sigma_e^2,\sigma_s^2,\mathbf{y}\right) \quad (10.26)$$

$$-(\mathbf{X}'\mathbf{X})^{-1}\mathbf{X}'\mathbf{H}_{(-)}E(\mathbf{b}_{(-)}|\sigma_e^2,\sigma_s^2,\mathbf{y}) \quad (10.27)$$

$$\text{where } \mathbf{H}_{(-)} = \left[\begin{array}{c|c} \mathbf{Z}_1\mathbf{K} & \mathbf{0} \\ \mathbf{0} & \mathbf{Z}_2\mathbf{K} \end{array}\right] \text{ and } \mathbf{b}'_{(-)} = (b_3,\ldots,b_{2m}). \quad (10.28)$$

From equation (10.4), $E(\mathbf{b}|\sigma_e^2,\sigma_s^2,\mathbf{y}) = (\mathbf{H}'\mathbf{P}^c\mathbf{H} + r\mathbf{I}_{2m})^{-1}\mathbf{H}'\mathbf{P}^c\mathbf{y}$, where $r = \sigma_e^2/\sigma_s^2$ and $\mathbf{P}^c = I_{2m} - \mathbf{X}(\mathbf{X}'\mathbf{X})^{-1}\mathbf{X}'$ is the projection onto the residual space of the fixed effects in an ordinary least squares analysis that ignores clustering. It is not hard to show (so it is an exercise) that

$$E\left(\begin{bmatrix} b_1 \\ b_2 \end{bmatrix}|\sigma_e^2,\sigma_s^2,\mathbf{y}\right) = \mathbf{0} \quad (10.29)$$

and

$$E(\mathbf{b}_{(-)}|\sigma_e^2,\sigma_s^2,\mathbf{y}) = \begin{bmatrix} (r\mathbf{I}_{m-1} + \mathbf{K}'\mathbf{Z}_1'\mathbf{P}_1\mathbf{Z}_1\mathbf{K})^{-1}\mathbf{K}'\mathbf{Z}_1'\mathbf{P}_1\mathbf{y}_1 \\ (r\mathbf{I}_{m-1} + \mathbf{K}'\mathbf{Z}_2'\mathbf{P}_2\mathbf{Z}_2\mathbf{K})^{-1}\mathbf{K}'\mathbf{Z}_2'\mathbf{P}_2\mathbf{y}_2 \end{bmatrix}, \quad (10.30)$$

where $\mathbf{P}_l = \mathbf{I}_{n^{(l)}} - \frac{1}{n^{(l)}}\mathbf{J}_{n^{(l)}}$ for $\mathbf{J}_{n^{(l)}}$ an $n^{(l)} \times n^{(l)}$ matrix of 1's, and $\mathbf{y}_l$ is the part of $\mathbf{y}$ from group l. $\mathbf{P}_l$ centers $\mathbf{y}_l$ without regard for clustering.

These results identify two places where collinearity can occur, first and most obviously between $\mathbf{X}$ and the completely confounded first two columns of $\mathbf{H}$ in equation (10.26), and second between $\mathbf{X}$ and $\mathbf{H}_{(-)}$, the rest of $\mathbf{H}$ in equation (10.27). It turns out that the obvious former collinearity is innocuous, while the latter non-obvious collinearity actually produces the change in $E(\beta|\sigma_e^2,\sigma_s^2,\mathbf{y})$ from including the random effect in equation (10.30), as we'll now see.

As noted in equation (10.29), the coefficients (b_1,b_2) of the two columns of $\mathbf{H}$ that are exactly collinear with the fixed effects in $\mathbf{X}$ have conditional posterior mean $E((b_1,b_2)|\sigma_e^2,\sigma_s^2,\mathbf{y}) = \mathbf{0}$, i.e., the machinery assigns all the relevant variation in $\mathbf{y}$ to the fixed effect irrespective of σ_e^2 and σ_s^2. Also, $\text{cov}((b_1,b_2)|\sigma_e^2,\sigma_s^2,\mathbf{y})$

$$= q^2\sigma_e^2\begin{bmatrix} a_1+a_2 & a_1-a_2 \\ a_1-a_2 & a_1+a_2 \end{bmatrix}, \quad (10.31)$$

$$\text{for } a_1 = \sum_{i=1}^{m} \frac{1}{n_i + r} + \left(n^{(1)} - \sum_{i=1}^{m} \frac{n_i^2}{n_i + r}\right)^{-1} \left(\sum_{i=1}^{m} \frac{n_i}{n_i + r}\right)^2 \text{ and} \tag{10.32}$$

$$a_2 = \sum_{i=m+1}^{2m} \frac{1}{n_i + r} + \left(n^{(1)} - \sum_{i=m+1}^{2m} \frac{n_i^2}{n_i + r}\right)^{-1} \left(\sum_{i=m+1}^{2m} \frac{n_i}{n_i + r}\right)^2 \tag{10.33}$$

where $r = \sigma_e^2/\sigma_s^2$ (the derivation is an exercise). If σ_s^2 is fixed at a positive, finite value and $\sigma_e^2 \to 0$, so $r \to 0$ and the intra-class correlation goes to 1, then the random effects become effectively unconstrained and (10.31) becomes infinite. However, as shown below, while the conditional posterior variance of β_1, the treatment difference, changes as $\sigma_e^2 \to 0$, it stays finite and appears largely if not completely unaffected by the perfect collinearity of $\mathbf{X}$ with these first two columns of $\mathbf{H}$. Thus, this particular collinearity appears innocuous; these two columns of $\mathbf{H}$ are in effect eliminated from the model when the intra-class correlation is large.

Now for the second collinearity, which does affect the posterior of β and which, it turns out, can be interpreted as informative cluster size. First, note in (10.27) that the change in $E(\beta|\sigma_e^2, \sigma_s^2, \mathbf{y})$ is, as in the spatial-confounding problem, the estimated coefficient in a regression on the fixed-effect design matrix $\mathbf{X}$ of the fitted value of the random effect, $\mathbf{H}_{(-)}E(\mathbf{b}_{(-)}|\sigma_e^2, \sigma_s^2, \mathbf{y})$. Let's isolate two key pieces of (10.27), $\mathbf{X}'\mathbf{H}_{(-)}$ and $E(\mathbf{b}_{(-)}|\sigma_e^2, \sigma_s^2, \mathbf{y})$, and consider them separately.

At first glance, it might seem that $\mathbf{X}'\mathbf{H}_{(-)}$ should be zero, because $\mathbf{H} = \mathbf{ZC}$, $\mathbf{C}$ is an orthogonal matrix, and the two columns of $\mathbf{H}$ excluded to form $\mathbf{H}_{(-)}$ are exactly collinear with $\mathbf{X}$. However, $\mathbf{X}'\mathbf{H}_{(-)}$ is zero only if each group has clusters that are all the same size, i.e, $n_1 = \ldots = n_m$ and $n_{m+1} = \ldots n_{2m}$. To see this, note that

$$\mathbf{X}'\mathbf{H}_{(-)} = \begin{bmatrix} \mathbf{1}'_{n^{(1)}}\mathbf{Z}_1\mathbf{K} & \mathbf{1}'_{n^{(2)}}\mathbf{Z}_2\mathbf{K} \\ \mathbf{1}'_{n^{(1)}}\mathbf{Z}_1\mathbf{K} & -\mathbf{1}'_{n^{(2)}}\mathbf{Z}_2\mathbf{K} \end{bmatrix}. \tag{10.34}$$

Now $\mathbf{1}'_{n^{(1)}}\mathbf{Z}_1\mathbf{K} = [n_1, n_2, \ldots, n_m]\mathbf{K}$, and recall that the columns of $\mathbf{K}$ are $m-1$ contrasts forming an orthonormal basis for the orthogonal complement of $\mathbf{1}_m$. Then $\mathbf{1}'_{n^{(1)}}\mathbf{Z}_1\mathbf{K} = \mathbf{0}$ if and only if $n_1 = n_2 = \ldots = n_m$; the analogous result follows for $\mathbf{1}'_{n^{(2)}}\mathbf{Z}_2\mathbf{K}$. Thus, $E(\beta|\sigma_e^2, \sigma_s^2, \mathbf{y})$ can change as a result of adding the random effect only if the cluster sizes vary within groups, in which case $\mathbf{X}$ and $\mathbf{H}_{(-)}$ are, to some extent, collinear.

Label the columns of $\mathbf{K}$ as $\mathbf{K}_k, k = 1, \ldots, m-1$, and suppose that adding the random effect makes a big change in the estimate of β. Then the kth contrast in cluster sizes, $[n_1, n_2, \ldots, n_m]\mathbf{K}_k$ and $[n_{m+1}, n_{m+2}, \ldots, n_{2m}]\mathbf{K}_k$ for groups 1 and 2 respectively, must be large in absolute value for some k while $E(\mathbf{b}_k|\sigma_e^2, \sigma_s^2, \mathbf{y})$ is also large. This too is analogous to the spatial-confounding problem. (And as in the spatial-confounding problem, if the foregoing is true for more than one k, the effects of the different k can cancel or reinforce each other.)

So when is $E(\mathbf{b}_k|\sigma_e^2, \sigma_s^2, \mathbf{y})$ large? Equation (10.30) has two sub-vectors each of length $m-1$, one each for Group 1 and Group 2. Each of these subvectors is the coefficient estimate in a regression of $\mathbf{y}_l$, the vector of outcomes in Group l, on $\mathbf{P}_l\mathbf{Z}_l\mathbf{K}, l = 1, 2$, where the estimates are shrunk toward zero by adding $r\mathbf{I}_{m-1}$ to the information matrix in the manner of ridge regression. So $E(\mathbf{b}_k|\sigma_e^2, \sigma_s^2, \mathbf{y})$ is large if r is *not* large, i.e., the coefficients in $\mathbf{b}_{(-)}$ are not shrunk much, and if either of $\mathbf{y}_1$ or $\mathbf{y}_2$ "is associated with" its respective $(\mathbf{P}_l\mathbf{Z}_l)\mathbf{K}_k$. That is, $E(\mathbf{b}_k|\sigma_e^2, \sigma_s^2, \mathbf{y})$ is large if the contrast $\mathbf{K}_k$ in cluster averages of $\mathbf{y}_l$ is large in magnitude. (The matrix $\mathbf{P}_l$ is a distraction here — it just subtracts out the group average ignoring clustering — so I'm ignoring it for the present heuristic purpose.)

This is the sense in which informative cluster size produces a change in the estimate of β from adding the random effect. If the same contrast $\mathbf{K}_k$ is large in magnitude for both cluster sizes and outcomes $\mathbf{y}_l$, and $r = \sigma_e^2/\sigma_s^2$ is not large, then the estimate of β changes when the random effect is added unless, as noted, there is more than one such k and their effects cancel. In this sense, cluster size is informative.

The same results can be produced in a more explicit form by treating the random effect $\mathbf{Zu}$ as part of the error term, as will now be shown. This too follows the development in the preceding section about spatial confounding.

Re-write the mixed linear model (10.20) as

$$\begin{aligned} \mathbf{y} &= \mathbf{X}\beta + \xi, \text{ where } \xi = \mathbf{Zu} + \varepsilon \qquad (10.35)\\ \text{so } \operatorname{cov}(\xi) &= \sigma_s^2\mathbf{ZZ}' + \sigma_e^2\mathbf{I}_n \\ &= \text{blockdiag}(\sigma_s^2\mathbf{J}_{n_i} + \sigma_e^2\mathbf{I}_{n_i}), \end{aligned}$$

where blockdiag$(\bullet)$ is a block-diagonal matrix and the $\mathbf{J}_{n_i}$ are order-n_i square matrices of 1's. Let $\operatorname{cov}(\xi)$ have spectral decomposition $\operatorname{cov}(\xi) = \Gamma\mathbf{D}\Gamma'$; it is easy to show that

$$\begin{aligned} \Gamma &= \text{blockdiag}(n_i^{-0.5}\mathbf{1}_{n_i}|\mathbf{K}_i), \qquad (10.36)\\ \mathbf{D} &= \text{blockdiag}\left(\begin{bmatrix} \sigma_e^2 + n_i\sigma_s^2 & \mathbf{0} \\ \mathbf{0} & \sigma_e^2\mathbf{I}_{n_i-1} \end{bmatrix}\right), \end{aligned}$$

where $\mathbf{K}_i$ is an $n_i \times (n_i - 1)$ matrix with orthonormal columns that are orthogonal to $\mathbf{1}_{n_i}$.

Transform the problem in equation (10.35) by pre-multiplying it by Γ', giving

$$\Gamma'\mathbf{y} = \Gamma'\mathbf{X}\beta + \xi^*, \text{ where } \xi^* = \Gamma'\xi \text{ and } \operatorname{cov}(\xi^*) = \mathbf{D}. \qquad (10.37)$$

This can be written more explicitly as

$$\begin{bmatrix} n_1^{0.5}\bar{y}_{1.} \\ \mathbf{K}_1\mathbf{y}_1 \\ \hline n_2^{0.5}\bar{y}_{2.} \\ \mathbf{K}_2\mathbf{y}_2 \\ \hline \vdots \\ \hline n_{2m}^{0.5}\bar{y}_{2m.} \\ \mathbf{K}_{2m}\mathbf{y}_{2m} \end{bmatrix} = \begin{bmatrix} n_1^{0.5} & n_1^{0.5} \\ \mathbf{0}_{n_1-1} & \mathbf{0}_{n_1-1} \\ \hline n_2^{0.5} & n_2^{0.5} \\ \mathbf{0}_{n_2-1} & \mathbf{0}_{n_2-1} \\ \hline \vdots & \\ \hline n_{2m}^{0.5} & -n_{2m}^{0.5} \\ \mathbf{0}_{n_{2m}-1} & \mathbf{0}_{n_{2m}-1} \end{bmatrix} \beta + \xi^*, \qquad (10.38)$$

where $\bar{y}_{i.}$ is the average of the outcomes y_{ij} in cluster i and $\mathbf{y}_i = (y_{i1}, \ldots, y_{in_i})'$. Not surprisingly, all of the information about β comes from the cluster averages. If the model in equation (10.38) is now pre-multiplied by

$$\text{blockdiag}\left(\begin{bmatrix} n_i^{-0.5} & \mathbf{0} \\ \mathbf{0} & \mathbf{I}_{n_i-1} \end{bmatrix}\right), \tag{10.39}$$

the model becomes

$$\begin{bmatrix} \bar{y}_{1.} \\ \mathbf{K}_1\mathbf{y}_1 \\ \hline \bar{y}_{2.} \\ \mathbf{K}_2\mathbf{y}_2 \\ \hline \vdots \\ \hline \bar{y}_{2m.} \\ \mathbf{K}_{2m}\mathbf{y}_{2m} \end{bmatrix} = \begin{bmatrix} 1 & 1 \\ \mathbf{0}_{n_1-1} & \mathbf{0}_{n_1-1} \\ \hline 1 & 1 \\ \mathbf{0}_{n_2-1} & \mathbf{0}_{n_2-1} \\ \hline \vdots & \\ \hline 1 & -1 \\ \mathbf{0}_{n_{2m}-1} & \mathbf{0}_{n_{2m}-1} \end{bmatrix} \beta + \xi^{**}, \tag{10.40}$$

$$\text{where } \text{cov}(\xi^{**}) = \text{blockdiag}\left(\begin{bmatrix} \sigma_s^2 + \sigma_e^2/n_i & \mathbf{0} \\ \mathbf{0} & \sigma_e^2\mathbf{I}_{n_i-1} \end{bmatrix}\right).$$

With the model in this form, it is now easy to show (so these are exercises) that

$$E(\beta_1|\mathbf{y},\sigma_s^2,\sigma_e^2) = 0.5\left[\sum_{i=1}^{m}\bar{y}_i\frac{w_i}{\sum_{j=1}^{m}w_j} - \sum_{i=m+1}^{2m}\bar{y}_i\frac{w_i}{\sum_{i=m+1}^{2m}w_j}\right] \text{ and} \tag{10.41}$$

$$\text{var}(\beta_1|\mathbf{y},\sigma_s^2,\sigma_e^2) = \sum_{j=1}^{2m}w_j\Big/\left[4\left(\sum_{j=1}^{m}w_j\right)\left(\sum_{i=m+1}^{2m}w_j\right)\right],$$

where $w_i = (\sigma_s^2 + \sigma_e^2/n_i)^{-1}$.

These expressions behave as expected at extreme values of $r = \sigma_e^2/\sigma_s^2$. If σ_e^2 is fixed at a finite, positive value and $\sigma_s^2 \to 0$ so $r \to \infty$, then $w_i \to n_i/\sigma_e^2$,

$$E(\beta_1|\mathbf{y},\sigma_s^2,\sigma_e^2) \to 0.5\left(\sum_{i=1}^{m}n_i\bar{y}_i/n^{(1)} - \sum_{i=m+1}^{2m}n_i\bar{y}_i/n^{(2)}\right) \text{ and} \tag{10.42}$$

$$\text{var}(\beta_1|\mathbf{y},\sigma_s^2,\sigma_e^2) \to 0.25\sigma_e^2(1/n^{(1)} + 1/n^{(2)}),$$

as in the analysis that ignores clusters and omits the random effect. On the other hand, if σ_s^2 is fixed at a finite, positive value and $\sigma_e^2 \to 0$ so $r \to 0$, then $w_i \to 1/\sigma_s^2$,

$$E(\beta_1|\mathbf{y},\sigma_s^2,\sigma_e^2) \to 0.5\left(\sum_{i=1}^{m}\bar{y}_i/m - \sum_{i=m+1}^{2m}\bar{y}_i/m\right) \text{ and} \tag{10.43}$$

$$var(\beta_1|\mathbf{y},\sigma_s^2,\sigma_e^2) \to \sigma_s^2/2m.$$

The difference between these two extremes is that when $r \to \infty$, individual observations y_{ij} are treated equally, while when $r \to 0$, clusters are treated equally regardless of their n_i. For small r, then, large clusters are down-weighted and small clusters are up-weighted relative to their sizes n_i.

Equation (10.41) can be used to show a more explicit sense in which the change in the estimate of β_1 arises from informative cluster size. Recalling that the analysis without the random effect is equivalent to letting $r \to \infty$ in the random-effect analysis, then

$$2(\hat{\beta}_{1,RE} - \hat{\beta}_{1,OLS}) = \left(\sum_{i=1}^{m} n_i \bar{y}_i / n^{(1)} - \sum_{i=m+1}^{2m} n_i \bar{y}_i / n^{(2)} \right) \tag{10.44}$$

$$- \left[\sum_{i=1}^{m} \bar{y}_i \frac{w_i}{\sum_{j=1}^{m} w_j} - \sum_{i=m+1}^{2m} \bar{y}_i \frac{w_i}{\sum_{i=m+1}^{2m} w_j} \right]$$

$$= \sum_{i=1}^{m} \bar{y}_i \left(\frac{w_i}{\sum_{j=1}^{m} w_j} - \frac{n_i}{n^{(1)}} \right) - \sum_{i=m+1}^{2m} \bar{y}_i \left(\frac{w_i}{\sum_{i=m+1}^{2m} w_j} - \frac{n_i}{n^{(2)}} \right). \tag{10.45}$$

Because $w_i = (\sigma_s^2 + \sigma_e^2/n_i)^{-1}$, $w_i/(\sum_j w_j) = (1 + r/n_i)^{-1}/(\sum(1 + r/n_i)^{-1})$ depends only on r, not on the individual variances. For given r,

$$\frac{w_i}{\sum_{j=1}^{m} w_j} - \frac{n_i}{n^{(1)}} \tag{10.46}$$

is negative for larger n_i and positive for smaller n_i, with the spread growing as r grows. Thus

$$\sum_{i=1}^{m} \bar{y}_i \left(\frac{w_i}{\sum_{j=1}^{m} w_j} - \frac{n_i}{n^{(1)}} \right) \tag{10.47}$$

is large in absolute value if the cluster averages $\bar{y}_i$ are correlated with the cluster sizes n_i. The same is true for Group 2. Note that the contributions to (10.45) from the two groups can cancel each other if the association between cluster size and cluster average is similar in the two groups, in a particular sense; the next subsection shows an example.

So the change in the estimated treatment effect from adding a random effect for clusters arises from collinearity between **X** and **Z**, the design matrices for the fixed and random effects, and this collinearity can be interpreted as informative cluster size, i.e., cluster size is associated with the outcome in a specific way.

10.2.2 Illustrative Examples

This section presents examples illustrating the previous section's results. The examples are extreme to make the points obvious.

Before beginning, we need to distinguish between *the models used to analyze the data* and *the models used to generate the data*. Chapter 13 discusses this distinction in detail. If you object to it, the argument is, briefly, that we *choose* the model used to analyze the data and rarely if ever know that it reasonably describes the mechanism that produced the data, although for models involving new-style random effects it certainly does not, because the random-effect model is only an analytical convenience. (More on this in Chapter 13.) When shrinkage estimation was introduced in the 1950s, nobody was confused about the distinction between the machinery that

Table 10.1: True Cluster Means in the Four Cases

		True Cluster Mean for	
Case	Group	Big Clusters	Small Clusters
A(i)	1	0	10
	2	10	0
A(ii)	1	0	0
	2	10	0
A(iii)	1	10	0
	2	10	0
B(iv)	1	10	0
	2	10	0

produced a shrunken estimate and the process that produced the data. People started becoming confused about this distinction only when shrinkage estimators were expressed as estimates for models with random effects.

The models used below for analysis are the random effect and no-random-effect models considered in the preceding section. The models used to generate the data are different. In each model used to generate data, each cluster's true mean is specified and iid error ε is added to the cluster means. To illustrate the previous section's results, I consider the expected values of $\hat{\beta}_1$, half the difference between groups, for the analyses without and with random effects, where the expectations are taken with respect to repeated draws of the error ε. For example, for the difference between the two estimators, this expectation is obtained by substituting the true cluster means into equation (10.45) in place of the $\bar{y}_{i.}$. I use different data-generation models — different cluster sizes and different true cluster means — to illustrate properties of this informative cluster size phenomenon.

Specifically, consider 4 data-generation models labeled A(i), A(ii), A(iii), and B(iv). For cases A(i), A(ii), A(iii), the cluster sizes are the same in the two Groups, with n_i as follows. Each of Group 1 and Group 2 has 1 big cluster with $n_i = 50$ and 25 small clusters with $n_i = 2$, so $m = 26$ and $n^{(1)} = n^{(2)} = 100$. These three cases differ according to the true means in each cluster, which depend on cluster size and are given in Table 10.1. For case B(iv), the cluster sizes differ in the two groups. Group 1 has 3 big clusters with $n_i = 50$ and 25 small clusters with $n_i = 2$, while Group 2 has only 1 big cluster with $n_i = 50$ and 27 small clusters, 24 with $n_i = 6$ and 3 with $n_i = 2$, so $m = 28$ and $n^{(1)} = n^{(2)} = 200$. Table 10.1 shows the true cluster means for case B(iv).

Table 10.2 shows how the expectation of the estimated difference between groups changes from adding the random effect, each computed for one of three fixed values of r: $r = 0$, corresponding to positive finite σ_s^2 and a tiny σ_s^2 (little shrinkage), and two moderate r. (The estimate from the analysis without a random effect corresponds to $r = \infty$.)

In case A(i), the analysis without the random effect has expected group difference zero, while the analysis with the random effect has a positive expected difference

Table 10.2: Expected Value of $\hat{\beta}_1$ in the Four Cases

Case	r	No-RE Est	RE Est	RE Est Minus No-RE Est
A(i)	0	0	4.62	4.62
	10		3.33	3.33
	100		0.95	0.95
A(ii)	0	–2.50	–0.19	2.31
	10		–0.83	1.67
	100		–2.02	0.48
A(iii)	0	0	0	0
	10		0	0
	100		0	0
B(iv)	0	2.50	0.36	–2.14
	10		1.47	–1.03
	100		2.40	–0.10

Note: "No-RE Est" is the expectation for the analysis without a random effect; "RE Est" is the expectation for the analysis including a random effect, for the given value of r.

between groups; the size of the expected difference depends on r. Adding the random effect makes a group difference appear. By contrast, in case A(ii), adding the random effect makes the group difference *dis*appear or rather become smaller in magnitude, with the extent of reduction again depending on r. In case A(iii), adding the random effect does not change the expected group difference. Cases A(i) and A(ii) show that informative cluster size can increase or decrease the estimated group difference even though the two groups have the same distribution of cluster sizes, i.e., cluster size is not associated with group. However, case A(iii) shows that this can only happen if the association between cluster size and cluster average is not the same in the two groups. In case A(iii), even though cluster size is associated with true cluster mean, the contributions to (10.45) from Groups 1 and 2 cancel each other because the association between cluster size and cluster average is identical in the two groups.

The result of adding the random effect can be interpreted in terms of the old-fashioned idea of "unit of analysis": When the random effect is absent, the big clusters dominate because the unit of analysis is an individual, but when the random effect is introduced, loosely speaking the small clusters dominate because the unit of analysis is clusters and each group has only one big cluster. (This is exactly correct only for $r = 0$.)

Case B(iv) differs from case A(iii) only in the distribution of cluster sizes in Groups 1 and 2; with different cluster sizes in Groups 1 and 2, adding the random effect makes the estimated group difference much smaller. Using case B(iv)'s cluster sizes, I believe it is impossible to specify true cluster means to create a case in which adding the random effect makes the estimated group difference larger in expectation. To construct such a case, I needed to drop one of the two constraints I have adopted

Table 10.3: Cross-Tabulation of Children by Crown Type and Number of Crowns

		Percent Children with			
Crown Type	N Children	1 Crown	2 Crowns	3 Crowns	4 Crowns
I	71	2.8%	36.6%	14.1%	46.5%
III	20	15.0%	25.0%	10.0%	50.0%
IV	76	4.0%	10.5%	21.1%	64.5%

Note: Table entries are percents within row (crown type).

for simplicity, i.e., requiring the two groups to have the same number of clusters m or requiring $n^{(1)} = n^{(2)}$. (But maybe this is not necessary; checking this is an exercise.)

I have described this effect as "informative cluster size" because cluster size does not confound the group comparison in the usual sense. If cluster size confounded groups in the usual sense, then cluster size would have to be associated both with groups and with the outcome y. However, for the cases labeled A, cluster size is not associated with groups; rather, cluster size is associated with true mean differently in the two groups. Case B(iv) fits the usual definition of confounding: Cluster size is associated with groups and with the true cluster means, while the association of cluster size and true mean is the same in the two groups.

10.2.3 Some Simple Analyses of the Kids-and-Crowns Data

We now return to the kids-and-crowns data. Unfortunately, I haven't found a tidy story that connects the analyses of this dataset to the theory and examples just presented. I give some simple analyses for what they are worth, including simple-minded statistical tests to calibrate some differences. Explaining the kids-and-crowns results is thus an open question, which I offer as an exercise. As noted, however, I can't see how the change in group differences from adding the random effect could be anything but an informative cluster size effect: The dataset has no other feature that *could* cause the change in estimated group difference. But if I am right, this is not a simple effect as in the preceding section's admittedly extreme examples.

Table 10.3 is a cross-tabulation of children according to the type of crown and the number of crowns they received. The crown types (rows) differ in distribution of crown counts per child (P = 0.003 by Pearson's chi-square) although the differences do not have a simple pattern. Type III differs from Type I mainly by having more children with 1 crown and fewer with 2 crowns, though the difference is not striking. Type IV differs from Type I by having fewer children with 2 crowns and more with 4 crowns. Average numbers of crowns per child are 3.04, 2.95, and 3.46 for Types I, III, and IV respectively (P = 0.012 by one-way ANOVA).

For the analyses in Section 9.1.2, however, cluster size is not a child's number of crowns; it is a child's number of crown *lifetimes*, because a repairable failure gives a new lifetime to a crown and the random effect analysis included a child-specific but not a tooth-specific random effect. Children with different crown types differ according to their numbers of crown lifetimes. The average numbers of lifetimes per child are 3.5, 4.4, and 4.4 for Types I, III, and IV respectively (P = 0.014 by one-way

Table 10.4: Percent of Crown Lifetimes with a Failure, by Crown Type and Number of Crowns

Crown Type	1 Crown	2 Crowns	3 Crowns	4 Crowns
Type I	50.0%	38.3%	12.9%	34.6%
Type III	50.0%	16.7%	64.3%	43.1%
Type IV	25.0%	57.9%	61.4%	35.8%

ANOVA). All crown types have median 4 lifetimes; the respective 75th percentiles are 4, 7, and 5 and the 90th percentiles are 6, 9, and 7. Thus while children with Type III crowns tended to have fewer crowns than either of the other two groups, they tended to have more lifetimes per crown: 1.49 lifetimes per crown for Type III crowns, compared to 1.17 and 1.28 for Types I and IV respectively.

So cluster size differs between the crown types. Is cluster size also associated with the outcome, crown failure? I consider this using two outcome measures, a simple binary outcome (failure/no failure) ignoring time on test, and time-to-event.

If, for now, each lifetime is described by crown type and the child's number of crowns, then the binary failure/no failure outcome is related to number of crowns differently for the different crown types ($P = 0.0003$ for the interaction of crown type with number of crowns, in a logistic regression ignoring clustering by child). Table 10.4 shows the fraction of lifetimes with failures for each combination of a crown type and number of crowns per child. The pattern is not simple but it is clear why the interaction tested significant. Both Types III and IV crowns differ from Type I crowns, though not in the same way.

Again, however, cluster size is a child's number of crown lifetimes, not number of crowns. But this introduces a complication: A child's frequency of failures (failures per lifetime) is necessarily associated with the child's number of crown lifetimes, because a crown only gets an extra lifetime by having a repairable failure. And indeed, number of lifetimes is associated with the chance of failing: Considering all children together, number of lifetimes and fraction of lifetimes with a failure have Pearson correlation 0.37 ($P < 0.0001$). Table 10.5 shows the fraction of lifetimes ending in failure, tabulated by the number of lifetimes (combining crown types; the association is quite similar for the three crown types). The fraction failing increases with the number of lifetimes, with a jump between 4 and 5 lifetimes, as might be expected.

The simple time-to-event analog to Table 10.5 is a Cox regression where a "case" is a crown lifetime and the predictors are crown type, number of lifetimes for the child, and their interaction. (In this simple-minded analysis, I've ignored clustering by child.) Not surprisingly, the relative hazard rises steeply with the child's number of lifetimes (estimated log relative hazard 0.30 per lifetime, standard error 0.033), but this slope in number of lifetimes does not test different between crown types ($P = 0.11$ by likelihood ratio test). In this analysis, after adjusting for number of lifetimes Type III crowns have *lower* hazard of failure than Type I crowns; if number of lifetimes and the interaction are removed from the model, the analysis is the same as the no-random-effect analysis in Section 9.1.2, and Type III crowns have significantly

Table 10.5: Fraction of Crown Lifetimes Ending in Failure, by Number of Crown Lifetimes

N Lifetimes	N Children	Fraction of Lifetimes Ending in Failure
1	6	16.7%
2	34	27.9%
3	20	25.0%
4	63	22.2%
5	11	52.7%
6	13	52.6%
7	12	61.9%
8	3	66.7%
9	3	63.0%
11	2	77.3%

higher hazard of failure. In both analyses, however, Type IV crowns have estimated hazard very similar to Type I crowns.

Exercises

Regular Exercises

1. (Section 10.1.1.2) Derive $E(\beta|\tau_e,\tau_s,\mathbf{y})$ (equation 10.3) and $E(\mathbf{b}|\tau_e,\tau_s,\mathbf{y})$ (equation 10.4). Reich et al. (2006) gives derivations but they are not hard, so see if you can do them without looking.
2. (Section 10.1.4.1) Derive the point estimates of β and their expected values, conditional on τ_s and τ_e when appropriate, under Models 0, H, and S. I found this not too easy, but perhaps you are better at linear algebra than I am.
3. (Section 10.2.1) Derive equations (10.29) and (10.30). Lacking finesse, I used brute force but a more elegant derivation might be possible.
4. (Section 10.2.1) Derive the posterior covariance matrices for β and (b_1,b_2) conditional on σ_e^2 and σ_s^2. Hint: Start with the **Zu** parameterization, derive the conditional posterior covariance of $(\beta',\mathbf{u}')'$, use partitioned-matrix formulas to get covariance matrices for β and $\mathbf{u}$ — you don't need the rest of the joint posterior covariance matrix for the present purpose — and then transform $\mathbf{u}$ to (b_1,b_2).
5. (Section 10.2.1) Derive equation (10.45), the difference between the estimated group difference without and with the random effect, conditional on r.
6. (Section 10.2.2) Construct an example using the true group means for case B(iv) in which adding the random effect makes the expectation of the group difference become larger. I conjecture you need to drop one of the two constraints I've used — that the two groups have the same numbers of clusters or that $n^{(1)}=n^{(2)}$ — but perhaps you can disprove this conjecture.

Open Questions

1. (Section 10.1.5) For the spatial-confounding problem, derive the marginal posterior distribution of the smoothing ratio $r = \tau_s/\tau_e$. Explore how this posterior is affected by confounding like that present in the Slovenia data. When a canonical regressor implicit in **S**'s precision matrix $\tau_s\mathbf{Q}$ is highly collinear with a fixed effect in **X**, what happens to this posterior distribution?
2. (Section 10.2.1) For the clustered-data problem, derive the marginal posterior distribution of the smoothing ratio $r = \sigma_e^2/\sigma_s^2$. This can be derived in closed form except for the proportionality constant. Explore how this posterior is affected, if at all, by informative cluster size.
3. (Section 10.2.3). In the kids-and-crowns data, explain the change in the estimate of the difference between treatment groups that arises from including the random effect.

Chapter 11

Differential Shrinkage

This chapter explores the differential shrinkage puzzle in Section 9.1.3, in which several interaction effects had roughly the same estimates and posterior standard deviations in an unshrunk ANOVA but were shrunk to very different extents in a smoothed ANOVA. That example had a time-to-event outcome, which for the present purpose makes exact derivations impossible. Also, its analysis included three patient characteristics (factors) having a total of 7 interaction contrasts with shrunken coefficients and 4 main-effect contrasts with unshrunk coefficients. For simplicity, the present chapter examines a simplified model with normal errors and three predictors, two with shrunk coefficients and one with an unshrunk coefficient.

This chapter's object is to determine, in this simplified model, whether non-zero correlation among the three predictors can induce differential shrinkage like that observed in Section 9.1.3's colon cancer example. The latter was striking because the columns in the design matrix were scaled to have the same Euclidean length and the unshrunk analysis had 4 interaction coefficients with roughly equal estimates and 95% posterior intervals. Reproducing these conditions in the simplified example required some restrictions on the specification. However, as we'll see, it is impossible to reproduce all the conditions in the original time-to-event analysis without placing so many restrictions on the simplified normal-errors analysis that there is plainly no possibility of finding differential shrinkage. The normal-errors formulation thus sacrifices a potentially important aspect of the original problem; I discuss this in the present chapter's concluding section.

11.1 The Simplified Model and an Overview of the Results

This section gives an overview of the results because the derivations themselves are brute-force exercises with little if any intuitive content. I chose a model with three predictors because a 3×3 matrix has an explicit inverse if all else fails and as it turned out, all else did fail. It's hard to believe these derivations can't be done more elegantly, and more elegant derivations with intuitive content might permit the results to be generalized to more and perhaps all normal-errors problems. Searching for elegant derivations and generalization is an exercise.

11.1.1 The Simplified Model

The sample size is, as usual, n. There are three predictors, X_1, X_2, and X_3. The coefficients of X_1 and X_2 are treated as random effects and shrunk, while X_3's coefficient is treated as a fixed effect in all analyses and is never shrunk. To represent explicitly the presence or absence of centering, the X_j are defined as follows:

$$\begin{aligned} X_1 &= m_1\mathbf{1}_n + Z_1 \\ X_2 &= c(m_2\mathbf{1}_n + Z_2) \\ X_3 &= m_3\mathbf{1}_n + Z_3, \end{aligned} \tag{11.1}$$

where each $Z_j, j = 1,2,3$ is an n-vector with average zero satisfying $Z_j'Z_j = n$, i.e., each Z_j is centered and scaled. X_1 and X_3 have averages m_1 and m_3 respectively, while X_2 has average cm_2. Define $\mathbf{m} = (m_1, m_2, m_3)$. The Z_j and thus the X_j have correlation matrix

$$\mathbf{F} = \begin{bmatrix} 1 & \rho_{12} & \rho_{13} \\ \rho_{12} & 1 & \rho_{23} \\ \rho_{13} & \rho_{23} & 1 \end{bmatrix}; \tag{11.2}$$

$Z_1'Z_3 = n\rho_{13}$ and $Z_j'Z_2 = cn\rho_{j2}, j = 1,3$. Define the design matrix $\mathbf{X}$ to have 4 columns: $\mathbf{1}_n$ for the intercept followed by X_1, X_2, and X_3. (This is not entirely consistent with notation in the rest of this book, but the alternatives seemed more awkward.) The constant c in X_2's definition is needed so that in the unshrunk analysis, the coefficients of X_1 and X_1 have the same posterior variance (or just variance, in a non-Bayesian analysis); $c = \left[(1-\rho_{13}^2)/(1-\rho_{23}^2)\right]^{0.5}$, as will be shown below.

The outcome or dependent variable is, as usual, the n-vector $\mathbf{y}$. The linear model for $\mathbf{y}$ is

$$\mathbf{y} = \mathbf{X}\theta + \varepsilon, \tag{11.3}$$

where the error term $\varepsilon \sim N(\mathbf{0}, \sigma_e^2\mathbf{I}_n)$ and $\theta = (\theta_0, \theta_1, \theta_2, \theta_3)'$ contains the coefficients, which are of primary interest. The derivations below do not use $\mathbf{y}$ but rather functions of $\mathbf{y}$. These are

- The coefficient estimates in the unshrunk analysis, $\hat{\theta}^U = (\mathbf{X}'\mathbf{X})^{-1}\mathbf{X}'\mathbf{y}$. These are the posterior means arising from an improper flat prior on θ as well as the least-squares estimates. A key condition in the differential shrinkage puzzle is that $\hat{\theta}_1^U = \hat{\theta}_2^U$, i.e., X_1 and X_2 have the same coefficient estimates in the unshrunk analysis.
- The residual sum of squares in the unshrunk analysis, $RSS = \mathbf{y}'\mathbf{y} - \hat{\theta}^{U\prime}\mathbf{X}'\mathbf{X}\hat{\theta}^U$; and
- $\mathbf{X}'\mathbf{y} = \mathbf{X}'\mathbf{X}\hat{\theta}^U$.

The analysis with shrinkage is implemented by giving $(\theta_1, \theta_2)'$ a model (or prior, if you prefer) with $\theta_j \sim N(0, \sigma_{sj}^2), j = 1,2$. As usual, shrinkage is determined by the ratios $r_j = \sigma_e^2/\sigma_{sj}^2$. The point estimate of θ with shrunken coefficients is denoted $\hat{\theta}^S$. Below, this is always shown conditioning on (r_1, r_2), so it is both the conditional posterior mean of $\hat{\theta}$ and the conventional estimate of $\hat{\theta}$ given (r_1, r_2).

With this machinery, the dataset and a non-Bayesian analysis are completely specified by five items: n, RSS, $\mathbf{F}$, $\hat{\theta}^U$, and $\mathbf{m}$. A Bayesian analysis needs two more items, priors on (r_1, r_2) and σ_e^2. The sections below show, as a by-product, that these seven items suffice. The results presented below do not use an explicit prior for (r_1, r_2) but just assume a prior that is symmetric in r_1 and r_2. An example of such a prior specifies r_1 and r_2 to be independent with each having the distribution implied by a flat prior on the approximate degrees of freedom in $X_j\theta_j$, $j = 1, 2$. (Section 2.2.2.3 gives the formula for approximate DF. Showing this prior is symmetric in (r_1, r_2) is an exercise to the present chapter.) The results shown below use an inverse gamma prior for σ_e^2.

The form of c implies that if X_1 and X_2 are required to have the same Euclidean length, then $\rho_{13} = \rho_{23}$. Applying this restriction with the above assumptions and a prior on (r_1, r_2) that is symmetric in r_1 and r_2 defines a problem that is in all aspects symmetric in X_1 and X_2 and in θ_1 and θ_2, so differential shrinkage cannot occur. Thus, $c \neq 1$ must be allowed in this normal-errors formulation, deviating from the original colon cancer smoothed ANOVA.

11.1.2 Overview of the Results

First, consider the case in which the predictors are centered ($\mathbf{m} = \mathbf{0}$), the outcome is centered ($\mathbf{1}_n'\mathbf{y} = 0$), and the model (11.3) has no intercept ($\theta_0 = 0$).

Section 11.2.1 shows that under these assumptions, the marginal posterior of (r_1, r_2) — and thus the profiled restricted likelihood of (r_1, r_2), i.e., the restricted likelihood with σ_e^2 replaced by the value of σ_e^2 that maximizes the restricted likelihood given (r_1, r_2) — is symmetric in r_1 and r_2. That is, substitute r_1 for r_2 and r_2 for r_1 and the marginal posterior and profiled restricted likelihood do not change. Also $\hat{\theta}^S$, the conditional posterior mean or conventional estimate of θ given (r_1, r_2), has $\hat{\theta}_1^S = \hat{\theta}_2^S$ whenever $r_1 = r_2$. Thus differential shrinkage is impossible irrespective of $\mathbf{F}$ and of the dataset and priors. Also, the marginal posterior and profiled restricted likelihood of (r_1, r_2) are not a function of $\hat{\theta}_3^U$, the coefficient of X_3 in the unshrunk analysis. Thus the value of $\hat{\theta}_3^U$ does not affect the extent to which $\hat{\theta}_1^S$ and $\hat{\theta}_2^S$ are shrunk, for given RSS.[1]

Now let the model have a non-zero intercept and do not require the outcome $\mathbf{y}$ to be centered, but continue to center the predictors ($\mathbf{m} = \mathbf{0}$). The previous results extend to this case with only one change: The exponent of σ_e^2 in the marginal posterior and restricted likelihood changes by 0.5. Otherwise, this problem reduces straightforwardly to the no-intercept, centered-$\mathbf{y}$ case above. Section 11.2.2 gives detailed results.

Finally, relax the assumption that the predictors are centered, i.e., allow $\mathbf{m} \neq \mathbf{0}$. The marginal posterior (and thus restricted likelihood) reduce to an expression

[1]An aside for students: The two key results here — symmetry in (r_1, r_2) and no dependence on $\hat{\theta}_3^U$ — surprised me. I coded up the machinery and after debugging it I randomly generated 10,000 examples expecting to find differential shrinkage and an effect of $\hat{\theta}_3^U$, which I planned to explore using further simulated examples. However, neither effect was present, leaving me little choice but to work out the admittedly stupefying proofs shown below.

equivalent to that arising in the preceding case with centered predictors but including an intercept. This is shown in Section 11.2.3. Changes in other analytical summaries are similarly minimal, so the preceding results follow in this case as well.

Thus in this simplified problem, the marginal posterior of (r_1, r_2) and the profiled restricted likelihood are symmetric in r_1 and r_2, as is $\hat{\theta}^S$, because $\hat{\theta}_1^U = \hat{\theta}_2^U$, a key aspect of the differential-shrinkage puzzle. Also, the marginal posterior and profiled restricted likelihood of (r_1, r_2) are not functions of $\hat{\theta}_3^U$. Thus differential shrinkage is impossible and $\hat{\theta}_3^U$ does not affect the shrinkage of $\hat{\theta}_1^S$ and $\hat{\theta}_2^S$ for given *RSS*. Section 11.3 discusses other possible explanations of the differential-shrinkage puzzle.

11.2 Details of Derivations

11.2.1 Centered Predictors and Outcome and No Intercept

11.2.1.1 Expressions for the Inferential Summaries

Assume $\mathbf{m} = \mathbf{0}$, $\mathbf{y}'\mathbf{1}_n = 0$, and $\theta_0 = 0$, so $\mathbf{X}$ is $n \times 3$ with columns X_1, X_2, and X_3. Then $\mathbf{X}'\mathbf{X} = n\mathbf{HFH}$ for $\mathbf{F}$ defined in (11.2) and $\mathbf{H}$ a diagonal matrix with diagonal entries $(1, c, 1)$. $(\mathbf{X}'\mathbf{X})^{-1} = \mathbf{H}^{-1}\mathbf{F}^{-1}\mathbf{H}^{-1}/n$, where

$$\begin{aligned} \mathbf{F}^{-1} &= \frac{1}{d}\begin{bmatrix} 1-\rho_{23}^2 & \rho_{13}\rho_{23}-\rho_{12} & \rho_{12}\rho_{23}-\rho_{13} \\ \rho_{13}\rho_{23}-\rho_{12} & 1-\rho_{13}^2 & \rho_{12}\rho_{13}-\rho_{23} \\ \rho_{12}\rho_{23}-\rho_{13} & \rho_{12}\rho_{13}-\rho_{23} & 1-\rho_{12}^2 \end{bmatrix} \qquad (11.4) \\ \text{and } d &= 1-(\rho_{12}^2+\rho_{13}^2+\rho_{23}^2)+2\rho_{12}\rho_{13}\rho_{23}. \end{aligned}$$

(Derivation of $\mathbf{F}^{-1}$ is an exercise.) In the unshrunk analysis, the variance of the least-squares estimator of θ and the posterior covariance of θ are proportional to $(\mathbf{X}'\mathbf{X})^{-1}$. Thus, if the unshrunk coefficients of X_1 and X_2 have the same variance, then $1-\rho_{23}^2 = c^{-2}(1-\rho_{13}^2)$ or $c = \left[(1-\rho_{13}^2)/(1-\rho_{23}^2)\right]^{0.5}$, as stated above.

The shrunken estimate of θ — the posterior mean and the conventional estimate given (r_1, r_2) — is

$$\begin{aligned} \hat{\theta}^S &= (\mathbf{X}'\mathbf{X}+\text{diag}(r_1, r_2, 0))^{-1}\mathbf{X}'\mathbf{y} \\ &= (\mathbf{X}'\mathbf{X}+\text{diag}(r_1, r_2, 0))^{-1}(\mathbf{X}'\mathbf{X})\hat{\theta}^U \\ &= (\mathbf{HKH})^{-1}\mathbf{HFH}\hat{\theta}^U \qquad (11.5) \\ & \text{where } \mathbf{K} = \mathbf{F}+\text{diag}(r_1/n, r_2/c^2n, 0), \end{aligned}$$

and the dependence of $\hat{\theta}^S$ on (r_1, r_2) has been suppressed for convenience.

Later in this derivation, I need an explicit form for $\mathbf{K}^{-1}$:

$$\mathbf{K}^{-1} = \frac{1}{D}\begin{bmatrix} \left(1+\frac{r_2}{c^2n}\right)-\rho_{23}^2 & \rho_{13}\rho_{23}-\rho_{12} & \rho_{12}\rho_{23}-\rho_{13}\left(1+\frac{r_2}{c^2n}\right) \\ \rho_{13}\rho_{23}-\rho_{12} & \left(1+\frac{r_1}{n}\right)-\rho_{13}^2 & \rho_{12}\rho_{13}-\rho_{23}\left(1+\frac{r_1}{n}\right) \\ \rho_{12}\rho_{23}-\rho_{13}\left(1+\frac{r_2}{c^2n}\right) & \rho_{12}\rho_{13}-\rho_{23}\left(1+\frac{r_1}{n}\right) & \left(1+\frac{r_2}{c^2n}\right)\left(1+\frac{r_1}{n}\right)-\rho_{12}^2 \end{bmatrix} \qquad (11.6)$$

and

$$\begin{aligned} D &= \det(\mathbf{K}) \\ &= \left(1+\frac{r_2}{c^2n}\right)\left(1+\frac{r_1}{n}\right) - \rho_{12}^2 - \left(1+\frac{r_2}{c^2n}\right)\rho_{13}^2 - \left(1+\frac{r_1}{n}\right)\rho_{23}^2 + 2\rho_{12}\rho_{13}\rho_{23}; \end{aligned} \tag{11.7}$$

this was obtained in the same inelegant manner as $\mathbf{F}^{-1}$ and is also an exercise.

The marginal posterior and restricted likelihood of (σ_e^2, r_1, r_2) are most easily derived using Section 2.1's constraint-case formulation. Let the sans-serif font indicates items specific to the constraint-case formulation, and define

$$\mathsf{Y} = \begin{bmatrix} \mathbf{y} \\ 0 \\ 0 \end{bmatrix}, \mathsf{X} = \begin{bmatrix} & \mathbf{X} & \\ \hline 1 & 0 & 0 \\ 0 & 1 & 0 \end{bmatrix}, \text{ and } \Gamma = \begin{bmatrix} \sigma_e^2\mathbf{I}_n & \mathbf{0} & \mathbf{0} \\ \mathbf{0} & \sigma_{s1}^2 & 0 \\ \mathbf{0} & 0 & \sigma_{s1}^2 \end{bmatrix}. \tag{11.8}$$

In the constraint-case formulation, the model is $\mathsf{Y} = \mathsf{X}\theta + \mathsf{E}$ where E is multivariate normal with covariance Γ, and the marginal posterior of (σ_e^2, r_1, r_2) is

$$\begin{aligned} \pi(\sigma_e^2, r_1, r_2|\mathbf{y}) &\propto \pi(\sigma_e^2, r_1, r_2) \\ &\quad |\Gamma|^{-0.5}|\mathsf{X}\Gamma^{-1}\mathsf{X}|^{-0.5}\exp\left(-\frac{1}{2}\left[\mathsf{Y}'\Gamma^{-1/2}[\mathbf{I}_{n+2} - \mathbf{V}]\Gamma^{-1/2}\mathsf{Y}\right]\right) \\ &\quad \text{where } \mathbf{V} = \Gamma^{-1/2}\mathsf{X}(\mathsf{X}'\Gamma^{-1}\mathsf{X})^{-1}\mathsf{X}'\Gamma^{-1/2}. \end{aligned} \tag{11.9}$$

(Deriving this is an exercise.) The restricted likelihood is (11.9) with $\pi(\sigma_e^2, r_1, r_2)$ set to 1.

The marginal posterior/restricted likelihood in (11.9) can be made much more explicit. The determinants are $|\Gamma| = (\sigma_e^2)^n\sigma_{s1}^2\sigma_{s1}^2$ and $|\mathsf{X}\Gamma^{-1}\mathsf{X}| \propto (\sigma_e^2)^{-3}|\mathbf{K}| = (\sigma_e^2)^{-3}D$ for D in (11.7). The argument of the exponent, multiplied by -2, can be written as

$$\frac{1}{\sigma_e^2}\mathbf{y}'\mathbf{y} - \hat{\theta}^{S\prime}(\mathsf{X}'\Gamma^{-1}\mathsf{X})\hat{\theta}^S = \frac{1}{\sigma_e^2}\left(\mathbf{y}'\mathbf{y} - n\hat{\theta}^{U\prime}\mathbf{HFK}^{-1}\mathbf{FH}\hat{\theta}^U\right). \tag{11.10}$$

Given RSS, $\hat{\theta}^U$, and $\mathbf{X}$, $\mathbf{y}'\mathbf{y} = RSS + \hat{\theta}^{U\prime}\mathbf{X}'\mathbf{X}\hat{\theta}^U = RSS + n\hat{\theta}^{U\prime}\mathbf{HFH}\hat{\theta}^U$. Thus, the marginal posterior in (11.9) can be written as

$$\begin{aligned} &\pi(\sigma_e^2, r_1, r_2|\mathbf{y}) \propto \pi(\sigma_e^2, r_1, r_2) \\ &\quad (\sigma_e^2)^{-\frac{(n-1)}{2}}(r_1r_2)^{0.5}D^{-0.5}\exp\left(-\tfrac{1}{2}\left[RSS + n\hat{\theta}^{U\prime}\mathbf{H}(\mathbf{F} - \mathbf{FK}^{-1}\mathbf{F})\mathbf{H}\hat{\theta}^U\right]\right), \end{aligned} \tag{11.11}$$

and the restricted likelihood is again obtained by setting $\pi(\sigma_e^2, r_1, r_2)$ equal to 1.

If σ_e^2 has an inverse gamma prior with density $\pi(\sigma_e^2) \propto (\sigma_e^2)^{-(a+1)}\exp(-b/\sigma_e^2)$, independently of (r_1, r_2) *a priori*, then σ_e^2 is easily integrated out to give the marginal posterior of (r_1, r_2)

$$\begin{aligned} &\pi(r_1, r_2|y) \propto \pi(r_1, r_2) \\ &\quad (r_1r_2)^{0.5}D^{-0.5}\left[b + 0.5RSS + 0.5n\hat{\theta}^{U\prime}\mathbf{H}(\mathbf{F} - \mathbf{FK}^{-1}\mathbf{F})\mathbf{H}\hat{\theta}^U\right]^{-\frac{(n+2a-1)}{2}}. \end{aligned} \tag{11.12}$$

The profiled restricted likelihood is obtained by inserting into the restricted likelihood the value of σ_e^2 that maximizes the restricted likelihood for fixed (r_1, r_2). This is easily shown to equal the marginal posterior in equation (11.12) with $\pi(r_1, r_2) = 1$ and $a = b = 0$.

The foregoing establishes, for this special case, the earlier claim that the dataset and analysis are completely specified by n, RSS, $\mathbf{F}$, $\hat{\theta}^U$, and priors on (r_1, r_2) and σ_e^2. Equation (11.12) can be written even more explicitly, using (11.4) and (11.6). I do so below, as needed.

11.2.1.2 Proofs of the Claims

The claims to be proved are that the marginal posterior (11.12) is symmetric in r_1 and r_2, that the two shrunken coefficient estimates $\hat{\theta}_1^S$ and $\hat{\theta}_2^S$ are equal when $r_1 = r_2$, and that (11.12) is not a function of $\hat{\theta}_3^U$, the coefficient of X_3 in the unshrunk analysis. The proof of the first claim is partly given and partly outlined just below. It is by unrelieved brute force; many details have been left as exercises. This is followed by proofs of the remaining two claims, which are slightly more elegant.

To show (11.12) is symmetric in r_1 and r_2, I first show symmetry for the parts of (11.12) arising from the determinants and then for the part arising from the exponent. Regarding the determinants, obviously $(r_1 r_2)^{0.5}$ is symmetric in r_1 and r_2. As for D, the part of D involving r_1 and r_2 is

$$\begin{aligned} &\left(1+\frac{r_2}{c^2 n}\right)\left(1+\frac{r_1}{n}\right) - \left(1+\frac{r_2}{c^2 n}\right)\rho_{13}^2 - \left(1+\frac{r_1}{n}\right)\rho_{23}^2 \\ =\ &\frac{1}{c^2}\left[\left(c^2+\frac{r_2}{n}\right)\left(1+\frac{r_1}{n}\right) - \left(c^2+\frac{r_2}{n}\right)\rho_{13}^2 - c^2\left(1+\frac{r_1}{n}\right)\rho_{23}^2\right]. \end{aligned} \quad (11.13)$$

Write $c^2 = 1 + q$ for $q = (\rho_{23}^2 - \rho_{13}^2)/(1 - \rho_{23}^2)$. Then (11.13) is

$$\frac{1}{c^2}\left[\left(1+\frac{r_2}{n}\right)\left(1+\frac{r_1}{n}\right) - q\rho_{13}^2 - \left(1+\frac{r_2}{n}\right)\rho_{13}^2 - \left(1+\frac{r_1}{n}\right)(c^2\rho_{23}^2 - q)\right]. \quad (11.14)$$

But $c^2\rho_{23}^2 - q = \rho_{13}^2$. Thus D is symmetric in r_1 and r_2.

Now for the part of (11.12) arising from the exponent, which involves $\mathbf{y}$ through RSS and $\hat{\theta}^U$. Of $b + 0.5RSS + 0.5n\hat{\theta}^{U\prime}\mathbf{H}(\mathbf{F} - \mathbf{F}\mathbf{K}^{-1}\mathbf{F})\mathbf{H}\hat{\theta}^U$, the part involving (r_1, r_2) is $\hat{\theta}^{U\prime}\mathbf{H}\mathbf{F}\mathbf{K}^{-1}\mathbf{F}\mathbf{H}\hat{\theta}^U$. Note that $\mathbf{K}^{-1} = \mathbf{A}/D$, where $\mathbf{A} = D\mathbf{K}^{-1}$ is the matrix enclosed in square brackets in equation (11.6). D is symmetric in r_1 and r_2, so we need to show $\hat{\theta}^{U\prime}\mathbf{H}\mathbf{F}\mathbf{A}\mathbf{F}\mathbf{H}\hat{\theta}^U$ is symmetric in r_1 and r_2. Recall that by assumption, $\hat{\theta}_1^U = \hat{\theta}_2^U$. Write $\mathbf{A}$ as

$$\begin{bmatrix} a_{11} & a_{12} & a_{13} \\ a_{12} & a_{22} & a_{23} \\ a_{13} & a_{23} & a_{33} \end{bmatrix}. \quad (11.15)$$

Below I will use notation like a_{21} for convenience, but $a_{ij} = a_{ji}$ and similarly $\rho_{ij} = \rho_{ji}$. Then

$$\mathbf{HFAFH} = \begin{bmatrix} \sum_{ij}\rho_{1i}\rho_{1j}a_{ij} & c\sum_{ij}\rho_{1i}\rho_{2j}a_{ij} & \sum_{ij}\rho_{1i}\rho_{3j}a_{ij} \\ c\sum_{ij}\rho_{2i}\rho_{1j}a_{ij} & c^2\sum_{ij}\rho_{2i}\rho_{2j}a_{ij} & c\sum_{ij}\rho_{2i}\rho_{3j}a_{ij} \\ \sum_{ij}\rho_{3i}\rho_{1j}a_{ij} & c\sum_{ij}\rho_{3i}\rho_{2j}a_{ij} & \sum_{ij}\rho_{3i}\rho_{3j}a_{ij} \end{bmatrix}. \quad (11.16)$$

Because $\hat{\theta}_1^U = \hat{\theta}_2^U$ and $\hat{\theta}_3^U$ can take any value, showing that $\hat{\theta}^{U\prime}\mathbf{HFAFH}\hat{\theta}^U$ is symmetric in r_1 and r_2 requires showing that each of the following items is symmetric in r_1 and r_2:

- the (3,3) element of **HFAFH** in (11.16);
- the sum of the (1,3) and (2,3) elements of **HFAFH**;
- the sum of the (1,2) and (2,1) elements of **HFAFH**; and
- the sum of the (1,1) and (2,2) elements of **HFAFH**.

I consider each of these in turn. For each, I ignored a_{ij} not involving r_1 or r_2, wrote out the remaining terms, and showed they are symmetric in r_1 and r_2. The next few paragraphs sketch these proofs.

Consider the (3,3) element of **HFAFH** in (11.16). Writing out the terms of $\sum_{ij}\rho_{3i}\rho_{3j}a_{ij}$ that involve r_1 or r_2 (which is an exercise) gives

$$\left(1+\frac{r_2}{c^2 n}\right)\left(1+\frac{r_1}{n}\right) - \left(1+\frac{r_2}{c^2 n}\right)\rho_{13}^2 - \left(1+\frac{r_1}{n}\right)\rho_{23}^2. \qquad (11.17)$$

This is identical to (11.13), the parts of D involving r_1 or r_2, which were shown above to be symmetric in r_1 and r_2. Thus the (3,3) element of **HFAFH** is symmetric in r_1 and r_2.

Now consider the sum of the (1,3) and (2,3) elements of **HFAFH**. After eliminating terms that do not involve r_1 or r_2 and doing some canceling (which is an exercise), this is

$$(\rho_{13}+c\rho_{23})\left[\left(1+\frac{r_2}{c^2 n}\right)\left(1+\frac{r_1}{n}\right) - \left(1+\frac{r_2}{c^2 n}\right)\rho_{13}^2 - \left(1+\frac{r_1}{n}\right)\rho_{23}^2\right]. \qquad (11.18)$$

The expression in square brackets is again identical to (11.13), so the sum of the (1,3) and (2,3) elements of **HFAFH** is symmetric in r_1 and r_2.

Now consider the sum of the (1,2) and (2,1) elements of **HFAFH**, $2c\sum_{ij}\rho_{1i}\rho_{2j}a_{ij}$. This must be considered separately from the (1,1) and (2,2) elements because the (1,2) and (2,1) elements have a factor c that cannot be eliminated. Omitting terms that do not involve r_1 or r_2, this sum is

$$\frac{1}{c^2}\left[\left(c^2+\frac{r_2}{n}\right)\left(1+\frac{r_1}{n}\right)\rho_{13}\rho_{23} + \left(c^2+\frac{r_2}{n}\right)(\rho_{12}-\rho_{12}\rho_{13}^2-\rho_{23}\rho_{13}) + c^2\left(1+\frac{r_1}{n}\right)(\rho_{12}-\rho_{12}\rho_{23}^2-\rho_{23}\rho_{13})\right]. \qquad (11.19)$$

Unlike the previous two items, this does not reduce to the relevant parts of D, but the proof follows a similar sequence. Write $c^2 = 1+q$, pull out terms involving q, add and subtract $\left(1+\frac{r_1}{n}\right)(\rho_{12}-\rho_{12}\rho_{13}^2-\rho_{23}\rho_{13})$, gather up the extra terms involving $\left(1+\frac{r_1}{n}\right)$, and show that the latter sum to zero. The details are an exercise.

This leaves the sum of the (1,1) and (2,2) elements of **HFAFH**. I proved its symmetry by applying the foregoing meat grinder separately to the (1,1) and (2,2) terms, each of which left an asymmetric term in $\left(1+\frac{r_1}{n}\right)$, which canceled each other. This too is an exercise.

Proofs of the final two claims — that the shrunken coefficient estimates $\hat{\theta}_1^S$ and $\hat{\theta}_2^S$ are equal when $r_1 = r_2$, and that the marginal posterior (11.12) is not a function of $\hat{\theta}_3^U$, the coefficient of X_3 in the unshrunk analysis — use an algebraic identity given as equation (1) in Smith (1973). For matrices **W**, **S**, and **T** with appropriate dimensions for which $\mathbf{W}^{-1}$ exists, this identity is

$$(\mathbf{W}+\mathbf{ST})^{-1} = \mathbf{W}^{-1} - \mathbf{W}^{-1}\mathbf{S}(\mathbf{I}+\mathbf{TW}^{-1}\mathbf{S})^{-1}\mathbf{TW}^{-1}. \tag{11.20}$$

First, I'll show that (11.12) is not a function of $\hat{\theta}_3^U$. The relevant part of (11.12) is $\hat{\theta}^{U\prime}\mathbf{H}(\mathbf{F}-\mathbf{FK}^{-1}\mathbf{F})\mathbf{H}\hat{\theta}^U$. The result follows if $\mathbf{H}(\mathbf{F}-\mathbf{FK}^{-1}\mathbf{F})\mathbf{H}$, a 3×3 matrix, has all zeroes in its third row and column. Recall that

$$\begin{aligned}\mathbf{K} &= \mathbf{F}+\mathrm{diag}(r_1/n, r_2/c^2 n, 0)\\ &= \mathbf{H}^{-1}(\mathbf{HFH}+\mathrm{diag}(r_1/n, r_2/n, 0))\mathbf{H}^{-1}.\end{aligned} \tag{11.21}$$

Therefore

$$\begin{aligned}&\mathbf{H}(\mathbf{F}-\mathbf{FK}^{-1}\mathbf{F})\mathbf{H}\\ =\; &\mathbf{HFH}-\mathbf{HFH}(\mathbf{HFH}+\mathrm{diag}(r_1/n, r_2/n, 0))^{-1}\mathbf{HFH}.\end{aligned} \tag{11.22}$$

In Smith's identity, let $\mathbf{W} = \mathbf{HFH}$ and let $\mathbf{S} = \mathbf{T} = \mathrm{diag}(\sqrt{r_1/n}, \sqrt{r_2/n}, 0)$. Then (11.22) becomes

$$\mathbf{W}-\mathbf{W}+\mathbf{S}(\mathbf{I}+\mathbf{SW}^{-1}\mathbf{S})^{-1}\mathbf{S}. \tag{11.23}$$

By the definition of **S**, **SMS** has third row and column equal to zero for any matrix **M** of suitable dimensions. The result follows: (11.12) is not a function of $\hat{\theta}_3^U$.

Finally, to show that $\hat{\theta}_1^S = \hat{\theta}_2^S$ when $r_1 = r_2$, recall from equation (11.5) that

$$\begin{aligned}\hat{\theta}^S &= (\mathbf{HKH})^{-1}\mathbf{HFH}\hat{\theta}^U\\ &= (\mathbf{HFH}+\mathrm{diag}(r_1/n, r_1/n, 0))^{-1}\mathbf{HFH}\hat{\theta}^U,\end{aligned} \tag{11.24}$$

because we are now assuming $r_1 = r_2$. As in the previous proof, apply Smith's identity with $\mathbf{W} = \mathbf{HFH}$ and $\mathbf{S} = \mathbf{T} = \mathrm{diag}(\sqrt{r_1/n}, \sqrt{r_1/n}, 0)$. Then

$$\hat{\theta}^S = \hat{\theta}^U - \mathbf{W}^{-1}\mathbf{S}(\mathbf{I}+\mathbf{SW}^{-1}\mathbf{S})^{-1}\mathbf{S}\hat{\theta}^U. \tag{11.25}$$

By assumption, $\mathbf{S} = \sqrt{r_1/n}\,\mathrm{diag}(1,1,0)$ and $\hat{\theta}^U = f(1,1,g)'$ for scalar constants $f = \hat{\theta}_1^U = \hat{\theta}_2^U$ and $g = \hat{\theta}_3^U/f$. Then

$$\hat{\theta}^S = \hat{\theta}^U - (fr_1/n)\mathbf{W}^{-1}\mathrm{diag}(1,1,0)(\mathbf{I}+\mathrm{diag}(1,1,0)\mathbf{W}^{-1}\mathrm{diag}(1,1,0))^{-1}\begin{bmatrix}1\\1\\0\end{bmatrix}. \tag{11.26}$$

The result follows if the upper left 2×2 block of $\mathbf{W}^{-1}$ has the form $\begin{bmatrix} s & t\\ t & s\end{bmatrix}$. Using (11.4), the upper left 2×2 block of $\mathbf{W}^{-1}$ is proportional to

$$\begin{bmatrix} 1-\rho_{23}^2 & c^{-1}(\rho_{13}\rho_{23}-\rho_{12})\\ c^{-1}(\rho_{13}\rho_{23}-\rho_{12}) & c^{-2}(1-\rho_{13}^2)\end{bmatrix}. \tag{11.27}$$

But $c^2 = (1-\rho_{13}^2)/(1-\rho_{23}^2)$, so the result follows.

11.2.2 Centered Predictors with an Intercept

This section's burden is to show that if the outcome $\mathbf{y}$ is no longer centered, i.e., we no longer require $\mathbf{1}_n'\mathbf{y} = 0$, and an intercept is included in the model and θ_0 can be non-zero, but the predictors are still centered ($\mathbf{m} = \mathbf{0}$), then all the analysis summaries reduce to those in the previous case except for small, unimportant changes. The desired results then follow immediately.

With these assumptions, $\mathbf{X}$ is $n \times 4$ with columns $\mathbf{1}_n$, X_1, X_2, and X_3. Now

$$\mathbf{X}'\mathbf{X} = \begin{bmatrix} n & \mathbf{0}_3' \\ \mathbf{0}_3 & n\mathbf{HFH} \end{bmatrix} \tag{11.28}$$

where $\mathbf{F}$ and $\mathbf{H}$ are defined as for the previous simpler case. Thus $\mathbf{X}_2$ needs the same multiplier c as in the simpler case to ensure that in the unshrunk analysis, the posterior variance of θ_2, or the variance of $\hat{\theta}_2^U$ in a non-Bayesian analysis, equals the posterior variance of θ_1.

The unshrunk estimate of θ, $\hat{\theta}^U$ is again taken as given; write it as $\hat{\theta}^U = (\hat{\theta}_0^U, \hat{\theta}_{(-)}^U)'$, where $\hat{\theta}_0^U$ is the estimate of the intercept and $\hat{\theta}_{(-)}^U$ contains the estimates of the coefficients $(\theta_1, \theta_2, \theta_3)'$. The shrunken point estimate of θ — the posterior mean and the conventional estimate given (r_1, r_2) — is

$$\hat{\theta}^S = \left(\mathbf{X}'\mathbf{X} + \text{diag}(0, r_1, r_2, 0)\right)^{-1} (\mathbf{X}'\mathbf{X})\hat{\theta}^U \tag{11.29}$$

$$= \begin{bmatrix} 1 & \mathbf{0}_3' \\ \mathbf{0}_3 & (\mathbf{HKH})^{-1}\mathbf{HFH} \end{bmatrix} \hat{\theta}^U, \tag{11.30}$$

where $\mathbf{K}$ is as in the previous simpler case and again the dependence of $\hat{\theta}^S$ on (r_1, r_2) is suppressed.

The marginal posterior and restricted likelihood of (σ_e^2, r_1, r_2) are again derived using the constraint-case formulation of Section 2.1. Recalling that the sans-serif font indicates items specific to the constraint-case formulation, define

$$\mathsf{Y} = \begin{bmatrix} \mathbf{y} \\ 0 \\ 0 \end{bmatrix}, \mathsf{X} = \begin{bmatrix} & \mathbf{X} & & \\ \hline 0 & 1 & 0 & 0 \\ 0 & 0 & 1 & 0 \end{bmatrix}, \tag{11.31}$$

and define Γ as in the previous simpler case. The marginal posterior of (σ_e^2, r_1, r_2) is again given by (11.9), and the restricted likelihood is given by setting $\pi(\sigma_e^2, r_1, r_2)$ to 1. For the rest of this chapter, I do not explicitly refer to the restricted likelihood or profiled restricted likelihood.

In making the marginal posterior more explicit, the determinant $|\Gamma|$ is, as before, $(\sigma_e^2)^n \sigma_{s1}^2 \sigma_{s1}^2$. The determinant $|\mathsf{X}\Gamma^{-1}\mathsf{X}|$ is now proportional to $(\sigma_e^2)^{-4}|\mathbf{K}| = (\sigma_e^2)^{-4} D$ for D in (11.7), while in the simpler case, σ_e^2 was raised to the power -3. This turns out to be the only change in the inferential summaries compared to the simpler case.

As for the argument of the exponent, this is $-1/2\sigma_e^2$ times

$$\mathbf{y}'\mathbf{y} - n(\hat{\theta}_0^U)^2 - n\hat{\theta}_{(-)}^{U\prime}\mathbf{HFK}^{-1}\mathbf{FH}\hat{\theta}_{(-)}^U. \tag{11.32}$$

Now

$$\begin{aligned} \mathbf{y}'\mathbf{y} &= RSS + \hat{\theta}^{U\prime}\mathbf{X}'\mathbf{X}\hat{\theta}^{U} \\ &= RSS + n(\hat{\theta}_0^U)^2 + n\hat{\theta}_{(-)}^{U\prime}\mathbf{HFH}\hat{\theta}_{(-)}^{U}. \end{aligned} \tag{11.33}$$

Thus the argument of the exponent is $-1/2\sigma_e^2$ times

$$RSS + n\hat{\theta}_{(-)}^{U\prime}\mathbf{H}(\mathbf{F} - \mathbf{FK}^{-1}\mathbf{F})\mathbf{H}\hat{\theta}_{(-)}^{U} \tag{11.34}$$

which is identical to the analogous item in the simpler case.

Thus, all the necessary results follow: The marginal posterior and profiled restricted likelihood are symmetric in r_1 and r_2, the two shrunken coefficient estimates $\hat{\theta}_1^S$ and $\hat{\theta}_2^S$ are equal when $r_1 = r_2$, and the marginal posterior is not a function of $\hat{\theta}_3^U$, the coefficient of X_3 in the unshrunk analysis. Incidentally, this also shows that in the present case, the dataset and the analysis are again completely specified by *n*, *RSS*, $\mathbf{F}$, $\hat{\theta}^U$, and priors on (r_1, r_2) and σ_e^2.

11.2.3 Predictors Not Centered

This section's burden, like its predecessor's, is to show that if the predictors are no longer required to be centered, i.e., we don't require $\mathbf{m} = 0$, then the analysis reduces to the previous case except for small, unimportant changes, so the desired results follow immediately.

We need one additional piece of notation: Define $\mathbf{m}_c = (m_1, cm_2, m_3)' = \mathbf{Hm}$, where, as before, $\mathbf{H} = \text{diag}(1, c, 1)$ and c is the scaling applied to $\mathbf{X}_2$ so that in the unshrunk analysis, θ_1 and θ_2 have the same posterior variance (or variance in a non-Bayesian analysis).

$\mathbf{X}$ now has the most general form given in equation (11.1);

$$\mathbf{X}'\mathbf{X} = \begin{bmatrix} n & n\mathbf{m}_c' \\ n\mathbf{m}_c & n(\mathbf{HFH} + \mathbf{m}_c\mathbf{m}_c') \end{bmatrix} \tag{11.35}$$

and

$$(\mathbf{X}'\mathbf{X})^{-1} = \frac{1}{n}\begin{bmatrix} 1 + \mathbf{m}_c'(\mathbf{HFH})^{-1}\mathbf{m}_c & -\mathbf{m}_c'(\mathbf{HFH})^{-1} \\ -(\mathbf{HFH})^{-1}\mathbf{m}_c & (\mathbf{HFH})^{-1} \end{bmatrix}. \tag{11.36}$$

Thus c has the same value in this most general case as it had in the two previous simpler cases.

The unshrunk estimate of θ is again written as $\hat{\theta}^U = (\hat{\theta}_0^U, \hat{\theta}_{(-)}^U)'$; write $\hat{\theta}^S$ with the analogous partition and notation. The shrunken point estimate of θ — the posterior mean and the conventional estimate given (r_1, r_2) — is

$$\begin{aligned} \hat{\theta}^S &= \left(\mathbf{X}'\mathbf{X} + \text{diag}(0, r_1, r_2, 0)\right)^{-1}(\mathbf{X}'\mathbf{X})\hat{\theta}^U \\ &= \begin{bmatrix} 1 & (1 + \mathbf{m}_c'(\mathbf{HKH})^{-1}\mathbf{m}_c)\mathbf{m}_c' - \mathbf{m}_c'(\mathbf{HKH})^{-1}(\mathbf{HFH} + \mathbf{m}_c\mathbf{m}_c') \\ 0 & (\mathbf{HKH})^{-1}\mathbf{HFH} \end{bmatrix}\hat{\theta}^U, \end{aligned} \tag{11.37}$$

where **K** is the same as in the previous simpler cases and the dependence of $\hat{\theta}^S$ on (r_1, r_2) is suppressed. Note that $\hat{\theta}^S_{(-)} = (\mathbf{HKH})^{-1}\mathbf{HFH}\hat{\theta}^U_{(-)}$ is the same as in the two preceding simpler cases.

Now for the marginal posterior. The determinant $|\Gamma|$ is, as before, $(\sigma_e^2)^n \sigma_{s1}^2 \sigma_{s1}^2$. As for the other determinant,

$$\mathsf{X}\Gamma^{-1}\mathsf{X} = n\sigma_e^{-2}\left(\begin{bmatrix} 1 & \mathbf{m}_c' \\ \mathbf{m}_c & \mathbf{HFH} + \mathbf{m}_c\mathbf{m}_c' \end{bmatrix} + \text{diag}(0, r_1/n, r_2/n, 0)\right), \tag{11.38}$$

and by Theorem 8.2.1 of Graybill (1983), the determinant of (11.38) is

$$(n\sigma_e^{-2})^{-4}|1||\mathbf{HKH} + \mathbf{m}_c\mathbf{m}_c' - \mathbf{m}_c\mathbf{m}_c'| \propto (\sigma_e^2)^{-4}D \tag{11.39}$$

for D in (11.7). Thus the contribution of the determinants is identical to the preceding case in which the predictors were centered but the outcome $\mathbf{y}$ was not.

Finally, for the argument of the exponent, proceeding as before this is $-1/2\sigma_e^2$ times

$$RSS + \hat{\theta}^{U\prime}\mathbf{X}'\mathbf{X}\hat{\theta}^U - \hat{\theta}^{U\prime}\mathbf{X}'\mathbf{X}(\mathsf{X}\Gamma^{-1}\mathsf{X})^{-1}\mathbf{X}'\mathbf{X}\hat{\theta}^U. \tag{11.40}$$

Brute force shows that

$$\hat{\theta}^{U\prime}\mathbf{X}'\mathbf{X}\hat{\theta}^U = n(\hat{\theta}_0^U)^2 + 2n\hat{\theta}_0^U\mathbf{m}_c'\hat{\theta}^U_{(-)} + n\hat{\theta}^{U\prime}_{(-)}(\mathbf{m}_c\mathbf{m}_c' + \mathbf{HFH})\hat{\theta}^U_{(-)} \tag{11.41}$$

and

$$\begin{aligned}&\hat{\theta}^{U\prime}\mathbf{X}'\mathbf{X}(\mathsf{X}\Gamma^{-1}\mathsf{X})^{-1}\mathbf{X}'\mathbf{X}\hat{\theta}^U \\ &\quad = n(\hat{\theta}_0^U)^2 + 2n\hat{\theta}_0^U\mathbf{m}_c'\hat{\theta}^U_{(-)} + n\hat{\theta}^{U\prime}_{(-)}(\mathbf{m}_c\mathbf{m}_c' + \mathbf{HFK}^{-1}\mathbf{FH})\hat{\theta}^U_{(-)}.\end{aligned} \tag{11.42}$$

Thus the exponent in the present case is identical to the exponent in the preceding case. All the desired results follow immediately.

11.3 Conclusion: What Might Cause Differential Shrinkage?

The preceding results leave three possible explanations for the differential shrinkage observed in smoothed ANOVA applied to the colon-cancer data (Section 9.1.3).

The first and least interesting possible explanation is a coding error or MCMC failure in the original analysis. For various reasons, I am not going to try to write new code to reproduce Yufen Zhang's analyses. However, Dr. Zhang used her code to do a great number and variety of analyses in which we found no oddities besides this one. Differential shrinkage might also have arisen from some feature of the posterior distribution, for example a second mode, which Dr. Zhang's MCMC routine failed to detect. Again, however, we found no irregularities of this kind in many varied examples and in view of the results derived above, it seems unlikely that the posterior would have a mode consistent with this chapter's results and a second mode with the differential-shrinkage property. Thus, I assign low personal probability to this possible explanation, though I cannot rule it out and unfortunately I cannot make the dataset available.

A second possible explanation is that differential shrinkage can happen in normal-errors problems but only if the model has more predictors. This may seem implausible but recall that Stein's historic result (Stein 1956), that shrinkage estimation of p group means dominates simple averages for certain loss functions, requires $p \geq 3$. The analogy is obviously imperfect; my point is that equally implausible things have happened in normal-theory problems, so that possibility cannot be dismissed here. One approach to examining this possible explanation is an exercise.

A third possible explanation is that differential shrinkage cannot happen in normal-errors problems even with more predictors and that smoothed ANOVA on the colon-cancer data shrank coefficients differentially because of some aspect of the time-to-event analysis that is absent in the normal-errors analysis. We have already seen one difference between the two types of analyses: In the normal-error analysis, it is impossible for the two predictors X_1 and X_2 to have the same Euclidean length and for their coefficients to have the same variance in the unshrunk analysis unless the two predictors also have the same correlation with X_3, $\rho_{13} = \rho_{23}$, a restriction that was absent in the analysis of the colon-cancer dataset. The obvious difference between the time-to-event and normal-errors analyses is that the former in effect assigns different weights to individual cases depending on their estimated relative hazard and censoring status, while the latter analysis gives equal weight to all cases. It may be that differential shrinkage requires differential case weighting. If so, it might be possible to produce differential shrinkage in other models with non-normal errors, such as Poisson or logistic regression. Exploration of this possibility is an exercise.

Exercises

Regular Exercises

1. (Section 11.1.1) Show that the prior on (r_1, r_2) with r_1 and r_2 independent and each having the distribution implied by a flat prior on the approximate degrees of freedom in $X_j\theta_j$, $j = 1, 2$, is symmetric in r_1 and r_2. Hint: The definition of c makes this work.
2. (Section 11.2.1.1) Derive $\mathbf{F}^{-1}$ and $\mathbf{K}^{-1}$.
3. (Section 11.2.1.1) Derive the marginal posterior and restricted likelihood of (σ_e^2, r_1, r_2) in the constraint-case formulation, equation (11.9).
4. (Section 11.2.1.2) Show that the (3,3) element of **HFAFH** reduces to the expression in (11.17) and that the sum of the (1,3) and (2,3) elements of **HFAFH** reduces to the expression in (11.18).
5. (Section 11.2.1.2) Complete the proof that the sum of the (1,2) and (2,1) elements of **HFAFH** is symmetric in r_1 and r_2.
6. (Section 11.2.1.2) Show that the sum of the (1,1) and (2,2) elements of **HFAFH** is symmetric in r_1 and r_2; I gave a hint in Section 11.2.1.2.

Open Questions

1. Derive this chapter's results without resorting to brute-force algebra, like I did. If you can do this for the 3-predictor case, you can probably extend that to models with general numbers of shrunk and unshrunk coefficients, e.g., by replacing X_1, X_2, and X_3 with blocks of columns.
2. (Section 11.3) Can you produce differential shrinkage in this simplified normal-errors model if you have more predictors with shrunk or unshrunk coefficients? Unless you can solve the preceding problem, the only way I can see to approach this problem is by generating predictors with a known correlation matrix, adjusting their lengths as necessary so that their unshrunk coefficients have the same posterior variance (or variance, in a non-Bayesian analysis), and doing a computer search for examples of differential shrinkage.
3. (Section 11.3) Can you produce differential shrinkage in a Poisson regression or logistic regression? Remember that the key feature of the problem is that the predictors with shrunk coefficients must have equal coefficient estimates with equal variances in the unshrunk analysis, which will probably make it non-trivial to construct predictors having these properties in the unshrunk analysis and also having known correlations with each other.

Chapter 12

Competition between Random Effects

Sections 6.3 and 9.1.4 described a puzzle in which a dynamic linear model (DLM) with a signal component and quasi-cyclic components for heartbeat and respiration was fit to the optical-imaging dataset, and when a third quasi-cyclic component, the "mystery" component, was added to the model, the fit changed radically. Adding the mystery component caused this effect because it captured variation the signal and respiration components had previously captured. When the model was elaborated further by adding second harmonics to the respiration and mystery components, the signal and respiration components were no long shrunk so drastically. The puzzle to be considered here is twofold: Why did the fit change so dramatically after the mystery term was added, and why was the heartbeat component largely unaffected? Section 9.1.4 proposed two hypotheses to explain this puzzle: Collinearity of the design matrices of the random effects (the signal and three quasi-cyclic components) or a likelihood function that was ill-behaved for some reason other than collinearity. This chapter explores these hypotheses.

I have made little analytic progress with the full DLMs Michael Lavine used in producing the puzzle. Instead, this chapter proceeds indirectly. First, I analyze simpler models explicitly. Apart from its inherent interest, this yields specific hypotheses about the DLM puzzle, which are then tested to some extent using Prof. Lavine's original data and models. The resulting argument is preliminary in that it is supported only by "engineer's proofs," as my professors called them, but the intuition is fairly compelling.

Section 12.1 explores three simpler models, developing and testing a conjecture about how the DLM puzzle arises. Section 12.2 then poses specific hypotheses about the DLM puzzle and tests them to some extent with the original data and models. It appears that to produce the phenomena in the DLM puzzle, the random effects must be highly collinear with each other and the first model must fail to fit the data in a specific way. Further, even though each of the four components in the second DLM — signal, heartbeat, respiration, and mystery — is a saturated model, the mystery component competes more strongly with the signal and respiration components than with the heartbeat component because, in a sense developed in Section 12.2, signal, respiration, and mystery are more highly collinear with each other than they are with heartbeat.

12.1 Collinearity between Random Effects in Three Simpler Models

This section considers a spatial model with clustering and heterogeneity effects, then a model with two crossed random effects, and finally Chapter 11's regression in which two of three coefficients are shrunk. The first model has two random effects with saturated and thus perfectly collinear design matrices; the second model has two random effects with design matrices that are orthogonal in a particular sense; in the third model, the two random effects have design matrices with one column each and can have any desired correlation.

12.1.1 Model with Clustering and Heterogeneity

This example is like our DLMs in that its components have as many unknowns as there are observations. It is unlike our DLMs in that it has only two such components, so it cannot show differential competition among components. It does permit explicit expressions, to which we now turn.

12.1.1.1 Theory

Cui et al. (2010, Section 3.1) analyzed a spatial model for periodontal measurements from a twin study of the heritability of periodontal disease. The following uses a simplified version of that model. (The data from Cui et al. 2010 is available on this book's web site.)

Consider a model for one person's attachment loss measurements, taken on 7 adjacent teeth (a so-called quadrant) with 6 measurement sites on each tooth and each site measured twice, giving $7 \times 6 \times 2 = 84$ total measurements. Let y_{ik} be the kth measurement of site i for $i = 1, \ldots, N$ for $N = 42$ and $k = 1, 2$, so the total sample size is $n = 2N = 84$. Model y_{ik} as

$$y_{ik} = \delta_i + \xi_i + \varepsilon_{ik}, \tag{12.1}$$

where the vector $\delta = (\delta_1, \ldots, \delta_{42})'$ captures spatial clustering and the vector $\xi = (\xi_1, \ldots, \xi_{42})'$ captures heterogeneity. This is a normal-errors version of a model proposed by Besag et al. (1991) and commonly used with Poisson errors for smoothing disease maps. The clustering component δ has a normal ICAR model or prior as in Section 5.2.2, with mean zero and precision matrix $\mathbf{Q}/\sigma_c^2$, where the $N \times N$ matrix $\mathbf{Q}$ encodes neighbor pairings. The analyses shown below use the neighbor pairings shown in Figures 5.1 and 5.4. (Cui et al. 2010 used different neighbor pairings.) The heterogeneity component ξ is iid, so $\xi \sim N(\mathbf{0}_N, \sigma_h^2 \mathbf{I}_N)$. Sort the observations with the 42 first measurements on each site followed by the 42 second measurements on each site; then this model can be written as

$$\mathbf{y} = \begin{bmatrix} \mathbf{I}_N \\ \mathbf{I}_N \end{bmatrix} \delta + \begin{bmatrix} \mathbf{I}_N \\ \mathbf{I}_N \end{bmatrix} \xi + \varepsilon. \tag{12.2}$$

Each effect has one parameter for each of the 42 measurement sites, and the clustering and heterogeneity effects have identical design matrices. This dataset is somewhat unusual for areal spatial data in that it has two measurements at each site, permitting identification of σ_e^2 and σ_h^2.

The derivation proceeds by re-parameterizing the ICAR effect δ as in Section 5.2.2. Let $\mathbf{Q}$ have spectral decomposition $\mathbf{Q} = \Gamma \mathbf{D} \Gamma'$, where $\mathbf{D}$ is diagonal with diagonal entries $d_1 \geq d_2 \geq \ldots \geq d_{N-1}$ and $d_N = 0$ (we are thus assuming the spatial map has one island) and Γ is an orthogonal matrix with N^{th} column (with the zero eigenvalue) proportional to $\mathbf{1}_N$. Reparameterize δ and ξ to $\theta = \Gamma'\delta$ and $\phi = \Gamma'\xi$. The new heterogeneity parameter ϕ is still distributed as $N(\mathbf{0}_N, \sigma_h^2 \mathbf{I}_N)$. The new clustering parameter θ has mean zero and precision matrix $\mathbf{D}/\sigma_c^2$, so $\theta_N = N^{-0.5}\mathbf{1}'\delta$ has precision zero, i.e., the intercept of model (12.2) is a fixed effect implicit in the ICAR model. This makes ϕ_N redundant and as the derivations in Section 10.2.1 suggest, ϕ_N will have posterior mean zero. (The proof is an exercise.) Delete ϕ_N from the model; the re-parameterized model can then be written as

$$\mathbf{y} = \mathbf{1}_{2N}\beta_0 + \begin{bmatrix} \Gamma_- \\ \Gamma_- \end{bmatrix} \theta_- + \begin{bmatrix} \Gamma_- \\ \Gamma_- \end{bmatrix} \phi_- + \varepsilon, \tag{12.3}$$

where β_0 is the intercept, the $N \times (N-1)$ matrix Γ_- is the left-most $N-1$ columns of Γ (i.e., without the column proportional to $\mathbf{1}_N$), and θ_- and ϕ_- are $N \times 1$ and equal to θ and ϕ without their respective N^{th} elements.

The restricted likelihood arising from this model and dataset is

$$RL(\sigma_e^2, \sigma_c^2, \sigma_h^2) \;\propto\; (\sigma_e^2)^{-N/2} \prod_{j=1}^{N-1} \left(\sigma_c^2/d_j + \sigma_h^2 + \sigma_e^2/2\right)^{-1/2} \tag{12.4}$$

$$\exp\left[-\frac{1}{2\sigma_e^2} SSE - \frac{1}{2}\sum_{j=1}^{N-1} \hat{\theta}_j^2 \left(\sigma_c^2/d_j + \sigma_h^2 + \sigma_e^2/2\right)^{-1}\right], \tag{12.5}$$

where $SSE = \sum_{i,k}(y_{ik} - \bar{y}_{i.})^2$, $\bar{y}_{i.}$ being the average of y_{i1} and y_{i2}, and $\hat{\theta}_j$ is the jth element of

$$\hat{\theta}_- = \left[\; \Gamma'_- \quad \Gamma'_- \;\right] \mathbf{y}/2, \tag{12.6}$$

the unshrunk estimate of θ_-. This can be derived straightforwardly and is an exercise; if each site has m measurements, $\sigma_e^2/2$ is replaced by σ_e^2/m. This form for a restricted likelihood is used frequently in later chapters. Here, it gives an immediate proof that the factor

$$\prod_{j=1}^{N-1} \left(\sigma_c^2/d_j + \sigma_h^2 + \sigma_e^2/2\right)^{-1/2} \exp\left[-\frac{1}{2}\sum_{j=1}^{N-1} \hat{\theta}_j^2 \left(\sigma_c^2/d_j + \sigma_h^2 + \sigma_e^2/2\right)^{-1}\right] \tag{12.7}$$

identifies σ_c^2 and $\sigma_h^2 + \sigma_e^2/2$ as long as at least two d_j are distinct; multiplying by the factor

$$(\sigma_e^2)^{-N/2} \exp\left[-\frac{1}{2\sigma_e^2} SSE\right] \tag{12.8}$$

identifies σ_e^2 and σ_h^2, so the variances are identified without a prior distribution $\pi(\sigma_e^2, \sigma_c^2, \sigma_h^2)$.

To explore our puzzle, we also need the restricted likelihood for a model including only the heterogeneity effect. This is the balanced one-way random effect model with N groups and 2 observations per group; its restricted likelihood can be written

$$RL(\sigma_e^2, \sigma_h^2) \propto (\sigma_e^2)^{-N/2} \left(\sigma_h^2 + \sigma_e^2/2\right)^{-(N-1)/2} \tag{12.9}$$

$$\exp\left[-\frac{1}{2\sigma_e^2} SSE - \frac{1}{2}\left(\sigma_h^2 + \sigma_e^2/2\right)^{-1} \sum_{j=1}^{N} (\bar{y}_{i.} - \bar{y}_{..})^2\right] \tag{12.10}$$

where $\bar{y}_{..} = \mathbf{1}_n'\mathbf{y}/n$ is the grand mean of $\mathbf{y}$.

For the model with both components, the degrees of freedom in the fit are

$$\text{DF clustering} = \sum_{N-1}^{j=1} \frac{\sigma_c^2/d_j}{\sigma_c^2/d_j + \sigma_h^2 + \sigma_e^2/2} \tag{12.11}$$

$$\text{DF heterogeneity} = \sum_{N-1}^{j=1} \frac{\sigma_h^2}{\sigma_c^2/d_j + \sigma_h^2 + \sigma_e^2/2}; \tag{12.12}$$

the derivation is again straightforward and given as an exercise. For the model with only the heterogeneity component,

$$\text{DF heterogeneity} = (N-1)\sigma_h^2/(\sigma_h^2 + \sigma_e^2/2). \tag{12.13}$$

The restricted likelihood shows information coming from two sources: Information about σ_e^2 from replicate measurements at each site, the factor in equation (12.8); and information about all three variances from the $\hat{\theta}_j$, which are normal with mean zero and variance $\sigma_c^2/d_j + \sigma_h^2 + \sigma_e^2/2$ conditional on $(\sigma_e^2, \sigma_c^2, \sigma_h^2)$. The latter is the key to reproducing part of the DLM puzzle here.

In $\sigma_c^2/d_j + \sigma_h^2 + \sigma_e^2/2$, the conditional variance of $\hat{\theta}_j$, the contributions of the heterogeity and error variances σ_h^2 and σ_e^2 are the same for all j. The contribution from the clustering variance, σ_c^2/d_j, is large for j with small d_j, i.e., for constrasts $\hat{\theta}_j$ in the site averages on which the ICAR imposes a small precision (large variance). Also, σ_c^2/d_j is small for j with large d_j, i.e., for constrasts $\hat{\theta}_j$ in the site averages on which the ICAR imposes a large precision (small variance). By comparison, in the model with heterogeneity only, the model imposes the same precision (variance) on all contrasts $\hat{\theta}_j$. Thus, for spatial maps with some small d_j and datasets in which σ_h^2 and σ_e^2 are not large relative to σ_c^2, the heterogeneity-only model does not fit. Table 12.1 shows the positive d_j for the neighbor pairings used here. If $\sigma_c^2 \approx \sigma_h^2$, $\sigma_c^2/d_1 \approx 37\sigma_h^2$.

This suggests how our DLM puzzle might occur. If in fact the clustering variance is large and the heterogeneity variance is small but we fit the heterogeneity-only model, then $\hat{\sigma}_h^2$ will be large to accommodate differences between sites created by spatial clustering. But if we then fit the model with clustering and heterogeneity, the clustering effect will capture variation previously captured by the heterogeneity effect and $\hat{\sigma}_h^2$ will be greatly reduced. Below, I show this effect in simulated data. Thus, in this simple model at least, the phenomena in the DLM puzzle require two

Table 12.1: Positive d_j from the ICAR Component of the Clustering and Heterogeneity Model

5.562	5.562	5.562	5.562	5.562	5.540	5.478	5.382	5.343	5.343	5.261
5.137	5.039	5.000	5.000	5.000	5.000	5.000	3.471	3.471	3.000	2.935
2.756	2.502	2.207	1.897	1.604	1.438	1.438	1.438	1.438	1.438	1.186
1.186	1.000	0.856	0.625	0.412	0.236	0.106	0.027			

conditions: The two effects must be collinear, so adding the second effect *can* capture the variability originally captured by the first effect, and the simpler model must have a specific lack of fit. Below, I use simulated data to show that for this model, other failures of fit do not produce effects like those in the DLM puzzle. Later, the example of crossed random effects provides a further engineer's proof that the DLM puzzle does not occur without collinearity between the random effects.

The expressions for degrees of freedom (DF) also give insight into the competition between effects in a multiply saturated model like this one and our DLMs. Compare the DF in the heterogeneity component in the two models, equations (12.12) and (12.13). Each expression is a sum with $N-1$ terms; the jth summand is the DF in the fit for the contrast in site averages $\hat{\theta}_j$. If we first fit the heterogeneity-only model and plug estimated variances $\hat{\sigma}_h^2$ and $\hat{\sigma}_e^2$ into (12.13), each contrast contributes $\sigma_h^2/(\sigma_h^2+\sigma_e^2/2)$ DF to the heterogeneity component. Now add the clustering component to the model by increasing σ_c^2 from 0 to the estimate obtained in fitting the model with clustering and heterogeneity, $\hat{\sigma}_c^2$, while leaving σ_h^2 and σ_e^2 unchanged. Although neither σ_e^2 nor σ_h^2 has changed, the heterogeneity component loses DF to the clustering component because in going from (12.13) to (12.12), the denominator has increased in each summand while the numerator is unchanged. This is a consequence of collinearity of the clustering and heterogeneity effects: If the two effects had orthogonal design matrices, this could not happen. (The proof is an exercise.) But of course adding the clustering effect to the model also causes the estimates $\hat{\sigma}_h^2$ and $\hat{\sigma}_e^2$ to change. If we now reduce $\hat{\sigma}_h^2$ and leave $\hat{\sigma}_e^2$ unchanged, the DF in the heterogeneity effect becomes smaller still while the DF in the clustering effect becomes larger, even though $\hat{\sigma}_c^2$ has not changed.

12.1.1.2 Illustrations Using Simulated Data

I simulated data from a model with large σ_c^2 and small σ_h^2, and the fits for these fake datasets did indeed behave as in the DLM puzzle. Then I simulated data from a model with small σ_c^2 and large σ_h^2 and the fits did *not* behave as in the DLM puzzle. Finally, I simulated data from a rather different model that I could still analyze using the model above and again the fits did not behave as in the DLM puzzle. These examples support the hypothesis that collinearity and a specific lack of fit are, for this model at least, necessary and sufficient to produce the effect in the DLM puzzle.

First, I simulated datasets from model (12.3) with $N=42$, $n=2$, $\sigma_e^2=1$, $\sigma_h^2=0.001$, and $\sigma_c^2=30$. For each fake dataset, new draws were made from each random effect. These are not realistic simulations of attachment-loss data, but that's

Table 12.2: Variance Estimates (Maximum Restricted Likelihood) for 14 Datasets Simulated with Large σ_c^2 and Small σ_h^2

Dataset	$\hat{\sigma}_h^2$, H Only	$\hat{\sigma}_h^2$, C+H	$\hat{\sigma}_c^2$, C+H
12	103.4	0.00001	40.0
1	30.0	0.00009	28.7
4	27.6	0.0002	29.2
5	22.8	0.0002	32.1
7	58.6	0.02	30.4
14	24.1	1.3	26.7
2	28.1	1.6	18.0
13	30.0	2.0	33.9
6	39.8	2.1	23.8
3	33.5	3.0	18.3
10	18.4	3.0	19.3
9	40.0	3.8	18.1
11	49.5	4.7	21.8
8	27.6	5.6	10.6

Note: "H Only" is the heterogeneity-only model; "C+H" is the model with clustering and heterogeneity. Datasets are sorted by $\hat{\sigma}_h^2, C+H$.

OK, the same effects could be produced with a smaller σ_c^2; they would just be less distinct. In the course of debugging and other blundering, I generated 14 fake datasets and because some results are surprising, I show results from all 14 datasets. The maximum restricted likelihood estimate $\hat{\sigma}_e^2$ ranged from 0.86 to 1.41 and hardly changed when clustering was added to the model: Each site has replicate measurements, so each simulated dataset provides ample information about σ_e^2. Thus, I do not discuss $\hat{\sigma}_e^2$ further.

Table 12.2 shows estimates of $\hat{\sigma}_h^2$ for the model with heterogeneity only ("H only") and for the model with clustering and heterogeneity ("C+H") and estimates of $\hat{\sigma}_c^2$ for the model that includes it ("C+H"). As expected, for all 14 datasets, $\hat{\sigma}_h^2$ is large in the heterogeneity-only model and becomes much smaller when clustering is added to the model. The surprise is that when clustering is included in the model, $\hat{\sigma}_h^2$ is often not so small.

Table 12.3 shows degrees of freedom (DF) computed using Table 12.2's variance estimates. The second column from the left shows DF in the heterogeneity component of the fit of the heterogeneity-only model; this cannot be larger than 41. The next column shows DF for the model with both components, with σ_c^2 set to its estimate for the C+H model but with the other two variances, σ_h^2 and σ_e^2, kept at their estimates in the heterogeneity-only model. This shows the reduction in DF for heterogeneity caused solely by increasing σ_c^2 from 0. Finally, the two right-most columns show DF for heterogeneity and clustering in the fit of the model including both effects, showing the further effect on DF for heterogeneity from reducing $\hat{\sigma}_h^2$. The sum of the two right-most columns is bounded above by 41. For all datasets, reducing $\hat{\sigma}_h^2$

Table 12.3: DF in Heterogeneity and Clustering Components for 14 Datasets Simulated with Large σ_c^2 and Small σ_h^2, Computed Using Variance Estimates in Table 12.2

Dataset	Het, H Only	Het, C+H, Est from H Only	Het, C+H	Clust, C+H
12	40.8	33.1	0.00004	39.4
1	40.1	27.4	0.0004	38.1
4	40.3	26.8	0.0007	39.1
5	39.9	24.5	0.0008	38.6
7	40.6	31.5	0.07	38.8
14	40.1	26.4	4.9	34.0
2	40.4	30.2	7.9	30.9
13	40.1	26.3	5.9	33.2
6	40.5	30.6	7.8	31.4
3	40.3	31.0	11.8	26.8
10	39.9	26.7	11.5	27.4
9	40.5	32.1	14.1	25.3
11	40.6	32.3	14.3	25.1
8	40.2	32.7	21.3	17.6

Note: "H Only" is the heterogeneity-only model; "C+H" is the model with clustering and heterogeneity. "Het, C+H, Est from H Only" are DF computed using $\hat{\sigma}_h^2$ and $\hat{\sigma}_e^2$ from the model with heterogeneity only but using $\hat{\sigma}_c^2$ from the model with both components. Datasets are sorted as in Table 12.2.

induces a larger part of the reduction in heterogeneity DF than does simply introducing the clustering component. The surprise here is that heterogeneity retains a substantial number of DF, 4.9 or higher, in 9 of the 14 datasets; in dataset 8, it has higher DF than clustering. (Incidentally, dataset 8 reminds us that σ_h^2 and σ_c^2 are not directly comparable: For this dataset, $\hat{\sigma}_c^2$ is almost twice $\hat{\sigma}_h^2$ but heterogeneity has more DF. The incomparability arises here because the precision matrix for the heterogeneity component of model (12.3) has trace $41/\sigma_h^2$, while the precision matrix for the clustering component has trace $132/\sigma_c^2$.)

It seems strange that we can generate data from a model with $\sigma_h^2 = 0.001$ and $\sigma_c^2 = 30$ and get even one dataset out of 14 in which heterogeneity has more DF in the fit than clustering. To examine these results further, consider Figures 12.1 and 12.2, showing the fits to datasets 12 and 14. (I use dataset 14 because it had the largest DF in heterogeneity among the few datasets I didn't accidentally destroy.) Figure 12.1, showing results for dataset 12, is as expected. In the heterogeneity-only fit (Panel a), $\hat{\sigma}_h^2$ is inflated to allow the model to fit the data closely; no shrinkage is visible. The fit with both components (Panel b) again fits the data with no visible shrinkage. Finally, Panel c shows that effectively the full model's entire fit is in the clustering component.

Now consider Figure 12.2 for dataset 14. At first glance, it seems hardly distinguishable from Figure 12.1 for dataset 12. On a closer look, Panel (c) of Figure 12.2

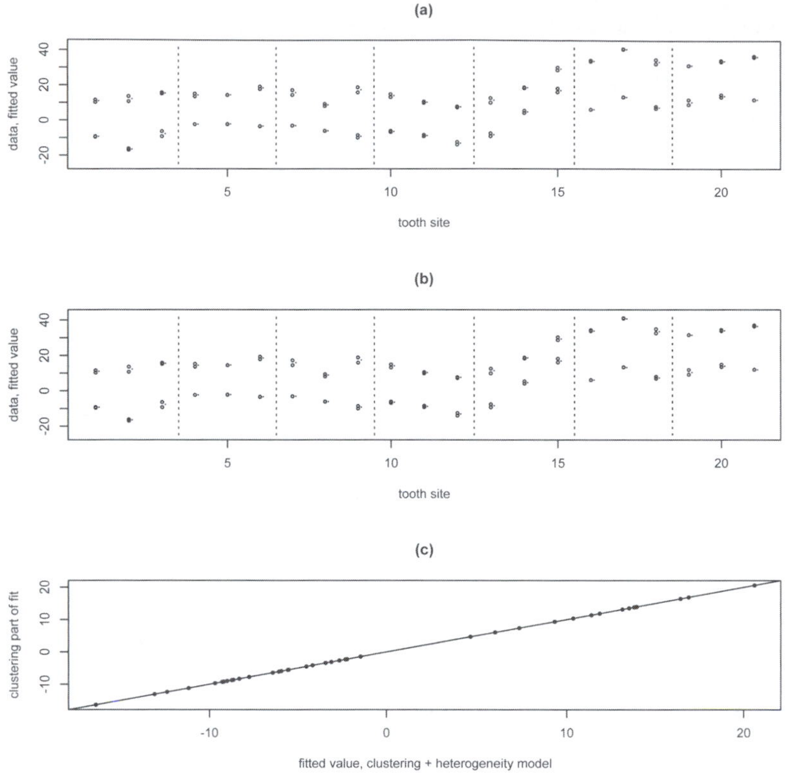

Figure 12.1: Data and fits for dataset 12. Panel (a): Open circles are data, with sites on one side of the dental arch shifted up by 20 units for clarity; solid circles are fitted values from the heterogeneity-only model. Panel (b): Open circles are data; solid circles are fitted values from the model with clustering and heterogeneity. Panel (c): The horizontal axis is fitted values from the full model, the vertical axis is the portion of the fit due to clustering, and the line plots $y = x$.

shows that the clustering component does not account for the whole fit to dataset 14: The heterogeneity component accounts for its largest portion of the fit at the sites with the most extreme combined fitted values. Still, heterogeneity's contribution to the fit seems rather small given that it has nearly 5 DF out of 38.9 total DF in the fit. Table 12.4 shows the fitted values for the heterogeneity component; by comparison, the fitted values for the clustering component range from -7.4 to 17.2.

This is an aspect of variance estimates and DF which I don't fully understand. The fitted clustering values are much larger than the fitted heterogeneity values — absolute values average 3.3 for clustering and 0.16 for heterogeneity, about 20 times as large — while clustering has not quite 7 times as many DF in the fit. It is true that in ordinary linear models, each design matrix column receives a full DF even if

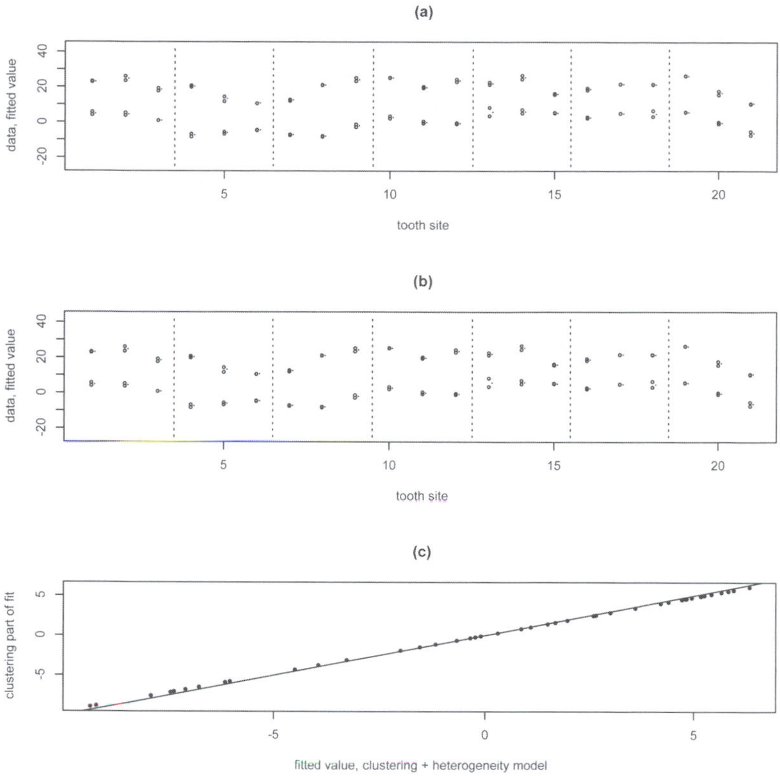

Figure 12.2: Data and fits for dataset 14. The three panels are as in Figure 12.1.

Table 12.4: Dataset 14, Fitted Values for the Heterogeneity Component in the Full Model

–0.83	–0.36	–0.36	–0.34	–0.27	–0.22	–0.18	–0.17	–0.12	–0.11	–0.09
–0.07	–0.05	–0.04	–0.04	–0.03	0.02	0.03	0.03	0.03	0.04	0.05
0.06	0.07	0.08	0.09	0.10	0.10	0.10	0.10	0.12	0.12	0.14
0.17	0.17	0.18	0.20	0.23	0.26	0.31	0.56			

its coefficient's estimate is zero, but in mixed linear models the DF in a component are determined through estimation of its smoothing/shrinkage variances, a feedback loop that is absent in an ordinary linear model. Something else seems to be happening here, something connected with the gross collinearity of the two random effects' design matrices. Cui et al. (2010, Section 3.1) examined the clustering and heterogeneity model using different neighbor pairings and simulated datasets having 12 persons instead of 1, and found that when the full model was fit to data generated with one of the effects truly absent, i.e., either $\sigma_c^2 = 0$ or $\sigma_h^2 = 0$, on average the

Table 12.5: Variance Estimates (Maximum Restricted Likelihood) for 5 Datasets Simulated with Large σ_h^2 and Small σ_c^2

Dataset	$\hat{\sigma}_h^2$, H Only	$\hat{\sigma}_h^2$, C+H	$\hat{\sigma}_c^2$, C+H
1	29.0	24.5	4.1
5	31.4	25.0	6.0
2	25.7	25.7	0.0004
4	40.4	40.4	0.0005
3	44.6	44.6	0.0002

Note: "H Only" is the heterogeneity-only model; "C+H" is the model with clustering and heterogeneity.

truly absent effect "stole" some DF from the truly present effect. The present simulation shows the same thing. Using the true values of the three variances, the DF in heterogeneity and clustering are 0.0004 and 39.0 respectively but using estimated variances, the averages over the 14 simulated datasets are 7.1 and 31.8. Even though the model is identified, the machinery cannot cleanly distinguish the two effects even when, as in Cui et al. (2010), the datasets are 12 times as large as the ones considered here. Cui et al. (2010) found this "theft" of DF in a Bayesian analysis and speculated that it arose because of the prior distribution on DF; the present results show it is not just a consequence of the prior.

Now consider a new set of artificial datasets simulated from model (12.3) with a large heterogeneity variance and small clustering variance, so that the heterogeneity-only model fits: $\sigma_e^2 = 1$, $\sigma_h^2 = 30$, and $\sigma_c^2 = 0.001$. Again, for each fake dataset, new draws were made from each of the random effects, and this time I only simulated 5 datasets. The variance estimates and DF from fits of the two models are presented in Tables 12.5 and 12.6. Again, $\hat{\sigma}_e^2$ was close to the true σ_e^2 in all cases and was hardly changed by adding clustering to the model.

Because the heterogeneity-only model fits these simulated datasets, it is not surprising that $\hat{\sigma}_h^2$ is large for the heterogeneity-only model and changes little when clustering is added to the model. However, even though the true σ_c^2 is quite small, for 2 of these 5 datasets $\hat{\sigma}_c^2$ is substantial. The results for DF (Table 12.6) are as expected given Table 12.5. The DF evaluated at the true variances are 40.3 for heterogeneity and 0.002 for clustering, while the respective averages for these 5 datasets are 38.3 and 2.0, respectively. "Theft" shows its face again, though this time clustering is "stealing" some of the variation in $\mathbf{y}$ that properly belongs to heterogeneity.

For the present chapter's main purpose, this example's implication is that, as the theory suggested, we do not see the DLM puzzle here because the heterogeneity-only model fits the data.

The final example is one in which the simulated data do not fit the heterogeneity-only model or the model with clustering and heterogeneity, but this failure to fit the heterogeneity-only model does not produce the effect seen in the DLM puzzle. Data were simulated from a non-standard model where $\mathbf{y} = \mathbf{X}\beta + \mathbf{Z}\mathbf{u} + \varepsilon$ was as in equation (12.3), but where $(\theta_{-} + \phi_{-})_i$ was drawn from a normal distribution with mean

Table 12.6: DF in Heterogeneity and Clustering Components for 5 Datasets Simulated with Large σ_h^2 and Small σ_c^2, Computed Using Variance Estimates in Table 12.5

Dataset	Het, H Only	Het, C+H, Est from H Only	Het, C+H	Clust, C+H
1	40.5	36.5	35.9	4.5
5	40.4	35.5	34.7	5.7
2	40.1	40.1	40.1	0.001
4	40.4	40.4	40.4	0.0009
3	40.6	40.6	40.6	0.0003

Note: "H Only" is the heterogeneity-only model; "C+H" is the model with clustering and heterogeneity. "Het, C+H, Est from H Only" are DF computed using $\hat{\sigma}_h^2$ and $\hat{\sigma}_e^2$ from the model with heterogeneity only but using $\hat{\sigma}_c^2$ from the model with both components. Datasets are sorted as in Table 12.5.

Table 12.7: Variance Estimates (Maximum Restricted Likelihood) for 5 Datasets Simulated from the Non-standard Model

Dataset	$\hat{\sigma}_h^2$, H Only	$\hat{\sigma}_h^2$, C+H	$\hat{\sigma}_c^2$, C+H
4	13.2	9.6	1.3
5	13.4	13.1	0.2
1	13.8	13.5	0.2
2	14.4	13.8	0.3
3	16.5	16.5	0.000006

Note: "H Only" is the heterogeneity-only model; "C+H" is the model with clustering and heterogeneity.

zero and variance 30 for odd i and 1 for even i. The sites were labeled $i = 1, \ldots, 42$ in such a way that for sites on one side of the dental arch, the variances appeared in the sequence $30, 1, 30, 30, 1, 30, 30, 1, 30, \ldots$, while on the other side of the arch, the variances appeared in the sequence $1, 30, 1, 1, 30, 1, 1, 30, 1, \ldots$. The resulting data have no spatial clustering but also do not fit the heterogeneity-only model. Again, for this structure I simulated only 5 datasets.

Since the heterogeneity term in effect acts as an error term for the site-level model, it can capture any kind of variation so it is not surprising that in fitting these datasets, $\hat{\sigma}_h^2$ is much larger than $\hat{\sigma}_c^2$ for all 5 simulated datasets (Table 12.7). The DF in these fits (Table 12.8) also have no surprises.

This example supports the hypothesis that a specific kind of lack of fit is needed to produce the effect seen in the DLM puzzle, but obviously this one example is far from conclusive. An exercise challenges readers to find a model for generating data that is *not* the model with clustering and heterogeneity — and, to avoid trivial counterexamples, is also not similar to it — but that still consistently produces the effect shown in Tables 12.2 and 12.3. This seems unlikely but intuition may not be trustworthy here.

Table 12.8: DF in Heterogeneity and Clustering Components for 5 Datasets Simulated from the Non-standard Model, Computed Using Variance Estimates in Table 12.7

Dataset	Het, H Only	Het, C+H, Est from H Only	Het, C+H	Clust, C+H
4	39.5	36.5	35.4	3.8
5	39.0	38.3	38.3	0.8
1	39.8	39.1	39.1	0.8
2	39.6	38.5	38.4	1.2
3	39.4	39.4	39.4	0.00002

Note: "H Only" is the heterogeneity-only model; "C+H" is the model with clustering and heterogeneity. "Het, C+H, Est from H Only" are DF computed using $\hat{\sigma}_h^2$ and $\hat{\sigma}_e^2$ from the model with heterogeneity only but using $\hat{\sigma}_c^2$ from the model with both components. Datasets are sorted as in Table 12.7.

12.1.2 Two Crossed Random Effects

This section considers a balanced ANOVA with two crossed random effects and multiple replications per design cell. It is like the DLMs in our puzzle because it has two random effects but unlike those DLMs in that each random effect has far fewer unknowns than the total number of design cells and the design matrices of the two random effects are uncorrelated, though in a less-than-obvious sense. As for the preceding example, we begin with some theory and illustrate the points with simulated examples.

12.1.2.1 Theory

The data can be displayed in an array with l rows and c columns and m replications in each of the $l \times c$ design cells. The only fixed effect is the intercept; the random effects are the row and column effects. Let the observations be y_{ijk}, where $i = 1, \ldots, l$ indexes rows, $j = 1, \ldots, c$ indexes columns, and $k = 1, \ldots, m$ indexes replicates within cells, and suppose the observations are ordered in $\mathbf{y}$ with k varying fastest and i slowest, so $\mathbf{y} = (y_{111}, \ldots, y_{11m}, y_{121}, \ldots, y_{12m}, \ldots, y_{lcm})'$. Then if $\otimes$ is the Kronecker product with $\mathbf{A} \otimes \mathbf{B} = (a_{ij}\mathbf{B})$, this model can be written in mixed-linear-model form $\mathbf{y} = \mathbf{X}\beta + \mathbf{Z}\mathbf{u} + \varepsilon$ with

$$\mathbf{X} = \mathbf{1}_{lcm} = \mathbf{1}_l \otimes \mathbf{1}_c \otimes \mathbf{1}_m \tag{12.14}$$

$$\mathbf{Z} = \left[\begin{array}{c|c} \mathbf{I}_l \otimes \mathbf{1}_c \otimes \mathbf{1}_m & \mathbf{1}_l \otimes \mathbf{I}_c \otimes \mathbf{1}_m \end{array}\right] \tag{12.15}$$

$$\beta = \beta_0, \quad \mathbf{u} = \left[\begin{array}{c|c} u_{11} \ldots u_{1l} & u_{21} \ldots u_{2c} \end{array}\right]' \tag{12.16}$$

$$\mathbf{G} = \left[\begin{array}{c|c} \sigma_l^2 \mathbf{I}_l & \mathbf{0} \\ \hline \mathbf{0} & \sigma_c^2 \mathbf{I}_c \end{array}\right] \tag{12.17}$$

and the ε_{ijk} are iid $N(0, \sigma_e^2)$. The column effect is $(u_{11} \ldots u_{1l})'$ and the row effect is $(u_{21} \ldots u_{2c})'$ with variances σ_l^2 and σ_c^2 respectively.

The restricted likelihood for this model is

$$\begin{aligned} RL(\sigma_e^2, \sigma_l^2, \sigma_c^2) \propto\ & (\sigma_e^2)^{-(lcm-l-c+1)/2} \exp\left[-\frac{1}{2\sigma_e^2} SSE\right] \\ & (\sigma_l^2 + \sigma_e^2/cm)^{-(l-1)/2} \exp\left[-\frac{1}{2}(\sigma_l^2 + \sigma_e^2/cm)^{-1} \sum_i (\bar{y}_{i..} - \bar{y}_{...})^2\right] \\ & (\sigma_c^2 + \sigma_e^2/lm)^{-(c-1)/2} \exp\left[-\frac{1}{2}(\sigma_c^2 + \sigma_e^2/lm)^{-1} \sum_i (\bar{y}_{.j.} - \bar{y}_{...})^2\right], \end{aligned} \tag{12.18}$$

where $SSE = \sum_{i,j,k}(y_{ijk} - \bar{y}_{...})^2 - cm\sum_i(y_{i..} - \bar{y}_{...})^2 - lm\sum_j(y_{.j.} - \bar{y}_{...})^2$, $\bar{y}_{...}$ is the grand mean, and $\bar{y}_{i..}$ and $\bar{y}_{.j.}$ are the averages for row i and column j respectively. The derivation is an exercise. For balanced models like this one, it is well known that the same restricted likelihood is obtained if the design matrices for the random effects in equation (12.15) are replaced by an orthogonal re-parameterization with one less column per random effect, i.e.,

$$\mathbf{Z} = \left[\ \mathbf{I}_l \otimes \mathbf{H}_l \otimes \mathbf{1}_m \ \middle|\ \mathbf{1}_l \otimes \mathbf{H}_c \otimes \mathbf{1}_m\ \right], \tag{12.19}$$

where $\mathbf{H}_l$ and $\mathbf{H}_c$ are $l \times (l-1)$ and $c \times (c-1)$ respectively and each has orthonormal columns. In Bayesian terms, these two posteriors are identical because of this model's implicit flat (improper) prior on the intercept β_0. The intercept's design matrix ($\mathbf{1}_{lcm}$) is exactly collinear with a linear combination of the design-matrix columns of each of the two random effects. Thus, because of the flat prior on the intercept, the posterior mean of each of those two linear combinations of $\mathbf{u}$ is identically zero and (switching back to non-Bayesian language) the corresponding item in the restricted likelihood vanishes. In this sense, the two random effects are perfectly non-collinear, at least as they contribute information to the restricted likelihood or posterior. The parameterization does matter if the design is unbalanced or if a Bayesian analysis does not put a flat prior on the intercept.

To see whether this model can produce the DLM puzzle, we also need a model that omits one of the random effects. I omitted the column effect; the row-effect-only model has restricted likelihood

$$\begin{aligned} RL(\sigma_e^2, \sigma_l^2) \propto\ & (\sigma_e^2)^{-(lcm-l)/2} \exp\left[-\frac{1}{2\sigma_e^2} SSE\right] \\ & (\sigma_l^2 + \sigma_e^2/cm)^{-(l-1)/2} \exp\left[-\frac{1}{2}(\sigma_l^2 + \sigma_e^2/cm)^{-1} \sum_i (\bar{y}_{i..} - \bar{y}_{...})^2\right], \end{aligned} \tag{12.20}$$

where now $SSE = \sum_{i,j,k}(y_{ijk} - \bar{y}_{...})^2 - cm\sum_i(y_{i..} - \bar{y}_{...})^2$. This is just the balanced one-way random effects model; the restricted likelihood can be obtained from (12.18) by setting $\sigma_c^2 = 0$.

In the model with both effects, the DF for the row and column effects are

$$\text{DF for rows} = (l-1)\frac{\sigma_l^2}{\sigma_l^2 + \sigma_e^2/cm} \tag{12.21}$$

$$\text{DF for columns} \quad = \quad (c-1)\frac{\sigma_c^2}{\sigma_c^2 + \sigma_e^2/lm}. \tag{12.22}$$

For the model with only the row effect ("rows-only model"), the DF in the fit for rows has exactly the same form as (12.21). However, the DF in the fit for the row effect will change when the column effect is added to the model if either σ_l^2 or σ_e^2 changes.

12.1.2.2 Illustrations Using Simulated Data

Information about the row and column random effects comes from projecting **y** onto the respective orthogonal subspaces of fitted-value space so adding the column effect can only change the estimate of the row-effect variance $\hat{\sigma}_l^2$ indirectly, by reducing $\hat{\sigma}_e^2$. We'll consider simulated examples to examine the size of this indirect effect. The results also let us see that the error term is a random effect that is collinear with every other random effect. Although we routinely see other fixed and random effects take variation away from the error term, we think nothing of it because we are trained to see the error term as a trash can for variation that isn't captured by more interesting effects. For many models, however, particularly complex ANOVAs, the error term is an interaction random effect and is distinguished from other random effects only by convention.

In the first simulated example, fake data were generated from the two-crossed random effects model with $\sigma_l^2 = 0.001$, $\sigma_c^2 = 30$, and $\sigma_e^2 = 1$. I used $l = c = 5$ row and column levels and $m = 3$ replications per cell for a total sample size of $n = 75$ and drew 5 simulated datasets, each with new draws on the random effects. Table 12.9 shows the estimates of the variances in fits of the row-effect-only model ("R only") and of the model with both random effects ("R+C"). As might be expected, the estimates $\hat{\sigma}_l^2$ are small in both the rows-only and the full-model fits, although for dataset 1, $\hat{\sigma}_l^2$ increased from 1×10^{-6} to 0.06 when the column effect was added to the model. In the rows-only model, the error variance estimate $\hat{\sigma}_e^2$ is large because the variation from the missing column effect is relegated to error; in the full model, $\hat{\sigma}_e^2$ is close to 1 for all five datasets, while $\hat{\sigma}_c^2$ is large. The effect seen in the DLM puzzle does not occur here because the row and column random effects are in effect orthogonal to each other. If we see the error term as a random effect, we can see that it behaves like the signal and respiration components in the DLM puzzle.

Table 12.10 shows the DF in the fits to these five datasets and presents a small surprise. The maximum DF in each of the fitted row and column effects is 4 and in the full model fit, the column effect gets almost 4 DF, as would be expected. The surprise is in dataset 1, where the row effect has 6×10^{-6} DF in the rows-only model, but 1.8 DF in the full model. This happens because adding the column effect to the model reduces $\hat{\sigma}_e^2$ and increases $\hat{\sigma}_l^2$, as noted above.

Now for a second simulated example, in which the row-effect variance is large, $\sigma_l^2 = 30$, while the column-effect variance is small, $\sigma_c^2 = 0.001$. Table 12.11 shows the variance estimates for five simulated datasets, each with new draws on the random effects, while Table 12.12 shows the DF in each fit. These datasets present no surprises: With two effectively orthogonal random effects and with the row-only model

Table 12.9: Variance Estimates (Maximum Restricted Likelihood) for 5 Datasets Simulated with Small σ_l^2 and Large σ_c^2

Dataset	$\hat{\sigma}_e^2$ R Only	$\hat{\sigma}_l^2$ R Only	$\hat{\sigma}_e^2$ R+C	$\hat{\sigma}_l^2$ R+C	$\hat{\sigma}_c^2$ R+C
1	10.7	< 0.00001	1.1	0.06	11.7
2	27.6	< 0.00001	1.1	< 0.00001	32.8
3	30.6	< 0.00001	1.0	< 0.00001	36.4
4	72.4	< 0.00001	0.9	0.0006	88.3
5	7.2	< 0.00001	1.2	< 0.00001	7.4

Note: "R Only" is the model with the row effect only; "R+C" is the model with row and column effects.

Table 12.10: DF in Row and Column Components for 5 Datasets Simulated with Small σ_l^2 and Large σ_c^2, Computed Using Variance Estimates in Table 12.9

Dataset	Row Effect R Only	Row Effect R+C	Column Effect R+C
1	< 0.00001	1.80	3.97
2	< 0.00001	< 0.00001	3.99
3	< 0.00001	< 0.00001	3.99
4	< 0.00001	< 0.00001	4.00
5	< 0.00001	< 0.00001	3.96

Note: "R Only" is the model with the row effect only; "R+C" is the model with row and column effects.

fitting the data, there would seem to be no way to produce an effect like the DLM puzzle or indeed for the column effect to capture much variation. However, by the symmetry in this model, we should expect that with enough simulated datasets, one dataset would give the column effect substantial DF in the fit, as the row effect did in Table 12.10's dataset 1.

Table 12.11: Variance Estimates (Maximum Restricted Likelihood) for 5 Datasets Simulated with Large σ_l^2 and Small σ_c^2

Dataset	$\hat{\sigma}_e^2$ R Only	$\hat{\sigma}_l^2$ R Only	$\hat{\sigma}_e^2$ R+C	$\hat{\sigma}_l^2$ R+C	$\hat{\sigma}_c^2$ R+C
1	1.0	35.5	1.0	35.5	< 0.00001
2	0.8	44.3	0.8	44.3	< 0.00001
3	0.8	12.4	0.8	12.4	< 0.00001
4	1.2	49.5	1.2	49.6	< 0.00001
5	1.0	30.5	1.0	30.5	< 0.00001

Note: "R Only" is the model with the row effect only; "R+C" is the model with row and column effects.

Table 12.12: DF in Row and Column Components for 5 Datasets Simulated with Large σ_l^2 and Small σ_c^2, Computed Using Variance Estimates in Table 12.11

Dataset	Row Effect R Only	Row Effect R+C	Column Effect R+C
1	3.99	3.99	< 0.00001
2	4.00	4.00	< 0.00001
3	3.98	3.98	< 0.00001
4	3.99	3.99	< 0.00001
5	3.99	3.99	< 0.00001

Note: "R Only" is the model with the row effect only; "R+C" is the model with row and column effects.

Table 12.13: Variance Estimates (Maximum Restricted Likelihood) for 5 Datasets Simulated from the Non-standard Model

Dataset	$\hat{\sigma}_e^2$ R Only	$\hat{\sigma}_l^2$ R Only	$\hat{\sigma}_e^2$ R+C	$\hat{\sigma}_l^2$ R+C	$\hat{\sigma}_c^2$ R+C
3	38.8	0.0006	38.0	0.0001	0.9
4	10.6	0.8	7.7	1.0	3.4
2	10.2	1.3	9.9	1.3	0.3
5	13.4	1.7	11.8	1.8	1.9
1	22.5	13.7	19.3	13.8	3.7

Note: "R Only" is the model with the row effect only; "R+C" is the model with row and column effects. Datasets are sorted in order of $\hat{\sigma}_l^2$.

The foregoing examples were fairly predictable, so I tried to concoct non-standard data-generating mechanisms which could not be expressed in the form of the two-crossed-random-effects model but which could be analyzed using it, to see if they produced the behavior in the DLM puzzle. Adding a row-by-column interaction would not work: The effective orthogonality of the effects would deliver nearly all of the interaction's variation to the error term in both the rows-only fit and the fit with row and column effects. I tried various ways of inducing confounding between the row and column effects, but failed. I show results for one of these non-standard models; the others gave qualitatively identical results. An exercise suggests a different approach to constructing examples.

This non-standard data-generating model had an intercept and $l \times c$ cells indexed by i and j and with m replicates per cell, as before. The cell means were the intercept plus a draw from a normal distribution with mean 0 and variance 30 for cells where $i+j$ was even and 0.001 for cells where $i+j$ was was odd. Again I used $l = c = 5$ and $m = 3$ and simulated 5 datasets with new draws of the cell means for each dataset. Table 12.13 shows the variance estimates from fits of the rows-only and full models, while Table 12.14 shows the DF in each fit. In all five datasets, $\hat{\sigma}_l^2$ and the DF in the row effect were largely unchanged when the column effect was added to the model.

Table 12.14: DF in Row and Column Components for 5 Datasets Simulated from the Non-standard Model, Computed Using Variance Estimates in Table 12.13

Dataset	Row Effect R Only	Row Effect R+C	Column Effect R+C
3	0.0010	0.0002	1.04
4	2.18	2.68	3.48
2	2.62	2.66	1.34
5	2.63	2.79	2.80
1	3.60	3.66	2.96

Note: "R Only" is the model with the row effect only; "R+C" is the model with row and column effects.

Instead, adding the column effect simply reduced $\hat{\sigma}_e^2$ by an amount roughly equal to $\hat{\sigma}_c^2$.

In summary, the results for this example are consistent with the hypothesis that the DLM puzzle requires collinear random effects and a specific kind of lack of fit.

12.1.3 Three Predictors, Two with Shrunk Coefficients

The last simple example explores our hypothesis using a model with two random effects, the design matrices of which have correlation that can be adjusted continuously. This example is the model used in Chapter 11 to explore differential shrinkage. The necessary theory is in Chapter 11; this section defines data-generating and analysis models and gives results from the simulations.

Chapter 11 used a linear model with an intercept θ_0 and three predictors X_1, X_2, and X_3 having coefficients θ_1, θ_2, and θ_3, where the estimates of θ_1 and θ_2 were shrunk toward zero by means of distributions $\theta_1 \sim N(0, \sigma_1^2)$ and $\theta_2 \sim N(0, \sigma_2^2)$. The correlation between X_i and X_j is ρ_{ij}. In the present section, the data are generated from a model in which the true $\theta_2 = 0$. The two fitted models, labeled by analogy with the DLM example, are Model 1, which includes X_2 as a predictor but not X_1, and Model 2, which includes both X_1 and X_2 as predictors:

$$
\begin{array}{llllllll}
\text{Data-generating:} & \mathbf{y} = & \mathbf{1}_n\theta_0 & +X_1\theta_1 & & +X_3\theta_3 & +\varepsilon & \\
\text{Model 1:} & \mathbf{y} = & \mathbf{1}_n\theta_0 & & +X_2\theta_2 & +X_3\theta_3 & +\varepsilon & \\
\text{Model 2:} & \mathbf{y} = & \mathbf{1}_n\theta_0 & +X_1\theta_1 & +X_2\theta_2 & +X_3\theta_3 & +\varepsilon. &
\end{array}
\tag{12.23}
$$

In this example we can see whether increasing ρ_{12} from zero to large values produces the behavior seen in the DLM puzzle. Thus, 5 artificial datasets were simulated for each of several values of ρ_{12}. Using Chapter 11's notation, the other aspects of the data-generating mechanism were as follows: $n = 100$, true $\theta = (1, 3, 0, -4)'$ consistent with (12.23), $\mathbf{m} = (10, 10, 10)$, $\rho_{13} = 0.6$, $\rho_{23} = 0.5$, $c = \left[(1-\rho_{13}^2)/(1-\rho_{23}^2)\right]^{0.5}$, and error variance $\sigma_e^2 = 1$. The correlations ρ_{13} and ρ_{23} were chosen so the correlation matrix of the X_j was positive definite for all ρ_{12}

Table 12.15: Maximum Restricted Likelihood Estimates $\hat{\sigma}_2^2$ for 5 Datasets Simulated from (12.23)

Model 1			Dataset		
ρ_{12}	1	2	3	4	5
0	1.6	1.8	1.3	2.2	1.2
0.2	0.1	0.0	0.2	0.0	0.2
0.4	0.2	0.1	0.1	0.1	0.0
0.6	1.2	1.6	0.9	1.4	1.2
0.8	4.6	4.3	5.2	3.6	5.0
0.9	7.6	6.6	6.9	6.8	6.1
0.95	8.6	7.7	8.1	7.3	7.7
0.99	9.4	11.1	7.5	8.4	8.6
Model 2			Dataset		
ρ_{12}	1	2	3	4	5
0	0.000	0.000	0.000	0.032	0.013
0.2	0.000	0.026	0.000	0.003	0.000
0.4	0.000	0.000	0.000	0.000	0.000
0.6	0.009	0.000	0.071	0.000	0.005
0.8	0.000	0.000	0.035	0.143	0.000
0.9	0.127	0.000	0.000	0.000	0.000
0.95	0.000	0.000	0.000	0.229	0.000
0.99	3.292	0.000	0.364	4.646	0.000

considered but these choices are otherwise arbitrary and probably contribute to some of the results that follow.

Tables 12.15 and 12.16 show results for X_2, $\hat{\sigma}_2^2$ and DF in the fit for $X_2\theta_2$, which has a maximum possible value of 1. As expected, $\hat{\sigma}_2^2$ and DF were always smaller in Model 2 than in Model 1. For ρ_{12} between 0 and 0.6, $\hat{\sigma}_2^2$ in Model 1 seems to drop down and then increase. As ρ_{12} rises from 0.6, however, $\hat{\sigma}_2^2$ in Model 1 becomes much larger. In Model 2, $\hat{\sigma}_2^2$ is quite small except when $\rho_{12} = 0.99$, though even here it is smaller than $\hat{\sigma}_2^2$ from Model 1. The change in $\hat{\sigma}_2^2$ from Model 1 to Model 2 is consistent with our hypothesis about what causes the DLM puzzle: The change is small for small ρ_{12} because although Model 1 doesn't fit, the included variable (X_2) is not highly correlated with the correct but excluded X_1. However, as collinearity becomes more pronounced, X_2's part of the fit becomes larger in Model 1 and is reduced more by adding X_1 to the model.

The DF in the fit for X_2 behave analogously. For $\rho_{12} = 0$ the DF for X_2 in Model 1's fit are large; as ρ_{12} increases, the DF for X_2 in Model 1's fit first drop and then rise. The DF for X_2 in Model 2's fit are always smaller than in Model 1's fit. The interesting thing is that for each ρ_{12} except 0.4, at least one dataset out of five has a substantial DF for X_2 in Model 2's fit.

Table 12.17 shows results for X_1 for Model 2 fits; in Model 1, $\hat{\sigma}_1^2$ and the DF for X_1 are necessarily zero. In Model 2, $\hat{\sigma}_1^2$ and the DF in the fit for X_1 change little for ρ_{12} between 0 and 0.95: $\hat{\sigma}_1^2$ is large and DF is very close to the maximum possible

Table 12.16: DF in $X_2\theta_2$ Computed Using Variance Estimates in Table 12.2

Model 1			Dataset		
ρ_{12}	1	2	3	4	5
0	0.95	0.95	0.93	0.96	0.93
0.2	0.52	0.001	0.61	0.0001	0.64
0.4	0.64	0.46	0.40	0.43	0.18
0.6	0.93	0.95	0.91	0.94	0.93
0.8	0.99	0.99	0.99	0.98	0.99
0.9	0.99	0.99	0.99	0.99	0.99
0.95	0.996	0.996	0.997	0.996	0.997
0.99	0.998	0.999	0.998	0.999	0.998
Model 2			Dataset		
ρ_{12}	1	2	3	4	5
0	0.000	0.000	0.000	0.608	0.404
0.2	0.000	0.668	0.000	0.174	0.002
0.4	0.000	0.000	0.000	0.000	0.000
0.6	0.296	0.000	0.802	0.000	0.198
0.8	0.000	0.000	0.514	0.829	0.000
0.9	0.617	0.000	0.000	0.000	0.000
0.95	0.000	0.000	0.000	0.618	0.000
0.99	0.724	0.000	0.207	0.862	0.000

Table 12.17: Results for $X_1\theta_1$ and $\hat{\sigma}_e^2$ for 5 Datasets Simulated from (12.23)

ρ_{12}	$\hat{\sigma}_1^2$	DF in $X_1\theta_1$	$\hat{\sigma}_e^2$
0	8.5-9.7	0.9979-0.9984	5.6-6.1
0.2	7.9-8.8	0.9978-0.9985	6.1-6.7
0.4	8.1-9.9	0.9981-0.9986	6.3-7.5
0.6	8.2-9.1	0.9977-0.9984	5.5-6.0
0.8	8.1-10.2	0.9971-0.9989	3.3-4.2
0.9	8.0-9.4	0.9930-0.9985	2.3-2.6
0.95	8.8-11.1	0.9915-0.9984	1.6-2.0
0.99	3.2, 11.3, 6.4, 1.4, 8.7	0.73, 0.999, 0.95, 0.56, 0.998	0.8-1.1

Note: Second and third columns from the left: Maximum restricted likelihood estimates $\hat{\sigma}_1^2$ and DF. For ρ_{12} between 0 and 0.95, table entries are the ranges over the 5 datasets of $\hat{\sigma}_1^2$ or DF in the fit; for $\rho_{12} = 0.99$, the table entries are $\hat{\sigma}_1^2$ or DF for the 5 datasets, in order, to allow comparison with Tables 12.15 and 12.16. Right-most column: Range of $\hat{\sigma}_e^2$ in the 5 datasets.

value, as we would expect. The surprising results are for $\rho_{12} = 0.99$. Here, $\hat{\sigma}_1^2$ and the DF in the fit for X_1 vary substantially between datasets; $\hat{\sigma}_1^2$ is small for datasets with large $\hat{\sigma}_2^2$ and large for datasets with small $\hat{\sigma}_2^2$. It seems that when $\rho_{12} = 0.99$, X_1 and X_2 are so close to a linear function of each other that tiny aspects of the simulated data tip the balance one way or the other. This is not a counterexample to our hypothesis: With $\rho_{12} = 0.99$, Model 1 *does* fit; in Model 2, the machinery cannot differentiate the portions of **y** attributable to X_1 and X_2.

Table 12.17 also shows estimates for the error variance $\hat{\sigma}_e^2$ for Model 1. (For Model 2, $\hat{\sigma}_e^2$ was close to the true value for all ρ_{12}.) For ρ_{12} between 0 and 0.6, $\hat{\sigma}_e^2$ was in the interval 5.5 to 6: The error term captured most of the variation arising from the omitted X_1 because X_2 was not correlated with X_1 highly enough to do so. These values of ρ_{12} resemble the two-crossed-random-effects example. As ρ_{12} increased beyond 0.6, X_2 captured more of the variation arising from the omitted X_1 and accordingly $\hat{\sigma}_e^2$ in Model 1 became smaller.

12.2 Testing Hypotheses on the Optical-Imaging Data and DLM Models

I have not been able to produce a re-expression of the likelihood or restricted likelihood for Prof. Lavine's DLMs like the one used above in examining the simple models, so instead I have used indirect methods. This section presents the results. The simple examples suggested a hypothesis for the DLM puzzle: To produce such an effect, the model's components must have highly collinear design matrices and the first fitted model (Model 1, without the mystery component) must have a specific lack of fit. This section formulates and tests, to some extent, specific hypothesis addressing both elements of this general hypothesis.

Section 12.2.1 shows two sets of results relevant to the hypothesis about collinearity. The first set decomposes the effect of adding the mystery component to Model 1 by first increasing the mystery component's variance from zero to its Model-2 estimate while keeping the variances for the other components at their Model-1 estimates, and then adjusting the other components' variances to their Model-2 estimates. At both steps of this process the signal and respiration components are selectively affected while the heartbeat component is affected much less. This suggests that the design matrices of the mystery, signal, and respiration terms are more highly collinear with each other, in some sense, than they are with the design matrix for the heartbeat component. The second set of results uses canonical correlation to show that this is the case. Each component forms a saturated model so their design matrices must be perfectly collinear, but shrinkage of each component's fitted values differentially mitigates the effect of this perfect collinearity. The idea of differential collinearity developed below relies on this differential mitigation.

Section 12.2.2 then considers the hypothesis of lack of fit. It does so by constructing artificial datasets that selectively omit parts of the fit of Model 3, which included signal, heartbeat, respiration, and mystery components with two harmonics for respiration and mystery. Models 1 and 2 are fit to each artificial dataset to see if the phenomenon in the DLM puzzle is reproduced. The object is to identify the aspect of Model 1's lack of fit that produced the DLM puzzle.

Table 12.18: DF in Each Component of the DLM Fit, Changing Variance Estimates One at a Time from Model 1's Estimates to Model 2's Estimates

Step	Variance Estimates	Signal	Heartbeat	Respiration	Mystery
1.	Model 1	32.9	183.5	433.6	—
2.	Model 2 $\hat{W}_m^2$	12.7	126.5	237.1	273.8
3.	Model 2 $\hat{W}_m^2$, $\hat{W}_h^2$, $\hat{W}_e^2$	12.7	120.3	239.5	277.6
4a.	*cf* 3, Model 2 $\hat{W}_m^2$, $\hat{W}_h^2$, $\hat{W}_e^2$, $\hat{W}_s^2$	2.07	120.3	240.6	287.0
4b.	*cf* 3, Model 2 $\hat{W}_m^2$, $\hat{W}_h^2$, $\hat{W}_e^2$, $\hat{W}_r^2$	15.0	147.5	20.0	467.4
5.	*cf* 3, all Model 2 estimates	2.07	147.7	20.3	480.0

Note: Each component's DF count includes 2 DF for the implicit fixed effects.

12.2.1 *The Hypothesis about Collinearity of the Design Matrices*

In this section, "Model 1" refers to the DLM including only error, signal, heartbeat, and respiration with one harmonic, while "Model 2" adds the mystery term with a single harmonic. The variances are called, respectively, W_v, W_s, W_h, W_r, and W_m. In Prof. Lavine's fit of Model 1 their estimates were $9.5 \times 10^{-8}, 1.6 \times 10^{-3}, 1.8 \times 10^{-2}, 2.3 \times 10^{-1}$, and 0, respectively; in the fit of Model 2 their estimates were $6.5 \times 10^{-9}, 3.1 \times 10^{-8}, 1.6 \times 10^{-2}, 3.9 \times 10^{-3}$, and 3.8×10^{-1}.

First we parse the effect on Model 1's fit from changing its variance estimates, one at a time, to the variance estimates from Model 2. Table 12.18 shows the results. Step 1 shows DF in the various components using the variance estimates from Model 1. Step 2 retains the Model-1 estimates for error, signal, heartbeat, and respiration but adds the mystery term with its Model-2 variance estimate. Omitting the 2 DF for fixed effects in each component, the random-effect DF in the signal, heartbeat, and respiration components have declined by 65%, 31%, and 46%, respectively. In Step 3, the variances for error and heartbeat are changed to their Model-2 estimates; these changes are small and have very little effect. Step 4a then changes the signal variance to its Model-2 estimate. The signal component is reduced to its 2 DF for fixed effects and almost nothing for random effects; the lost 10.6 DF move mostly to mystery with a bit to respiration. Step 4b changes Step 3's variances by changing only the estimate for respiration, leaving $\hat{W}_s^2$ at its Model-1 estimate. With this change, the random-effects DF in the respiration component are reduced from 237.5 to 18.0; most of these lost DF move to the mystery component, although 27 move to heartbeat and a bit to signal. Finally, Step 5 sets all variances at their Model-2 estimates, showing that the changes to $\hat{W}_s^2$ and $\hat{W}_r^2$ (Steps 4a and 4b) act more or less independently.

When the signal component loses DF, they move almost entirely to the mystery component. When the respiration component loses DF, they move mostly to the mystery component though some go to the heartbeat and signal components. This suggests that in some sense, the signal and mystery components have design matrices that are especially highly correlated, while respiration is less highly correlated with mystery and signal and all three of these are less highly correlated with heartbeat. I now attempt to formulate this vague idea of differential collinearity between random-effect design matrices that are full-rank or nearly full-rank. This is a crude first attempt, presented for the heuristic value it might have.

The following development re-expresses Model 2's DLM in a manner like Section 5.2.2's re-expression of the ICAR model. First each component of Model 2 is expressed as a mixed linear model, then each component's random effects are re-expressed with a canonical kind of design matrix.

The signal component is re-expressed using the linear growth model in equations (2.38) and (2.39), with local-intercept evolution terms $w_{1t} = 0$, i.e., the smoothing variance for the local intercept is fixed at zero. The implicit fixed effects have the design matrix $\mathbf{X}_s$ given in (2.38), with one column each for the initial intercept and slope. The implicit random effects have the design matrix $\mathbf{Z}_s$ given in (2.39), with one column for each of the $T-1$ steps by which the local slope evolves.

Using the approach in Section 6.2, it is straightforward to express each quasi-cyclic DLM component in mixed-linear-model form, so the details are an exercise. The state equation for each quasi-cyclic component can be written as in equation (6.19), with each component having its own period and thus step-length $\delta = 2\pi/\text{period}$. Using the results in Section 6.1, quasi-cyclic component j for $j \in \{h,r,m\}$ can be written as a mixed linear model with

$$\mathbf{X}_j = \begin{bmatrix} 1 & 0 \\ \cos\delta & \sin\delta \\ \cos 2\delta & \sin 2\delta \\ \cos 3\delta & \sin 3\delta \\ & \vdots & \\ \cos T\delta & \sin T\delta \end{bmatrix}, \quad \boldsymbol{\beta}_j = \begin{bmatrix} \beta_{10} \\ \beta_{20} \end{bmatrix}, \tag{12.24}$$

$$\mathbf{Z}_j = \begin{bmatrix} 0 & 0 & 0 & 0 & & 0 & 0 \\ 1 & 0 & 0 & 0 & & 0 & 0 \\ \cos\delta & \sin\delta & 1 & 0 & \ldots & 0 & 0 \\ \cos 2\delta & \sin 2\delta & \cos\delta & \sin\delta & & 0 & 0 \\ \vdots & & \vdots & & \ddots & \vdots & \\ \cos(T-1)\delta & \sin(T-1)\delta & \cos(T-2)\delta & \sin(T-2)\delta & \ldots & 1 & 0 \end{bmatrix}, \tag{12.25}$$

$$\mathbf{u}'_j = \left[\, w_{j11} \;\; w_{j21} \,\middle|\, w_{j12} \;\; w_{j22} \,\middle|\, \ldots \,\middle|\, w_{j1T} \;\; w_{j2T} \,\right], \tag{12.26}$$

where the pairs (w_{j1t}, w_{j2t}) are independent bivariate normals with mean 0 and covariance $W_j\mathbf{I}_2$.

In the re-expression that follows, I gather the 8 fixed effect columns, 2 per component, into a single fixed-effect design matrix $\mathbf{X}$ but keep separate the random-effect design matrices $\mathbf{Z}_j, j \in \{s,h,r,m\}$. Then, if $\mathbf{Q}_X = \mathbf{I}_{T+1} - \mathbf{X}(\mathbf{X}'\mathbf{X})^{-1}\mathbf{X}'$ is the residual projection for the fixed effects, I re-express component j's random effects as $\mathbf{Q}_X\mathbf{Z}_j\mathbf{u}_j$. The rationale is as follows.

First, in computing DF (Section 2.2.2.2), DF for a random effect is computed for the projection of that random effect onto the orthogonal complement of the fixed effects' fitted-value space. In other words, fixed effects have priority over random effects. Re-expressing the random effect as $\mathbf{Q}_X\mathbf{Z}_j\mathbf{u}_j$ makes this explicit. Second, recall from Section 10.2 that because the fixed effects had a flat (improper) prior, the linear combination of the random effects having the same span as the fixed effects necessarily had posterior mean zero. Although I have not proved this in generality, I have seen it in other guises and conjecture it is true for the present models. (This is consistent with the just-noted fact about DF, which arises from the flat prior on fixed effects.)

Expressing the random effect implicit in component j this way implies that each $\mathbf{u}_j$ has 8 redundant dimensions. We rectify that by re-parameterizing the random effect for the jth component. Let $\mathbf{Q}_X\mathbf{Z}_j$ have singular value decomposition $\Gamma_j\Delta_j\mathbf{V}'_j$, where all three matrices are $(T+1)\times(T+1)$ even though the last 8 diagonal entries in Δ_j are zero. Write $\mathbf{Q}_X\mathbf{Z}_j\mathbf{u}_j = \Gamma_j\mathbf{v}_j$, where $\mathbf{v}_j = \Delta_j\mathbf{V}'_j\mathbf{u}_j$ is multivariate normal with mean $\mathbf{0}$ and covariance $W_j\mathbf{D}_j$ for diagonal $\mathbf{D}_j = \Delta_j^2$. The last 8 diagonal entries of $\mathbf{D}_j$ are zero, so the corresponding columns of Γ_j and entries in $\mathbf{v}_j$ can be dropped, and the following does so while retaining the symbols Γ_j and $\mathbf{v}_j$.

With this re-parameterization of the random effects, by convention for each component j the columns of the new random effect design matrix Γ_j are sorted so the first entry in the new random effect, v_{j1}, has the largest variance and thus is shrunk least of all the v_{jk} for a given value of the smoothing variance W_j. The extent of shrinkage of v_{jk} increases as k increases because the diagonal elements of $\mathbf{D}_j$ decline as k increases. For the three quasi-cyclic components, the diagonal entries in $\mathbf{D}_j$ are very similar: The ranges of positive diagonal entries for heartbeat, respiration, and mystery are 0.31 to 2.15×10^4, 0.26 to 2.16×10^4, and 0.25 to 2.18×10^4 respectively, and for all three pairs of components, Pearson's correlation of these diagonal entries exceeds 0.999. Thus differences among these components arise almost entirely from the eigenvectors in the columns of their Γ_j.

Now we can define the sense in which the signal and mystery components are especially highly correlated. As we will see below, the first few columns of the signal component's design matrix Γ_s — the ones with the least shrunk coefficients $v_{s,k}$ — have extremely high canonical correlation with the first few columns of the mystery component's design matrix Γ_m, which have this component's least shrunk coefficients $v_{m,k}$. By definition, the first canonical correlation between the first m columns of Γ_s and the first m columns of Γ_m is the maximal correlation between a linear combination of the first m columns of Γ_s and a linear combination of the first m columns of Γ_m (Seber 1984, Section 5.7). If this canonical correlation is large, it means the two components' random effects are almost perfectly collinear in those columns for which the coefficients $v_{j,k}$ are least shrunk. The design matrices of signal and respiration and of respiration and mystery are also highly correlated in this sense, though less so than signal and mystery. Finally, the design matrix of heartbeat is not, in this sense, highly correlated with any of the other components' design matrices.

Table 12.19 shows canonical correlations between the first m columns of pairs of design matrices for m = 10, 20, 30, 40, 50, 100, and 200. The correlations between signal and mystery are especially large. Considering smaller m, for $m = 1,\ldots,8$ the first canonical correlation is 0.07, 0.22, 0.24, 0.36, 0.53, 0.73, 0.94, 0.9996 respectively. The first 8 columns of these two design matrices are almost perfectly collinear. It is thus hardly surprising that these two components compete so vigor-

Table 12.19: First Canonical Correlations of the First m Columns of Random-Effect Design Matrices of DLM Components

Design matrices		m 10	20	30	40	50	100	200
signal	mystery	0.999999	1	1	1	1	1	1
signal	respiration	0.14	0.28	0.44	0.64	1	1	1
respiration	mystery	0.10	0.21	0.36	0.56	1	1	1
heartbeat	signal	0.03	0.05	0.08	0.10	0.13	0.25	0.50
heartbeat	respiration	0.02	0.03	0.05	0.07	0.09	0.17	0.41
heartbeat	mystery	0.02	0.04	0.06	0.08	0.10	0.19	0.39

Note: "1" means "≥ 0.99999995."

ously. The question is why the mystery component wins for Prof. Lavine's dataset; Section 12.2.2 below suggests this happens because the data fit the mystery component better.

Table 12.19 shows a similar but less dramatic collinearity between the signal and respiration components and between respiration and mystery. Both pairs of components jump to essentially perfect collinearity between $m = 40$ and $m = 50$. By contrast, the heartbeat component's design matrix does not have very high canonical correlation with any of the other components' design matrices even as far out as $m = 200$ columns.

I now offer a sort of explanation; I apologize for its arm-waving quality. If you plot columns of the Γ_j with time on the horizontal axis (this is an exercise), you will find that the columns with least-shrunk v_{jk} look like relatively low-frequency sine curves, and that the frequency increases as k increases. I said "relatively low-frequency sine curves": For each quasi-cyclic component, the fixed effect columns have period equal to the component's nominal period (e.g., 117 for the mystery component) and the columns in Γ_j have frequencies that increase from that starting point. The signal component's fixed effects are an intercept and a straight line, and the first column in Γ_s resembles a half-cycle of a sine curve (or a quadratic). For the mystery component, the first column in Γ_m is roughly 6 cycles of a sine curve. The first few columns of these two components' Γ_j have high canonical correlation because both have low frequencies in the columns with least-shrunk v_{jk}. By contrast, for the heartbeat component the columns of Γ_h with least-shrunk $v_{h,k}$ have quite high frequencies; the columns of Γ_s with similarly high frequencies have $v_{s,k}$ that are shrunk almost to zero for the $\hat{W}_s^2$ considered here. Thus, heartbeat and signal do not compete in these fits to any noteworthy extent. Heartbeat competes with respiration to a small extent because these two components have somewhat closer nominal periods (2.78 and 18.75 time steps respectively), and because, in Table 12.18, respiration was not shrunk much until Steps 4b and 5.

I stop here lest I dislocate my shoulder from all the arm-waving. Obviously there is plenty of room to explore and make rigorous this idea of collinearity among full-rank or nearly full-rank design matrices for random effects. This is given as an exercise.

12.2.2 The Hypothesis about Lack of Fit

Section 12.1.1 hypothesized that to produce our DLM puzzle, Model 1's fit must be defective in a particular way. That hypothesis was motivated by the restricted likelihood for the clustering and heterogeneity model, which was expressed in a simple form. Unfortunately, neither of the DLMs that produced our puzzle seems to allow such a simple re-expression of the likelihood or restricted likelihood, so I cannot identify an analogous lack of fit. In particular, for the clustering and heterogeneity model the lack of fit problem was poor fit of the variances of specific contrasts of the site averages. I have not been able to devise analogous contrasts for the DLMs and thus cannot show that Prof. Lavine's Model 1 has the same variety of lack of fit. Thus for the DLMs, I have taken a different approach to constructing artificial datasets with known lack of fit, explained below. Also, although for the simpler models I used the restricted likelihood, in the following I use the likelihood because I chose to rely on the R package dlm (Petris 2010) rather than writing my own code (if you had my computing skills, you would too), and because maximizing the likelihood appears to be customary in non-Bayesian DLM analyses.

This section uses two terms loosely for brevity. First, any component of these DLMs is a saturated model if unshrunk, so it can fit literally anything and in that sense "lack of fit" cannot occur. When I say Model 1 does not fit a dataset where the respiration component has a second harmonic, I mean that in order to capture this feature of the data, the respiration component must have a large evolution variance W_r. This is not the same sense of "lack of fit" that I used in discussing the clustering-with-heterogeneity model. Second, I say "the second harmonic in respiration" to refer to that part of our artificial data arising from the fit of the second harmonic of Model 3's respiration component. This is not the same as a second harmonic in a Fourier decomposition; rather, it is a portion of the fit of a quasi-cyclic model component. Given the large DF in this fitted component, it must deviate considerably from a cyclic shape with a fixed frequency.

To construct artificial data with known lack of fit, I began with the fit of Prof. Lavine's Model 3 (Section 9.1.4, Figure 9.3), which included second harmonics for the respiration and mystery components. Model 3's fit had effectively zero error variance, so the fit is nearly identical to the data. This was the most sensible of the fits considered in this book and has three pieces that do not fit Model 1: the second harmonic for respiration and the first and second harmonics for mystery. Thus, the artificial data considered here are built by including Model 3's fitted signal and heartbeat components and the fitted first harmonic of the respiration term, and then including one or more of these three pieces that do not fit Model 1. Table 12.20 shows the artificial datasets considered. Each artificial dataset was constructed by adding together the included components of Model 3's fit and then adding iid error with variance $\hat{W}_v$ from Model 3's fit, though $\hat{W}_v$ was so small that the time series with error is indistinguishable from the time series without it.

Artificial Data A. For fits of Models 1 and 2, Table 12.21 shows the estimated variances and component DF. Model 1's fit has a complication: I found two local maxima

Table 12.20: Artificial DLM Datasets Considered

	Respiration	Mystery	
Artificial Dataset	2nd Harmonic	1st Harmonic	2nd harmonic
A	X		
B		X	
C		X	X

Note: In the row for a given artificial dataset, an "X" under a piece of Model 3's fit means that piece was included in the artificial dataset.

Table 12.21: Fits for Artificial Data A, Including the Second Harmonic in the Respiration Component but No Mystery Component

Variance Estimates						
Model	Obj fcn	Signal	Heartbeat	Respiration	Mystery	Error
1	103.2	0.13	0.014	0.0012	—	3.5×10^{-8}
1	31.7	9.5×10^{-6}	0.013	0.11	—	9.6×10^{-8}
2		1.0×10^{-5}	0.013	0.11	2.3×10^{-6}	1.5×10^{-8}
DF in components of fit						
Model	Obj fcn	Signal	Heartbeat	Respiration	Mystery	Error
1	103.2	133.8	42.0	6.2	—	458.0
1	31.7	13.1	213.5	423.4	—	0.00005
2		12.3	213.3	422.4	2.06	0.06

Note: Variance estimates are from maximizing the likelihood; DF are computed using the variance estimates. "Obj fcn" is the value of the objective function minimized by the dlm package. For each component, 2 DF are for fixed effects.

in the likelihood with quite different variance estimates and DF. Table 12.21's second fit of Model 1 has the higher likelihood (the objective function in dlm is minimized to obtain the maximum-likelihood fit, so lower objective function means higher likelihood). The less-preferred local maximum (objective function 103.2) has 458 DF for error and shrinks the respiration component greatly (6.2 DF, 2 for fixed effects), while the signal component is greatly enlarged (133.8 DF). The more-preferred local maximum (objective function 31.7) resembles Model 1's fit to the real data (Figure 6.1). For both the real data and this preferred fit to the artificial data, the fitted respiration component clearly shows the two harmonics. For Artificial Data A and the preferred local maximum, the fitted signal component looks like Model 3's signal fit to the real data. (To avoid pages and pages of plots, I haven't shown plots for this fit. Fitting these models and drawing plots are exercises.)

Fitting Model 2 to Artificial Data A, I found one local maximum for the likelihood after trying a few different starting values. This fit is very similar to the preferred local maximum for Model 1: The mystery component is shrunk almost to its fixed effects (2.06 DF, 2 of which are fixed effects) while the other components are almost unchanged from Model 1's fit. The mystery component's 2 DF for fixed ef-

Table 12.22: Fits for Artificial Data B, Including the First Harmonics of the Respiration and Mystery Components

More Preferred Local Maximum						
Variance Estimates						
Model	Obj fcn	Signal	Heartbeat	Respiration	Mystery	Error
1	−176.3	3.1×10^{-4}	0.018	0.011	—	1.2×10^{-2}
2	−174.3	5.1×10^{-5}	0.018	0.011	2.9×10^{-3}	1.2×10^{-2}
DF in Components of Fit						
Model	Obj fcn	Signal	Heartbeat	Respiration	Mystery	Error
1	−176.3	41.5	312.8	173.6	—	122.0
2	−174.3	17.3	317.4	173.8	38.5	103.1
Less Preferred Local Maximum						
Variance Estimates						
Model	Obj fcn	Signal	Heartbeat	Respiration	Mystery	Error
1	−171.1	3.0×10^{-4}	0.021	0.017	—	9.6×10^{-8}
2	−171.0	3.5×10^{-5}	0.021	0.013	6.2×10^{-3}	9.6×10^{-8}
DF in Components of Fit						
Model	Obj fcn	Signal	Heartbeat	Respiration	Mystery	error
1	−171.1	38.9	374.8	236.2	—	0.0009
2	−171.0	12.4	367.8	200.8	69.0	0.0009

Note: Variance estimates are from maximizing the likelihood; DF are computed using the variance estimates. For each component, 2 DF are for fixed effects.

fects had to come from somewhere; respiration and signal lost nearly a full DF each from their Model 1 fits. Thus although the second harmonic in respiration does not fit Model 1, this aspect of lack of fit did not produce the DLM puzzle.

Artificial Data B. This artificial dataset included only the first harmonic of Model 3's respiration component fit and added the first harmonic of Model 3's mystery component fit. Table 12.22 shows the results; again, plots are an exercise. For this artificial dataset, the likelihood for both Model 1 and Model 2 had two local maxima which had similar objective function (i.e., likelihood) values. To reduce confusion, Table 12.22 has separate sections for the (slightly) more-preferred and less-preferred local maxima.

For each of Models 1 and 2, the more-preferred local maximum had large error variance and about 100 DF for error, while the less-preferred local maximum had effectively 0 DF for error. For Model 1, for both local maxima, the signal component captured much of the variation arising from the mystery component, which was in the data but not in the model. For the more preferred local maximum (objective function −174.3 for Model 2), adding the mystery component reduced DF only in the signal component and in error, but the fitted signal component in Model 2 still captured the declining-then-rising curve used in constructing these artificial datasets. For the less-preferred local maximum (objective function −171.0 for Model 2), adding the mystery component took DF from signal, error, respiration, and a bit from heartbeat

Table 12.23: Fits for Artificial Data C, Including the First Harmonic of the Respiration Component and Both Harmonics of the Mystery Component

Variance Estimates						
Model	Obj fcn	Signal	Heartbeat	Respiration	Mystery	Error
1	49.2	0.0032	0.019	0.019	—	0.054
1	60.3	0.0025	0.024	0.060	—	9.6×10^{-8}
2		3.0×10^{-5}	0.020	0.0068	0.10	4.8×10^{-7}
DF in Components of Fit						
Model	Obj fcn	Signal	Heartbeat	Respiration	Mystery	Error
1	49.2	62.6	217.3	137.3	—	232.8
1	60.3	48.6	297.2	304.1	—	0.0005
2		6.6	248.6	56.3	338.5	0.002

Note: Variance estimates are from maximizing the likelihood; DF are computed using the variance estimates. For each component, 2 DF are for fixed effects.

but again, the signal component still captured the signal curve that was built into the artificial data, and the respiration component's fit changed little from Model 1. Thus, this specific failure in Model 1's fit — omitting the first harmonic of the mystery term — did not produce the phenomena observed in the DLM puzzle.

Artificial Data C. This artificial dataset included, from Model 3's fit to the real data, the first harmonic of the respiration component and both harmonics from the mystery component. This artificial dataset reproduced some aspects of the phenomena in our DLM puzzle. Table 12.23 and Figures 12.3, 12.4, and 12.5 show the results. Figures 12.3 and 12.4 have the same format as Figure 6.1, which shows Model 1's fit to the real data, while Figure 12.5 has the same format as Figures 9.2 and 9.3, which show the fits of Models 2 and 3 to the real data.

Again, Model 1's likelihood had two local maxima, with the more-preferred local maximum giving 232.8 DF to error (objective function 49.2; Figure 12.3) and the less-preferred one giving almost zero DF to error (objective function 60.3; Figure 12.4). However, the fits arising from the two local maxima don't differ much visually. The 170 extra DF in the respiration component's less-preferred fit are reflected in somewhat more jaggedness but the gross outlines of the two fits are quite similar. At both local maxima, the fitted signal component is capturing a large piece of the omitted mystery component.

In fitting Model 2 I found just one local maximum (Figure 12.5). This fit reduced the signal component to 6.6 DF, the lowest in any model fit to these artificial datasets. However, Model 3's fit to the real data gave 6.5 DF to the signal component, and in Figure 12.5 the signal component's fit still captures the down-then-up shape that was built into this artificial dataset, although it is oversmoothed at the data's lowest point (Figure 12.5a). The respiration component's fit is also reduced to the fewest DF in any model fit to these artificial datasets and shows little of the detail that was present in Model 1's fit (Figure 12.5c). However, this component is also not reduced as severely as it was in the real data's Model 2 fit.

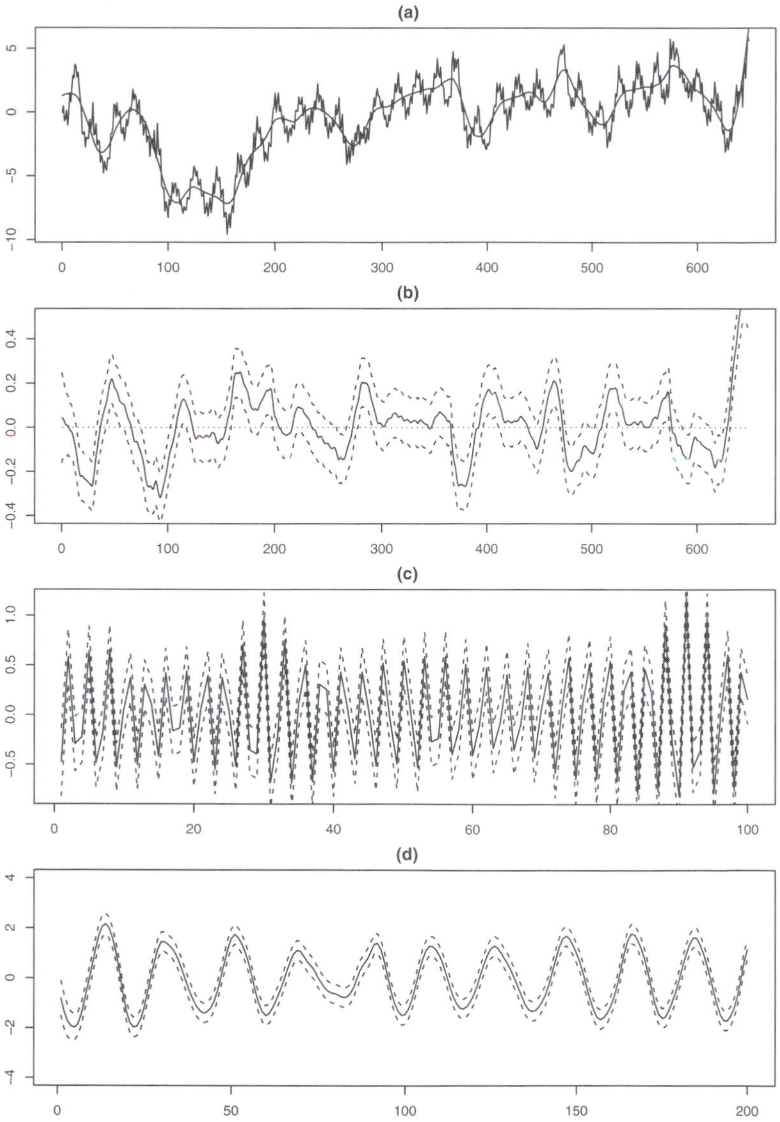

Figure 12.3: Model 1 fit to Artificial Data C, more preferred fit (objective function 49.2). Panel (a): Data and full fit (jagged line) and the smoothed response s_t (smooth line). Panel (b): Fitted local slope, slope_t. Panel (c): Fitted heartbeat component, first 100 time steps. Panel (d): Fitted respiration component, first 200 time steps. Panels b, c, and d show pointwise 90% posterior regions, treating the variance estimates as known.

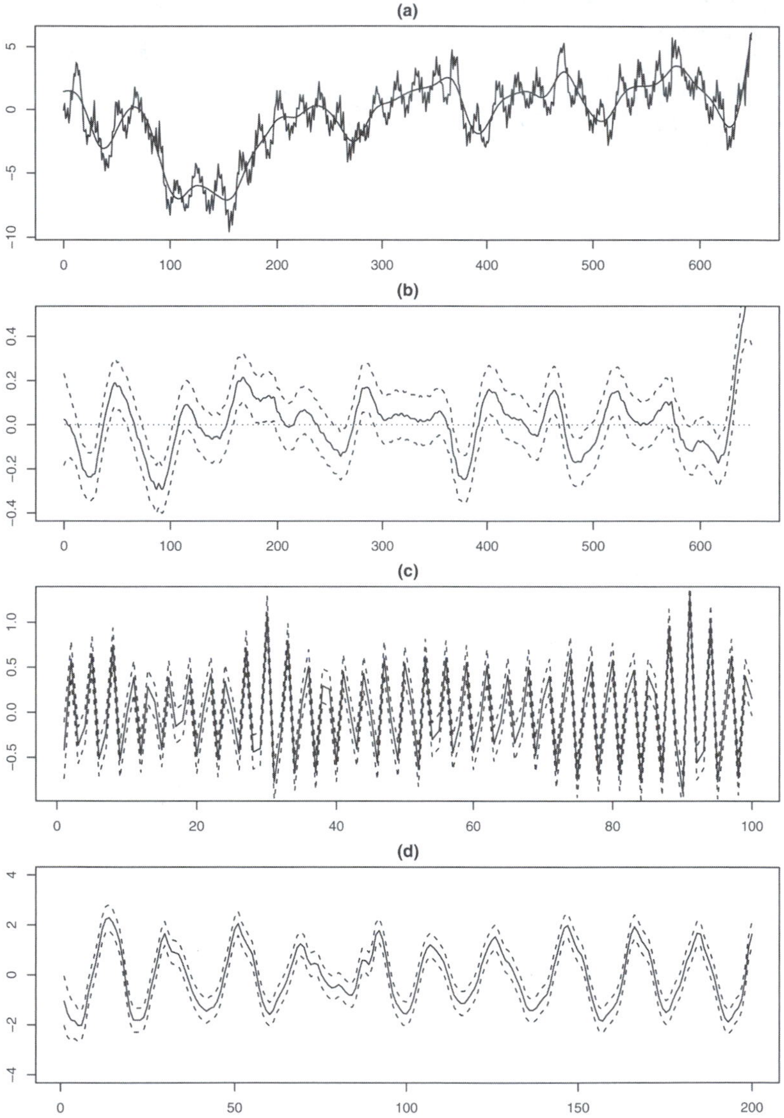

Figure 12.4: Model 1 fit to Artificial Data C, less preferred fit (objective function 60.3). Panel (a): Data and full fit (jagged line) and the smoothed response s_t (smooth line). Panel (b): Fitted local slope, slope_t. Panel (c): Fitted heartbeat component, first 100 time steps. Panel (d): Fitted respiration component, first 200 time steps. Panels b, c, and d show pointwise 90% posterior regions, treating the variance estimates as known.

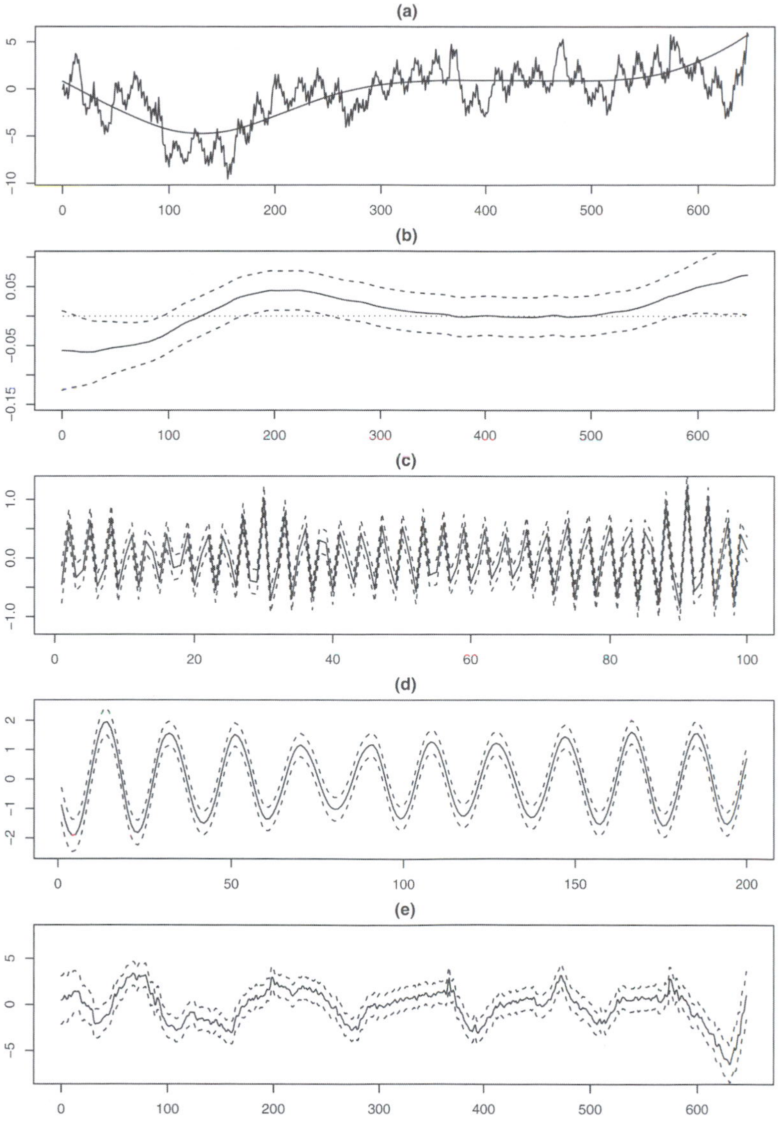

Figure 12.5: Model 2 fit to Artificial Data C. Panel (a): Data and full fit (jagged line) and the smoothed response s_t (smooth line). Panel (b): Fitted local slope, slope$_t$. Panel (c): Fitted heartbeat component, first 100 time steps. Panel (d): Fitted respiration component, first 200 time steps. Panel (e): Fitted mystery component. Panels b through e show pointwise 90% posterior regions, treating the variance estimates as known.

Thus, to produce the full effect seen in the DLM puzzle, it appears that the necessary lack of fit is contained in the second harmonic of the respiration component — though Artificial Data A showed that this alone is not sufficient — *and* both harmonics of the mystery component. Including only the first harmonics of respiration and mystery produced nothing like our DLM puzzle (Artificial Data B), and adding the second harmonic of mystery but not respiration (Artificial Data C) only partly reproduced the phenomena DLM puzzle.

12.3 Discussion

Because I couldn't analytically "crack" the DLM models, I approached this problem indirectly. I analyzed simpler models to produce hypotheses that might generalize to the DLM models, tested them using simulated data, and formulated more specific hypotheses for the DLM models. I then tested the latter to some extent, first by showing a sense in which the mystery component is more highly correlated with the signal and respiration components than with the heartbeat component and second by analyzing artificial datasets with specific kinds of lack of fit to Model 1. This supports some modest conclusions: It appears that both collinearity and a specific lack of fit are needed to produce dramatic effects of the sort seen in the DLM model. This was shown for simpler models and suggested strongly for the DLM model itself. None of this is well developed as yet so much work and no doubt more surprises remain.

For the simpler problems in Section 12.1, the restricted likelihood and thus the marginal posterior of the variances could be rewritten in a simple form permitting near-total understanding. This form is used extensively in later chapters of this book. It can be understood as arising from a linear transformation of the data $\mathbf{Ky}$ such that $\mathbf{K}$ does not depend on any unknowns and $\mathbf{Ky}$ has mean zero and a diagonal covariance matrix conditional on the mixed linear model's covariance matrices $\mathbf{R}$ and $\mathbf{G}$. Such a $\mathbf{K}$ does not exist for mixed linear models in general; Chapter 17 proves this for a counterexample and this chapter's DLMs are almost certainly another counterexample. The present chapter suggests it could still be enlightening to extend this approach by constructing $\mathbf{K}$ so that $\mathbf{Ky}$ has a covariance matrix that is not diagonal but which has illuminating diagonal entries. This might permit us to identify the specific lack of fit in Prof. Lavine's Model 1 that produced the DLM puzzle, in the same way we could see the necessary lack of fit in the clustering and heterogeneity problem. Chapter 17 explores one way to do this. Imagining such an extension raises the question of what exactly the transformation $\mathbf{Ky}$ is intended to accomplish if it cannot produce a diagonal covariance matrix. It may be that different $\mathbf{K}$ are useful for different purposes. Also, $\mathbf{K}$ need not be of full rank in the residual space of the mixed linear model's fixed effects, though presumably giving up dimensions in that space sacrifices information. Exploring this idea is given as an exercise.

Exercises

Regular Exercises

1. (Section 12.1.1) Show that in the re-parameterized model with clustering and heterogeneity, ϕ_N necessarily has posterior mean zero. This is a consequence of the flat (improper) prior on $\theta_N = N^{-0.5}\mathbf{1}'\delta$ implicit in the ICAR model.

2. (Section 12.1.1) Derive the restricted likelihood arising from the model in equation (12.3). Hints: I used the constraint-case formulation for the restricted likelihood given in equation (11.9) and inversion and determinant theorems for partitioned matrices (e.g., Graybill 1983, Section 8.2).

3. (Section 12.1.1) Derive the degrees of freedom in the components of the clustering and heterogeneity model. I did this using equation (2.35) in Section 2.2.2.2, treating the fixed effect (the intercept) as the limit of a random effect. A direct approach is probably workable.

4. (Section 12.1.1) Consider a model with two random effects, and assume the design matrices of the random effects, $\mathbf{Z}_1$ and $\mathbf{Z}_2$, are orthogonal, so $\mathbf{Z}_1'\mathbf{Z}_2 = \mathbf{0}$. Consider also a second model with only the random effect having design matrix $\mathbf{Z}_1$. For fixed values of the error variance σ_e^2 and the random effect variances σ_1^2 and σ_2^2, show that the DF in the random effect with design matrix $\mathbf{Z}_1$ is the same in the two models.

5. (Section 12.1.1) See if you can find a non-trivial data-generating model such that when the heterogeneity-only and clustering-with-heterogeneity models are fit to data generated from it, fits to many datasets consistently show an effect like that in Tables 12.2 and 12.3. If you can do this, it might allow you to sharpen the notion of "specific lack of fit" that I've hypothesized is necessary to produce the DLM puzzle.

6. (Section 12.1.2) Derive the restricted likelihood for the two-crossed random effects model, first using the non-orthogonal parameterization in (12.15) and then using the orthogonal parameterization in (12.19). Hint: Use the constraint-case formulation and inversion and determinant theorems for partitioned matrices (e.g., Graybill 1983, Section 8.2).

7. (Section 12.1.2) Derive the DF for the row and column effects in the model with both effects and the DF for the row effect in the model with only that effect.

8. (Section 12.1.2) See if you can find a data-generating model such that when the rows-only and the two-crossed-random-effects models are fit to data generated from it, fits to many datasets consistently show the effect in the DLM puzzle. I conjecture it is impossible to do this with examples like my non-standard example, which simply manipulate the variances of the $l \times c$ cells. You might be able to construct a counter-example by directly specifying row and column means that are confounded, instead of trying to draw them randomly, as I have. However, if the row and column effects really are confounded, then it's not clear the row-only model gives a bad fit, even if adding the column effect reduces the estimated row effect by a lot, because as in any situation with confounded effects, the data do not distinguish the effects well.

9. (Section 12.2.1). Derive the re-expressed random-effect design matrices Γ_j for the DLM components and examine plots of their columns, in particular the first 8 columns of the signal and mystery components' design matrices. The high first canonical correlations arise from linear combinations with many zero entries.

10. (Section 12.2.2). Re-do the fits of the artificial datasets, or create your own artificial datasets and fit these models to them. Draw plots like Figures 12.3 and 12.5.

Open Questions

1. (Section 12.2.1). Can you make rigorous the vague notion of differential collinearity between random effects with full-rank or nearly full-rank design matrices?
2. (Section 12.3). Consider transformations $\mathbf{Ky}$ for the data in the DLM puzzle. I am $(100-\varepsilon)\%$ sure that for $\mathbf{K}$ of rank 644 (for Model 1) or 642 (for Model 2) no $\mathbf{K}$ exists such that $\mathbf{Ky}$ has a diagonal covariance matrix, though I haven't proved this. Assuming that's true, can you find full-rank $\mathbf{K}$ that illuminate any of the problems considered in this chapter? What properties do you think are desirable in such a $\mathbf{K}$? You can write $\text{cov}(\mathbf{y})$ as a sum of terms of the form $\sigma_j^2\mathbf{M}_j$, where $\mathbf{M}_j$ is a matrix arising from the jth component of the model, and then select $\mathbf{K}$ to induce desired properties in $\mathbf{KM}_j\mathbf{K}$. For example, it might be useful if the diagonal elements of the $\mathbf{KM}_j\mathbf{K}$ were maximally differentiated in some sense, as this would seem to highlight the kind of lack of fit we saw in considering the model with clustering and heterogeneity.

Chapter 13

Random Effects Old and New

The term *random effect* is now used more broadly than it was, say, 50 years ago. Many things that are now called random effects differ qualitatively from random effects as defined in, e.g., Scheffé's classic 1959 text on analysis of variance. Chapter 1 introduced this distinction between old- and new-style random effects and later chapters explored some of its consequences. The present chapter ties together the material about this distinction and summarizes its consequences, conceptual and practical. Few statisticians seem to recognize this distinction and our field does not have language to describe it. At least one prominent statistician has declared that the term *random effect* is now ill-defined to the point of being meaningless, so he does not use it and would abolish it (Gelman 2005a, Section 6; Gelman 2005b). I have some sympathy with this view, but following Gelman's advice would abolish not only the term *random effect* but also the distinction between old- and new-style random effects which, this chapter argues, would be a mistake.

The argument, briefly, is this: The traditional, old-style definition of a random effect (Section 13.1) is explicit and specific. As analysis of richly parameterized models becomes unified in the mixed linear model and other syntaxes, more analyses use the *form* of old-style random-effects, but many, perhaps most, such random effects do not meet the traditional definition. Section 13.2 describes how new-style random effects deviate from the old-style definition and gives three examples that capture the sense of new-style random effects. Section 13.3 argues that these deviations from the old-style definition are not mere niceties but imply distinct ways to do inference and prediction, to interpret analytical artifacts like spatial confounding, and to do simulations for evaluating statistical methods. In statistical theory and practice, therefore, we should identify whether a random effect is old-style or new-style and proceed accordingly. It may ultimately be desirable to abolish the term *random effect* but if so we should replace it with new terms for old- and new-style random effects.

13.1 Old-Style Random Effects

The traditional definition of a random effect is given, for example, in Scheffé (1959, p. 238): The levels of a random effect are draws from a population; the draws are not of interest in themselves but only as samples from the larger population, which *is* of interest. (I have changed Scheffé's definition a bit; I return to this below.) "Level" is ANOVA jargon, which I use to highlight the difference between new- and old-style random effects. In this traditional definition, the population can be real and finite or hypothetical and infinite; my example has a real finite population.

Example 13: Measuring epidermal nerve density. Dr. William Kennedy's lab at the University of Minnesota uses novel methods to count and measure nerve fibers in skin and mucosa as an objective way to measure the status and progress of neuropathies, especially among diabetics. A recent study (Panoutsopoulou et al. 2009) compared two skin sampling methods using 25 normal (not diabetic) subjects, from each of whom skin was sampled by each of two methods, biopsy and blister, on the calf and on the foot. The analysis included three old-style random effects: Subjects, the method-by-subject interaction ("method" being blister or biopsy), and the location-by-subject interaction ("location" being foot or calf). A fourth random effect, the method-by-location-by-subject interaction, is customarily treated as the residual error term because it is not identified without replicate measurements. In traditional jargon, it is called a variance component but not a random effect.

To represent this in a model, let factor A be methods, with levels $i = 1, 2$, factor B be location, with levels $j = 1, 2$, and factor C be subjects, with levels $k = 1, \ldots, 25$. Then the model representing this experimental situation is

$$y_{ijk} = \mu + \alpha_i^A + \alpha_j^B + \alpha_{ij}^{AB} + a_k^C + a_{ik}^{AC} + a_{jk}^{BC} + \varepsilon_{ijk} \tag{13.1}$$

where the αs are fixed effects, the as are random effects with mean zero and variances σ_C^2, σ_{AC}^2, and σ_{BC}^2 respectively, and ε_{ijk} is the error term. (I have used Scheffé's notation but not his parameterization of the random effects.)

These random effects arise from sampling subjects and describe how the average nerve density varies between subjects (subject main effect); how the difference in nerve density, blister minus biopsy, varies between subjects (subject-by-method interaction); and how the difference in nerve density, foot minus calf, varies between subjects (subject-by-location interaction). As in many, perhaps most, applications of old-style random effects, these subjects were not selected by a formal sampling mechanism. This is in contrast with Scheffé's (1959) definition of a random effect, in which the draws "can reasonably be regarded as a random sample from some population" (p. 238). Nonetheless, these are old-style random effects: The levels (subjects) are a sample, albeit not a random sample; the subjects are not interesting in themselves, but only as members of the population from which they were drawn (non-diabetic adults); and the object was to measure average differences in that population between methods and locations.

13.2 New-Style Random Effects

In discussing new-style random effects, it will be useful to distinguish three senses in which the notion of probability is used. One familiar sense involves draws using a random mechanism, either one we create and control (e.g., sampling using random numbers) or one we imagine is out there in the world (e.g., random errors in measuring a length). This conforms more or less to the frequentist notion of probability, the fraction of times an event occurs in a hypothetical infinite sequence of draws. A second familiar sense of probability describes uncertainty that does not arise from operation of a random mechanism but rather is a person's uncertainty about an unknown quantity. This is the subjective Bayesian notion of probability; de Finetti's slogan "Probability does not exist" summarized his argument that random mechanisms do not exist out in the world and that probability properly has *only* this second sense.

The third and least familiar sense of probability is that a probability distribution can be used as a descriptive device. (Draper et al. 1993 argue that a descriptive notion of exchangeability is more primitive than probability. Freedman & Lane 1983 give a way to use this descriptive sense of probability to justify a form of inference.) For example, if we measured the heights of all U.S.-born 52-year-old males employed by the University of Minnesota, those heights could be well described as following a normal distribution with some mean μ and variance σ^2. Describing the heights this way does not imply that anyone's height is a random draw from N(μ, σ^2): Each man's height is fixed (for now, at least). Rather, this is an aggregate statement describing the heights of a group of men. If we used a random-number generator to select one man from this group and report his height, the height we report could be represented as a draw from N(μ, σ^2) in the first sense of probability (random mechanism). However, the selected individual's height is fixed; we created the randomness by our method of selecting him.

We can now identify three varieties of new-style random effects, which are listed below and then described in detail with examples, some from earlier chapters. I use ANOVA jargon to emphasize how these random effects fail to meet Scheffé's definition and indeed how foreign the old-style jargon seems when used to describe new-style random effects. The varieties are as follows:

- The levels of the effect are not draws from a population because there is no population. The mathematical form of a random effect is used for convenience only.
- The levels of the effect come from a meaningful population but they are the whole population and these particular levels are of interest.
- A sample has been drawn and the sample is modeled as levels of a random effect but a new draw from that random effect could not conceivably be made even if it made sense to imagine the random effect was drawn in the first place.

Rendered in ANOVA jargon, these new-style random effects are almost unrecognizable but as we will see, they are familiar when described in contemporary jargon.

13.2.1 There Is No Population

In the first variety of new-style random effect, the levels of the effect are not draws from a population because there is no population. The mathematical form of a random effect is used for convenience only.

The mixed-model representation of penalized splines (Chapter 3; Ruppert et al. 2003) is an example of random effects with levels that are not draws from any conceivable population. In the case of a one-dimensional penalized spline with a truncated-line basis, the random effect's levels are the changes in the slope at the

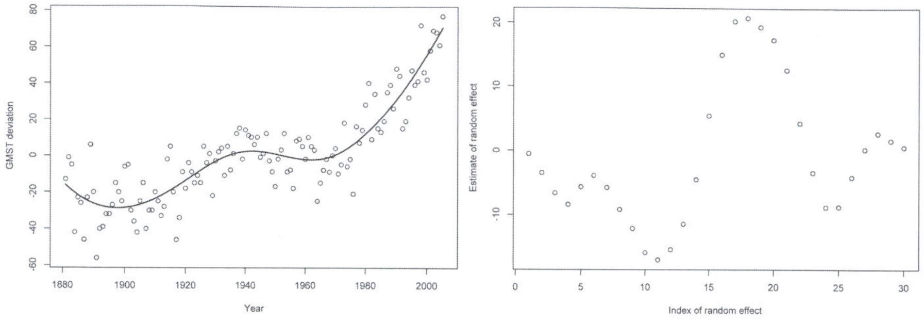

(a) Data and spline fit (b) Fitted $\hat{u}_k$

Figure 13.1: Global mean surface temperature deviation data. Panel (a): The data and spline fit. Panel (b): Fitted values (EBLUPs) for the random effects u_i.

knots and the random effect's distribution is just a device for constraining those changes. The senselessness of imagining further draws of such random effects is clearest for the examples in Ruppert et al. (2003) in which penalized splines are used to estimate smooth functions in the physical sciences. As Ruppert et al. (2003) put it, "we used the mixed model formulation of penalized splines as a convenient fiction to estimate smoothing parameters. The mixed model is a reasonable (though not compelling) Bayesian prior for a smooth curve, and ... [maximized restricted-likelihood] estimates of variance components give estimates of the smoothing parameter that generally behave well" (p. 138). To make this concrete, we recapitulate the mixed-linear-model formulation of penalized splines.

Example 5 revisited. Global mean surface temperature (GMST). Consider smoothing the annual series of global mean surface temperature deviations from 1881 to 2005, plotted in Figure 13.1a. Call the GMST measurements y_i, $i = 1, \ldots, 125$ (units are 0.01C) and let $\mathbf{y}' = (y_1, \ldots, y_{125})$. In the analyses presented here, the years are centered and scaled so they have finite-sample variance 1.

The object is to draw a smooth line through the data capturing its general shape. Penalized splines do this by fitting a linear model with many right-hand-side variables and constraining their coefficients. For the right-hand-side variables, we use a truncated quadratic basis, so the fitted spline for year i has the form

$$\beta_0 + x_i\beta_1 + x_i^2\beta_2 + \sum_{j=1}^{30} u_j([x_i - \kappa_j]_+)^2 \tag{13.2}$$

where x_i is year i after centering and scaling, $[z]_+ = 0$ if $z < 0$ and z otherwise, and the κ_j are 30 knots, the years 1880 + $\{4, 8, \ldots, 120\}$, centered and scaled in the same manner as the x_i. The spline is fit by choosing values for $(\beta, \mathbf{u})$, where $\beta' = (\beta_0, \beta_1, \beta_2)$ and $\mathbf{u}' = (u_1, \ldots, u_{30})$. We can write (13.2) in the form $\mathbf{X}\beta + \mathbf{Z}\mathbf{u}$, where the ith row of $\mathbf{X}$ is $(1, x_i, x_i^2)$ and the ith row of $\mathbf{Z}$ is $(([x_i - \kappa_1]_+)^2, \ldots, ([x_i - \kappa_{30}]_+)^2)$.

If we fit a conventional linear model $\mathbf{y} = \mathbf{X}\beta + \mathbf{Z}\mathbf{u} + \varepsilon$, it would give a wiggly fit, wildly over-fitting the data, as we saw in Chapter 3. Penalized splines avoid this by constraining $\mathbf{u}$. This was originally done by selecting a positive semi-definite matrix $\mathbf{D}$ and then fitting the spline — selecting $(\beta, \mathbf{u})$ — by solving this optimization problem:

$$\text{minimize} \quad (\mathbf{y} - \mathbf{X}\beta - \mathbf{Z}\mathbf{u})'(\mathbf{y} - \mathbf{X}\beta - \mathbf{Z}\mathbf{u})) \quad \text{subject to} \quad \mathbf{u}'\mathbf{D}\mathbf{u} \leq K \tag{13.3}$$

for some constant K. This optimization problem is equivalent to another optimization problem:

$$\text{minimize} \quad (\mathbf{y} - \mathbf{X}\beta - \mathbf{Z}\mathbf{u})'(\mathbf{y} - \mathbf{X}\beta - \mathbf{Z}\mathbf{u})) + \alpha\mathbf{u}'\mathbf{D}\mathbf{u}, \tag{13.4}$$

where α is a function of K that can be chosen by any of several methods. The optimization problem (13.4) can in turn be re-cast using the *form* of a mixed linear model. For simplicity, let $\mathbf{D}$ be the identity matrix; then the vector $\mathbf{y}$ is modeled as

$$\mathbf{y} = \mathbf{X}\beta + \mathbf{Z}\mathbf{u} + \varepsilon, \quad \varepsilon_i \overset{iid}{\sim} N(0, \sigma_e^2), u_j \overset{iid}{\sim} N(0, \sigma_s^2), \tag{13.5}$$

and $(\beta, \mathbf{u})$ is selected by solving the generalized least squares problem that arises from (13.5) when the variances σ_e^2 and σ_s^2 are fixed at particular values. The optimization problem (13.3) is thus formally identical to computing the conventional estimate of β and the best linear unbiased predictors (BLUPs) of $\mathbf{u}$ for model (13.5) given σ_e^2 and σ_s^2. The penalized spline now has the form of a mixed linear model and as Ruppert et al. (2003) noted, we can use this convenient fiction as a source of tools for (implicitly) selecting the penalty K or α in (13.3) and (13.4) respectively, by using restricted likelihood to estimate the variances in (13.5).

Although the penalized spline now has the *form* of a mixed linear model, nowhere is there any suggestion that $\mathbf{Z}\mathbf{u}$ is a draw from a random mechanism. The so-called random effect $\mathbf{u}$, like the so-called fixed effect β, is simply unknown before the optimization problem is solved. The model is a device that *we choose*; it would be ridiculous to imagine that the data were produced, out there in the world, by a draw from this model. The representation (13.5) merely provides a flexible collection of smooth curves and some discipline to avoid overfitting. Even if we assumed that some random process produced year i's true GMST and measurement y_i, it would be silly to imagine making more draws from $\mathbf{u}$'s distribution, which is only an analytical convenience, a device to facilitate smoothing. Indeed, the GMST series clearly does not fit a model of iid errors around a smooth curve (see, e.g., Adams et al. 2000), but maximizing the restricted likelihood arising from (13.5) is just a way to pick α in (13.4) and thus an unobjectionable way to draw a smooth line through the data, which was the object of this exercise.

In many cases there cannot be *any* sense in which $\mathbf{X}\beta + \mathbf{Z}\mathbf{u}$ arose from a random process, even once. Many examples in Ruppert et al. (2003) come from the physical sciences, e.g, the LIDAR, Janka hardness, and NOx examples, in each of which the smooth function estimated by $\mathbf{X}\beta + \mathbf{Z}\mathbf{u}$ is implied by physical or chemical laws. Even more plainly, I have used splines to approximate complicated deterministic functions, to simplify computation or presentations. In these cases, $\mathbf{X}\beta + \mathbf{Z}\mathbf{u}$ may be considered a random variable in the Bayesian use of that term (the second sense of probability), but it did not come into being as a draw from a random mechanism.

Bayesian terminology is delicate here, making it is easy to fall into conceptual errors. In a Bayesian approach to a penalized spline analysis, Your uncertainty (as de Finetti put it) about $\mathbf{X}\beta + \mathbf{Zu}$ is described using a probability distribution, the second sense of "probability," but that distribution describes You, not $\mathbf{X}\beta + \mathbf{Zu}$.

It is easy to see that that $\mathbf{u}$ is not an iid draw from $N(0, \sigma_s^2)$ by fitting (13.5) to the GMST data and examining the BLUPs for $\mathbf{u}$. Figure 13.1a shows the fit obtained using (13.5) and $\hat{\sigma}_s^2 = 947$ and $\hat{\sigma}_e^2 = 145$, which maximize the restricted likelihood. The fixed effect estimates are $\hat{\beta}' = (85.35, 176.40, 68.40)$; Figure 13.1b plots the EBLUPs for $\mathbf{u}$. Runs of positive and negative signs as in this figure are needed to fit a curve like the one in Figure 13.1a but they are spectacularly improbable if $\mathbf{u}$ is truly iid normal with mean zero. However, iid is not the wrong model for $\mathbf{u}$; rather, when we use (13.5) as a device to fit a smooth curve, there is no reason to expect the fitted $\mathbf{u}$ to look like an iid sample. (A colleague has objected that these EBLUPs are fitted values, not draws from $\mathbf{u}$'s distribution, so they wouldn't necessarily look like such a draw even if the model were literally true. Section 13.3.3 below gives my response.)

13.2.2 The Effect's Levels Are the Whole Population

In the second variety of new-style random effect, the levels of the effect come from a meaningful population but they are the whole population and these particular levels are of interest.

A common example here is smoothing disease maps, in which a geographical entity is partitioned into regions and some color-coded measure of a disease, e.g., incidence, is shown for each region. If the disease is rare or many regions have small populations, a map showing each region's raw disease measure will be noisy and of little use. A smoothed map replaces each region's raw disease measure with a measure obtained by "borrowing strength" from neighboring regions. (I use quotes because of the value judgment implicit in this expression, but it is so well-established as to be *de rigueur*.) Smoothing is often implemented by means of a random effect: The regions are the random effect's levels, the collection of regions is the entire population, and individual regions are of interest. Such analyses often include regressors as fixed effects.

Example 10 revisited. Stomach cancer in Slovenia. As described in earlier chapters, Dr. Vesna Zadnik collected counts of stomach cancers for the years 1995 to 2001 inclusive in the 194 municipalities that partitioned Slovenia in 2005, to study the possible association of stomach cancer with socioeconomic status. As part of this analysis, she used the model of Besag, York, & Mollié (1991), in which the counts of stomach cancers O_i are conditionally independent Poisson random variables with mean

$$\log\{E(O_i)\} = \log(E_i) + \beta SEc_i + S_i + H_i. \tag{13.6}$$

E_i is the expected number of cancers, the rate of stomach cancers per person in all of Slovenia multiplied by the population of municipality i (indirect standardization). SEc_i is the centered measure of socioeconomic status. The intercept is the sum of two random effects, $\mathbf{S} = (S_1, \ldots, S_{194})'$ capturing spatial clustering and $\mathbf{H} = (H_1, \ldots, H_{194})'$ capturing heterogeneity. The H_i were modeled as iid normal and the S_i were modeled using the ICAR model described in Sections 5.2.2 and 10.1.

In this example, the spatially correlated random effect $\mathbf{S}+\mathbf{H}$ does not meet Scheffé's definition of a random effect because the levels (municipalities) are the entire population and are themselves of interest. This example differs from penalized splines in that there is a meaningful population but as with splines, it is awkward to imagine $\mathbf{S}+\mathbf{H}$ as a draw from a random mechanism. It is not hard to imagine that some process with a stochastic element produced the $S_i + H_i$ and obviously counts of stomach cancers could be made for the following 7-year period. However, it would do serious violence to the subject matter to imagine or model these new counts as arising from an iid draw from the same mechanism that produced the counts for 1995–2001, or that it would even be possible to make a second draw from the same mechanism that produced the counts for 1995–2001. Those of a modeling inclination might model the *change* from 1995–2001 to the next 7-year period using a stochastic model and argue that the change from each 7-year period to the next could be considered an iid draw from some process. However, that has no relevance to Dr. Zadnik's problem, in which she only had data for 1995–2001. Thus it is hard to see how the random effects $\mathbf{S}+\mathbf{H}$ can be usefully described as arising from a random mechanism, the first sense of probability, or how we could usefully imagine making a second draw of either random effect.

It seems less problematic to consider the random-effect models for $\mathbf{S}$ or $\mathbf{H}$ as an example of the third sense of probability, a descriptive device. Just as we could say that the heights of the 52-year-old U.S.-born males at the University of Minnesota look like draws from a particular normal distribution, we could also say that the 194 Slovenian S_i, were we to observe them, would look like a draw from an ICAR with particular neighbor pairings. As with the heights of 52-year-old men, this would not imply that the S_i were actually drawn from any random mechanism, it would only be a convenient way to describe the ensemble of fixed but unknown constants S_i.

Using a probability distribution to describe $\mathbf{S}$ is decidedly less concrete than using a probability distribution to describe the heights of a group of men, which are at least in principle observable while $\mathbf{S}$ is necessarily unobservable. But the intuition that motivates use of a spatial model — that municipalities near each other tend to have more similar S_i than municipalities far from each other — is ultimately descriptive. Those of a subjective Bayesian turn of mind might say that it is equally natural to think of $\mathbf{S}$'s random-effect distribution as a probability statement in the second sense, describing Your uncertainty about how the S_i tend to be similar to each other. I have no objection to this but many people still find it awkward or impossible to think of probability statements as describing personal belief. Those of a conciliatory turn of mind might suggest that for the present problem, at least, there is no practical difference between the descriptive and subjective-Bayesian interpretations of this random effect: In neither of these senses of probability could we imagine making another draw from $\mathbf{S}$'s distribution. If we view $\mathbf{S}$'s distribution as descriptive, we might choose to deploy that description as part of a statistical method, the operating characteristics of which can be studied using simulations, for example. If instead we view **S**'s distribution in the subjective-Bayesian manner, we would use it like any other probability statement in the Bayesian calculus. Either way, the random effect **S** is a device that *we choose* for a particular analytic purpose and it would be absurd to imagine that this model generated the counts of stomach cancer O_i.

I have belabored this point because in the subculture of spatial analysis, at least some people tend to ignore the nature of a random effect and proceed directly to mathematical specifications and calculations but as I argue below, the nature of the random effect has practical implications.

13.2.3 A Sample Has Been Drawn, but . . .

In the third variety of new-style random effect, a sample has been drawn and the sample is modeled as levels of a random effect but a new draw from that random effect could not conceivably be made even if it made sense to imagine the random effect was drawn in the first place.

Consider the classic problem of geostatistics, mineral exploration. Suppose that in the region being explored, we are interested in the fraction of iron in rock at a certain depth, called W. Each location $\mathbf{s}$ in the region has a particular true W, $W(\mathbf{s})$, which is fixed but unknown. $W(\mathbf{s})$ is not observed but is measured with error at a sample of locations $\{\mathbf{s}_i\}$, giving measurements $y(\mathbf{s}_i)$ which could be modeled as $y(\mathbf{s}_i) = W(\mathbf{s}_i) + error(\mathbf{s}_i)$, where the $error(\mathbf{s}_i)$ are independent of each other. We could estimate $W(\mathbf{s})$ from the $y(\mathbf{s}_i)$ using a two-dimensional penalized spline. If we did, $W(\mathbf{s})$ would be another instance of the new-style random effect in Section 13.2.1. However, I want to make a different point that strikes closer to the heart of spatial statistics as it is usually conceived.

The present example, mineral exploration, differs from the global-mean surface temperature and Slovenian stomach-cancer examples in a few ways. The measurement locations $\mathbf{s}_i$ are a sample of possible locations (in practice rarely selected by a formal sampling mechanism, but that is immaterial) and they are not of interest in themselves but for what they tell us about the region as a whole, which may be understood as the population from which the $\mathbf{s}_i$ were drawn. We could draw a new sample of locations $\mathbf{s}_i$ and we could make new measurements at our original locations $\mathbf{s}_i$, or at least very close to them.

So far this sounds like an old-style random effect, but it is not. If we were to take new measurements at old $\mathbf{s}_i$ or take measurements at new $\mathbf{s}_i$, it would be untenable to imagine that these would involve a new draw of the random effect $W(\mathbf{s})$. We may imagine that $W(\mathbf{s})$ came into being by some stochastic process but it is now fixed, if unknown, and no more draws will be made from that stochastic process. We have observed it (with error) at some $\mathbf{s}_i$ but its value is already determined at *all* $\mathbf{s}$, observed and unobserved. It may be convenient to *describe* $W(\mathbf{s})$ in aggregate by saying that at any finite set of locations $\{\mathbf{s}_i\}$, $\{W(\mathbf{s}_i)\}$ looks like a draw from a normal distribution with mean $\mu(\mathbf{s}_i)$ and covariance $C(\mathbf{s}_i, \mathbf{s}_j; \theta)$ with parameter θ, and we may understand this probability distribution in either the second (subjective Bayesian) or third (descriptive) senses. In these respects, $W(\mathbf{s})$ is identical to **S** in the Slovenian stomach-cancer example.

13.2.4 Comments on New-Style Random Effects

(a) With some exceptions noted in Section 13.2.1, one might argue that for all these examples of new-style random effects, when the true $\mathbf{X}\beta + \mathbf{Z}\mathbf{u}$ or $\mathbf{S}$ or $W(\mathbf{s})$ came into being out there in the world, it *did* involve a draw or draws from a random mechanism. In such cases, it may make sense to imagine that producing the now fixed but unknown $\mathbf{X}\beta + \mathbf{Z}\mathbf{u}$, $\mathbf{S}$, or $W(\mathbf{s})$ involved a random draw *on one occasion*, after which $\mathbf{X}\beta + \mathbf{Z}\mathbf{u}$, $\mathbf{S}$, or $W(\mathbf{s})$ became fixed. Having once made such a draw, however, it makes no sense to imagine further draws, and the value of that single draw is of intrinsic interest: Estimating it is often the entire purpose of the analysis. It is therefore a hazardous distraction to conceive of $\mathbf{X}\beta + \mathbf{Z}\mathbf{u}$, $\mathbf{S}$, or $W(\mathbf{s})$ arising as a draw from a random mechanism and better to think of them as simply fixed and unknown. As hypothesized *mechanisms* for producing the data, the random-effect models are silly. It is more accurate to think of them as superficial descriptions, which is not intended to deny their utility.

(b) Instead of being a draw from a random mechanism, a new-style random effect is more clearly seen as something we *choose* as an analytic tool. I do not mean to suggest that the things labeled $\mathbf{X}\beta + \mathbf{Z}\mathbf{u}$, $\mathbf{S}$, or $W(\mathbf{s})$ never have real physical existence; $W(\mathbf{s})$ certainly does in our example and $\mathbf{X}\beta + \mathbf{Z}\mathbf{u}$ and $\mathbf{S}$ do in at least some cases. Accordingly, some choices of analytical tools — random effect distributions — will be better suited than others to estimating the actual (but unknown) $\mathbf{X}\beta + \mathbf{Z}\mathbf{u}$, $\mathbf{S}$, or $W(\mathbf{s})$. But "better suited" here means they yield better estimates, not that they are better characterizations of any stochastic process that might have produced the actual $\mathbf{X}\beta + \mathbf{Z}\mathbf{u}$, $\mathbf{S}$, or $W(\mathbf{s})$. In other words, the probabilistic form of our chosen analytical tool does not correspond to a random mechanism in the world that produced $\mathbf{X}\beta + \mathbf{Z}\mathbf{u}$, $\mathbf{S}$, or $W(\mathbf{s})$.

It is an odd consequence of our attitude toward models that so many of us find this distinction hard to make. As noted in Chapter 10, when shrinkage estimation was introduced in the 1950s, it was certainly considered strange that shrunken estimates dominate unshrunk estimates in the decision-theoretic sense, but as far as I know nobody had trouble identifying shrunken estimates as merely a particular decision rule and thus distinct from the process that produced the data. It seems that statisticians became confused about this distinction — began mistaking the shovel for the process that produced the soil — only when shrinkage estimators were expressed as estimates arising from models with random effects.

(c) New-style random effects can all be understood as merely formal devices for implementing smoothing or shrinkage. This is explicitly the case for penalized splines, in which the random effect is a "convenient fiction." It is less explicit for the kinds of spatial models in Sections 13.2.2 and 13.2.3, perhaps because people in our discipline habitually think of these models in terms of covariance matrices and thus draws from random mechanisms.

(d) Regarding the Slovenian stomach-cancer example, I have heard it said that $S_i + H_i$ is only in the model as an error term, just as we include a simple error term in an ordinary linear regression. This raises a deep question, which is beyond this chapter's scope. The question is: What is "error," and is there only one kind? Briefly,

I would argue that $S_i + H_i$ is local lack of fit of the mean structure $\log(E_i) + \beta SEc_i$, which is error in the sense of bias, not measurement error. Labeling the $S_i + H_i$ as error, understood as bias, does not provide a rationale for imagining that a new draw could be made from their distribution or indeed that a draw ever was made.

(e) There are, of course, cases of spatial analysis that are properly conceived as involving old-style random effects. For example, suppose we are studying ozone in the Boston area and have a dataset with measurements taken each day for several years. It may be meaningful, depending on the questions being asked, to represent days as draws from a distribution with spatial correlation. More such draws could be made and for many questions the draws themselves have no intrinsic interest, the interest lying instead in the population of days from which they were drawn.

Suppose however we are interested in a specific week because, for example, a particular kind of temperature inversion occurred that week. In this case, Boston has, for this week, a smooth spatial ozone gradient around which daily measurements varied. It violates the substantive question to treat that smooth spatial ozone gradient as a draw from a random mechanism. Rather, it is a fixed feature of Boston for the study's week, in which we have a specific interest but which we happen not to know. We may choose to make an aggregate description of this fixed but unknown feature of Boston using a probability distribution as a descriptive tool. There is a meaningful sense in which this fixed but unknown feature of Boston was drawn from a probability distribution, but that sense is not relevant to this particular question.

13.3 Practical Consequences

The distinction between old- and new-style random effects has implications for how we should do inference and prediction, interpret analytical artifacts, and design simulation experiments. Some of these were discussed at length in earlier chapters; here I will summarize the earlier discussion and emphasize new points.

13.3.1 Inference and Prediction

By "inference," I mean analyses focused on the present set of observations and on the models we posit as having generated them. By "prediction," I mean analyses focused on other observations related to the present set but as yet unobserved or unknown to us.

13.3.1.1 Inference

In a Bayesian analysis, all inferences are based solely on the posterior distribution of unknowns in the model being entertained and in manipulating the Bayesian machinery it makes no difference whether a random effect is old- or new-style. However, the old- vs. new-style distinction is relevant to prior specification.[1] For old-style random effects as in the nerve density example, data from earlier studies can be used to rationalize a prior distribution for variances, if it is plausible that a new dataset's random

[1]This observation is due to Jon Wakefield.

effects, e.g., for subjects or batch effects, are drawn from distributions like those in the earlier studies. No such rationale is possible for the variances of new-style random effects and in my experience, it is impossible to develop intuition about them. This is one advantage of specifying a prior distribution on the degrees of freedom in the fit instead of on the variances (Chapter 2): Such a prior is specified on quantities about which some intuition or prior information is possible.

For conventional analyses, the old-style notion of a random effect is deeply ingrained in the concepts and terminology. Consider best linear unbiased predictions (BLUPs). The notion of unbiasedness here refers to the expectation over random effects as well as error terms. Outside of a Bayesian analysis, computing an expectation over a random effect seems to require that repeated sampling of the random effect be meaningful. While it is defensible to imagine that new draws can be made from old-style random effects, the idea of new draws makes no sense for new-style random effects, as I have argued. Thus, it makes no sense to consider BLUPs unbiased for new-style random effects and in fact it is well known that penalized splines — which are BLUPs in the mixed-model formulation — flatten peaks and fill valleys (e.g., Ruppert et al. 2003, p. 141), as do spatial random-effect models. These tendencies are plainly biases if we understand new-style random effects simply as tools for estimating ensembles of fixed but unknown quantities. Such an understanding was present in early work on shrinkage estimation, which was called "biased estimation" among other things although that term has died out. I do not mean to impugn BLUPs, which are immensely useful (Robinson 1991), but it does not make sense to consider them unbiased for new-style random effects.

The old-style notion of random effects is so deeply embedded in non-Bayesian theory that cataloging its effects is beyond the scope of this chapter; two examples will have to suffice. The first example is pointwise confidence intervals for a penalized spline fit when the spline is implemented as a mixed linear model. This was discussed in Section 3.3.1. To summarize, if a random effect $\mathbf{u}$ is new-style, the confidence interval should be computed conditioning on — treating as fixed, although unknown — the realized value of the random effect, just as we do in simpler cases when for example the standard error of an estimate depends on unknowns. On the other hand, if the random effect is old-style, confidence intervals should not condition on $\mathbf{u}$'s realized value.

The second example of an effect on inference involves a spatial random effect; specifying the correct bootstrap depends on whether the spatial random effect is old- or new-style.[2]

Example 14: Health effects of pollutants. The object of this work is to estimate the effect on health outcomes of exposure to airborne pollutants. A health outcome y_i is measured on persons $i = 1, \ldots, n$ who are assigned spatial locations s_i, usually the map coordinates of their residences. (This type of analysis has many issues including assignment of locations to persons, but for the present purpose they are irrelevant.)

[2]This example is taken from an early version of Szpiro & Paciorek (2014).

The health-outcome model is

$$y_i = \beta_0 + x_i\beta_x + \mathbf{z}_i\beta_z + \varepsilon_i, \tag{13.7}$$

where x_i is a measure of person i's exposure to the pollutant of interest, β_x is its average effect on the health outcome, $\mathbf{z}_i\beta_z$ captures the effect of other personal characteristics (e.g., age), and ε_i is iid $N(0, \sigma_e^2)$. Often x_i cannot be measured for person i and must be estimated using measurements x_j^* made by monitors at locations $s_j^*, j = 1, \ldots, N$. This is a spatial misalignment problem: We have exposure measurements x_j^* at the s_j^*, but we want the exposures x_i at the s_i. One solution is to model the x_j^* as a function $\mu(s_j^*)$ of spatially referenced predictors like traffic and land use, capture residual spatial correlation in the x_j^* using a spatial random effect, and then predict the missing x_i using the model fit to the x_j^*. Specifically, jointly model the vectors of observed $\mathbf{x}^*$ and desired $\mathbf{x}$ as

$$\begin{bmatrix} \mathbf{x} \\ \mathbf{x}^* \end{bmatrix} = \begin{bmatrix} \mu(s;\gamma) \\ \mu(s^*;\gamma) \end{bmatrix} + \begin{bmatrix} \eta \\ \eta^* \end{bmatrix}, \tag{13.8}$$

where $\mu(s;\gamma)$ and $\mu(s^*;\gamma)$ are the vectors of fixed-effect predictions for $\mathbf{x}$ and $\mathbf{x}^*$, which depend on unknowns γ, and η and η^* are the vectors of spatial random effects, modeled as

$$\begin{bmatrix} \eta \\ \eta^* \end{bmatrix} \sim N(\mathbf{0}, \Sigma_{\eta,\eta^*}(\theta)), \tag{13.9}$$

where $\Sigma_{\eta,\eta^*}(\theta)$ is a function of unknowns θ. The unmeasured exposures $\mathbf{x}$ are predicted as

$$\hat{\mathbf{w}} = E(\mathbf{x}|\mathbf{x}^*; \hat{\gamma}, \hat{\theta}). \tag{13.10}$$

for estimates $(\hat{\gamma}, \hat{\theta})$ of the unknowns in the exposure model (13.8).

The standard bootstrap analysis for this problem (e.g., Szpiro et al. 2011) has these steps for a given method of fitting (13.7) – (13.10):

1. Use $\mathbf{x}^*$ to estimate exposure parameters $(\hat{\gamma}, \hat{\theta})$ in (13.8, 13.9).
2. Compute exposure predictions $\hat{\mathbf{w}}$ from (13.10) using $(\hat{\gamma}, \hat{\theta}, \mathbf{x}^*)$.
3. Estimate the unknowns $(\hat{\beta}_0, \hat{\beta}_x, \hat{\beta}_z, \hat{\sigma}_e^2)$ in the disease model (13.7), using $\hat{\mathbf{w}}$ in place of the unobserved $\mathbf{x}$.
4. For $k = 1, \ldots, B$ bootstrap samples
   - Draw a bootstrap realization of $(\mathbf{x}_k, \mathbf{x}_k^*, \mathbf{y}_k)$ from the joint model with spatial random effect (13.9), using parameter values $(\hat{\gamma}, \hat{\theta}, \hat{\beta}_0, \hat{\beta}_x, \hat{\beta}_z, \hat{\sigma}_e^2)$.
   - Estimate $(\hat{\gamma}_k, \hat{\theta}_k)$ using $\mathbf{x}_k^*$.
   - Compute the exposure prediction $\hat{\mathbf{w}}_k$ using $(\hat{\gamma}_k, \hat{\theta}_k, \mathbf{x}_k^*)$.
   - Estimate $\hat{\beta}_{x,k}$ using $\hat{\mathbf{w}}_k$ as the exposure.
5. Compute bootstrap standard errors and bias estimates from the moments of $\hat{\beta}_{x,k} - \hat{\beta}_x$.

In the first step of the bootstrap loop (Step 4), a draw is made from the joint spatial random effect model (13.9). This makes sense if (13.9) is an old-style random effect. But is it? As Szpiro and Paciorek say, the motivation for including the spatial random effect was to capture unexplained spatial structure in the residuals around the exposure prediction model $\mu(s;\gamma)$. The bootstrap above assigns the unexplainable spatial structure in **x** to the *variance* portion of model (13.7). However, the spatial random effect (η, η^*) was included to capture unmodeled *mean* structure in **x**; that is, the spatial random effect is new-style, not old-style. The bootstrap above makes sense if the mechanism that generated the spatial exposure data is a predictable component $\mu(s;\gamma)$ plus a spatial random effect from which new draws can be and are made out there in the world. However, it is more plausible that the mechanism generating the spatial exposure data is a fixed function of space that is predictable in principle but which is incompletely captured by $\mu(s;\gamma)$, and the spatial random effect is used to capture (some of) the bias in $\mu(s;\gamma)$. The bootstrap above answers the question "how would the results have changed if the spatial pattern of air pollution in this city was different?" Instead, the appropriate question is "how would the results have changed if I had picked different subjects and monitor locations?" That question implies a different bootstrap, in which subjects and monitor locations are resampled but the spatial random effect is not.

This new bootstrap, which is appropriate for this new-style random effect, raises its own set of technical issues, for example about bias in the estimate of β_x. For this chapter's purpose, however, the point is that defining the appropriate bootstrap depends on whether the spatial random effect is understood to be old- or new-style.

13.3.1.2 Prediction

For old-style random effects, everyone distinguishes two cases, which I will describe using the nerve density example. The first case is a prediction that involves drawing a new level of each random effect; in the nerve density example, this would be predicting biopsy and blister nerve density measurements from a new subject's calf and foot. Because a new subject is drawn, each new measurement's variance is the sum of all the components of variation, the three customarily called random effects (subjects, method-by-subject interaction, and location-by-subject interaction) and the one customarily called residual error. The second case is prediction of a new measurement for levels of the random effects that have already been drawn. In the nerve density example, this would be a new skin sample from, say, the foot of an already-sampled subject. Because no new subjects are sampled, each new measurement's variance is simply the residual error variance. The proper measure of prediction uncertainty for this new measurement also accounts for uncertainty about the true nerve density of this specific subject's foot in the area being sampled, i.e., of some function of the fixed effects and the already-sampled random effects. But the only variance component that contributes to the new draw is residual error.

For *all* new-style random effects, predictions are like the second case, if predictions are sensible at all. For all new-style random effects, the device of a random effect is used to estimate an ensemble of quantities that are fixed but unknown, like the true foot nerve density in the already-sampled subject from whom we take a new skin sample. For some problems, like the global-mean surface temperature data, predictions of new measurements may be impossible or ill-advised: It is not possible to make a new observation on the years 1881–2005 and using a spline fit to extrapolate past 2005 is a bad idea. However, when a new measurement can be made (e.g., some physical-science examples in Ruppert et al. 2003), predictions about these new measurements are simply inferences about the fitted spline surface, as in Section 13.3.1.1, plus residual error. In this respect, the new-style random effects in Sections 13.2.1 and 13.2.2 are the same.

Now consider the mineral-exploration example. It is possible to take a second measure at an existing location $\mathbf{s}_i$ (or very close to it) or to take a measurement at a new location $\mathbf{s}_0$. In either case, the only interpretation that does not do violence to the subject matter is that the random effect $W(\mathbf{s}_i)$ or $W(\mathbf{s}_0)$ has already been drawn and is simply unknown; otherwise, making a prediction would involve re-drawing the process that produced the ore seam. Rather, the spatial process used as a model for $W(\mathbf{s})$ simply facilitates spatial smoothing.

13.3.2 Interpretation of Analytical Artifacts

Section 10.1 discussed spatial confounding, as in the Slovenian stomach-cancer example, which has a different interpretation depending on whether the random effect **S** is new- or old-style. If **S** is a new-style random effect — a mere formal device to capture local lack of fit and not to be taken literally as a draw from a random mechanism — then in general adding it to the model introduces a bias into the estimate of β. If **S** is intended to adjust the estimate of β to account for a missing covariate, the adjustment from adding **S** to the model is biased, perhaps highly. If, on the other hand, **S** is an old-style random effect, as in the example in comment (e) of Section 13.2.4, and the fixed effect is treated as measured without error, then adding the spatial random effect **S** does not introduce a bias into the estimate of β, but simply inflates its posterior variance or standard error. Thus, this analytical artifact cannot be interpreted or handled properly without first determining whether the random effect is old- or new-style.

13.3.3 Simulation Experiments to Evaluate Statistical Methods

This section's premise is that in any simulation experiment, artificial data should be simulated in a manner that is appropriate for the question the experiment is intended to answer. This may seem breathtakingly obvious but based on the papers and talks I see, it is not. It sometimes seems that a contemporary statistical paper or talk is not complete without a simulation, even if that simulation is inane. I could go on (and on and on) but that is a rant for a different occasion.

So: A simulation experiment is intended to address a specific question or questions. This section argues that for simulations involving methods with new-style random effects, the question to be addressed determines whether it is permissible or meaningful, in simulating the data, to make draws from the random effect. I admit this consideration does not always give a clear-cut answer and that I have not always been consistent, as my students have thoughtfully pointed out. I return to this point below, after the main argument.

Consider a simulation experiment intended to evaluate methods that involve old-style random effects. What questions might such an experiment be intended to answer? Genuine questions would be about the performance of estimators and intervals and sometimes other kinds of decision rules (in the decision theoretic sense). For example, such an experiment might compare point estimates of variances computed from the posterior distribution, or the robustness of certain estimates or intervals in the presence of outliers or other contamination, or prediction accuracy.

To make this concrete, consider a simulation experiment to evaluate questions like these in the context of the nerve-density example. In such an experiment, simulated datasets must be drawn by first drawing the random effects for new subjects, a_k^C, then drawing the interaction random effects for each subject, a_{ik}^{AC} and a_{jk}^{BC}, then drawing a residual error ε_{ijk}, and finally adding these random effect draws to the fixed effects μ, α_i^A, α_j^B, and α_{ij}^{AB}. This way of simulating data reflects the nature of the old-style random effects, which exist because the subjects are a sample, and permits the questions to be answered as posed.

The foregoing is not controversial. However, based on talks, journal articles, and informal communications, many statisticians appear to think that for simulation experiments evaluating a method involving *new-style* random effects, data should be simulated in the same manner, with repeated draws from the random effects. Views on this matter do not seem to correspond to any Bayes-frequentist divide; it seems to be the default position of most statisticians of both persuasions. I will argue that for questions about a method involving new-style random effects, repeated draws never *need* to be made from those new-style random effects and for many questions is it incorrect and self-defeating to make draws from new-style random effects.

To be specific, consider a simulation experiment intended to compare different penalized splines for fitting smooth functions to datasets like the global-mean surface temperature data, e.g., to compare different bases or methods of choosing knots. Questions suitable to such an experiment include the following: What is the fit's bias or mean-squared error at various predictor values? What is the coverage of a given type of confidence interval? In experiments like this, I will argue, it is *never* appropriate to make draws from the random effect in the penalized spline's mixed-linear-model formulation. I give two arguments, one from first principles and one pragmatic. To summarize, the argument from first principles is that new-style random effects are only convenient fictions used to implement smoothing; it is a conceptual error to take these fictions literally and draw from them. The pragmatic argument is that draws from the penalized spline's random effect do not behave the way our intuition, developed from *fitting* such models, would suggest. Simulating from such a new-style random effect does not produce data with features relevant to questions like those given above and is thus self-defeating.

When fitting a penalized spline, we assume there is a fixed but unknown $f(x)$ and the object is to estimate it. In the mixed-model representation $f(x) = \mathbf{X}\beta + \mathbf{Z}\mathbf{u}$ with u_j iid $N(0, \sigma_s^2)$, $\mathbf{u}$'s distribution is simply a device that penalizes overfitting. The argument from first principles is that in evaluating a penalized spline procedure,

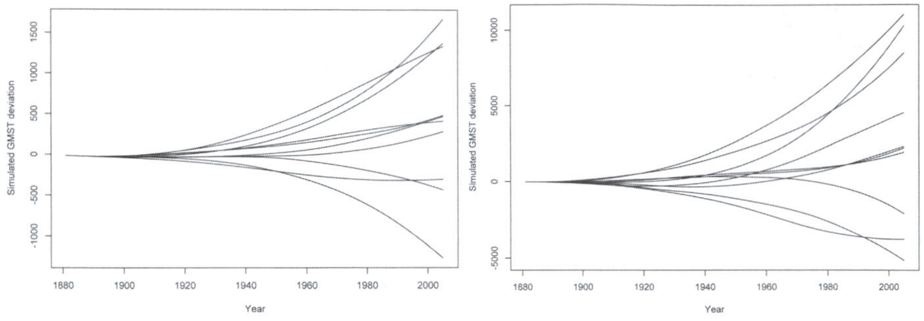

(a) $\sigma_u^2 = 947$ (b) $\sigma_u^2 = 94700$

Figure 13.2: Draws from the convenient fiction used to fit a penalized spline to the GMST data. Panel (a): $\sigma_u^2 = 947$; Panel (b): $\sigma_u^2 = 94700$.

the question is how well the procedure captures features of $f(x)$ such as sharp turns or valleys. Therefore, data should be simulated by adding residual error to specific true $f(x)$ having such interesting features. Simulating true $f(x)$ by making repeated draws from $\mathbf{X}\beta + \mathbf{Zu}$ with u_j iid $N(0, \sigma_s^2)$ omits precisely the most relevant features.

The pragmatic argument applied to penalized splines is about the effect of increasing the variance σ_s^2 of the random effect $\mathbf{u}$. In *fitting* a penalized spline, larger σ_s^2 implies a wigglier fit: Larger σ_s^2 means changes at the knots are penalized less, so larger changes will be accepted in the fit. But *draws* from $\mathbf{X}\beta + \mathbf{Zu}$ do not behave this way, as is easily verified. Consider, for example, the penalized spline model with the truncated quadratic basis fit to the GMST data in Sections 3.3.1 and 13.2.1. Figure 13.2a shows 10 draws from this spline fit's $\mathbf{X}\hat{\beta} + \mathbf{Zu}$, using the estimated fixed effects $\hat{\beta}$ and the REML estimate of the smoothing variance, $\hat{\sigma}_s^2 = 947$. None of these curves shows anything like the key feature of the spline fit to the GMST data, namely the turn down about 1940 and back up about 1970. In hundreds of draws from this model, I have seen no shapes deviating notably from the ones in Figure 13.2a. (See for yourself, in an exercise to this chapter.) Indeed, we should not expect any: The u_j fitted to the GMST data, with long runs of negative and positive u_j (Figure 13.1b), do not look anything like draws from the iid model used for a penalty and the chance of drawing such a collection of u_j from an iid normal centered at zero is vanishingly small. (Note also that most of these simulated $\mathbf{X}\beta + \mathbf{Zu}$ span a far wider vertical range than the actual GMST data, and Figure 13.2a does not even include the error term.)

Figure 13.2b shows 10 draws from a model identical to Figure 13.2a's model except that σ_s^2 is 94,700, larger by a factor of 100. Contrary to intuition, Figure 13.2b's draws are no more wiggly than Figure 13.2a's; the two plots differ only in the vertical scale. Thus, if a simulation experiment is intended to compare different penalized splines according to how well they estimate $f(x)$ at peaks, valleys, and turns, such an experiment can only be done by specifying particular $f(x)$ with peaks, valleys, or turns and simulating datasets by adding residual error to them. Drawing curves from $\mathbf{X}\beta + \mathbf{Zu}$ with large σ_s^2 and adding residual error is simply incorrect; it defeats the experiment's purpose.

The same arguments apply to the new-style spatial random effects in Sections 13.2.2 and 13.2.3. In *fitting* models using either of these random effects, increasing the variance of the random effect makes the fitted curve or surface rougher. In *generating data from* these random effects, increasing the random effect's variance does not qualitatively change the shape of the generated curve or surface, it merely changes the scale. Thus, for the same arguments given above, in a simulation experiment intended to compare the performance of different spatial models for the mineral-exploration example, the experiment should generate data from a variety of fixed true $W(\mathbf{s})$, with the shapes of those fixed true $W(\mathbf{s})$ chosen to serve the purposes of the simulation experiment.

When I present this chapter's material in seminars or in my class, the argument just given elicits the strongest response. Apparently some of us *really like* making draws from random effects, or else really dislike being told our simulations are ill-conceived. If the audience includes Gaussian process mavens, invariably one points out that if you use such-and-such obscure stochastic process you *can* get draws that have the up-then-down-then-up shape of the GMST fit. But why bother? What possible advantage could that have, if the purpose of the simulation is to answer questions about bias, MSE, or interval coverage? There is no advantage. It is better to take the fit of a Gaussian process to data, use that as the true $f(x)$, and simulate data by adding residual error to it. Such fits, unlike draws from the same Gaussian processes, often have relevant features such as peaks.

For some questions it is not so clear whether it is acceptable to simulate data by making draws from new-style random effects. In some simulations I see — though this is rarely stated explicitly — the purpose seems merely to be determining whether a new method can be computed and whether its results are not outlandish. In spatial analysis, I have seen simulations that simulated a single dataset and then computed for a week to produce the analyses under consideration. Such an exercise can be rationalized as a way to test computer code and for this purpose it seems harmless to simulate data by drawing from the random effects. However, as an evaluation of a new method such a simulation yields so little information that it hardly seems worth reporting.

In Section 12.1's simulations — the only ones in this book — I made new draws from the random effects for each simulated dataset. These simulations were intended to determine what types of data could produce effects like those seen in the DLM puzzle. I simulated data this way because I knew it would produce datasets with the specific kinds of lack of fit that I wanted to test, and because it was easy. But simulating data this way did have a disadvantage, as I noted in a comment buried in one of Chapter 12's exercises. In the example of two crossed random effects (Section 12.1.2), I am pretty sure it is impossible to draw from any kind of random effect and produce phenomena like the DLM puzzle. However, as I suggested in Exercise 8 to Chapter 12, it may be possible to produce such phenomena by simulating data in the manner of the present chapter: Specify the $l \times c$ cell means directly and make repeated draws of residual error. I left this as an exercise because I have a book deadline to make and I didn't think it was too interesting but I admit I have not been consistent on this point.

In Section 12.1, I could have specified data-generation mechanisms that did not draw from the random effects. For example, for the clustering-and-heterogeneity model of Section 12.1.1, to produce datasets with large $\hat{\sigma}_c^2$ and small $\hat{\sigma}_h^2$, I could have specified the site means directly as a function of the lattice locations, where the function had fairly broad peaks and valleys; this would give results like those I reported. Similarly, I could have produced cases with small $\hat{\sigma}_c^2$ and large $\hat{\sigma}_h^2$ by randomly permuting the peaks-and-valleys site means I just described. If the purpose of my simulation experiments had been to measure bias, then these alternative data-generating models would have been both correct and necessary, and the simulations I actually did would have been incorrect and self-defeating. But my simulation experiment had a different purpose, which could be and (mostly) was served by making draws from the random effects.

13.4 Conclusion

This chapter and the earlier material it drew on are an attempt to define two distinct types of random effects, an older type consistent with the definition given in Scheffé (1959), and a newer type that does not fit Scheffé's definition and can generally be described as a formal device to implement smoothing or shrinkage. While this distinction is not novel, it is not well-developed or consistently observed even among authors who recognize it (including me, I admit), so that sometimes statisticians make the error of taking literally the convenient fiction of a random mechanism. I've tried to develop some language and a body of examples to show how the two types of random effects differ in interpretation and how their differences affect inference and prediction, interpretation of analytical artifacts, and design of simulation experiments. I do not hope that I have completely cataloged new-style random effects or the implications of their difference from old-style random effects. Rather, this is a first step. In particular, as analyses involving new-style random effects become more common, new analytical artifacts will no doubt turn up and call for interpretations that old-style random effects cannot provide.

Exercises

Regular Exercises

1. (Section 13.3.3) For the penalized spline model fit to the global-mean surface temperature data, draw several $\mathbf{u}$ from $N(\mathbf{0}, \sigma_s^2)$ for each of several smoothing variances σ_s^2 and plot the resulting $\mathbf{X}\hat{\beta} + \mathbf{Z}\mathbf{u}$. If you ignore the vertical axis, you will find that the curves look pretty similar unless σ_s^2 is quite small, in which case the curves are close to the quadratic fixed effects.

Part IV

Beyond Linear Models: Variance Structure

Part IV Introduction

Part IV's primary question is this: What functions of the data supply information about each of the unknowns in the mixed linear model's covariance matrices **G** and **R**?

For example, in the balanced one-way random effects model with $y_{ij} = \mu + u_i + \varepsilon_{ij}, i = 1, \ldots, N, j = 1, \ldots, m$, where $u_i \sim N(0, \sigma_s^2)$ and $\varepsilon_{ij} \sim N(0, \sigma_e^2)$, how do the data provide distinct information about σ_s^2 and σ_e^2? For σ_e^2, $y_{ij} - \bar{y}_{i.} \sim N(0, \frac{m-1}{m}\sigma_e^2)$, where $\bar{y}_{i.}$ is the ith group's average, so these functions of the data provide "clean" information about σ_e^2. What about σ_s^2? It appears that no function of the data provides information about σ_s^2 that is not "contaminated" by σ_e^2. But the plot thickens: Because $\text{var}(\bar{y}_{i.}) = \sigma_s^2 + \sigma_e^2/m$, the group averages appear to provide some information about σ_e^2. So even this simplest case is not so simple.[1]

But now consider the ICAR model. If $y_i = \theta_i + \varepsilon_i$ where the θ_i follow an ICAR model with variance parameter σ_s^2 and the ε_i are iid $N(0, \sigma_e^2)$, it's not obvious the two variances are even identified. It turns out they are identified except for certain extreme spatial maps, but it is not simple to determine which functions of the data provide information about which variance.

In Part III there were obvious ways to borrow ideas from ordinary linear models like case deletion, collinearity, etc. Here we are in new territory because while ordinary linear models have a single unknown variance, a mixed linear model's two covariance matrices have a total of at least two unknowns. And indeed, our analytical workhorses for mixed linear models, the restricted likelihood and marginal posterior of ϕ, are black boxes. We can manipulate them numerically but we hardly understand them at all. When they produce mysterious, inconvenient, or plainly wrong results — and Chapter 14 gives examples of each — we can rarely explain them and can usually only guess about how to fix or avoid them. In the words of this book's epigraph, when we use mixed linear models to analyze data, we believe in things that we don't understand.

Chapters 15 to 19 are one attempt to begin prying open the black boxes of the restricted likelihood and marginal posterior. They use a re-expression of the restricted likelihood that provides insight for a variety of problems. As it turns out, this approach allows us to borrow some tools from generalized linear models and thus from ordinary linear models, although at one further remove than in Part III. This approach is quite underdeveloped; I hope Part IV will stimulate you to go forth and do better.

---

[1] It is easy to show that no non-trivial *linear* function of the data has distribution depending on σ_s^2 but not σ_e^2. I believe but have not proved that no function of the data has distribution depending on σ_s^2 but not σ_e^2, though there may be a function of the data with distribution depending on σ_e^2/σ_s^2 but not on either variance individually. An exercise to Chapter 14 asks you to examine these questions.

Chapter 14

Mysterious, Inconvenient, or Wrong Results from Real Datasets

This chapter describes five examples in which my colleagues and I got mysterious, inconvenient, or plainly wrong results using mixed linear models to analyze real datasets. All these examples involve the data's information about ϕ, the unknowns in the mixed linear model's **R** or **G**, and thus lead to examination of the restricted likelihood and its close relative, the marginal posterior of ϕ.

14.1 Periodontal Data and the ICAR Model

Example 14: Periodontal measurements. Reich & Hodges (2008a) analyzed one dental patient's clincal attachment loss (CAL) measurements, shown in Figure 14.1a, using the ICAR model described in Chapter 5. We began with the simple no-covariates form of the ICAR in equation (5.14) with sites joined as neighbor pairs as shown in Figure 5.1. The analysis was Bayesian. The two island means — one each for the maxilla and mandible, i.e., the upper and lower jaws — had flat (improper) Gaussian priors implicit in the ICAR model and σ_e^2 and σ_s^2, the error and smoothing variance respectively, had inverse gamma priors, IG(0.01,0.01). The posterior medians of σ_e^2 and σ_s^2 were 1.25 and 0.25 respectively, and their ratio σ_e^2/σ_s^2, which determines the degree of smoothness (a higher ratio implies a smoother fit), had posterior median 4.0. Figure 14.1a shows the posterior means of the site-specific true CAL, δ in the notation of equation (5.14). These posterior means are highly smoothed.[1]

A diagnostic statistic suggested that the direct sites — the sites in the middle of each tooth, either on the lingual (tongue) or buccal (cheek) side of the tooth — tend to have lower attachment loss than their neighboring interproximal (distal or mesial) sites, especially for teeth farther back in the mouth. This tendency can be seen fairly clearly in Figure 14.1a. For example, in the bottom echelon of the plot (the mandibular buccal measurements), for the 6 teeth on the left of the plot the direct site is prominently lower than its neighbors. In the mixed-linear-model expression of the

[1] My colleague Bryan Michalowicz, a periodontologist, said of this fit, "I spend all my time worrying about the 6 and 7 mm sites and you're telling me I should ignore them and pay attention to this smooth line instead." Um … no. This was our first fit to periodontal data; Reich & Hodges (2008b) gives a much more sensible analysis.

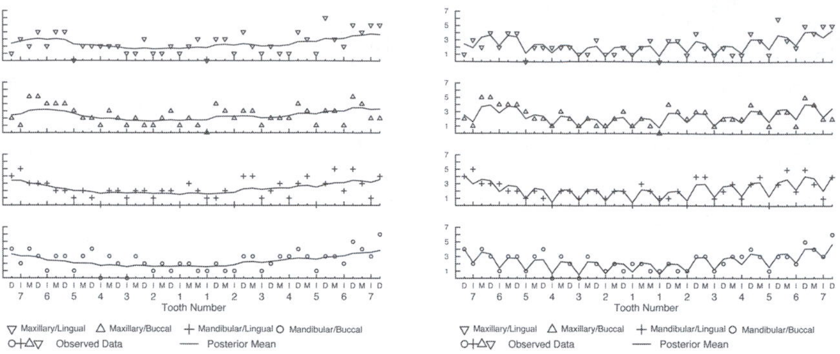

(a) Simple ICAR model (b) Modified ICAR

Figure 14.1: One person's attachment loss measurements with two fits. Vertical axes are clinical attachment loss (CAL; mm). Horizontal axes are labeled by tooth number (7 is the second molar, 1 the central incisor) and site ("D," "I," and "M" refer to distal, direct, and mesial sites, respectively). "Maxillary" and "mandibular" refer to upper and lower jaws while "lingual" and "buccal" refer to tongue and cheek side of the teeth, respectively. Panel (a): Simple ICAR model. Panel (b): Modified ICAR with maxillary and mandibular direct vs. interproximal contrasts treated as fixed effects.

ICAR model for these data, two columns in the canonical random-effects design matrix **Z**, one each for the upper and lower jaws, are precisely the difference between the average of the direct sites and the average of the interproximal sites; these canonical columns were shown in Chapter 5's Figure 5.4b.

The contrasts in the data describing the difference between the average of the direct sites and the average of the interproximal sites would seem to provide information about the smoothing variance σ_s^2 because these contrasts are columns of the random effect's design matrix. Thus, we would expect that if we moved these two columns from the random effects design matrix **Z** to the fixed effects design matrix **X** and gave their coefficients flat (improper) priors, making them fixed instead of random effects, then less variation in the data would be attributed to the random effects and the estimate of σ_s^2 would become smaller. However, when we made this change in the model, we got a surprise. Figure 14.1b shows the fitted values in the new model, which have a pronounced average difference between the direct and interproximal sites. The posterior median of σ_s^2, instead of becoming smaller, increased from 0.25 to 0.41; the posterior median of the *error* variance σ_e^2 decreased from 1.25 to 0.63, and the posterior median of the ratio σ_e^2/σ_s^2 declined from 4.0 to 1.6, indicating *less* smoothing of the random effect in the new fit, not more. (The posterior median of DF in the fit increased from 23.5 to 45.6.)

Chapter 16 shows that this puzzle is quite straightforwardly explained. Some — sometimes most — contrasts in the data that lie in the column space of the random-effects design matrix **Z** provide information mostly about the *error* variance σ_e^2, not

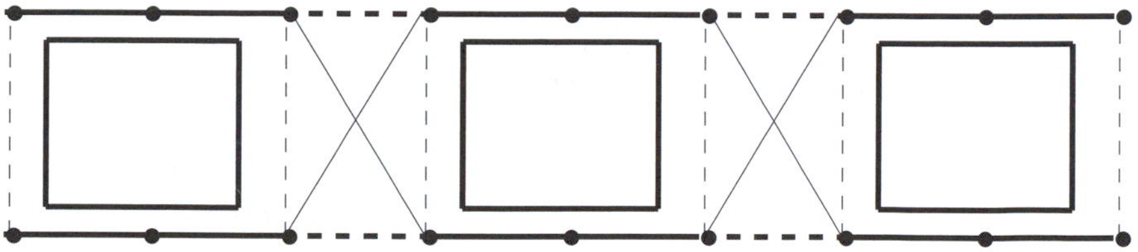

Figure 14.2: Four types of neighbor pairs for periodontal data. Rectangles are teeth, circles are sites where AL is measured, and different line types are different types of neighbor pairs (from Reich et al. 2007).

about the smoothing variance σ_s^2. This may appear paradoxical but turns out to have a very simple explanation.

14.2 Periodontal Data and the ICAR with Two Classes of Neighbor Pairs

Example 14 again: Periodontal measurements. Other analyses of periodontal data in Brian Reich's doctoral dissertation used the fact that pairings of neighboring measurement sites are of four distinct types, as shown in Figure 14.2. The simple ICAR model in the preceding section assumes attachment loss is equally similar within all neighbor pairs but in fact different types of neighbor pairs have different degrees of within-pair similarity. For example, Reich et al. (2007, Table 1) show Pearson's correlations of attachment loss measurements of 0.47 for Type I pairs and 0.60 for Type IV pairs, based on over 5,200 and 2,600 pairs respectively. Therefore, we considered (Reich et al. 2004, Reich et al. 2007) ICAR-like models in which the four types of neighbor pairs were partitioned into two classes, with different degrees of similarity permitted in the two classes. (This model was introduced by Besag & Higdon 1999.) For the present purpose, consider these two classifications: Classification A, side pairs (Types I, II) versus interproximal pairs (Types III, IV); and Classification B, direct sites (Type I) vs. interproximal sites (Types II, III, IV).

The two-neighbor-relation CAR (2NRCAR) model (Reich et al. 2007) can be written as follows. Let n be the number of measurement sites; for a single patient with no missing teeth, $n = 168$. In the simplest 2NRCAR model, the n-vector of attachment loss measurements, $\mathbf{y}$, is modeled as $\mathbf{y} = \delta + \varepsilon$, where $\varepsilon \sim N_n(\mathbf{0}, \sigma_e^2 \mathbf{I}_n)$ and δ has an improper normal density specified in terms of its precision matrix:

$$f(\delta|\sigma_{s1}^2, \sigma_{s2}^2) \propto \exp\left[-\frac{1}{2}\delta' \left(\mathbf{Q}_1/\sigma_{s1}^2 + \mathbf{Q}_2/\sigma_{s2}^2\right)\delta\right], \tag{14.1}$$

where σ_{sk}^2 controls the similarity of class k neighbor pairs, $k = 1, 2$, with larger values allowing less similarity. The $n \times n$ matrices $\mathbf{Q}_k = (q_{ij,k})$ encode the class-k neighbor pairs of measurement sites with $q_{ii,k}$ being the number of site i's class-k neighbors and $q_{ij,k} = -1$ if sites i and j are class-k neighbors and 0 otherwise.

This model can be rewritten as a mixed linear model using the lemma (Newcomb 1961) that there is a non-singular matrix $\mathbf{B}$ such that $\mathbf{Q}_k = \mathbf{B}'\mathbf{D}_k\mathbf{B}$, where $\mathbf{D}_k$ is diagonal with non-negative diagonal elements. $\mathbf{D}_k$ has I_k zero diagonal elements, where I_k is the number of $\mathbf{Q}_k$'s zero eigenvalues and the number of islands in the spatial map specified by $\mathbf{Q}_k$. $I_k \geq I$, where I is the number of zero eigenvalues of $\mathbf{Q}_1 + \mathbf{Q}_2$ and also the number of zero diagonal entries in $\mathbf{D}_1 + \mathbf{D}_2$. $\mathbf{B}$ is non-singular but is orthogonal if and only if $\mathbf{Q}_1\mathbf{Q}_2$ is symmetric (Graybill 1983, Theorem 12.2.12).

Define $\mathbf{v} = \mathbf{B}\delta = (\mathbf{u}', \beta')'$, where $\mathbf{u}$ and β have dimensions $(n-I) \times 1$ and $I \times 1$ respectively, and write $\mathbf{B}^{-1} = (\mathbf{Z}|\mathbf{X})$, where $\mathbf{Z}$ is $n \times (n-I)$ and $\mathbf{X}$ is $n \times I$. Because $\delta = \mathbf{B}^{-1}\mathbf{v}$, the model for $\mathbf{y}$ becomes $\mathbf{y} = \delta + \varepsilon = \mathbf{X}\beta + \mathbf{Z}\mathbf{u} + \varepsilon$ and the quadratic form in the exponent of equation (14.1) becomes

$$\begin{aligned} \delta' \left(\mathbf{Q}_1/\sigma_{s1}^2 + \mathbf{Q}_2/\sigma_{s2}^2\right) \delta &= \mathbf{v}' \left(\mathbf{D}_1/\sigma_{s1}^2 + \mathbf{D}_2/\sigma_{s2}^2\right) \mathbf{v} \\ &= \mathbf{u}' \left(\mathbf{D}_{1+}/\sigma_{s1}^2 + \mathbf{D}_{2+}/\sigma_{s2}^2\right) \mathbf{u} \qquad (14.2) \end{aligned}$$

where $\mathbf{D}_{k+}$ is the upper-left $(n-I) \times (n-I)$ submatrix of $\mathbf{D}_k$. The 2NRCAR model is now a mixed linear model with random-effects covariance matrix $\mathbf{G} = (\mathbf{D}_{1+}/\sigma_{s1}^2 + \mathbf{D}_{2+}/\sigma_{s2}^2)^{-1}$. I have assumed and will continue to assume that the proportionality constant that's missing in equation (14.1) is chosen by the same convention I have used for the usual ICAR model since Section 5.2.2. As noted in that section, other choices also give legal probability distributions for $\mathbf{u}$.

This model seemed straightforward enough. Because we thought it would simplify the MCMC, Prof. Reich put a Gamma(0.01,0.01) prior (mean 1, variance 100) on $1/\sigma_e^2$, derived the joint marginal posterior of the log variance ratios $z_k = \log(\sigma_e^2/\sigma_{sk}^2)$, which control the shrinkage of class k neighbor pairs, and ran a Metropolis-Hastings chain for (z_1, z_2). (Deriving this density is a special case of Chapter 1's Regular Exercise 6 and is an exercise to the present chapter.) However, despite extensive tinkering with the Metropolis-Hastings candidates, Prof. Reich found extremely high lagged autocorrelations in the MCMC series. I told him there must be an error in his MCMC code, but he proved me wrong with contour plots of (z_1, z_2)'s log marginal posterior.

Figure 14.3 shows contour plots of (z_1, z_2)'s log marginal posterior for one person's attachment loss data for classifications A and B, with independent Uniform(-15,15) priors on the z_j (the plots truncate the axes for clarity). For classification A, the posterior is quite flat along horizontal and vertical "legs" extending out from the modest global maximum at about $(z_1, z_2) = (2, -2)$. (Note the grey scale legend at the right of each plot.) For classification B, the contour plot does not show a distinct global maximum though it also has two "legs" of nearly equal height. Our naïve (in retrospect) MCMC had high lagged autocorrelations because it moved into one of the "legs" and meandered there for many iterations before moving into the other "leg" and meandering *there* for many iterations.

Increasing the sample size does not make this posterior better-behaved. Figure 14.4 shows contour plots for analyses including three persons' data, assuming

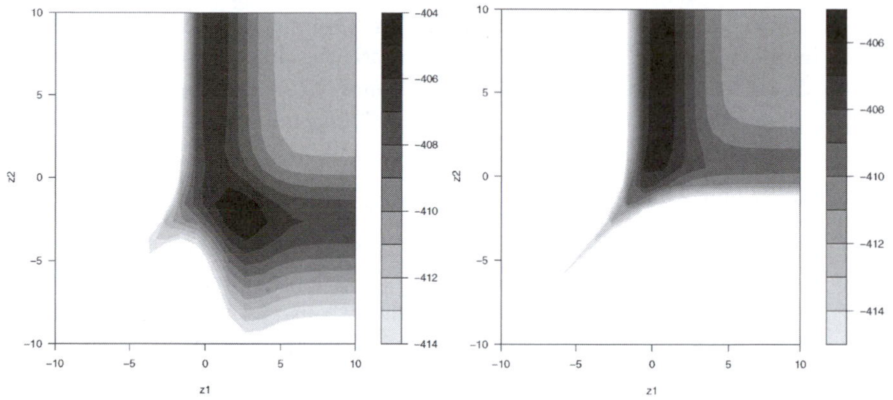

(a) Classification A (b) Classification B

Figure 14.3: Contour plots of the log marginal posterior of (z_1, z_2) for one person's attachment loss measurements. Panel (a): Classification A. Panel (b): Classification B. (From Reich et al. 2007).

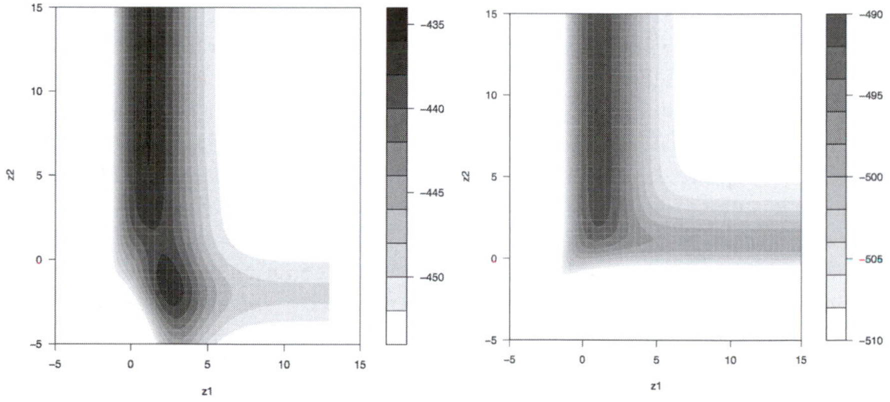

(a) Classification A (b) Classification B

Figure 14.4: Contour plots of the log marginal posterior of (z_1, z_2) for three persons' attachment loss measurements. Panel (a): Classification A. Panel (b): Classification B. (From Reich et al. 2004).

$(\sigma_e^2, \sigma_{s1}^2, \sigma_{s2}^2)$ is the same for all three persons. For classification A, the two "legs" now have rather more distinct heights but the posterior has two modes; the three individuals, analyzed separately, had posterior medians for z_2 of -1.75, 0.08, and 6.09. The assumption of common $(\sigma_e^2, \sigma_{s1}^2, \sigma_{s2}^2)$ apparently does not fit these three persons, but mere lack of fit is not enough to produce a bi-modal posterior in most common models. For classification B, the posterior still has no distinct peak.

Why do the posteriors have this strange shape, and can shapes like this be predicted? Chapter 17 shows the odd shapes in Figure 14.3 can be explained and predicted fairly well, at least for models with 3 variances like this one. Reich et al. (2004) shows even weirder posteriors arising from 2NRCAR models, one of which is identical to a second-order Markov chain model for a time series.

14.3 Two Very Different Smooths of the Same Data

Example 5 revisited: Global mean surface temperature. Earlier chapters showed fits of a penalized spline to the global mean surface temperature data. Using a truncated-quadratic basis with 30 knots and estimating the two variances by maximizing the restricted likelihood gave the entirely reasonable fit shown in Figures 3.5 and 3.7 and reproduced in Figure 14.5 (the dashed line).

We can also smooth this series using the ICAR model with adjacent years as the single type of neighbor pair, so the first and last years have one neighbor and other years have two. Fitting this model to the GMST data by maximizing the restricted likelihood gives a fit differing greatly from the penalized-spline fit. Figure 14.5a shows this ICAR fit (solid line), which has 26.5 DF compared to 6.7 DF in the penalized spline fit (dashed line). I thought this might be an instance in which, as Ruppert et al. (2003, p. 177) warned, maximizing the restricted likelihood does not give a reasonable degree of smoothing. Thus, I re-fit the ICAR model and increased the ratio σ_e^2/σ_s^2 from the value that maximized the restricted likelihood until the ICAR fit also had 6.7 DF. Figure 14.5b shows the 6.7 DF ICAR fit (solid line) as well as the 6.7 DF penalized spline fit (dashed line). Although the two fits have identical DF (i.e., complexity), the ICAR and penalized spline allocate those DF differently. The ICAR cannot capture the decline in GMST between about 1940 and 1970 but at the same time it is more rough than the penalized spline, e.g., the years 1890–1910.

Why do these smoothers give such different fits? Chapter 16 shows that although these are usually understood as local smoothers, when the variances σ_e^2 and σ_s^2 are estimated as part of the fit, both fits are in fact determined by *global* features of the data and the ICAR and penalized spline differ in the global features that determine their respective fits. Differences in the fixed- and random-effects design matrices implicit in their formulations turn out to be the key here. For this dataset, these differences force the ICAR either to undersmooth so it can capture the downward-then-upward trend or to miss that trend while still capturing noise.

14.4 Misleading Zero Variance Estimates

Example 13 revisited: Measuring epidermal nerve density. The Kennedy lab's new blister method for measuring epidermal nerve density was used in a second study in normal children, to compare it to the older biopsy method and to collect normative data in children. (This study is as yet unpublished. The data analyzed below are available on this book's web site.) In this study, the investigators had 19 subjects and again measured two locations (calf and thigh) with two blisters at each location on each subject. A blister was measured using two non-overlapping images, each

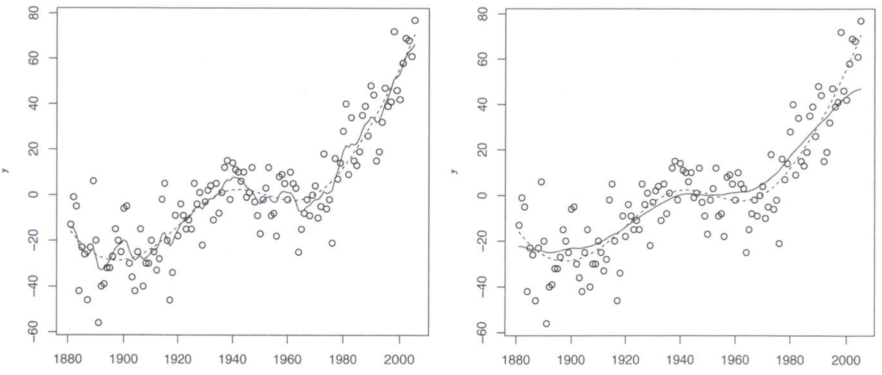
(a) Maximizing the restricted likelihood
(b) Forcing DF = 6.7

Figure 14.5: Global mean surface temperature data with ICAR and penalized spline fits. Panel (a): ICAR fit (solid line) with 26.5 DF; penalized-spline fit with 6.7 DF (dashed line). Panel (b): ICAR fit forced to have 6.7 DF (solid line); penalized-spline fit (dashed line) with 6.7 DF.

Table 14.1: Epidermal Nerve Density: Estimates of Variance Components for Blister Measurements from Maximizing the Restricted Likelihood, with Confidence Intervals from the Satterthwaite Approximation as Described in Section 1.2.4.3

Variance Component	Variance Estimate	Confidence Interval	
Subject	18,031	8,473	61,169
Subject-by-Location	9,561	4,684	29,197
Blister within Subject/Location	0	0	0
Residual	6,696	5,181	8,992

covering a large portion of the blister. The outcome measure was axons/mm^2; the investigators had measurements on 136 images (some blisters yielded 0, 1, or 3 images). Besides the fixed effect of location, the analysis included these components of variation: between subjects in overall level (subject main effect); between subjects in the difference between locations (subject-by-location interaction); between blisters at a given location (blisters within subject-by-location); and between images within blister (residual). The investigators were interested in these components of variation because the skin sample in a blister covers a larger area than the skin sample in a biopsy and they expected this would reduce some components in blisters compared to biopsies.

When I analyzed these data using the model just described, the JMP package's conventional analysis gave estimates and confidence intervals shown in Table 14.1. (JMP uses the same statistical methods as SAS's MIXED procedure but different numerical routines.)

The between-blister variance was of particular interest; it was estimated to be zero and JMP reported a 95% confidence interval containing only zero. Obviously this confidence interval is wrong, though in fairness to JMP's authors this is only slightly worse than the missing values that other software (e.g., SAS) would give for the same confidence interval. The zero estimate is also obviously wrong, because blisters do vary. When I show this to people, the most common first reaction, sometimes from very smart people, is "maybe the variance really is zero." However, the restricted likelihood itself contradicts this optimistic view. If you fix the subject, subject-by-location, and residual variances at the values that maximize the restricted likelihood and then increase the between-blister variance from zero, it reaches 1,550 before the log restricted likelihood declines by 2 log units. That is, the data are quite consistent with non-trivial values of this variance; the maximizing value just happens to be zero. If instead the other three variances were profiled out (replaced by their maximizing values given the between-blister variance), an even larger between-blister variance would be needed to reduce the profiled restricted likelihood by 2 log units from its maximum. Thus there is no reason to conclude that the between-blister variance is zero, but I had to change software (to SAS) and do rather more work than usual to establish that.

In this problem, the restricted likelihood is quite flat as a function of the between-blister variance for values of that variance that are, loosely speaking, rather less than the estimated residual variance. A between-blister variance of zero maximizes the restricted likelihood, but not by much. Software that handles this problem responsibly would give some warning to this effect, perhaps even a one-sided confidence interval for this variance, but none does and a naïve JMP user could draw a seriously mistaken conclusion from Table 14.1. Bayesians should not gloat: I learned that this restricted likelihood was flat by computing posterior distributions for this variance for many different priors and happening to notice that the left tail of each posterior hardly differed from the left tail of its respective prior. A Bayesian analysis here is very sensitive to its prior but there is currently no way to know this without trying several priors and specifically looking for this type of evidence of a flat tail. Thus, both approaches to this analysis are subject to error or misinterpretation unless the analyst makes an unusual effort, which I happened to do in this problem only because the standard method gave a silly answer to a question that was important to my collaborators.

Chapter 18 explores zero variance estimates in the balanced one-way random effects model and in a balanced version of the model used above to analyze the blister data. I have only seen zero estimates for the residual variance in fits of dynamic linear models with multiple components (e.g., Michael Lavine's dataset discussed in Chapters 6, 9, and 12 and in Chapter 17). Apart from that example, in thousands of analyses of a great variety of data, I have never seen a zero estimate for the residual variance, while I routinely see zero variance estimates for other variance components. For the balanced one-way random-effect model, flat left tails in the restricted likelihood or posterior occur when the experimental design or data provide poor resolution. This occurs when the error variance is large or clusters have few replications.

The situation is more complicated for designs with more components of variance although broadly speaking it seems that the problem is still poor resolution. Chapter 18 briefly considers some approaches to diagnostics for a flat left tail.

14.5 Multiple Maxima in Posteriors and Restricted Likelihoods

Example 9 revisited: HMO premium data. Section 5.2 of Hodges (1998) reported an analysis of the HMO premium data using two state-level predictors suggested by that paper's diagnostics, state average expenses per hospital admission (centered and scaled) and an indicator for the New England states. As in the analyses described in Chapter 8, I used a flat prior on the fixed effects, a flat prior on $1/\sigma_e^2$ (which I would not do now), and "a low information gamma prior for $[1/\sigma_s^2]$, with mean 11 and variance 110." I described my 1000-iteration Gibbs sampler output[2] by saying, "The Gibbs sampler was unstable through about 250 iterations; from iteration 251 to 1000 the sampler provided satisfactory convergence diagnostics, and these [750] iterations were used for the computations that follow." "Unstable" refers to the trace plot for σ_s^2, which was stable for the first 250 iterations then dropped to a distinctly lower level, where it was stable for the remaining 750 iterations. Such behavior means the posterior is bimodal but when I did this analysis I had never entertained the possibility that this posterior might have two modes.

Wakefield's discussion of this paper (Wakefield 1998) touched on several topics before re-running the above analysis with 10,000 Gibbs iterations. Figure 14.6, Wakefield's scatter plot of the Gibbs draws on the two variances, clearly shows the posterior's two modes, one with tiny σ_s^2 (between-state variance) and one with rather large σ_s^2; the former has more mass than the latter. Wakefield commented that the bimodal posterior showed "there are two competing explanations for the observed variability in the data" (p. 525).

This was embarrassing, as you can imagine, but I was naïve and deserved it, and Wakefield exposed my error as gently as possible. My next PhD student, Jiannong Liu, studied bimodality in mixed linear models; Chapter 19 summarizes his results (Liu & Hodges 2003), among others. I hope it does not seem churlish when Chapter 19 shows that the mode with more mass in Figure 14.6 is not present in the restricted likelihood but arises entirely from the prior on $1/\sigma_s^2$. Thus, it is not quite true that "there are two competing explanations for the observed variability in the data"; rather, the data's information about σ_s^2 is weak enough to be punctured easily by a thoughtlessly chosen prior.

Bimodal posteriors arising from a strong prior are not mysterious but they can be inconvenient and given that many priors are chosen by convention without careful substantive consideration, in a given problem one might decide that a bimodal posterior is wrong. Anti-Bayesians should not gloat: The restricted likelihood itself can have more than one maximum and as we have seen, even when it has a single maximum that is not necessarily a sensible estimate. The likelihood appears to be even

[2] I now know that such a short run is malpractice, but I didn't know it then and oddly, none of the referees pointed it out although one of the discussants did.

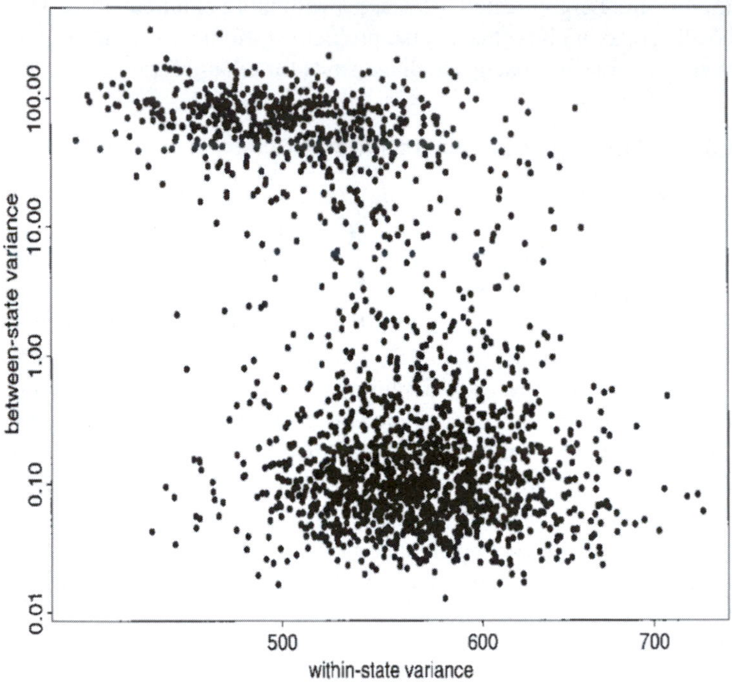

Figure 14.6: Wakefield's (1998) Gibbs draws on the two variances for the model with two state-level predictors. The vertical and horizontal scales are logarithmic. (This figure is reprinted from Wakefield 1998.)

more prone to multiple maxima than the restricted likelihood; Chapter 12 showed several examples.

14.6 Overview of the Remaining Chapters

The five mysterious, inconvenient, or wrong results described above can be explained, at least to some extent. My current foothold on all these problems is a particular re-expression of the restricted likelihood and thus the marginal posterior of ϕ (for a flat prior on the fixed effects β) in which all the square matrices are diagonalized so matrix expressions become scalar expressions. We saw examples of this re-expression for the relatively simple models considered in Section 12.1. A large sub-class of mixed linear models can be re-expressed in this manner but by no means all of them. This is therefore a place to start, not the last word. Some have argued to me that this foothold has little value because it is not as general as one might hope. I would say, to the contrary, that we will make progress toward a better theory by working problems with the modest tools we have, not by waiting for someone to deliver a general theory. If you have a more general place to start, I would be delighted to see it. As far as I can tell, however, nobody has yet developed such a general theory, so I present this modest beginning for what it may be worth.

This book's remaining chapters develop and use the re-expressed restricted likelihood, proceeding as follows. Chapter 15 introduces the re-expression for a small but interesting sub-class of mixed linear models called two-variance models and applies it to penalized splines, ICAR models, and a dynamic linear model with one quasi-cyclic component. Chapter 15 also introduces some analytic tools derived using the re-expression. Chapter 16 uses this re-expression to examine which functions of the data provide information about the smoothing and error variances in a two-variance model and applies the results to penalized splines and ICAR models to explain the first two mysteries discussed above. Chapter 17 extends the re-expression to models that are more general than two-variance models. I cannot explicitly characterize the class of models for which the re-expression can be done, but this chapter proves that it cannot be done for all mixed linear models though useful extensions might be possible, as suggested at the end of Chapter 12. Chapter 17 chapter gives two expedients that can be used for some models that cannot be re-expressed in the diagonalized form, one of which allows us to explain the third mystery, for 2NRCAR models. Chapter 18 uses the re-expressed restricted likelihood for an as-yet too brief consideration of zero variance estimates, while Chapter 19 uses it to examine multiple local maxima in the restricted likelihood and marginal posterior.

Two other problems would fit here in Part IV but so far I do not even have a foothold on explanations for them. These problems were mentioned in connection with Chapter 1's Example 2 (vocal folds) and Example 3 (glomerular filtration rate). Example 2 exhibited unpredictable convergence of numerical algorithms, which is inconvenient and currently mysterious. Example 3 exhibited unpredictable convergence and also correlation estimates of ± 1, which are inconvenient and wrong. Estimates of ± 1 are somewhat like zero variance estimates but as yet I have no insight into why they occur. I guess they are more complicated than zero variance estimates, which (Chapter 18 argues) generally arise from poor resolution in the data or design, but that is only a guess.

Regarding unpredictable convergence, one reader noted that the solution may simply be a smarter optimization algorithm or running the same optimization routine many times with different starting values. The latter is certainly the sensible thing to do when you encounter this problem. However, if we *really* understood our mixed-linear-model tools, it would be possible to determine which aspects of the data and design cause the restricted likelihood or posterior to be ill-behaved and perhaps to avoid the difficulty by using a different design or by selecting a numerical algorithm more suited to the problem.

Bad behavior of this sort is fairly well understood for generalized linear models. For example, we know we'll have convergence problems with a logistic regression if a categorical predictor has a category in which all the outcomes are either "No" or "Yes." Wedderburn (1976, p. 31)[3] gave a catalog of generalized linear models organized by error distribution and link function, identifying which models give

[3]Thanks to Lisa Henn for pointing out this under-appreciated gem.

maximum-likelihood estimates that are finite or unique or in the parameter space's interior. Ideally, a similar catalog could be constructed for mixed linear models or at least a large subset of them. This is too much to hope for now, of course, but the following chapters may suggest a place to start.

Exercises

Regular Exercises

1. (Section 14.2). Derive the joint marginal posterior of (z_1, z_2) in the 2NRCAR model, for the assumptions used in Section 14.2. Use a gamma prior for σ_e^2 and an independent and non-specific prior $\pi(z_1, z_2)$ for (z_1, z_2).

Open Questions

1. (Introduction to Part IV) Prove that for the balanced one-way random-effects model, no function of the data **y** has a distribution that depends on the between-group variance σ_s^2 but not the within-group variance σ_s^2. This is easy to show for linear functions of **y** and for a second-order approximation to an arbitrary continuous function, but I haven't been able to show it for more general functions. Are there functions of the data that depend on σ_e^2/σ_s^2 but not on either variance individually?
2. Re-analyze the available datasets from this chapter to see whether you can figure out why the mysterious, inconvenient, or wrong results happened, before you read the following chapters and see what my colleagues and I did.

Chapter 15

Re-expressing the Restricted Likelihood: Two-Variance Models

This chapter and the next consider two-variance models, which can be defined as models in mixed-linear-model form (1.1) with $\mathbf{R} = \sigma_e^2 \Sigma_e$ and $\mathbf{G} = \sigma_s^2 \Sigma_s$ for known Σ_e and Σ_s and unknown σ_e^2 and σ_s^2. I assume Σ_e and Σ_s are positive definite, though this may not be necessary. For every model in this class, the restricted likelihood can be re-expressed in a simple form that this chapter and the next develop and explore. Chapter 17 extends this re-expression to a larger class of models. The class of two-variance models may seem small but it includes examples of most models considered in Part II of this book (not additive models) and its simplicity permits considerable insight.

In this chapter, the re-expressed restricted likelihood is derived for two-variance models, its key features are identified, and it is shown to be identical to a particular generalized linear model. The re-expression involves a rather obscure re-parameterization so it is then applied to several examples, for which the re-parameterization is more or less interpretable. The chapter concludes by introducing some tools based on the re-expression.

15.1 The Re-expression

As far as I know, this re-expression was first presented by Reich & Hodges (2008a) and independently presented for a special case by Welham & Thompson (2009), though it generalizes a representation used in analysis of variance (McCullagh & Nelder 1989, Section 8.3.5). Reich & Hodges (2008a) expressed two-variance models in a hierarchical form using precision matrices. Their re-expression is the same as the one given below, which begins with two-variance models expressed in mixed-linear-model form. Besides being consistent with the rest of this book, the derivation below avoids some untidy aspects of the derivation in Reich & Hodges (2008a) and thus supersedes it.

First, in the definition of two-variance models given above, without loss of generality set $\Sigma_e = \mathbf{I}_n$ and $\Sigma_s = \mathbf{I}_q$. If Σ_e is not the identity, the data $\mathbf{y}$ can be transformed to $\Sigma_e^{-0.5}\mathbf{y}$ so the transformed problem has $\Sigma_e = \mathbf{I}_n$. Similarly, if Σ_s is not the identity, the random effects can be re-parameterized as $\mathbf{Zu} = \mathbf{Z}^*\mathbf{u}^*$ with $\mathbf{u}^* = \Sigma_s^{-0.5}\mathbf{u}$, which has covariance proportional to the identity.

The derivation proceeds by re-parameterizing the mixed linear model's $(\beta, \mathbf{u})$ to a canonical parameterization that immediately gives the desired simple form for the restricted likelihood. To begin, define $s_X = \text{rank}(\mathbf{X}) \in \{1, 2, \ldots, p\}$ and define $s_Z = \text{rank}(\mathbf{X}|\mathbf{Z}) - s_X \in \{1, 2, \ldots, q\}$; I assume $s_X, s_Z > 0$ and necessarily $s_X + s_Z \leq p + q$. These definitions are needed because $(\mathbf{X}|\mathbf{Z})$ commonly has rank less than $p+q$. Now let Γ_X be an $n \times s_X$ matrix with columns forming an orthonomal basis for the column space of $\mathbf{X}$, e.g., Γ_X is the matrix $\mathbf{U}$ in the singular value decomposition $\mathbf{X} = \mathbf{U}\Delta\mathbf{V}'$, where Δ has dimension s_X. Define the matrix Γ_Z to be $n \times s_Z$ such that $\Gamma_Z'\Gamma_X = \mathbf{0}$ and the columns of $(\Gamma_X|\Gamma_Z)$ are an orthonormal basis for the column space of $(\mathbf{X}|\mathbf{Z})$. (Construction of such a Γ_Z is straightforward and given as an exercise.) Finally, define Γ_c to be $n \times (n - s_X - s_Z)$ so that $\Gamma_c'\Gamma_X = \mathbf{0}$, $\Gamma_c'\Gamma_Z = \mathbf{0}$, and $(\Gamma_X|\Gamma_Z|\Gamma_c)$ is an orthonormal basis for real n-space.

Finally, define the matrix $\mathbf{M}$ such that $(\mathbf{X}|\mathbf{Z}) = (\Gamma_X|\Gamma_Z)\mathbf{M}$ and partition $\mathbf{M}$ conformably as

$$\mathbf{M} = \begin{bmatrix} \mathbf{M}_{XX} & \mathbf{M}_{XZ} \\ \mathbf{0} & \mathbf{M}_{ZZ} \end{bmatrix}, \tag{15.1}$$

where $\mathbf{M}_{XX}$ is $s_X \times p$, $\mathbf{M}_{XZ}$ is $s_X \times q$, and $\mathbf{M}_{ZZ}$ is $s_Z \times q$ so $\mathbf{X} = \Gamma_X\mathbf{M}_{XX}$ and $\mathbf{Z} = \Gamma_X\mathbf{M}_{XZ} + \Gamma_Z\mathbf{M}_{ZZ}$. Let $\mathbf{M}_{ZZ}$ have singular value decomposition $\mathbf{M}_{ZZ} = \mathbf{P}\mathbf{A}^{0.5}\mathbf{L}'$, where $\mathbf{P}$ is $s_Z \times s_Z$ and orthogonal, $\mathbf{A}$ is $s_Z \times s_Z$ and diagonal, and $\mathbf{L}'$ is $s_Z \times q$ with orthonormal rows, so $\mathbf{M}_{ZZ}\mathbf{M}_{ZZ}' = \mathbf{PAP}'$. Now the mixed linear model can be re-parameterized as

$$\begin{aligned} \mathbf{y} &= (\mathbf{X}|\mathbf{Z}) \begin{bmatrix} \beta \\ \mathbf{u} \end{bmatrix} + \varepsilon \\ &= (\Gamma_X|\Gamma_Z)\mathbf{M} \begin{bmatrix} \beta \\ \mathbf{u} \end{bmatrix} + \varepsilon \\ &= (\Gamma_X|\Gamma_Z\mathbf{P}) \begin{bmatrix} \beta^* \\ \mathbf{v} \end{bmatrix} + \varepsilon \end{aligned} \tag{15.2}$$

where $\beta^* = \mathbf{M}_{XX}\beta + \mathbf{M}_{XZ}\mathbf{u}$ and $\mathbf{v} = \mathbf{A}^{0.5}\mathbf{L}'\mathbf{u}$. Note that β^* is indeed a fixed effect: It has a normal prior with mean zero and infinite variance. This can be shown by giving β a proper normal distribution with covariance $\lambda\Sigma$ for some fixed positive definite Σ and letting λ go to infinity.[1] Also — and of greater interest — $\mathbf{v}$ is s_Z-variate normal with mean zero and covariance $\sigma_s^2\mathbf{A}$, where $\mathbf{A}$ is diagonal with diagonal elements $a_j > 0, j = 1, \ldots, s_Z$. The canonical random-effects design matrix $\Gamma_Z\mathbf{P}$ and the a_j are the keys to understanding the restricted likelihood.

The restricted likelihood can be obtained from this re-parameterized model (15.2) in at least two ways. First, I'll use the definition of the restricted likelihood as the likelihood of a transformed observation $\mathbf{w} = \mathbf{K}'\mathbf{y}$, for $\mathbf{K}$ a known full-rank $n \times (n - s_X)$ matrix chosen so that $\mathbf{w}$'s distribution is independent of β, as in Section 1.2.3.2's equation (1.13). After that, it will be useful also to derive the restricted likelihood as a marginal likelihood, as in equation (1.15).

---

[1] Alternatively, $\mathbf{M}_{XX}\beta$ has prior precision zero and since β^*'s precision can be no larger than $\mathbf{M}_{XX}\beta$'s, β^* must also have precision zero.

In the first definition of the restricted likelihood, define $\mathbf{K} = (\Gamma_Z \mathbf{P} | \Gamma_c)$. Pre-multiply equation (15.2) by $\mathbf{K}'$ to give

$$\mathbf{w} = \mathbf{K}'\mathbf{y} = \left[\begin{array}{c} \mathbf{v} \\ \mathbf{0}_{(n-s_X-s_Z)\times 1} \end{array} \right] + \xi, \tag{15.3}$$

where $\xi \sim N(\mathbf{0}, \sigma_e^2 \mathbf{I}_{n-s_X})$ and, as before, $\mathbf{v} \sim N_{s_Z}(\mathbf{0}, \sigma_s^2 \mathbf{A})$. The log restricted likelihood for (σ_s^2, σ_e^2) is the log likelihood arising from (15.3):

$$\begin{aligned} \log RL(\sigma_s^2, \sigma_e^2 | \mathbf{y}) &= B - \frac{n - s_X - s_Z}{2} \log(\sigma_e^2) - \frac{1}{2\sigma_e^2} \mathbf{y}'\Gamma_c \Gamma_c' \mathbf{y} \\ & \quad - \frac{1}{2} \sum_{j=1}^{s_Z} \left[\log(\sigma_s^2 a_j + \sigma_e^2) + \frac{\hat{v}_j^2}{\sigma_s^2 a_j + \sigma_e^2} \right], \end{aligned} \tag{15.4}$$

for B an unimportant constant and $\hat{\mathbf{v}} = (\hat{v}_1, \ldots, \hat{v}_{s_Z})' = \mathbf{P}'\Gamma_Z' \mathbf{y}$. Note that $\mathbf{y}'\Gamma_c \Gamma_c' \mathbf{y}$ is the residual sum of squares from the simple linear model with outcome $\mathbf{y}$ and design matrix $(\mathbf{X}|\mathbf{Z})$, or equivalently from a fit of the mixed linear model with $\sigma_s^2 = \infty$. I call this the unshrunk fit of $\mathbf{y}$ on $(\mathbf{X}|\mathbf{Z})$ because the $\mathbf{u}$ are not shrunk toward zero. Also, $\hat{\mathbf{v}}$ is a known linear function of $\mathbf{y}$, the estimate of $\mathbf{v}$ in a fit of model (15.2) with $\sigma_s^2 = \infty$, which I'll call the unshrunk estimate of $\mathbf{v}$. The $\hat{v}_j$ capture the information in the data that is specific to the model's random effects, that is, not confounded with the fixed effects.

The choice of Γ_Z would appear to affect the $\hat{v}_j$, but in fact (15.4) is invariant to the choice of Γ_Z. (The proof is straightforward and given as an exercise.) If some a_j has multiplicity greater than 1, the terms in $\sum_{j=1}^{s_Z} \hat{v}_j^2 (\sigma_s^2 a_j + \sigma_e^2)^{-1}$ having the same a_j become $(\sigma_s^2 a_j + \sigma_e^2)^{-1} \sum_k \hat{v}_k^2$, and $\sum_k \hat{v}_k^2$ is invariant to the choice of Γ_Z.

Two comments: Note that $\mathbf{y}'\Gamma_c \Gamma_c \mathbf{y} = \mathbf{y}'(\mathbf{I}_n - \mathbf{X}(\mathbf{X}'\mathbf{X})^{-1}\mathbf{X}')\mathbf{y} + \sum_{j=1}^{s_Z} \hat{v}_j^2$, where $\mathbf{y}'(\mathbf{I}_n - \mathbf{X}(\mathbf{X}'\mathbf{X})^{-1}\mathbf{X}')\mathbf{y}$ is the residual sum of squares for a fit of $\mathbf{y}$ on $\mathbf{X}$ only, i.e., a fit of the mixed linear model in which the random effects $\mathbf{u}$ are shrunk to $\mathbf{0}$. This allows (15.4) to be rewritten in terms of the precisions $1/\sigma_e^2$ and $1/\sigma_s^2$, possibly to some advantage. Also, note the importance of the flat prior on β. If β is instead given a proper normal prior, this becomes a three-variance model even if that proper prior has no unknowns, and the marginal posterior of (σ_s^2, σ_e^2) has a simple form like (15.4) only under certain conditions on $\mathbf{X}$ and $\mathbf{Z}$.

It is useful also to derive the restricted likelihood as a marginal likelihood. Begin with (15.2) and pre-multiply both sides of the equation by $(\Gamma_X | \Gamma_Z \mathbf{P} | \Gamma_c)'$ to give

$$\left[\begin{array}{c} \Gamma_X' \\ \mathbf{P}'\Gamma_Z' \\ \Gamma_c' \end{array} \right] \mathbf{y} = \left[\begin{array}{cc} \mathbf{I}_{s_X} & \mathbf{0} \\ \mathbf{0} & \mathbf{I}_{s_Z} \\ \mathbf{0} & \mathbf{0} \end{array} \right] \left[\begin{array}{c} \beta^* \\ \mathbf{v} \end{array} \right] + \varepsilon, \tag{15.5}$$

where as before $\mathbf{v} \sim N_{s_Z}(\mathbf{0}, \sigma_s^2 \mathbf{A})$ and the distribution of ε is unchanged because the equation was pre-multiplied by an orthogonal matrix. If $\pi(\sigma_e^2, \sigma_s^2)$ is the prior distribution for (σ_e^2, σ_s^2), the joint posterior distribution of all the unknowns is easily shown to be

$$\begin{aligned}
\pi(\beta^*, \mathbf{v}, \sigma_e^2, \sigma_s^2|\mathbf{y}) \propto\ & \pi(\sigma_e^2, \sigma_s^2) \\
& (\sigma_e^2)^{-s_X/2} \exp\left(-(\beta^* - \Gamma_X'\mathbf{y})'(\beta^* - \Gamma_X'\mathbf{y})/2\sigma_e^2\right) && (15.6) \\
& \prod_{j=1}^{s_Z} \left(\sigma_e^2 \frac{a_j}{a_j+r}\right)^{-0.5} \exp\left(-\sum_{j=1}^{s_Z} \left(2\sigma_e^2 \frac{a_j}{a_j+r}\right)^{-1} (v_j - \tilde{v}_j)^2\right) && (15.7) \\
& (\sigma_e^2)^{-(n-s_X-s_Z)/2} \exp\left(-\mathbf{y}'\Gamma_c\Gamma_c'\mathbf{y}/2\sigma_e^2\right) && (15.8) \\
& \prod_{j=1}^{s_Z} (\sigma_s^2 a_j + \sigma_e^2)^{-0.5} \exp\left(-\sum_{j=1}^{s_Z} \hat{v}_j^2/2(\sigma_s^2 a_j + \sigma_e^2)\right), && (15.9)
\end{aligned}$$

where $\tilde{v}_j = \frac{a_j}{a_j+r}\hat{v}_j$ and $r = \sigma_e^2/\sigma_s^2$, as usual. (The proof is an exercise.) Equation (15.6) is the conditional posterior of the re-parameterized fixed effects, β^*, given (σ_e^2, σ_s^2). Equation (15.7) is the conditional posterior of the re-parameterized random effects, $\mathbf{v}$, given (σ_e^2, σ_s^2). Conditionally, the v_j are independent with mean $\tilde{v}_j$, variance $\sigma_e^2 a_j/(a_j + r)$, and DF $a_j/(a_j + r)$. Thus given $r = \sigma_e^2/\sigma_s^2$, v_j is shrunk more for j with smaller a_j. Integrating (15.6) and (15.7) out of this joint posterior leaves (15.8) and (15.9); they are the re-expressed restricted likelihood as in (15.4).

It is convenient to break the log restricted likelihood (15.4) into two pieces. The first piece, $-(n - s_X - s_Z)\log(\sigma_e^2)/2 - \mathbf{y}'\Gamma_c\Gamma_c'\mathbf{y}/2\sigma_e^2$, is a function only of σ_e^2 and will be called the *free terms* for σ_e^2 because it is free of σ_s^2. The second piece is

$$-\frac{1}{2}\sum_{j=1}^{s_Z}\left[\log(\sigma_s^2 a_j + \sigma_e^2) + \frac{\hat{v}_j^2}{\sigma_s^2 a_j + \sigma_e^2}\right], \tag{15.10}$$

which will be called *mixed terms* because each summand (mixed term) involves both variances. As we will see, the free terms are absent for some models, e.g., a model including an ICAR random effect. When present, they provide strong information about σ_e^2, information that is in a sense uncontaminated by information about σ_s^2. The data's information about σ_s^2 enters the restricted likelihood entirely through the mixed terms and is thus always mixed with information about σ_e^2.

Several features make this re-expression interpretable and useful. First, the $\hat{v}_j$ are known functions of the data; whenever the fixed effects include an intercept, they are contrasts in the data. Second, the form of (15.4) implies that the $\hat{v}_j$ are independent conditional on (σ_e^2, σ_s^2); that's why the restricted likelihood decomposes in this simple way. (The following paragraph gives another way to see this.) In particular, the conditional distribution of $\hat{v}_j$ given (σ_e^2, σ_s^2) is normal with mean zero and variance $\sigma_s^2 a_j + \sigma_e^2$, a simple linear function of the unknown variances. This means the a_j — which are functions of the design matrices $\mathbf{X}$ and $\mathbf{Z}$, albeit obscure ones at this point — determine how the canonical "observations" $\hat{v}_j$ and $\mathbf{y}'\Gamma_c\Gamma_c'\mathbf{y}$ provide information about σ_e^2 and σ_s^2 and do so in a fairly simple way that is explored in this and the next chapter. Third, the v_j are independent *a posteriori* conditional on (σ_e^2, σ_s^2) and have simple conditional posterior mean, variance, and DF. Finally, by assumption $s_Z > 0$

Table 15.1: The Restricted Likelihood as in Equation (15.4), Interpreted as a Generalized Linear Model

GLM Notation	j^{th} Mixed Term $j = 1, \ldots, s_Z$	Free Terms $j = s_Z + 1$
Data y_i	$\hat{v}_j^2$	$\hat{v}_{s_Z+1}^2 = \mathbf{y}'\Gamma_c\Gamma_c'\mathbf{y}/(n - s_X - s_Z)$
Canonical parameter θ_i	$-1/(\sigma_s^2 a_j + \sigma_e^2)$	$-1/\sigma_e^2$
Gamma shape parameter ν_i	1/2	$(n - s_X - s_Z)/2$
$E(y_i) = -1/\theta_i$	$\sigma_s^2 a_j + \sigma_e^2$	σ_e^2
$\text{Var}(y_i) = [E(y_i)]^2/\nu_i$	$2(\sigma_s^2 a_j + \sigma_e^2)^2$	$2(\sigma_e^2)^2/(n - s_X - s_Z)$

so there is always at least one mixed term. Thus (15.4) implies that σ_e^2 and σ_s^2 are identified if and only if either (a) there are free terms (i.e., $n - s_X - s_Z > 0$) or (b) there are at least two distinct a_j.

The restricted likelihood (15.4) can be interpreted as a generalized linear model (GLM; Henn & Hodges 2013, Welham & Thompson 2009). Table 15.1 maps the free terms and mixed terms into the GLM terminology and notation of McCullagh & Nelder (1989). Specifically, (15.4) has gamma errors and the identity link (which is not the canonical link), the unknown coefficients in the linear predictor are σ_e^2 and σ_s^2, each mixed term provides one "observation," and the free terms provide one "observation." Here the coefficients in the linear predictor, σ_e^2 and σ_s^2, are constrained to be positive, unlike more typical uses of generalized linear models. McCullagh & Nelder (1989) noted (p. 287, Section 8.3.5) that GLMs of this sort provide a method for estimating variance components "where the observations are sums of squares of Normal variables"; interpreting the re-expressed restricted likelihood as a GLM extends this.

The free terms can alternatively be interpreted not as an "observation" in this GLM but rather as an informative inverse gamma prior for σ_e^2, with the mixed terms supplying all the "observations." If an informative inverse gamma prior is added for σ_s^2, it has the same form as the free terms for σ_e^2.

15.2 Examples

Although the re-expressed restricted likelihood (15.4) is simple, its key components, the $\hat{v}_j$ and a_j, are obscure functions of the original model's design matrices $\mathbf{X}$ and $\mathbf{Z}$. This section's examples give some intuition for these functions while providing preliminaries for Chapter 16. Besides these examples, the class of two-variance models also includes random-intercept models (e.g., all the models fit in Hodges 1998), multiple-membership models (e.g., Browne et al. 2001, McCaffrey et al. 2004), and no doubt others.

15.2.1 Balanced One-Way Random Effect Model

The balanced one-way random effect model can be written as $y_{ij} = \beta_0 + u_i + \varepsilon_{ij}$ for $i = 1,\ldots,N$ and $j = 1,\ldots,m$, $u_i \sim N(0,\sigma_s^2)$ and $\varepsilon_{ij} \sim N(0,\sigma_e^2)$. Order the observations in $\mathbf{y}$ with j varying fastest, as $\mathbf{y} = (y_{11},\ldots,y_{1m},y_{21},\ldots,y_{Nm})'$. Then in the mixed-linear-model form, $n = Nm$, $\mathbf{X} = \mathbf{1}_n$, $\mathbf{Z} = \mathbf{I}_N \otimes \mathbf{1}_m$, where the Kronecker product $\otimes$ is defined for matrices $\mathbf{A} = (a_{ij})$ and $\mathbf{B}$ as $\mathbf{A} \otimes \mathbf{B} = (a_{ij}\mathbf{B})$, $\mathbf{G} = \sigma_s^2\mathbf{I}_N$, and $\mathbf{R} = \sigma_e^2\mathbf{I}_{Nm}$. $\mathbf{X} = \mathbf{Z}\mathbf{1}_N$, so $s_Z = N-1$ while $s_X = 1$.

The matrix Γ_X in the re-expression is $\frac{1}{\sqrt{n}}\mathbf{1}_n$. Γ_Z can be any matrix of the form $\mathbf{F} \otimes \frac{1}{\sqrt{m}}\mathbf{1}_m$ where $\mathbf{F}$ is $N \times (N-1)$ satisfying $\mathbf{F}'\mathbf{F} = \mathbf{I}_{N-1}$, i.e., $\mathbf{F}$'s columns are an orthonormal basis for the orthogonal complement of $\mathbf{1}_N$. Then $\mathbf{M}_{ZZ} = \Gamma_Z'\mathbf{Z} = \sqrt{m}\mathbf{F}'$, so $\mathbf{M}_{ZZ}\mathbf{M}_{ZZ}' = m\mathbf{I}_{N-1}$, $a_j = m, j = 1,\ldots,N-1$, and $\mathbf{P}$ is I_{N-1}. Finally, $\hat{v} = \Gamma_Z'\mathbf{y}$ so $\hat{v}' = \sqrt{m}(\bar{y}_{1.},\ldots,\bar{y}_{N.})\mathbf{F}$, where $\bar{y}_{i.} = \sum_j y_{ij}/m$ is the average of group i's observations.

Plugging all this into the re-expressed log restricted likelihood (15.4) gives the familiar form

$$-\frac{n-N}{2}\log(\sigma_e^2) - \frac{1}{2\sigma_e^2}\sum_{i,j}(y_{ij}-\bar{y}_{i.})^2 - \frac{N-1}{2}\log(\sigma_s^2 m + \sigma_e^2)$$
$$-\frac{m}{2}\sum_i(\bar{y}_{i.}-\bar{y}_{..})^2/(\sigma_s^2 m + \sigma_e^2), \qquad (15.11)$$

where $\bar{y}_{..} = \sum_{ij} y_{ij}/n$ is the average of all $n = Nm$ observations. Because there is only one distinct a_j, the two variances are identified if and only if $n > N$, i.e., $m > 1$, not exactly a novel finding. Interpreted as a GLM with unknown coefficients σ_s^2 and σ_e^2, this GLM has two "observations," with one each arising from the residual and between-group mean squares.

15.2.2 Penalized Spline

Example 5 again, global mean surface temperature. Consider yet again the penalized spline fit to this dataset in several previous chapters. The dataset has n = 125 observation. The spline has a truncated quadratic basis and 30 knots at years $1880 + \{4,\ldots,120\}$. The predictor (year) and knots are centered and scaled as before. The fixed-effects design matrix $\mathbf{X}$ has $s_X = p = 3$ columns for the intercept, linear, and quadratic terms, while the random-effects design matrix $\mathbf{Z}$ has $s_Z = q = 30$ columns, one for each knot. The combined design matrix $(\mathbf{X}|\mathbf{Z})$ has rank 33 though it is very close to rank-deficient: The first canonical correlation of $\mathbf{X}$ and $\mathbf{Z}$ is greater than 0.99999995.

Table 15.2 lists the eigenvalues a_j in the restricted likelihood's re-expression. The a_j decline quickly: $a_1/a_6 = 1841$, and the last 18 a_j are all smaller than a_1 by a factor of at least 100,000. In Chapter 16 we will see that this rapid decline in the a_j implies that the first few $\hat{v}_j$ provide almost all of the information in the data about σ_s^2, with the remaining $\hat{v}_j$ providing information almost exclusively about σ_e^2.

Table 15.2: Global Mean Surface Temperature Data: Penalized Spline, a_j

36.0	3.15	0.562	0.147	0.0493	0.0195
8.76×10^{-3}	4.32×10^{-3}	2.30×10^{-3}	1.30×10^{-3}	7.68×10^{-4}	4.75×10^{-4}
3.05×10^{-4}	2.01×10^{-4}	1.37×10^{-4}	9.51×10^{-5}	6.76×10^{-5}	4.89×10^{-5}
3.61×10^{-5}	2.71×10^{-5}	2.06×10^{-5}	1.60×10^{-5}	1.26×10^{-5}	1.01×10^{-5}
8.32×10^{-6}	7.00×10^{-6}	6.06×10^{-6}	5.42×10^{-6}	5.06×10^{-6}	3.75×10^{-6}

Figure 15.1 shows several columns of the canonical random-effect design matrix $\Gamma_Z\mathbf{P}$, i.e., canonical predictors or regressors. The first canonical predictor (Figure 15.1a) — which has the largest a_j, so its coefficient's unshrunk estimate $\hat{v}_1$ has variance $\sigma_s^2 a_j + \sigma_e^2$ conditional on σ_s^2 and σ_e^2, largest of all the $\hat{v}_j$ — is shaped like a cubic polynomial; alternatively, it is roughly sinusoidal with 1.5 cycles. As j increases and the a_j decline, the jth canonical predictor resembles a polynomial of progressively higher order, quartic, quintic, etc., or alternatively a sine curve with progressively more cycles, 2, 2.5, etc. Based on inspection of plots, these canonical predictors have increasing numbers of cycles and at least roughly constant amplitude up to about $j = 20$, at which point the amplitude starts to change visibly over the 125 years of the data. Figure 15.1c shows the last 3 canonical predictors, which are still cyclic but have distinctly non-constant amplitude.

In other words, this penalized spline can be understood as a quadratic regression with unshrunk coefficients plus a regression on higher-order polynomials in which the coefficients are shrunk, with the extent of shrinkage increasing with the order of the polynomial. Shrinkage of all the latter coefficients is controlled by σ_s^2 but the a_j control the relative degrees of shrinkage of their respective v_j. I have briefly explored penalized splines with other truncated polynomial bases and these observations appear to generalize in the obvious way: With a truncated cubic basis, the first canonical predictor in the random effect is roughly quartic, etc.[2]

Turning to the restricted likelihood itself, consider its interpretation as a generalized linear model with gamma observables $\hat{v}_j^2$ and linear predictor with intercept σ_e^2 and σ_s^2 the slope on a_j. If the $\hat{v}_j^2$ are more or less constant in j, that suggests σ_s^2 is close to zero. On the other hand, if the $\hat{v}_j^2$ decline as j increases, i.e., as a_j decreases, that suggests σ_s^2 is large. In Figure 15.2, panel (a) plots the 30 $\hat{v}_j^2$ in order of j, while panel (b) plots the 30 a_j. The three largest $\hat{v}_j^2$ occur for $j = 1,2,3$, which have the three largest a_j. For this model, the a_j decline so quickly in j that most a_j are quite small, which suggests again that for this model at least, most of the information about σ_s^2 is contained in the first few $\hat{v}_j^2$. Chapter 16 explores this idea in detail.

The free terms and mixed terms make distinct contributions to the restricted likelihood for this model. Figure 15.3 shows contour plots of the log restricted likelihood (middle row left) and the parts of it contributed by the mixed terms involving both

[2]This characterization of penalized splines is probably in the spline literature somewhere but apart from a few heroes like Ruppert et al. (2003) and Green & Silverman (1994), this literature is so user-hostile that I am not willing to look for and decode the relevant papers.

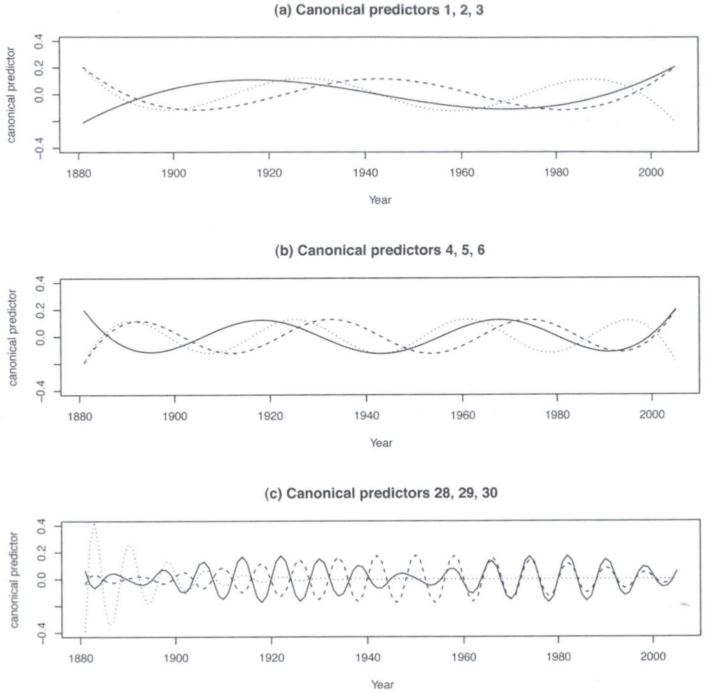

Figure 15.1: Global mean surface temperature data: Penalized spline model, canonical predictors corresponding to several a_j. The horizontal axis is the row in $\Gamma_Z \mathbf{P}$ (labeled by years), while the vertical axis is the value of the predictor for that row. Panel (a): Predictors corresponding to a_1 (solid line), a_2 (dashed line), a_3 (dotted line). Panel (b): Predictors corresponding to a_4 (solid line), a_5 (dashed line), a_6 (dotted line). Panel (c): Predictors corresponding to a_{28} (solid line), a_{29} (dashed line), a_{30} (dotted line).

variances (top row and middle row right) and the free terms for σ_e^2 (bottom row). The log restricted likelihood is very well-behaved as a function of the log variances.

These plots suggest, like the re-expressed restricted likelihood itself, that all the information about σ_s^2 is in the mixed terms. Comparing the two plots in the middle row, the contours in the two plots appear to align perfectly in the vertical direction. In fact, however, they do not, although the differences are small enough to be invisible in plots of this size. Analyses presented in Chapter 16 show that the free terms affect both the maximizing value of σ_s^2 and the restricted likelihood's curvature in the σ_s^2 direction. In that sense, the free terms indirectly provide some information about σ_s^2.

Clearly both the free and mixed terms provide information about σ_e^2, which can be seen by comparing the three plots in Figure 15.3's left column. The free terms in effect arise from a fit of the same mixed linear model with unshrunk random effects, which overfits the data and thus has a relatively small residual mean square. The value of σ_e^2 that maximizes the free terms (between the two very close vertical lines)

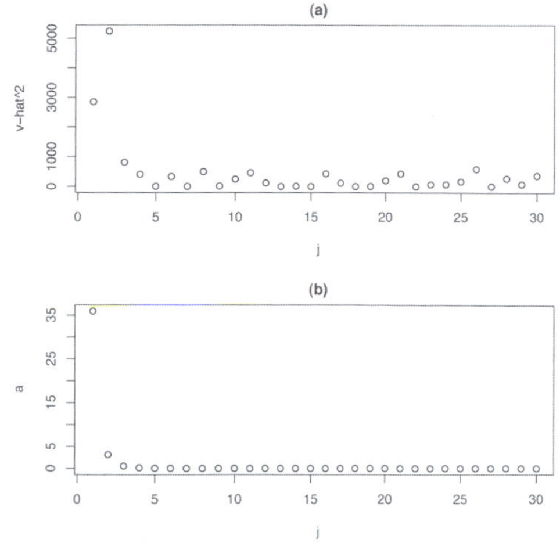

Figure 15.2: Global-mean surface temperature data: Penalized spline. Panel (a), $\hat{v}_j^2$; Panel (b), a_j.

is therefore smaller than the value that maximizes the entire restricted likelihood, which in turn is smaller than the value that maximizes the mixed terms. The value of σ_e^2 that maximizes the restricted likelihood is determined by a contest between the free and mixed terms. In the present case, the free terms dominate because they are worth $n - s_X - s_Z = 92$ observations compared to 30 for the mixed terms, and accordingly the restricted likelihood is maximized at $\hat{\sigma}_e^2 = 145.3$, while the free and mixed terms are maximized at 137.4 and 173.9 respectively. Besides shifting the maximizing value of σ_e^2, the mixed terms also make the restricted likelihood more peaked, as can be seen by comparing the contours of the restricted likelihood and the free terms.

If we had used a full-rank penalized spline here, with a knot at every distinct predictor value (year), it is easy to see that the restricted likelihood would have no free terms for σ_e^2. This would seem to imply that the restricted likelihood would be less well-behaved for the full-rank spline than for the low-rank spline I actually fit. However, further examination suggests that the restricted likelihoods for the two splines are quite similar. Table 15.3 shows the first 30 a_j for the full-rank spline for the GMST data. The first 15 a_j for the full-rank spline are almost exactly 4 times the analogous a_j for the low-rank spline. For the next 6 a_j, the multiple declines but is still greater than 3.9. The first 9 canonical predictors in the full- and low-rank splines are visually indistinguishable although as j increases the canonical predictors in the two models become progressively more different. Regarding the loss of the free terms, the last 92 mixed terms in the full-rank model have $a_j < 8.5 \times 10^{-6}$, small enough that these mixed terms are effectively free terms for σ_e^2. Thus the only real difference between the restricted likelihoods for the full- and low-rank splines is that the former's restricted likelihood is shifted lower in σ_s^2 by a factor of about 4 because the first 15 a_j are larger by a factor of 4. (An exercise allows you to see this for yourself.)

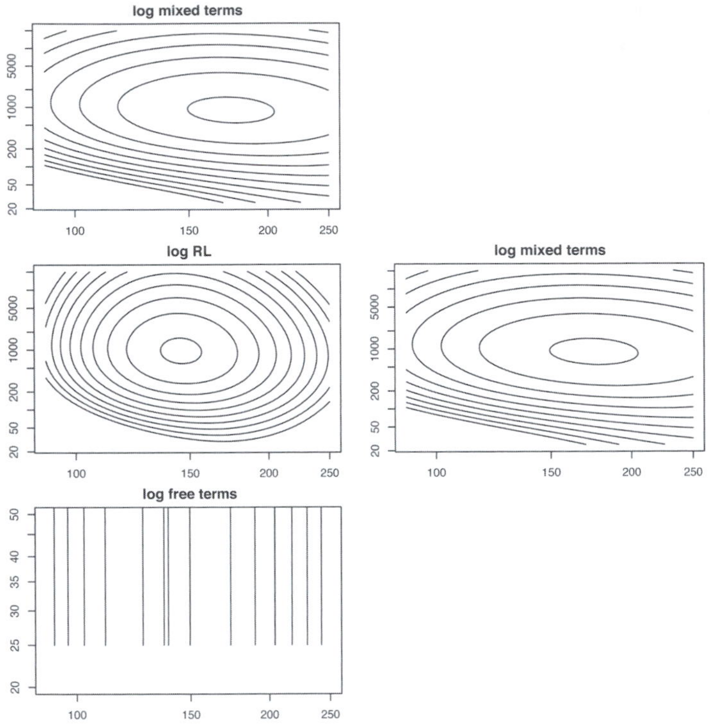

Figure 15.3: Global mean surface temperature data: Contour plots of the log restricted likelihood (middle row left), mixed terms (top row and middle row right), and free terms (bottom row). In all panels the vertical axis is σ_s^2, the horizontal axis is σ_e^2, the axes have logarithmic scales, and the contours are lower than their respective maxima by 0.16 + $\{0, 1, \ldots 9\}$ logs, except that the contour plot of the free terms has an extra pair of contours very close to the maximizing value $\sigma_e^2 = 137.4$.

Table 15.3: Global Mean Surface Temperature Data

145	12.6	2.25	0.589	0.197	0.0781
0.0350	0.0173	9.18×10^{-3}	5.18×10^{-3}	3.07×10^{-3}	1.90×10^{-3}
1.22×10^{-3}	8.05×10^{-4}	5.46×10^{-4}	3.79×10^{-4}	2.69×10^{-4}	1.94×10^{-4}
1.43×10^{-4}	1.07×10^{-4}	8.06×10^{-5}	6.17×10^{-5}	4.78×10^{-5}	3.74×10^{-5}
2.95×10^{-5}	2.35×10^{-5}	1.89×10^{-5}	1.53×10^{-5}	1.25×10^{-5}	1.02×10^{-5}

Note: The first 30 eigenvalues a_j in the re-expressed restricted likelihood for a full-rank penalized spline.

15.2.3 ICAR Model; Spatial Confounding Revisited

First we consider Section 14.1's simple ICAR model for the attachment-loss data, contrasting it with the penalized spline model just discussed. Then we consider the ICAR model with one fixed-effect predictor that was used to explore spatial confounding in Sections 9.1.1 and 10.1. This allows us to say more about some questions that were left open at the end of Section 10.1.

15.2.3.1 Attachment-Loss Data: Simple ICAR Model

Section 14.1 fit a simple ICAR model, with no explicit fixed-effect predictors, to the attachment-loss data shown in Figure 14.1. For this dataset, the spatial map has $I = 2$ islands, one each for the maxillary (upper) and mandibular (lower) arches, so the ICAR model has $p = s_X = 2$ implicit fixed effects, intercepts or means for the two islands. This person had $n = 168$ measurement sites, so the ICAR model's random effects have $q = s_Z = 166$. The neighbor pairings of measurement sites are as shown in Figure 14.2, although the present analysis does not distinguish types of neighbor pairs. The data are sorted as in Figure 14.1: Maxilla then mandible; within arch, lingual (tongue side) then buccal (cheek side); and within arch and side, from left to right as in Figure 14.1. Before reading further, it may be helpful to review Section 5.2.2.

In the mixed-linear-model formulation of the ICAR model given in Section 5.2.2, the random effects θ_1 have covariance matrix $\sigma_s^2\mathbf{D}_1^{-1}$, where $\mathbf{D}_1$ is a diagonal matrix containing the positive eigenvalues of $\mathbf{Q}$, the matrix that encodes the ICAR's neighbor pairings (and which is proportional to the prior precision of the ICAR random effect). This covariance matrix, $\sigma_s^2\mathbf{D}_1^{-1}$, is not proportional to the identity so before we begin, the ICAR model's random effects $\mathbf{V}_1\theta_1$, as written in equation (5.14), must be re-parameterized. Using the notation of Section 5.2.2, for the present purpose the fixed effect design matrix is $\mathbf{X} = \mathbf{V}_2$, of which the two columns are indicators for the maxillary and mandibular islands; the random effect design matrix is $\mathbf{Z} = \mathbf{V}_1\mathbf{D}_1^{-0.5}$, where $\mathbf{V}_1$'s columns are the eigenvectors of $\mathbf{Q}$ that have positive eigenvalues; and the random effects are $\mathbf{u} = \mathbf{D}_1^{0.5}\theta_1$ with covariance matrix $\mathbf{G} = \sigma_s^2\mathbf{I}_{n-2}$.

Define $\mathbf{W}_m$ as an $m \times m$ matrix with 1's down the diagonal from upper right to lower left and 0's elsewhere; post-multiplying a matrix by $\mathbf{W}_m$ reverses the order of its columns, pre-multiplying by $\mathbf{W}_m$ reverses the order of its rows. If we follow the steps in re-expressing the restricted likelihood, $\Gamma_Z = \mathbf{V}_1$, $\mathbf{A} = \mathbf{W}_{s_Z}\mathbf{D}_1^{-1}\mathbf{W}_{s_Z}$, and $\mathbf{P} = \mathbf{W}_{s_Z}$ so the canonical observations are $\hat{\mathbf{v}} = (\mathbf{V}_1\mathbf{W}_{s_Z})'\mathbf{y}$. These $\hat{\mathbf{v}}$ are the same contrasts in the data that were examined in Section 5.2.2, but their order has been reversed: $\mathbf{V}_1$'s columns were sorted in decreasing order of the *precisions* of the random effects θ_1, while the canonical predictors are sorted in decreasing order of the *variances* of the canonical random effects $\mathbf{v}$.[3]

Consider the a_j in the re-expressed restricted likelihood, summarized in Table 15.4. Because this subject has no missing teeth, the maxillary and mandibular

[3] I apologize for this notational kluge. ICARs were largely developed by Bayesians, who like precisions, while mixed-linear-model theory was largely developed by non-Bayesians, who prefer variances. Sometimes the translation is awkward.

Table 15.4: Simple ICAR Model for the Attachment-Loss Data: Summary of the 166 a_j

a_j	Multiplicity	Number of Distinct a_j
149.0	2	1
37.33	2	1
16.64	2	1
9.40 to 1	2	11
0.843	4	1
0.695	24	1
0.672 to 0.333	2	14
0.288	4	1
0.200	24	1
0.1996 to 0.1872	2	7
0.1872	4	1
0.1858 to 0.1800	2	6
0.1798	24	1

arches have identical spatial structures so the multiplicity of each a_j is an even number split equally between the two arches. Some a_j have multiplicity 24, 12 per arch, referring to the 12 interior teeth in an arch; others have multiplicity 4, 2 per arch, referring to the 2 "edge" teeth in each arch, labeled as tooth number 7 in Figure 14.1.

Compared to the a_j for the penalized spline, the ICAR's a_j decrease much more slowly in j. For the penalized spline, $a_1/a_6 = 1,841$; for the ICAR, $a_1/a_6 = 9.0$. (Arguably $a_1/a_{12} = 35.2$ is a more fair comparison because the ICAR's spatial structure is identical in the two arches, but my point stands.) For the spline, $a_1/a_{15} = 263,083$; for the ICAR, $a_1/a_{15} = 61.4$ ($a_1/a_{30} = 177$). The ratio of largest to smallest a_j is 9,593,165 for the spline and 829 for the ICAR, and the ICAR has 166 a_j compared to the spline's 30. This means that shrinkage of the canonical random effects v_j is much less differentiated for the ICAR than for the spline. This appears to be true of ICAR models generally and as we will see in Chapter 16, this fact helps explain why the penalized spline and ICAR models smooth the global mean surface temperature data so differently (Section 14.3).

The ICAR's canonical predictors, on the other hand, are fairly similar to the spline's canonical predictors. Figure 15.4 shows several canonical predictors. The two arches have identical sets of 83 canonical predictors and for each arch, the first few canonical predictors are identical on the buccal and lingual sides of the arch. Thus, Figure 15.4 shows the values of the first few canonical predictors for one side of one arch. The ICAR's first 8 canonical predictors, i.e., the first 4 in each arch, strongly resemble sine curves. The first canonical predictor in each arch (Panel (a)) is close to a half-cycle of a sine curve. The second canonical predictor in each arch (Panel (b)) resembles a full cycle of a sine curve; each of the next two canonical predictors (Panels (c) and (d)) adds a half-cycle. Each canonical predictor deviates from

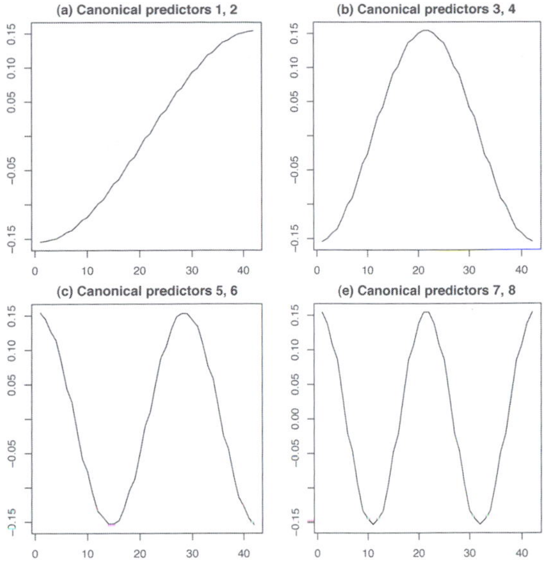

Figure 15.4: Simple ICAR model for periodontal data: Canonical predictors corresponding to several a_j. The horizontal axis is the measurement site, running along one side of an arch (the two sides of the arch are identical for these predictors); the vertical axis is the value of the predictor. Panel (a): The first canonical predictor for each arch, corresponding to a_1 and a_2. Panel (b): The second canonical predictor for each arch, corresponding to a_3 and a_4. Panel (c): The third canonical predictor for each arch, corresponding to a_5 and a_6. Panel (d): The fourth canonical predictor for each arch, corresponding to a_7 and a_8.

a sine curve by making a little jerk at interproximal measurement sites (i.e., those on a gap between two teeth), showing the effect of the neighbor pairings across the interproximal area with measurement sites on the other side of the arch.

Some of the ICAR's canonical predictors were shown in Figure 5.4, where the model included only a single arch. Figure 5.4a showed the last canonical predictor for each arch as it was computed by my then-student, Brian Reich (Reich & Hodges 2008a). The 12 last canonical predictors in each arch have the same a_j and as computed by Prof. Reich, each one had the pattern shown in Figure 5.4a for one of the 12 interior teeth. I re-did these calculations for this chapter and my last 12 canonical predictors (per arch) differ from Prof. Reich's because this a_j has multiplicity 12 in each arch, so the corresponding canonical predictors are uniquely specified only up to the subspace they span. Mine span the same subspace as Prof. Reich's Figure 5.4a but by a lucky choice of software, his are more interpretable.

Now consider again the interpretation of the re-expressed restricted likelihood as a generalized linear model with gamma observables $\hat{v}_j^2$ and with linear predictor having intercept σ_e^2 and σ_s^2 as the slope on a_j. Because the a_j are positive and decline

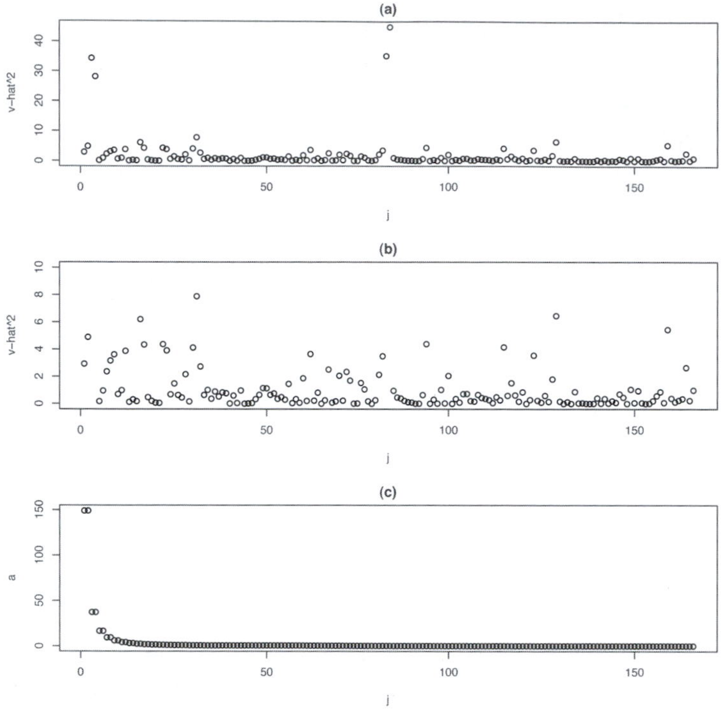

Figure 15.5: ICAR fit to the attachment loss data: Panel (a), $\hat{v}_j^2$ with full vertical scale; Panel (b), $\hat{v}_j^2$ with vertical scale restricted to 0 to 10; Panel (c), a_j.

as j increases, if this model fits the $\hat{v}_j^2$ should generally decline as j increases. For the attachment-loss data, Figure 15.5 shows the $\hat{v}_j^2$ (Panels (a) and (b), identical except for their vertical scales) and the a_j (Panel (c)). Figure 15.5a reveals a feature of these data that does not fit the expected pattern of $\hat{v}_j^2$ decreasing in j, namely the two prominent points in the middle of Panel (a). Their canonical predictors are the contrasts describing the difference between the average of the direct sites and the average of the interproximal sizes, one for each arch. In the puzzle described in Section 14.1, these two canonical predictors produced an unexpected effect when they were changed from random effects to fixed effects. One way to make this change is to change these two a_j to infinity; in the second derivation of the re-expressed restricted likelihood, equations (15.6) to (15.9), this shifts these coefficients v_j from equation (15.7) into equation (15.6) and deletes the corresponding terms from (15.9). Chapter 16 explores the effect of this shift and explains its unexpected effect.

Apart from these two $\hat{v}_j^2$, Figure 15.5b shows that the other $\hat{v}_j^2$ behave more or less as expected: At the plot's left, with large a_j, a majority of the $\hat{v}_j^2$ are relatively large, while for j larger than about 30 the $\hat{v}_j^2$ are only sporadically large. (I admit this is all quite informal as yet, but for now my purpose is just to illustrate suggestively the re-expression of the restricted likelihood.)

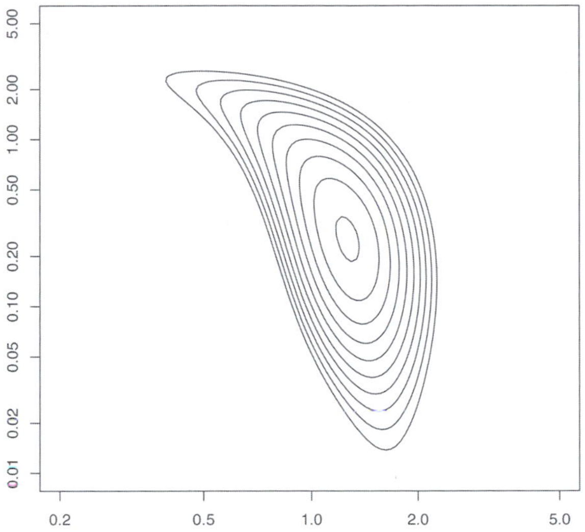

Figure 15.6: ICAR fit to the attachment loss data: Contour plot of the log restricted likelihood. The horizontal axis is σ_e^2, the vertical axis is σ_s^2, both axes have logarithmic scales, and the contours are lower than the maximum by 0.16 + $\{0, \ldots, 9\}$ logs.

Despite Figure 15.5a's evidence that the ICAR model has a lack-of-fit problem and despite the absence of free terms, the log restricted likelihood is not badly behaved (Figure 15.6). Interpreting the log restricted likelihood as a log marginal posterior (which it is), Figure 15.6 shows some negative posterior correlation between the two variances: As σ_s^2 is increased, smaller values of σ_e^2 are favored. Given that each mixed term in the re-expressed restricted likelihood has the form $\sigma_s^2 a_j + \sigma_e^2$ for $a_j > 0$ and that the ICAR's restricted likelihood has only mixed terms, this is not surprising. By contrast, for the penalized spline the restricted likelihood, again interpreted as a marginal posterior, shows essentially no posterior correlation between the two log variances (Figure 15.3); even the mixed terms show little sign of such correlation. This would seem to be another effect of having free terms or having many mixed terms with very small a_j: These tend to reduce the posterior correlation of the two variances.

A final note about identification of the simple ICAR model: As noted in Section 15.1, because this model has no free terms, the two variances are identified if and only if there are at least two distinct a_j. Reich & Hodges (2008a, Appendix A) shows that this condition holds *except* when all I islands in the spatial map have the same number of areas (regions, nodes) and on each island, each area is a neighbor of every other area.

15.2.3.2 Spatial Confounding: ICAR with Predictors

The re-expressed restricted likelihood allows us to examine some aspects of the spatial confounding problem that Section 10.1.5 described as underdeveloped. I present results for a one-dimensional fixed-effect predictor **F** as in Section 10.1, but extension to higher-dimensional **F** is straightforward. As before, **F** is centered and scaled so $\mathbf{F}'\mathbf{F} = n-1$.

Following the notation in Chapters 5 and 10, the model (before re-expression) is $\mathbf{y} = \mathbf{F}\beta + \mathbf{I}_n\mathbf{S} + \varepsilon$, where **S** is an n-dimensional ICAR random effect with smoothing variance σ_s^2 and adjacency matrix **Q**, i.e., with precision matrix $\mathbf{Q}/\sigma_s^2$. For simplicity, assume the spatial map is connected, with $I = 1$ island, so **Q**'s zero eigenvalue has multiplicity 1. (This also generalizes in the obvious way.) As before, **Q** has spectral decomposition $\mathbf{Q} = \mathbf{VDV}'$, where **V** is an orthogonal matrix that partitions as $(\mathbf{V}_1|\frac{1}{\sqrt{n}}\mathbf{1}_n)$, where $\mathbf{V}_1$ is $n \times (n-1)$, and **D** is diagonal with n^{th} diagonal entry zero; its upper left $(n-1)\times(n-1)$ partition $\mathbf{D}_1$ is a diagonal matrix with positive entries $d_1 \geq \ldots \geq d_{n-1} > 0$.

Then as in equation (5.15), the model can be rewritten as

$$\mathbf{y} = \mathbf{1}_n\beta_0 + \mathbf{F}\beta + \mathbf{V}_1\theta_1 + \varepsilon \tag{15.12}$$

where the now-explicit intercept β_0 has absorbed the factor $1/\sqrt{n}$ and $\theta_1 \sim N(\mathbf{0}, \sigma_s^2\mathbf{D}_1^{-1})$, where $\sigma_s^2\mathbf{D}_1^{-1}$ is a covariance matrix. To follow the derivation of the re-expressed restricted likelihood, the random effect must be re-parameterized so its covariance is proportional to the identity, so for the present purpose, $\mathbf{X} = (\mathbf{1}_n|\mathbf{F})$ and $\mathbf{Z} = \mathbf{V}_1\mathbf{D}_1^{-0.5}$.

Following the derivation of the re-expressed restricted likelihood, $\Gamma_X = (\frac{1}{\sqrt{n}}\mathbf{1}_n|\frac{1}{\sqrt{n-1}}\mathbf{F})$. As noted, Γ_Z is not unique but the restricted likelihood is invariant to the choice of Γ_Z. Recall, however, that Γ_Z is defined so that $(\Gamma_X|\Gamma_Z)$ is a basis for the column space of $(\mathbf{X}|\mathbf{Z})$ and $\Gamma_X'\Gamma_Z = \mathbf{0}$. $\mathbf{M}_{ZZ} = \Gamma_Z'\mathbf{Z} = \Gamma_Z'\mathbf{V}_1\mathbf{D}_1^{-0.5}$, so $\mathbf{M}_{ZZ}\mathbf{M}_{ZZ}' = \Gamma_Z'\mathbf{V}_1\mathbf{D}_1^{-1}\mathbf{V}_1'\Gamma_Z$. In the re-expression, the final key items are obtained from the spectral decomposition $\mathbf{M}_{ZZ}\mathbf{M}_{ZZ}' = \mathbf{PAP}'$. The canonical predictors are $\Gamma_Z\mathbf{P}$ and the canonical observations are $\hat{\mathbf{v}} = \mathbf{P}'\Gamma_Z'\mathbf{y}$. Unfortunately, **A**, $\Gamma_Z\mathbf{P}$, and $\hat{\mathbf{v}}$ do not have simple forms in general. I consider special cases below, some of which do yield simple forms.

The rest of this subsection considers two questions about the restricted likelihood. First, given the spatial map, which implies $\mathbf{V}_1$ and $\mathbf{D}_1$, how does the fixed effect **F** affect the restricted likelihood and thus, through σ_s^2 and σ_e^2, the amount of spatial smoothing and the potential severity of spatial confounding? Second, it is widely believed that adding the spatial random effect **S** will adjust the estimate of β for an omitted fixed-effect predictor **H**. Section 10.1.4.1 debunked this belief using derivations that assumed the data were in fact generated by Model H in which $\mathbf{y} = \mathbf{1}_n\beta_0 + \mathbf{F}\beta + \mathbf{H}\gamma + \varepsilon$. Those derivations treated the variances σ_s^2 and σ_e^2 as known but now we can, to some extent, examine the information about them in the restricted likelihood (and thus in the posterior distribution arising from a flat prior on (β_0, β). Thus the second question about the restricted likelihood is: If the data were in fact generated by Model H, how does the specific **H** affect the restricted likelihood and thus the amount of spatial smoothing and the potential for spatial confounding?

So: How does the fixed effect **F** affect the restricted likelihood? I can't answer this in general, but I present three illuminating special cases. In the first, **F** is exactly proportional to the kth column of $\mathbf{V}_1$. In this case, $\Gamma_Z = \mathbf{V}_1^{(-k)}$, where this notation indicates $\mathbf{V}_1$ with its kth column deleted. Then $\mathbf{M}_{ZZ}\mathbf{M}'_{ZZ} = (\mathbf{D}_1^{(-k)})^{-1}$, where $\mathbf{D}_1^{(-k)}$ is $\mathbf{D}_1$ with its kth row and column deleted, which implies $\mathbf{P} = \mathbf{W}_{n-2}$ ($\mathbf{W}_m$ was defined above as the $m \times m$ matrix with 1's down the diagonal from upper right to lower left and 0's elsewhere) and $\mathbf{A} = \mathbf{W}_{n-2}(\mathbf{D}_1^{(-k)})^{-1}\mathbf{W}_{n-2}$. The canonical predictors are the columns of $\mathbf{V}_1^{(-k)}\mathbf{W}_{n-2}$ and the canonical observations are $\hat{\mathbf{v}} = (\mathbf{V}_1^{(-k)}\mathbf{W}_{n-2})'\mathbf{y}$. The information about σ_s^2 comes from contrasts in the data corresponding to the columns of $\mathbf{V}_1$, the eigenvectors of the adjacency matrix **Q**, *excluding* the contrast proportional to **F**, which does not contribute to the restricted likelihood but rather informs only about the fixed effects.

This special case presents two distinct possibilities. If in fact the data were generated from the true model $\mathbf{y} = \mathbf{1}_n\beta_0 + \mathbf{F}\beta + \varepsilon$, so all variation in **y** apart from **F** is iid, then in truth $\text{var}(\hat{v}_j|\sigma_s^2, \sigma_e^2) = \sigma_e^2$. Considering the re-expressed restricted likelihood as a gamma regression with "data" $\hat{v}_j^2$ and linear predictor $\sigma_s^2 a_j + \sigma_e^2$, σ_s^2 will generally be estimated as being close to zero, so **S** will be shrunk close to zero, and spatial confounding cannot occur. This partially contradicts one of the hypotheses stated in Section 10.1.5. On the other hand, if the true data-generating model is $\mathbf{y} = \mathbf{1}_n\beta_0 + \mathbf{F}\beta + \mathbf{V}_1^{(-k)}\gamma + \varepsilon$ for some non-trivial γ, then σ_s^2 may have a large estimate. I am not sure whether spatial confounding can occur in this case. The key equations here are (10.3) and (10.4), which show the change in the estimate of β from adding the spatial random effect; the complication arises from the effect of $r\mathbf{D}$ in equation (10.4). Exploring this question is an exercise; I don't think it will be too hard but my deadline looms.

In transforming from the original model, equation (15.12), to the re-parameterized model that gives the re-expressed restricted likelihood, the random effects lose one dimension corresponding to the column space of the fixed effect **F**. In the very simple special case just discussed, where **F** is exactly proportional to the kth column of $\mathbf{V}_1$, this transition is simple: Γ_Z is simply $\mathbf{V}_1$ without its kth column. But what happens when **F** is not proportional to any column in $\mathbf{V}_1$?

To explore this, our second special case is the Slovenian stomach-cancer data considered in Chapters 5, 9, and 10. Here, $n = 194$ and **F** is the centered and scaled indicator of socioeconomic status. It has correlation 0.72 with $\mathbf{V}_{1,193}$, the random effect design-matrix column for which the random effect, $\theta_{1,193}$, has the largest variance $d_{193}^{-1}\sigma_s^2$ among all the random effects. Because the squared correlations between **F** and the columns of $\mathbf{V}_1$ must add to 1 (see Section 10.1), **F** cannot have a very large correlation with any other column of $\mathbf{V}_1$; the 2nd and 3rd largest correlations are 0.23 with $\mathbf{V}_{1,34}$ and -0.13 with $\mathbf{V}_{1,173}$. In deriving Γ_Z and the canonical predictors $\Gamma_Z\mathbf{P}$, we remove from $\mathbf{V}_1$'s columns the parts that lie in the column space of **F** and omit a column to obtain the 192 canonical predictors. Given the high correlation between

Table 15.5: Slovenian Stomach-Cancer Data

j	1	2	3	4	5	6
$\text{cor}(\mathbf{F}, \mathbf{V}_{1,194-j})$	0.72	0.12	-0.05	0.07	0.10	-0.04
$\text{cor}(\mathbf{V}_{1,194-j}, \mathbf{CP}_j)$	-0.69	-0.96	-0.99	0.98	-0.97	1.00
$(d_{194-j})^{-1}$	33.7	9.31	5.76	4.59	4.10	2.56
a_j	16.9	8.93	5.73	4.55	4.00	2.55

Note: The transition from the original to the transformed random effects; "cor" means Pearson's correlation, "CP" means "canonical predictor."

$\mathbf{V}_{1,193}$ and $\mathbf{F}$, this would suggest at least a radical change in $\mathbf{V}_{1,193}$ and $(d_{193})^{-1}$ in the re-parameterization and much smaller changes in other columns of $\mathbf{V}_1$.

Table 15.5 gives some sense of what the re-parameterization does by showing some information about the last six columns in $\mathbf{V}_1$ and the corresponding first six canonical predictors and their a_j. The correspondence cannot be perfect, of course, because in constructing the canonical predictors we must lose one column out of $\mathbf{V}_1$'s 193 and remove the bits that are in the column space of $\mathbf{F}$. Table 15.5's first row shows the correlation of $\mathbf{F}$ with each of the last six columns in $\mathbf{V}_1$, for which the corresponding random effects have the largest variances. Except for $\mathbf{V}_{1,193}$ (i.e., $j = 1$), these correlations are quite small. Table 15.5's second row shows the correlation between these same six columns in $\mathbf{V}_1$ and the corresponding first six canonical predictors in the re-parameterized model. Apart from sign changes, columns $\mathbf{V}_{1,192}, \ldots, \mathbf{V}_{1,188}$ (i.e., $j = 2,3,4,5,6$) are hardly changed at all in the reparameterization, while $\mathbf{V}_{1,193}$ ($j = 1$) is changed rather more, as indicated by its relatively small correlation with the first canonical predictor (-0.69). Table 15.5's third row shows $(d_{194-j})^{-1}$ for these six columns of $\mathbf{V}_1$ and the fourth row shows a_j for the analogous canonical predictors. Consistent with the changes from the columns of $\mathbf{V}_1$ to the canonical predictors, a_1 is smaller than d_{193}^{-1} by about half while $a_j, j = 2,3,4,5,6$, are barely reduced from the corresponding d_{194-j}^{-1}. Thus in the re-parameterization, the column of $\mathbf{V}_1$ that is most highly correlated with the fixed effect $\mathbf{F}$ is changed most, while the other 5 columns here are changed very little.

Figure 15.7 gives another view of how the re-parameterization changes $\mathbf{V}_1$ into the canonical predictors. Figure 15.7b, c, and d show that canonical predictors 2, 3, and 4 are indeed changed little in the re-parameterization; these columns of $\mathbf{V}_1$ had low correlations with the fixed effect $\mathbf{F}$. Figure 15.7a shows that the first canonical predictor is still almost perfectly correlated with $\mathbf{V}_{1,193}$ *within each of the five distinct values of socioeconomic status*, but considered as a whole Figure 15.7a looks very much like a plot of the residuals from a regression of $\mathbf{V}_{1,193}$ on $\mathbf{F}$.

Table 15.5 describes the six columns of $\mathbf{V}_1$ with least-shrunk random effects and the first six canonical predictors. In fact, the 33 columns of $\mathbf{V}_1$ with least-shrunk random-effects and the corresponding first 33 canonical predictors are very highly correlated with each other; after these 33 the correlations abruptly begin to vary much more uniformly between -1 and +1. Among these 33 pairs of vectors, there are 5 exceptions: $V_{1,193}$ and the first canonical predictor, as noted; the pairs including the

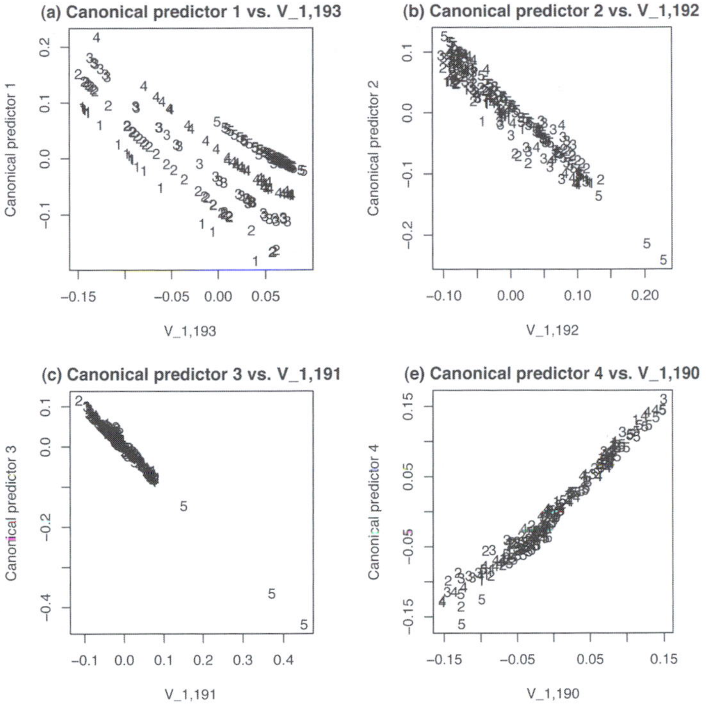

Figure 15.7: Slovenian stomach-cancer data: The first four canonical predictors (vertical axis) versus the corresponding four columns of $\mathbf{V}_1$ (horizontal axis). The plotting symbol (1, 2, 3, 4, 5) is the value of the socioeconomic status measure for each municipality; $\mathbf{F}$ is the centered and scaled version of this variable.

11th and 12th canonical predictors (correlation $+/-$ 0.59 with the analogous columns of $\mathbf{V}_1$); and the pairs including the 28th and 29th canonical predictors (correlation 0.39 with the analogous columns of $\mathbf{V}_1$). Except for $V_{1,193}$ and the first canonical predictor, I have no idea what is distinctive about these columns or why the consistent high correlations stop after the 33rd pair.

How does the re-parameterization affect columns of $\mathbf{V}_1$ with the smallest d_j^{-1}, which correspond to the canonical predictors with the smallest a_j? For the first 20 such pairs of columns starting with $j = 192$, the correlations are at least 0.94 in absolute value, except for the pairs including the 9th and 10th from last canonical predictors (correlation $+/-$ 0.44 with the analogous columns of $\mathbf{V}_1$). Again, I have no idea why this pattern occurs. Obviously, it and the analogous pattern for canonical predictors with small j arise from reducing the number of columns by one and removing the parts of $\mathbf{V}_1$'s columns that are in $\mathbf{F}$'s column space, but otherwise this is a minor mystery.

The two special cases just discussed had, respectively, $\mathbf{F}$ perfectly correlated with one column of $\mathbf{V}_1$ and $\mathbf{F}$ rather highly correlated with one column of $\mathbf{V}_1$. In our third special case, $\mathbf{F} = \mathbf{V}_1\mathbf{1}_{193}$, which has the same correlation (0.023) with all 193 columns of $\mathbf{V}_1$. We might expect all the columns of $\mathbf{V}_1$ to be affected about equally in the re-parameterization and they are. Reflecting this equal treatment and the need to lose one column in the re-parameterization, $\mathbf{V}_{1,194-j}$ and canonical predictor j have correlations with absolute value at least 0.94 for $j = 1,\ldots,27$; as j increases further, these correlations vary in a progressively wider band, though up to $j = 100$ the smallest absolute correlation is 0.52. Similarly, the a_j are reduced very little compared to d_{194-j}^{-1}: The ratio lies in the interval 0.982 to 1 for $j = 1,\ldots,183$ and overall the smallest ratio is 0.938.

Now for the second question about the restricted likelihood in the spatial-confounding case: If the data were in fact generated by Model H, in which $\mathbf{y} = \mathbf{1}_n\beta_0 + \mathbf{F}\beta + \mathbf{H}\gamma + \varepsilon$, how does the specific true $\mathbf{H}$ affect the restricted likelihood? Under Model H,

$$\begin{aligned}\hat{\mathbf{v}} &= \mathbf{P}'\Gamma_Z'\mathbf{y} \\ &= \mathbf{P}'\Gamma_Z'(\mathbf{1}_n\beta_0 + \mathbf{F}\beta + \mathbf{H}\gamma + \varepsilon) \\ &= \mathbf{P}'\Gamma_Z'\mathbf{H}\gamma + \mathbf{P}'\Gamma_Z'\varepsilon. \end{aligned} \tag{15.13}$$

Because $\mathbf{P}'\Gamma_Z'$ has orthonormal rows, $\mathbf{P}'\Gamma_Z'\varepsilon \sim N(\mathbf{0}, \sigma_e^2\mathbf{I}_{s_Z})$, so conditional on σ_e^2, $\hat{\mathbf{v}}$ is normal with covariance $\sigma_e^2\mathbf{I}_{s_Z}$ and mean $\mathbf{P}'\Gamma_Z'\mathbf{H}\gamma$. $\mathbf{P}'\Gamma_Z'\mathbf{H}\gamma$ is the vector of coefficient estimates in a regression of $\mathbf{H}\gamma$ on the canonical predictors $\Gamma_Z\mathbf{P}$. Thus, $\hat{v}_j$ will tend to be large if $\mathbf{H}$ is highly correlated with canonical predictor j. Recalling that the re-expressed restricted likelihood is a gamma regression with "data" $\hat{v}_j^2$ and linear predictor $\sigma_s^2 a_j + \sigma_e^2$, it is clear that for $\mathbf{H}\gamma$ of a given magnitude, the restricted likelihood will depend strongly on which canonical predictors j happen to be highly correlated with $\mathbf{H}$.

Because the a_j decline as j increases, if σ_s^2 is large, the $\sigma_s^2 a_j + \sigma_e^2$ decline as j increases, while if σ_s^2 is small or zero, the $\sigma_s^2 a_j + \sigma_e^2$ are more or less constant in j. Thus, if $\mathbf{H}$ is highly correlated with canonical predictor 1, $\hat{v}_1^2$ will be large, the restricted likelihood will tend to indicate a large σ_s^2, and the random effect will not be shrunk much. This creates the possibility of spatial confounding as well as the possibility that the spatial random effect will adjust $\mathbf{F}$'s coefficient β for the omitted predictor $\mathbf{H}$. On the other hand, if $\mathbf{H}$ is highly correlated with canonical predictor j for large j, then that $\hat{v}_j^2$ will be large and other $\hat{v}_j^2$ will be small and the restricted likelihood will tend to indicate a small σ_s^2. In this case, the random effect will be shrunk close to zero so spatial confounding cannot occur but the spatial random effect also cannot adjust $\mathbf{F}$'s coefficient β for the omitted predictor $\mathbf{H}$.

The actual results for a given $\mathbf{H}$ will depend not just on how $\mathbf{H}$ projects on the various canonical predictors but also on γ and σ_e^2. An exercise explores the effect of $\mathbf{H}$ in detail for the Slovenian stomach-cancer dataset.

15.2.4 Dynamic Linear Model with One Quasi-Cyclic Component

For the present purpose I consider a simplified version of the dynamic linear model (DLM) used for Michael Lavine's dataset on localizing epileptic activity in the brain, discussed in Chapters 6, 9, and 12. Specifically, consider a data series with 650 time steps and a model for it that includes an intercept and one quasi-cyclic component with nominal period (wavelength) 18.75 time steps. This is the single-harmonic respiration component in Prof. Lavine's Models 1 and 2 and the first harmonic of Model 3's respiration component. Below, I consider fits of this model to artificial data with various structures. First, however, we consider the re-expressed restricted likelihood.

The observation equation, the model for the data series $y_t, t = 0\ldots, T = 649$, is

$$y_t = \mu + c_t + v_t, \tag{15.14}$$

where μ is the intercept, c_t is the quasi-cyclic component (this is Section 6.3's respiration component r_t), and v_t is an error term distributed as iid $N(0, \sigma_e^2)$. The quasi-cyclic component has a two-dimensional state; equation (6.19) gives the state equation describing each time step's update of that state. In the notation of equation (6.19), the state is $(b_t \cos\alpha_t, b_t \sin\alpha_t)'$; for simplicity, call the state $(c_t, c_t^{(2)})'$. One step in the state's evolution consists of rotating it through the angle $\delta = 2\pi/18.75$ radians and adding an error $\mathbf{w}_t' = (w_{1,t}, w_{2,t}) \sim$ iid $N_2(0, \sigma_s^2 \mathbf{I}_2)$. It is easy to show (so it is an exercise) that this model can be written as a mixed linear model with

$$\mathbf{X} = \begin{bmatrix} 1 & 1 & 0 \\ 1 & \cos\delta & \sin\delta \\ 1 & \cos 2\delta & \sin 2\delta \\ \vdots & \vdots & \vdots \\ 1 & \cos T\delta & \sin T\delta \end{bmatrix}, \beta = \begin{bmatrix} \mu \\ c_0 \\ c_0^{(2)} \end{bmatrix},$$

$$\mathbf{Z} = \left[\begin{array}{cc|cc|c|cc} 0 & 0 & 0 & 0 & & 0 & 0 \\ 1 & 0 & 0 & 0 & & 0 & 0 \\ \cos\delta & \sin\delta & 1 & 0 & & 0 & 0 \\ \cos 2\delta & \sin 2\delta & \cos\delta & \sin\delta & \ldots & 0 & 0 \\ \cos 3\delta & \sin 3\delta & \cos 2\delta & \sin 2\delta & & 0 & 0 \\ \vdots & \vdots & \vdots & \vdots & & \vdots & \vdots \\ \cos(T-1)\delta & \sin(T-1)\delta & \cos(T-2)\delta & \sin(T-2)\delta & & 1 & 0 \end{array}\right],$$

$$\mathbf{u} = \begin{bmatrix} w_{1,1} \\ w_{2,1} \\ w_{1,2} \\ w_{2,2} \\ \cdots \\ w_{1,T} \\ w_{2,T} \end{bmatrix},$$

where $\mathbf{X}$ is $(T+1)\times 3$, β is 3×1, and $\mathbf{Z}$ is $(T+1)\times 2T$. The last column of $\mathbf{Z}$ is redundant, so it is omitted in the following, leaving $\mathbf{Z}$ with dimension $(T+1)\times(2T-1)$. This leaves $\mathbf{u}$ with dimension $(2T-1)\times 1$, i.e., $w_{2,T}$ is dropped, but this creates no problem because the $w_{k,t}$ are iid $N(0, \sigma_e^2)$.

The elements of the re-expressed restricted likelihood are as follows. We have $n = T+1 = 650$, $s_X = 3$ dimensions for the fixed effects, and $s_Z = 647$ dimensions

Table 15.6: DLM with One Quasi-Cyclic Component: The 20 Largest a_j

21577	21232	5396	5308	2399	2360	1350	1328	864	850
601	591	442	434	338	333	268	263	217	213

for the canonical random effects. Table 15.6 shows the 20 largest a_j. The largest a_j come in pairs; within a pair of a_j, the values are similar but not identical. As we will see, the corresponding canonical predictors also come in pairs. The last roughly 150 a_j are all between 0.30 and 0.26. The a_j decline in j somewhat more quickly than for the simple ICAR model for periodontal data but much more slowly than for the penalized spline for the GMST data: $a_1/a_6 = 9.1$, $a_1/a_{15} = 63.8$, $a_1/a_{166} = 8,072$, and $a_1/a_{647} = 83,908$.

The canonical predictors $\Gamma_Z \mathbf{P}$ are not easy to describe. Broadly, each canonical predictor looks like a combination of sinusoidal curves where the predominant curve generally increases in frequency as j increases. Figure 15.8 shows the first four canonical predictors, i.e., with the four largest a_j. The first two (Panels (a) and (b)) appear nearly identical but the ends of the curves show that they differ by about a half cycle of the predominant sinusoidal curve. The same is true for the third and fourth canonical predictors (Panels (c) and (d)). For each of these canonical predictors, the amplitude changes cyclically, one cycle for the first two canonical predictors and two cycles for the third and fourth. The canonical predictors progress in this manner for several j before changing their pattern.

Figure 15.9 shows the last four canonical predictors, i.e., with the four smallest a_j. These are no longer in similar-looking pairs but again each looks like a sine curve with a very high frequency, the amplitude of which oscillates through 4, 3, 2, and 1 cycles respectively.

To give some sense of how the canonical predictors change as j runs from 1 to 647, I summarized each canonical predictor using its power spectrum, which decomposes a canonical predictor's oscillations into the frequencies present in it, the "power" of a frequency being the extent to which that frequency is present. The purpose is to identify the most prominent frequency — and thus period or wavelength, which are 1/frequency — in canonical predictor j and to describe how it changes as j increases.

Figure 15.10 shows the period (wavelength) with maximum power for each canonical predictor j. Starting with $j = 119$ (the kink in the plot's lower line) this most-powerful wavelength declines monotonically as j increases, i.e., the most prominent frequency in canonical predictor j increases with j. For $j \leq 118$, however, the most powerful wavelength switches back and forth between the lower, decreasing line and the upper, increasing line. The upper line increases to a wavelength of 675 at $j = 118$; I've cut off the axis at 50 to avoid obscuring the lower line's shape. If you look at power spectra for individual canonical predictors for $j \leq 118$ (as an exercise incites you to do), each spectrum has two prominent peaks, one corresponding to each of the lower and upper lines in Figure 15.10. For each j, one of these two peaks is just barely higher than the other; the "winner" switches back and forth between the

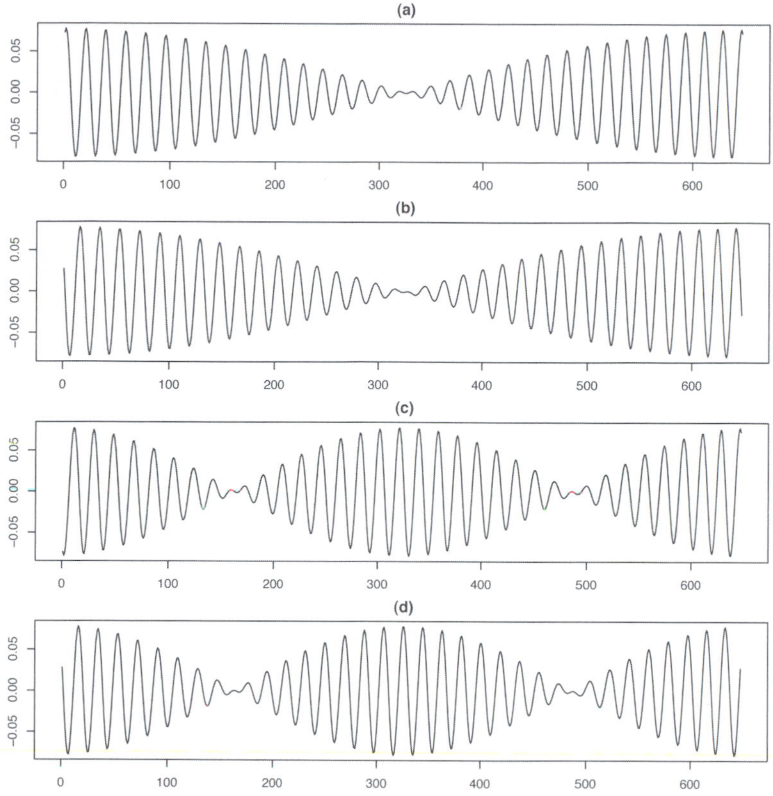

Figure 15.8: DLM with one quasi-cyclic component: Panels (a) through (d) are the first four canonical predictors in the order $j = 1,2,3,4$. The horizontal axis is the time index t, running from 0 to 649; the vertical axis is the value of the canonical predictor.

two peaks as j increases. For $j \geq 119$, however, the peak with the shorter wavelength (higher frequency) predominates and the other peak declines quickly. I have no idea why the break point occurs at $j = 119$.

A given observed $\mathbf{y}$ gives canonical observations $\hat{\mathbf{v}} = \mathbf{P}'\Gamma_Z'\mathbf{y}$. For the present purpose, I constructed artificial data from the fit of Michael Lavine's Model 3, described in Sections 9.1.4 and 12.2. Specifically, the artificial $\mathbf{y}$ is the sum of two pieces from the fit of Model 3 to Prof. Lavine's optical-imaging dataset, the first harmonic of the fitted respiration component, which should fit model (15.14) reasonably well, and the fitted heartbeat component, which has a nominal wavelength (period) of 2.75 time steps and does not fit model (15.14). I included the latter to see how it is manifested in the re-expressed restricted likelihood.

Figure 15.11 shows the resulting $\hat{v}_j$ and a_j. Both plots have logarithmic vertical scales, unlike the analogous plots above. Panel (a) shows that the $\hat{v}_j$ do not generally

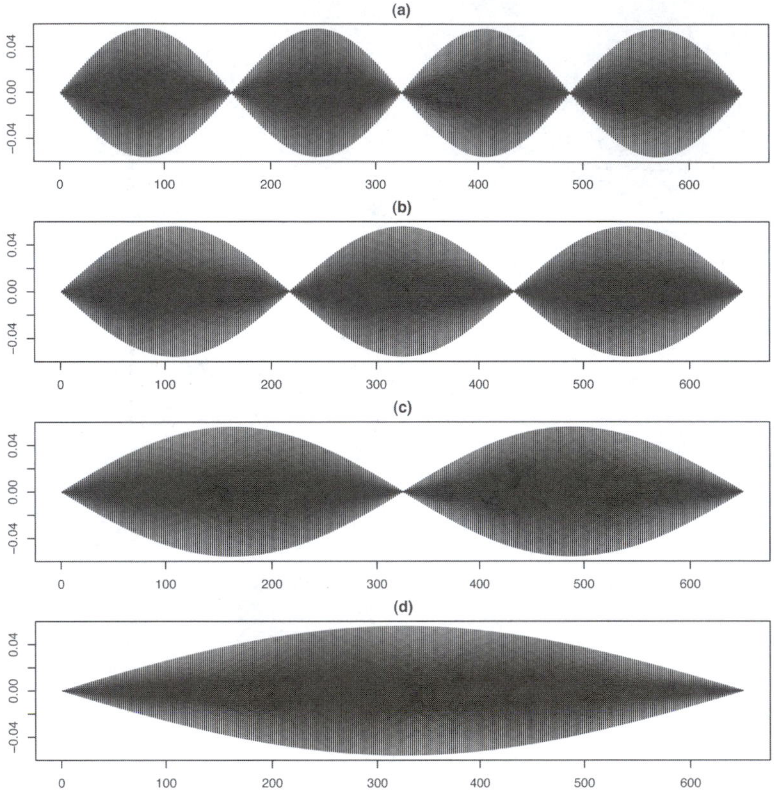

Figure 15.9: DLM with one quasi-cyclic component: Panels (a) through (d) are the last four canonical predictors in the order $j = 644, 645, 646, 647$. The horizontal axis is the time index t, running from 0 to 649; the vertical axis is the value of the canonical predictor.

decrease in j but rather rise to second peak for j about 450. This secondary peak's maximum occurs at a canonical predictor with maximum power at a wavelength of 2.87 time steps, very close to the nominal 2.75 time-step wavelength of the heartbeat component I included. If Panel (a) were plotted with an linear vertical scale instead of a logarithmic scale, it would show a narrow peak at the extreme left (small j) with very few prominent $\hat{v}_j$, and around $j = 450$ only two points would be visible above the row of points lying on the horizontal axis.

If instead of including the fitted heartbeat component in the artificial data, I included a sine curve with period 2.75, Panel (a) would have a more pronounced peak at about $j = 450$ but would otherwise appear similar. If instead the artificial data had included *only* the fitted first harmonic component for respiration, Panel (a) would show a marked *valley* — with a logarithmic vertical scale, though it would be invisible with a linear vertical scale — for j about 450, presumably because these aspects

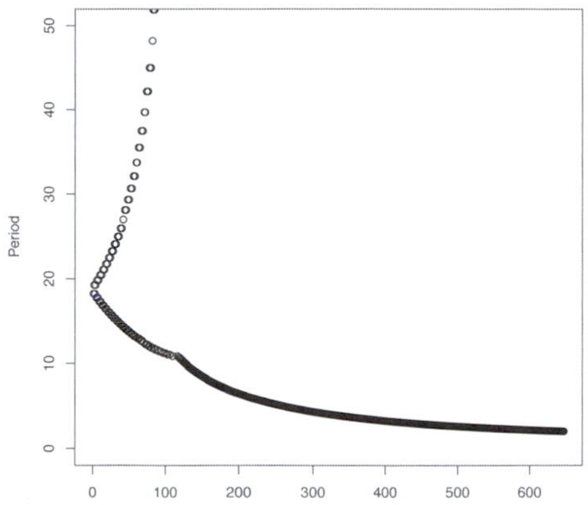

Figure 15.10: DLM with one quasi-cyclic component: The vertical axis is the period (wavelength) with the largest power in the jth canonical predictor's power spectrum.

of Prof. Lavine's data series had been captured in the fitted heartbeat component and were thus deficient in the respiration component's fit. (An exercise asks you to construct these two artificial datasets and draw the plots.)

These plots suggest how the restricted likelihood might be affected if the model does not fit because of an omitted quasi-cyclic component. Recalling that the re-expressed restricted likelihood is a generalized linear model with gamma errors, "observations" $\hat{v}_j^2$, and linear predictor $\sigma_s^2 a_j + \sigma_e^2$, Figure 15.11 suggests that if the omitted quasi-cyclic component has a higher nominal frequency than the included component, the lack of fit might tend to reduce the included component's estimated σ_s^2. But what if the omitted component has a *lower* nominal frequency than the included component? For artificial data consisting of Model 3's fitted first harmonic for respiration plus the fitted first harmonic for mystery (with nominal period 117 time steps), the analog to Figure 15.11's Panel (a) again shows a peak at the far left of the plot but also, surprisingly, a second and somewhat higher peak centered at about $j = 100$. Obviously a good deal of exploration remains to be done here.

15.3 A Tentative Collection of Tools

I learned that the re-expressed restricted likelihood for two-variance models is a generalized linear model just before I started writing this chapter, so I have had little time to work with that idea. Chapter 17 shows two classes of models for which the restricted likelihood can be re-expressed in a diagonalized or scalar-ized form as above but for which the re-expressed restricted likelihood is not a generalized linear model with linear predictor a linear function of the unknowns in ϕ. Thus, while the generalized linear model interpretation has been fruitful, it is probably not the last word here.

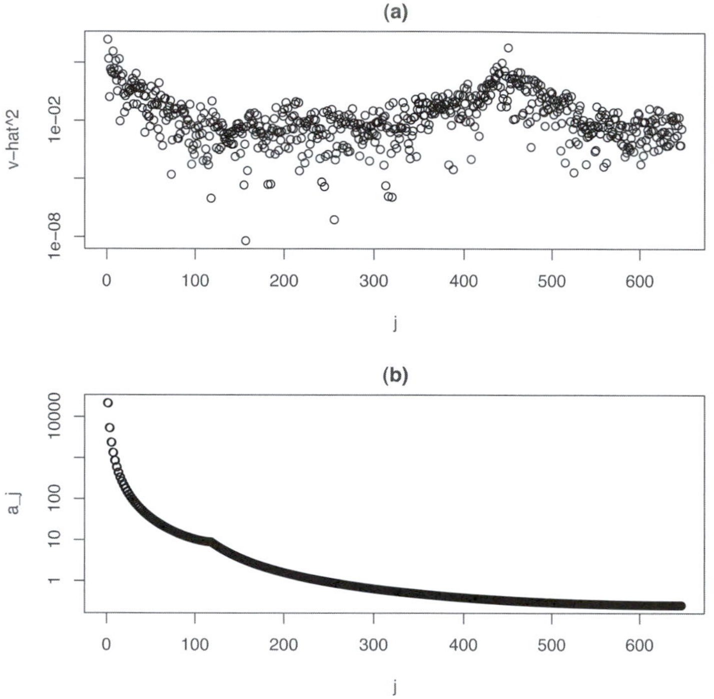

Figure 15.11: Artificial data fit with two quasi-cyclic components, fit using a DLM with one quasi-cyclic component: Panel (a), $\hat{v}_j^2$; Panel (b), a_j.

Tools from generalized linear models. Generalized linear models provide several tools: The deviance (a measure of goodness of fit, Section 2.3 in McCullagh & Nelder 1989), residuals defined in various ways (Section 2.4), and measures of leverage (Section 12.7.1) and case influence (Section 12.7.3) that generalize the analogous measures for ordinary linear models. In the following chapters, I use only residuals and case-influence measures, modifying the latter to explore which canonical observations $\hat{v}_j$ provide information primarily about σ_s^2 and which provide information primarily about σ_e^2. Regarding deviance, I generally don't find omnibus goodness-of-fit tests useful because the important question is not whether a model fits but whether its deficiencies of fit affect the answers to the subject-matter questions under study (*cf* Wakefield 1998, Critchley 1998), and goodness-of-fit tests do not reliably indicate this. However, deviance is used for other purposes and it could be that a goodness-of-fit test applied to the restricted likelihood can identify occasions when more detailed exploration is likely to be useful. An exercise considers this possibility.

The canonical observations $\hat{v}_j$ and their conditional variances $\sigma_s^2 a_j + \sigma_e^2$. Discrepant canonical observations $\hat{v}_j$ can be identified by computing $\hat{v}_j / \sqrt{\sigma_s^2 a_j + \sigma_e^2}$ with the two variances replaced by estimates. Also, because $\text{var}(\hat{v}_j | \sigma_s^2, \sigma_e^2) = \sigma_s^2 a_j + \sigma_e^2$, it seems obvious that mixed terms with very small a_j generally provide information only about σ_e^2 because σ_s^2 has a negligible effect on $\hat{v}_j$ unless σ_s^2 is very large or σ_e^2 is very small. By the same heuristic reasoning, mixed terms with large a_j provide information mainly about σ_s^2. These considerations are developed in Section 16.1.

A modified restricted likelihood that omits all $j > m$. To identify which mixed terms provide information about which variance, I have found it useful to consider, for various m, a modified restricted likelihood that omits canonical observations $j = m + 1, \ldots, s_Z + 1$, the last of these being the free terms. This is closely related to the case influence of the group of "cases" $j = m+1, \ldots, s_Z + 1$, though usually case influence analyses consider only the effects of one or two cases. Also, I have found it useful to consider the usual approximate standard errors computed from these modified restricted likelihoods, which is also not usually part of case influence analysis.

The degrees of freedom in the fit for v_j. Equation (15.7) is the conditional posterior distribution given (σ_e^2, σ_s^2) of the re-parameterized random effects $\mathbf{v}$. In this conditional posterior, the v_j are independent with mean $\tilde{v}_j = \hat{v}_j a_j/(a_j + r)$, variance $\sigma_e^2 a_j/(a_j + r)$, and DF $a_j/(a_j + r)$ for $r = \sigma_e^2/\sigma_s^2$. These DF tell us about the contributions the canonical predictors $(\Gamma_Z \mathbf{P})_j$ make to the fitted model, with larger DF indicating a greater contribution. Again, the a_j are key: v_j with large a_j contribute more to the fit than v_j with small a_j, with the absolute sizes of the contributions determined by σ_s^2. As we'll see in Chapter 16, these DF, along with an understanding of which $\hat{v}_j$ provide information about σ_s^2, explain why Section 14.3's two smoothers gave such different fits.

Reich & Hodges (2008a) emphasized two tools that are not listed here. Those tools were helpful in understanding the examples described in that paper but not helpful when applied to later examples, which led me to focus more on the $\hat{v}_j$ and their conditional variances and on the modified restricted likelihood. One of the two discarded tools, called $\hat{r}_i$ in Reich & Hodges (2008a), now appears entirely superseded. The other discarded tool is a plot with vertical axis σ_s^2 and horizontal axis σ_e^2 showing the lines $\sigma_s^2 = \hat{v}_j^2/a_j - \sigma_s^2/a_j$ (in the present chapter's notation). The jth such line describes the maximum of the jth mixed term; a vertical line for the free terms was also drawn. This plot may turn out to have uses other than the ones in Reich & Hodges (2008a), so I am not yet ready to consign it to the trash heap of history.

Exercises

Regular Exercises

1. (Section 15.1) Prove that Γ_Z exists and give an expression for it.
2. (Section 15.1) Show that the re-expressed restricted likelihood is invariant to the choice of Γ_Z.

3. (Section 15.1) Derive the joint posterior $\pi(\beta^*, \mathbf{v}, \sigma_e^2, \sigma_s^2|\mathbf{y})$ in equations (15.6) to (15.9).
4. (Section 15.2.2) Derive the re-expressed restricted likelihood for the full-rank spline for the global-mean surface data and draw a contour plot of the log restricted likelihood. Compare your contour plot to Figure 15.3's middle row left panel; your plot should look nearly identical except that it is shifted down by log(4).
5. (Section 15.2.4) Derive the mixed-linear-model form for the dynamic linear model given after equation (15.14).
6. (Section 15.2.4) Examine the canonical predictors for the dynamic linear model (15.14). You may wish to examine their power spectra, e.g., using the R function spectrum. I find it remarkable how much their shapes change within the general pattern of a gradually increasing frequency overlaid with other cyclic patterns, e.g., cyclically changing amplitude.
7. (Section 15.2.4) Re-draw Figure 15.11 for other artificial datasets. The text discusses artificial datasets with a sine curve added to the fitted first harmonic for respiration; consisting *only* of the fitted first harmonic for respiration; and consisting of Model 3's fitted first harmonics for respiration and mystery.

Open Questions

1. (Section 15.2.3.2) Consider spatial confounding in the special case in which $\mathbf{F}$ is exactly proportional to the kth column of $\mathbf{V}_1$. Suppose the true data-generating model is $\mathbf{y} = \mathbf{1}_n\beta_0 + \mathbf{F}\beta + \mathbf{V}_1^{(-k)}\gamma + \varepsilon$ and that some entries in γ are big enough in absolute value that σ_s^2 has a large estimate. Can addition of the spatial random effect $\mathbf{S}$ to the model $\mathbf{y} = \mathbf{1}_n\beta_0 + \mathbf{F}\beta + \varepsilon$ change the estimate of β much or indeed at all? As noted in the text, the key equations are (10.3) and (10.4) and the complication arises from the effect of $r\mathbf{D}$ in equation (10.4).
2. (Section 15.2.3.2) In the spatial-confounding problem, if the data are truly generated by Model H, in which $\mathbf{y} = \mathbf{1}_n\beta_0 + \mathbf{F}\beta + \mathbf{H}\gamma + \varepsilon$, how does $\mathbf{H}$ affect the restricted likelihood? The text discussed this qualitatively; examine this in detail by considering various $\mathbf{H}$, γ, and σ_e^2.
3. (Section 15.3) Interpreting the re-expressed restricted likelihood as a gamma-errors GLM, compute deviance-based goodness-of-fit tests for the penalized spline, simple ICAR, and DLM examples. If this goodness-of-fit test is worthwhile as a "gatekeeper" for deeper exploration, it should detect problems in the simple ICAR fit to the attachment-loss data and the simple DLM fit to the artificial datasets with quasi-cyclic components that are not in the model.

Chapter 16

Exploring the Restricted Likelihood for Two-Variance Models

Section 16.1 uses the re-expressed restricted likelihood to ask which $\hat{v}_j$ mostly provide information about σ_s^2 or σ_e^2 and Section 16.2 uses those results to explain the mysteries in Sections 14.1 and 14.3.

16.1 Which $\hat{v}_j$ Tell Us about Which Variance?

This section makes an argument that mixed terms with large a_j mostly provide information about σ_s^2 while mixed terms with smaller a_j and the free terms mostly provide information about σ_e^2. Thus, the relatively few mixed terms with relatively large a_j are especially important in determining the estimate of σ_s^2 and therefore the extent of smoothing. As will become clear, this generalization is oversimplified but I think it is useful as long as its limitations are not forgotten. I develop this material using two examples from Chapter 15, the penalized spline for the global mean surface temperature data and the simple ICAR (no explicit fixed effects) applied to the attachment-loss data. These examples differ in two ways: The a_j decline much more quickly in j for the spline than for the ICAR and the spline's restricted likelihood has fixed terms for σ_e^2 while the ICAR's does not. I begin with heuristic comments and become progressively more specific as the section proceeds.

This discussion is complicated by a fact about the restricted likelihood: When some change increases the maximizing value of σ_s^2, the maximizing value of σ_e^2 usually decreases, and vice versa. Considering the re-expressed restricted likelihood in equation (15.4), the intuition is that once the data are observed, they are fixed in the restricted likelihood and the two variances compete to capture the available variation. Thus, if the estimate (maximizing value) of one variance increases, the estimate of the other variance must decrease. This complicates consideration of which variance a particular $\hat{v}_j$ tells us about. I say more about this as the section proceeds.

16.1.1 Some Heuristics

The free terms are a function of σ_e^2 only — in effect the free terms are $n - s_X - s_Z$ mixed terms with $a_j = 0$ — so they can only indirectly affect the maximizing value of σ_s^2 and the restricted likelihood's curvature with respect to σ_s^2 (i.e., its approximate posterior precision). Also, the jth mixed term is a function of the variances only through $\sigma_s^2 a_j + \sigma_e^2$, so if a_j is small (which is admittedly vague), the jth mixed term is effectively a function of σ_e^2 only, while if a_j is very large (again, vague), the jth mixed term is dominated by σ_s^2.

We can make these observations somewhat less vague by considering derivatives of the mixed terms. Call the jth mixed term m_j; the first derivatives of m_j are

$$\begin{aligned} \frac{\partial m_j}{\partial \sigma_e^2} &= -0.5(\sigma_s^2 a_j + \sigma_e^2)^{-1} + 0.5\hat{v}_j^2(\sigma_s^2 a_j + \sigma_e^2)^{-2} \\ \frac{\partial m_j}{\partial \sigma_s^2} &= a_j \frac{\partial m_j}{\partial \sigma_e^2}. \end{aligned} \tag{16.1}$$

If a_j is small, $\partial m_j / \partial \sigma_s^2$ is much closer to zero than $\partial m_j / \partial \sigma_e^2$ and both derivatives are insensitive to changes in σ_s^2, so the jth mixed term m_j can have little effect on the value of σ_s^2 that maximizes the restricted likelihood. On the other hand, if a_j is large, $\partial m_j / \partial \sigma_s^2$ is much larger in magnitude than $\partial m_j / \partial \sigma_e^2$ and both derivatives are sensitive to changes in σ_s^2.

As for the second derivatives,

$$\begin{aligned} \frac{\partial^2 m_j}{\partial (\sigma_e^2)^2} &= 0.5(\sigma_s^2 a_j + \sigma_e^2)^{-2} - \hat{v}_j^2(\sigma_s^2 a_j + \sigma_e^2)^{-3} \\ \frac{\partial^2 m_j}{\partial \sigma_e^2 \partial \sigma_s^2} &= a_j \frac{\partial^2 m_j}{\partial (\sigma_e^2)^2} \\ \frac{\partial^2 m_j}{\partial (\sigma_s^2)^2} &= a_j^2 \frac{\partial^2 m_j}{\partial (\sigma_e^2)^2} \end{aligned} \tag{16.2}$$

Evaluating (16.2) at the maximum-restricted-likelihood estimates, multiplying by -1, and summing over j gives an approximate precision matrix for the maximum restricted likelihood estimates or an approximate posterior precision matrix for (σ_e^2, σ_s^2). Inverting this matrix approximates the covariance or posterior covariance though admittedly not very well. If a_j is small, m_j makes a negligible contribution to $\partial^2 m_j / \partial (\sigma_s^2)^2$. If a_j is large, m_j makes a much larger contribution to the approximate posterior precision for σ_s^2 than for σ_e^2. Section 16.1.3 below shows this effect in discussing the modified restricted likelihood that includes only terms with $j = 1, \ldots, m$.

16.1.2 Case-Deletion Using the Gamma GLM Interpretation

We can make the discussion more precise by using the gamma GLM interpretation of the restricted likelihood and considering the case influence of the "observations" $\hat{v}_j^2$, the mixed terms, and $\hat{v}_{s_Z+1}^2$, the free terms. (I will now drop the quotes around "observation.") Usually measures of leverage and case influence describe how an observation's removal affects the whole parameter vector (e.g., McCullagh & Nelder 1989, Sections 12.7.1, 12.7.3). Here, however, we are interested in an observation's separate effects on the two variances σ_s^2 and σ_e^2; I'll develop this using McCullagh & Nelder's approach to case influence in GLMs.

Omitting a mixed term from the restricted likelihood corresponds to fitting a new model in which that mixed term's canonical predictor is treated as a fixed effect instead of a random effect, i.e., moved from **Z** to **X** and given infinite prior variance. This was done for the simple ICAR model for the attachment-loss data in Section 14.1 and produced a puzzle that is resolved below. Omitting the free terms amounts to reducing the dataset by throwing away all but certain summaries. In simpler models this is equivalent to retaining design-cell means and throwing away within-cell replications. For models like the ICAR or penalized spline, it amounts to reducing the dataset to the $\hat{v}_j$ for mixed terms. Omitting a mixed term corresponds to something we will actually do sometimes, while omitting the free terms is interesting only as a way of understanding their effect on the fit.

If this were an ordinary linear regression, we could use the updating formulas given in Section 8.4 to derive the exact change in each variance's estimate from removing an observation. However, the restricted likelihood for a two-variance model is a gamma-errors regression, which differs from an ordinary linear regression in that, among other things, each observation's variance is a function of its mean, which is a function of the unknowns σ_s^2 and σ_e^2. Thus, to measure an observation's influence exactly we need to re-fit the model for each deleted observation, which usually involves several iterations of the iteratively re-weighted least squares (IRLS) fitting algorithm for GLMs. To reduce the computing burden, McCullagh & Nelder (1989, Section 12.7.3) suggest computing an approximation to the change from deleting an observation by making just one IRLS iteration, starting at the estimates from the complete dataset.

Because the re-expressed restricted likelihood is a GLM with the identity link, this one-step approximation is especially simple. Take the full-dataset estimates for σ_s^2 and σ_e^2 as if known to be true, and compute the corresponding variances of the observations $\hat{v}_j^2$, $j = 1, \ldots, s_Z + 1$, as in Table 15.1. The reciprocals of these variances are the weights in the single IRLS step we will take. But these are also the weights that were used in the *last* iteration of the IRLS algorithm that gave the full-dataset estimates, because that fully iterated IRLS algorithm converged. This allows us to compute the desired one-step approximation using the updating formula from linear regression. (Proof of this assertion is straightforward and given as an exercise.) A detailed derivation of the one-step approximation follows.

Let **W** be the diagonal weight matrix evaluated at the full-data estimates of σ_s^2 and σ_e^2, which for this purpose I'll call $(\tilde{\sigma}_s^2, \tilde{\sigma}_e^2)$. **W**'s diagonal entries are

$$\begin{aligned} W_j &= \frac{1}{2}(\tilde{\sigma}_s^2 a_j + \tilde{\sigma}_e^2)^{-2}, j = 1, \ldots, s_Z \\ W_{s_Z+1} &= \frac{n - s_X - s_Z}{2}(\tilde{\sigma}_e^2)^{-2}. \end{aligned} \tag{16.3}$$

Define the $(s_Z + 1) \times 2$ matrix **B** with rows $\mathbf{B}_j = [1, a_j]$, with $a_{s_Z+1} = 0$ for the free terms; this is the design matrix in the gamma GLM. Now apply the updating formula for linear regression to the approximate linear model

$$\mathbf{W}^{1/2}\hat{\mathbf{v}}^2 = \mathbf{W}^{1/2}\mathbf{B}\begin{bmatrix} \sigma_e^2 \\ \sigma_s^2 \end{bmatrix} + \varepsilon, \tag{16.4}$$

where the error term ε has constant variance as a consequence of pre-multiplying by $\mathbf{W}^{1/2}$. The one-step approximation to the change in the estimates of σ_s^2 and σ_e^2 from deleting $\hat{v}_j^2$ is then

$$W_j^{1/2}(\hat{v}_j^2 - \tilde{\sigma}_s^2 a_j - \tilde{\sigma}_e^2) \tag{16.5}$$

$$\times \quad (-W_j^{1/2}(\mathbf{B}'\mathbf{W}\mathbf{B})^{-1}\mathbf{B}_j')/(1 - W_j\mathbf{B}_j(\mathbf{B}'\mathbf{W}\mathbf{B})^{-1}\mathbf{B}_j'); \tag{16.6}$$

(the proof is straightforward and given as an exercise). Equation (16.5) is a scaled residual for the jth observation, which is being deleted; McCullagh & Nelder (1989, Section 2.4.1) call these Pearson residuals. Equation (16.6) is a 2-vector that multiplies the scaled residual to give the (approximate) case-deletion change for σ_s^2 and σ_e^2. Examining (16.6) gives some insight into whether $\hat{v}_j^2$ mainly provides information about σ_s^2 or σ_e^2.

The least explicit part of equation (16.6) is $(\mathbf{B}'\mathbf{W}\mathbf{B})^{-1}$, because inverting

$$\mathbf{B}'\mathbf{W}\mathbf{B} = 0.5\begin{bmatrix} (n - s_X - s_Z)(\sigma_e^2)^{-2} + \sum_i(\sigma_s^2 a_i + \sigma_e^2)^{-2} & \sum_i a_i(\sigma_s^2 a_i + \sigma_e^2)^{-2} \\ \sum_i a_i(\sigma_s^2 a_i + \sigma_e^2)^{-2} & \sum_i a_i^2(\sigma_s^2 a_i + \sigma_e^2)^{-2} \end{bmatrix} \tag{16.7}$$

is messy. Let

$$\mathbf{B}'\mathbf{W}\mathbf{B} = \begin{bmatrix} f & g \\ g & h \end{bmatrix} \text{ so that } (\mathbf{B}'\mathbf{W}\mathbf{B})^{-1} = \frac{1}{fh - g^2}\begin{bmatrix} h & -g \\ -g & f \end{bmatrix}. \tag{16.8}$$

Then equation (16.6) can be written as

$$\underbrace{\frac{-1}{fh - g^2}}_{\text{(i)}} \times \underbrace{\frac{W_j^{1/2}}{1 - W_j\mathbf{B}_j(\mathbf{B}'\mathbf{W}\mathbf{B})^{-1}\mathbf{B}_j'}}_{\text{(ii)}} \times \begin{bmatrix} (\; h \quad -g\;)\,\mathbf{B}_j' \\ (\; -g \quad f\;)\,\mathbf{B}_j' \end{bmatrix} \begin{matrix} \}\text{(iii)} \\ \}\text{(iv)} \end{matrix}. \tag{16.9}$$

In the equation above, item (i) is the same for all cases and for both σ_s^2 and σ_e^2. Item (ii) is specific to case j but is the same for σ_s^2 and σ_e^2. The product of items (i) and (ii) is negative for all j. Item (iii) is specific to case j and σ_e^2 and item (iv) is specific to case j and σ_s^2. Items (iii) and (iv) will now be written more explicitly and examined.

Item (iii) is

$$(\; h \quad -g\;)\,\mathbf{B}_j' = \frac{1}{2}\sum_i \frac{a_i(a_i - a_j)}{(\tilde{\sigma}_s^2 a_i + \tilde{\sigma}_e^2)^2} \tag{16.10}$$

(the derivation is an exercise). For $j = 1$ this is negative because a_1 is the largest of the a_j. As j increases, this increases; for $j = s_Z$ and for the free terms ($a_{s_Z+1} = 0$), it is positive. This allows us to see, qualitatively, how $\hat{\sigma}_e^2$ is changed by omitting an observation $\hat{v}_j^2$ from the restricted likelihood.

If the free terms are omitted, $\hat{\sigma}_e^2$ generally increases: The scaled residual (16.5) for the free terms is non-positive because $\hat{v}_{s_Z+1}^2$, the residual mean square from the unshrunk fit, is necessarily no larger than $\tilde{\sigma}_e^2$ and almost always rather smaller. The effect of omitting mixed term j depends on a_j and on the scaled residual $W_j^{1/2}(\hat{v}_j^2 -$

$\tilde{\sigma}_s^2 a_j - \tilde{\sigma}_e^2$). Suppose the jth scaled residual is negative, which means that $\hat{v}_j^2$ indicates less variation than expected in the direction of canonical predictor j. If a_j is small — small enough that (16.10) is positive — then $\hat{\sigma}_e^2$ increases when the jth mixed term is omitted, as for the free terms. This change is in the "right" direction: Omitting a relatively small $\hat{v}_j^2$ increases the estimate. However if a_j is large — large enough that (16.10) is negative — then omitting the jth mixed term has the opposite, "wrong" effect on $\hat{\sigma}_e^2$: Omitting this relatively small $\hat{v}_j^2$ *decreases* the estimate. (I'll say more on this below.) Now suppose the jth scaled residual is positive, so that $\hat{v}_j^2$ indicates more variation than expected in the direction of canonical predictor j. Omitting the jth mixed term again has the "right" effect on $\hat{\sigma}_e^2$ if a_j is small but the "wrong" effect if a_j is large.

Now consider equation (16.9)'s item (iv), which is

$$\begin{pmatrix} -g & f \end{pmatrix} \mathbf{B}'_j = \frac{1}{2}\left[a_j(n - s_X - s_Z)(\tilde{\sigma}_e^2)^{-2} + \sum_i \frac{a_j - a_i}{(\tilde{\sigma}_s^2 a_i + \tilde{\sigma}_e^2)^2} \right] \tag{16.11}$$

(the derivation is an exercise). The first summand in (16.11) is zero for the free terms ($a_{s_Z+1} = 0$); for mixed terms, it is positive and decreases as j increases. The second summand in (16.11) is positive for $j = 1$ and decreases as j increases, becoming negative when a_j is small enough, in particular for the free terms. Again, this allows us to see qualitatively how $\hat{\sigma}_s^2$ is changed by omitting an observation $\hat{v}_j^2$ from the restricted likelihood.

By the same logic as above, if the free terms are omitted from the restricted likelihood, $\hat{\sigma}_s^2$ decreases. The effect of omitting mixed term j from the restricted likelihood depends, again, on a_j and on the scaled residual. Suppose the scaled residual is negative, i.e., $\hat{v}_j^2$ indicates less variation than expected in the direction of canonical predictor j. Then if a_j is large — large enough that (16.11) is positive — $\hat{\sigma}_s^2$ increases when the jth mixed term is omitted. This change is in the "right" direction: Omitting a relatively small $\hat{v}_j^2$ increases the estimate. But if a_j is small — small enough that (16.11) is negative — then omitting the jth mixed term has the opposite, "wrong" effect on $\hat{\sigma}_s^2$: Omitting this relatively small $\hat{v}_j^2$ *decreases* the estimate. If the scaled residual is positive, the effects on $\hat{\sigma}_s^2$ are in the opposite directions but are still "right" for large a_j and "wrong" for small a_j.

We've noted that the estimates $\hat{\sigma}_e^2$ and $\hat{\sigma}_e^2$ react in the "right" or "wrong" way to omitting $\hat{v}_j^2$ depending on a_j. The "right" reactions seem fairly obvious; the odd thing is the "wrong" reactions. Heuristically, these "wrong" reactions happen because once the data are observed, they are fixed in the restricted likelihood and the two variances compete to capture the available variation, so if the estimate of one variance increases, the estimate of the other variance must decrease. Thus $\hat{\sigma}_s^2$ reacts backwardly to omitting $\hat{v}_j^2$ with small a_j by an indirect effect: Omitting a $\hat{v}_j^2$ with small a_j changes $\hat{\sigma}_e^2$ the "right" way, which induces the opposite, "wrong," change in $\hat{\sigma}_s^2$. If you accept this,[1] then the "wrong" reactions support the generalization that

[1]Because this argument is weak, though it's the best story I've been able to devise. An exercise asks you to devise a better explanation.

$\hat{v}_j^2$ with small a_j mostly provide information about σ_e^2 and $\hat{v}_j^2$ with large a_j mostly provide information about σ_s^2.

Now let's consider the one-step approximation for the two examples from Chapter 15. The approximation works well for both examples and largely supports the heuristic discussion above.

For the penalized spline applied to the global-mean surface temperature data, Table 16.1 shows the a_j, $\hat{v}_j^2$, scaled residuals, multipliers of the scaled residuals for σ_e^2 and σ_s^2, and approximate and exact changes in the estimates from case deletion. Comparing the one-step approximate changes to the exact changes (the four columns on the right of the table), the approximation does worst when the changes are largest, as you might expect, but the approximation is always close. (It may not seem close for $\hat{\sigma}_s^2$ when $j \geq 9$, but these changes are all tiny compared to the full-data estimate $\hat{\sigma}_s^2 =$ 947.7.) Table 16.1 supports the qualitative discussion about the effects of deleting different $\hat{v}_j^2$. For σ_s^2, the only big changes occur for $j < 6$ and for $j = 31$, the free terms. The rapid decline in influence as j increases is caused by the rapid decline in the a_j. The influence on $\hat{\sigma}_s^2$ is especially large for $j = 1,2$, on the order of half the full-data estimate of 948. Recall that these canonical predictors are respectively roughly cubic and quartic. Since the most interesting feature of this smooth is the decline-then-rise between about 1940 and 1970, you might expect the cubic predictor ($j = 1$) to be the most prominent, but in fact it has a negative scaled residual. Look back at Figure 14.5: For the earliest years, the smooth fit actually declines,[2] so the smooth's shape is more quartic than cubic, and the scaled residuals for $j = 1,2$ reflect this.

As for $\hat{\sigma}_e^2$, the influence of individual observations j is not large except for the free terms; omitting the free terms increases $\hat{\sigma}_e^2$ by about 20%. It is not surprising that the free terms are influential, because 92 contrasts in the data contribute to them and the restricted likelihood weights them accordingly. The changes in $\hat{\sigma}_e^2$ are especially small for $j = 1,\ldots,5$, which have large influence on $\hat{\sigma}_s^2$. However, the effect of any mixed term on $\hat{\sigma}_e^2$ is diluted by the other 29 mixed terms and the free terms. In this sense, the data provide a lot of information about $\hat{\sigma}_e^2$ for this model.

Finally, note that for $j = 4,\ldots,10$, the factor by which the scaled residual is multiplied to give case influence — in equation (16.9)'s labeling, these are (i)×(ii)×(iii) for σ_e^2 and (i)×(ii)×(iv) for σ_s^2 — is negative for both σ_e^2 and σ_s^2. Deleting these observations changes both variance estimates in the same direction. Omitting some of these induce rather large changes in $\hat{\sigma}_s^2$, as large as 10%.

The next example, the simple ICAR model fit to the attachment-loss data, contrasts with the preceding example in that there are no free terms and the a_j decline much more slowly. Table 16.2 shows the effects of deleting selected $\hat{v}_j^2$. (The selected j are also of interest for the modified restricted likelihood, considered in Section 16.1.3. An exercise allows you to see all of them.) As for the penalized spline, the one-step approximation generally works well. It is least accurate for $j = 83, 84$, for which the canonical predictors are contrasts between the direct and interproximal

[2]This is an artifact of the larger variation in earlier years in the 2006 version of this series, which I use in this book. Later versions of this series use more data in estimating GMST deviations in the early years and a fit of the same penalized spline does not have this artifact.

Table 16.1: Global Mean Surface Temperature Data, Penalized Spline Fit: Changes in Maximum-Restricted-Likelihood Estimates of σ_e^2 and σ_s^2 from Deleting Case j, Exact and One-Step Approximation

			Scaled	Multiplier for		Estimate's Change from			
						$\tilde{\sigma}_e^2 = 145.3$		$\tilde{\sigma}_s^2 = 947.7$	
j	a_j	$\hat{v}_j^2$	Resid	σ_e^2	σ_s^2	1-Step	Exact	1-Step	Exact
1	3.6(+1)	2856	–0.65	0.81	–727	–0.53	–0.45	471.1	448.2
2	3.2(+0)	5248	0.48	0.63	–666	0.30	0.92	–318.1	–453.8
3	5.6(–1)	815	0.14	0.05	–475	0.01	0.03	–68.2	–74.7
4	1.5(–1)	409	0.31	–0.69	–252	–0.21	–0.21	–77.5	–79.0
5	4.9(–2)	8	–0.68	–1.23	–116	0.83	0.76	78.4	94.7
6	2.0(–2)	335	0.74	–1.51	–51.0	–1.12	–1.11	–37.7	–38.2
7	8.8(–3)	6	–0.68	–1.64	–22.5	1.12	1.12	15.3	16.3
8	4.3(–3)	497	1.65	–1.71	–9.67	–2.81	–2.81	–15.9	–15.8
9	2.3(–3)	15	–0.63	–1.73	–3.58	1.10	1.10	2.3	1.8
10	1.3(–3)	248	0.49	–1.75	–0.51	–0.86	–0.86	–0.2	0.2
11	7.7(–4)	461	1.53	–1.76	1.13	–2.68	–2.69	1.7	3.3
12	4.8(–4)	119	–0.13	–1.76	2.04	0.23	0.23	–0.3	–0.4
13	3.0(–4)	2	–0.70	–1.76	2.57	1.23	1.23	–1.8	–2.6
14	2.0(–4)	12	–0.65	–1.77	2.90	1.15	1.15	–1.9	–2.7
15	1.4(–4)	1	–0.70	–1.77	3.10	1.24	1.24	–2.2	–3.0
16	9.5(–5)	429	1.38	–1.77	3.23	–2.44	–2.44	4.4	6.3
17	6.8(–5)	118	–0.13	–1.77	3.32	0.23	0.24	–0.4	–0.6
18	4.9(–5)	1	–0.70	–1.77	3.37	1.24	1.25	–2.4	–3.2
19	3.6(–5)	8	–0.67	–1.77	3.41	1.18	1.19	–2.3	–3.1
20	2.7(–5)	199	0.26	–1.77	3.44	–0.46	–0.46	0.9	1.2
21	2.1(–5)	428	1.38	–1.77	3.46	–2.43	–2.44	4.8	6.6
22	1.6(–5)	0	–0.71	–1.77	3.48	1.25	1.25	–2.5	–3.3
23	1.3(–5)	67	–0.38	–1.77	3.49	0.67	0.67	–1.3	–1.8
24	1.0(–5)	71	–0.36	–1.77	3.50	0.64	0.64	–1.3	–1.7
25	8.3(–6)	170	0.12	–1.77	3.50	–0.21	–0.21	0.4	0.6
26	7.0(–6)	588	2.15	–1.77	3.51	–3.81	–3.82	7.5	10.6
27	6.1(–6)	7	–0.67	–1.77	3.51	1.19	1.20	–2.4	–3.2
28	5.4(–6)	274	0.62	–1.77	3.51	–1.10	–1.11	2.2	3.0
29	5.1(–6)	78	–0.33	–1.77	3.51	0.58	0.58	–1.1	–1.6
30	3.7(–6)	365	1.07	–1.77	3.52	–1.90	–1.90	3.8	5.2
31	0	137	–0.37	–78.31	156.15	28.86	29.06	–57.5	–66.0

Note: Row $j = 31$ is the free terms. The a_j are in scientific notation with the power of 10 in parentheses.

sites (one for each arch). These observations have huge scaled residuals, 17.8 and 22.9, so it is hardly surprising that the approximation stumbles a bit, though it is quite good for σ_e^2. The approximation is also not so good for $j = 3,4$, omission of which induces the next-largest change in $\hat{\sigma}_s^2$ after $j = 83,84$. Canonical predictors $j = 3,4$ are, for each arch, roughly quadratic from one end of the arch to the other and capture the well-known phenomenon that attachment loss tends to be greater in the back of the mouth than in the front. However, although $\hat{v}_3^2$ and $\hat{v}_4^2$ are very large, their scaled residuals are not particularly large; a_3 and a_4 are large so $\sqrt{W_3}$ and $\sqrt{W_4}$, which scale the residuals, are small.

This example supports, though a bit tepidly, the generalization that observations with large a_j mostly provide information about σ_s^2 and observations with small a_j mostly provide information about σ_e^2. As for the penalized spline, apart from the spectacular outliers $j = 83,84$, the biggest changes in $\hat{\sigma}_s^2$ occur in the first 25 or so observations (though Table 16.2 only shows some of these). Of particular note are $j = 3,4$: Their scaled residuals are not especially large but omitting either of them causes a larger change in $\hat{\sigma}_s^2$ than omitting any other $\hat{v}_j^2$, even $j = 83,84$. By contrast, $\hat{\sigma}_e^2$ changes moderately and in the "wrong" direction when $j = 3,4$ are omitted, but changes a great deal and in the "right" direction when $j = 83,84$ are omitted.

Table 16.2 includes $j = 147,\ldots,166$ to show that the changes in $\hat{\sigma}_s^2$ from omitting these observations are all tiny. Apart from $j = 3,4,83,84$, the largest changes in $\hat{\sigma}_e^2$ are induced by omitting $j = 94,115,129$ and 159, but these have very little effect on $\hat{\sigma}_s^2$. In contrast to the penalized spline, the a_j decline rather slowly here, so the canonical predictors are less strongly differentiated.

16.1.3 Modified Restricted Likelihood

We can also examine the information provided by each $\hat{v}_j$ by considering modified restricted likelihoods that include only $j = 1,\ldots,m$ for various m. As noted, this is like case deletion but the calculations shown below are exact and they include the approximate standard errors (SEs) for $\hat{\sigma}_e^2$ and $\hat{\sigma}_s^2$ from the modified restricted likelihood (i.e., compute $-1\times$ the hessian of the log restricted likelihood, invert it, and take the square root of its diagonal). As noted in Chapter 1, these are generally poor approximations for the standard deviation of the variance estimates but they do describe the curvature of the modified restricted likelihood at its peak and tell us about how that curvature changes as we add mixed terms.

So consider the penalized spline fit to the global-mean surface temperature data, which has free terms ($j = 31$) and 30 mixed terms. A modified restricted likelihood must have at least two mixed terms to identify σ_e^2 and σ_s^2, so we consider $m = 2,\ldots,31$; $m = 31$ gives the entire restricted likelihood. Figure 16.1 plots the maximizing value of the modified restricted likelihood for each m (top row) and the approximate SE (bottom row). For $m = 5$, $\hat{\sigma}_e^2 = 0$ and the hessian is not negative definite so the approximate SE cannot be computed.

Regarding the estimates, the most striking thing is that increasing m (adding mixed terms) changes both estimates; if one estimate is increased, the other is decreased (Figure 16.1, upper panels). One might argue that the first few mixed terms

Table 16.2: Attachment Loss Data, Simple ICAR Fit: For Selected $\hat{v}_j^2$, Changes in Maximum-Restricted-Likelihood Estimates of σ_e^2 and σ_s^2 from Deleting $\hat{v}_j^2$, Exact and One-Step Approximation

						Estimate's Change from			
			Scaled	Multiplier for		$\tilde{\sigma}_e^2 = 1.26$		$\tilde{\sigma}_s^2 = 0.25$	
j	a_j	$\hat{v}_j^2$	Resid	σ_e^2	σ_s^2	1-Step	Exact	1-Step	Exact
1	149.0	2.9	–0.7	0.028	–0.057	–0.018	–0.018	0.037	0.037
2	149.0	4.9	–0.6	0.028	–0.057	–0.017	–0.017	0.035	0.035
3	37.3	34.3	1.6	0.023	–0.050	0.036	0.048	–0.078	–0.094
4	37.3	28.2	1.2	0.023	–0.050	0.027	0.033	–0.058	–0.067
5	16.6	0.2	–0.7	0.017	–0.041	–0.012	–0.013	0.028	0.030
6	16.6	1.0	–0.6	0.017	–0.041	–0.010	–0.011	0.024	0.026
7	9.4	2.4	–0.2	0.012	–0.033	–0.003	–0.003	0.008	0.009
8	9.4	3.2	–0.1	0.012	–0.033	–0.001	–0.001	0.003	0.003
9	6.1	3.6	0.2	0.007	–0.027	0.001	0.002	–0.006	–0.006
10	6.1	0.7	–0.5	0.007	–0.027	–0.004	–0.005	0.014	0.016
16	2.4	6.2	1.6	0.003	–0.013	–0.004	–0.005	–0.022	–0.021
17	1.9	4.3	1.0	–0.005	–0.011	–0.005	–0.005	–0.011	–0.011
30	0.8	4.1	1.3	–0.010	–0.003	–0.013	–0.013	–0.004	–0.004
31	0.8	7.9	3.1	–0.010	–0.003	–0.031	–0.031	–0.009	–0.009
83	0.3	35.3	17.8	–0.014	0.002	–0.242	–0.253	0.028	0.043
84	0.3	45.1	22.9	–0.014	0.002	–0.312	–0.329	0.036	0.060
85	0.3	1.0	–0.2	–0.014	0.002	0.003	0.003	0.000	0.000
94	0.2	4.4	1.7	–0.015	0.003	–0.024	–0.025	0.005	0.006
95	0.2	0.0	–0.7	–0.015	0.003	0.010	0.010	–0.002	–0.002
114	0.2	0.3	–0.6	–0.015	0.003	0.008	0.008	–0.002	–0.002
115	0.2	4.2	1.5	–0.015	0.003	–0.022	–0.023	0.005	0.005
128	0.2	1.8	0.3	–0.015	0.003	–0.004	–0.004	0.001	0.001
129	0.2	6.5	2.8	–0.015	0.003	–0.041	–0.042	0.009	0.010
147	0.2	0.5	–0.5	–0.015	0.003	0.007	0.007	–0.001	–0.002
148	0.2	0.0	–0.7	–0.015	0.003	0.010	0.011	–0.002	–0.002
149	0.2	1.0	–0.1	–0.015	0.003	0.002	0.002	0.000	0.000
150	0.2	0.1	–0.7	–0.015	0.003	0.010	0.010	–0.002	–0.002
151	0.2	0.9	–0.2	–0.015	0.003	0.003	0.003	–0.001	–0.001
152	0.2	0.1	–0.7	–0.015	0.003	0.010	0.010	–0.002	–0.002
153	0.2	0.0	–0.7	–0.015	0.003	0.010	0.011	–0.002	–0.002
154	0.2	0.0	–0.7	–0.015	0.003	0.010	0.010	–0.002	–0.002
155	0.2	0.2	–0.6	–0.015	0.003	0.009	0.009	–0.002	–0.002
156	0.2	0.6	–0.4	–0.015	0.003	0.006	0.006	–0.001	–0.001
157	0.2	0.9	–0.2	–0.015	0.003	0.004	0.004	–0.001	–0.001
158	0.2	0.1	–0.7	–0.015	0.003	0.010	0.010	–0.002	–0.002
159	0.2	5.4	2.2	–0.015	0.003	–0.033	–0.034	0.007	0.008
160	0.2	0.4	–0.5	–0.015	0.003	0.007	0.007	–0.002	–0.002
161	0.2	0.1	–0.6	–0.015	0.003	0.009	0.010	–0.002	–0.002
162	0.2	0.2	–0.6	–0.015	0.003	0.009	0.009	–0.002	–0.002
163	0.2	0.4	–0.5	–0.015	0.003	0.008	0.008	–0.002	–0.002
164	0.2	2.7	0.7	–0.015	0.003	–0.011	–0.011	0.002	0.003
165	0.2	0.2	–0.6	–0.015	0.003	0.009	0.009	–0.002	–0.002
166	0.2	1.0	–0.2	–0.015	0.003	0.003	0.003	–0.001	–0.001

Note: Horizontal lines in the table indicate omitted $\hat{v}_j^2$.

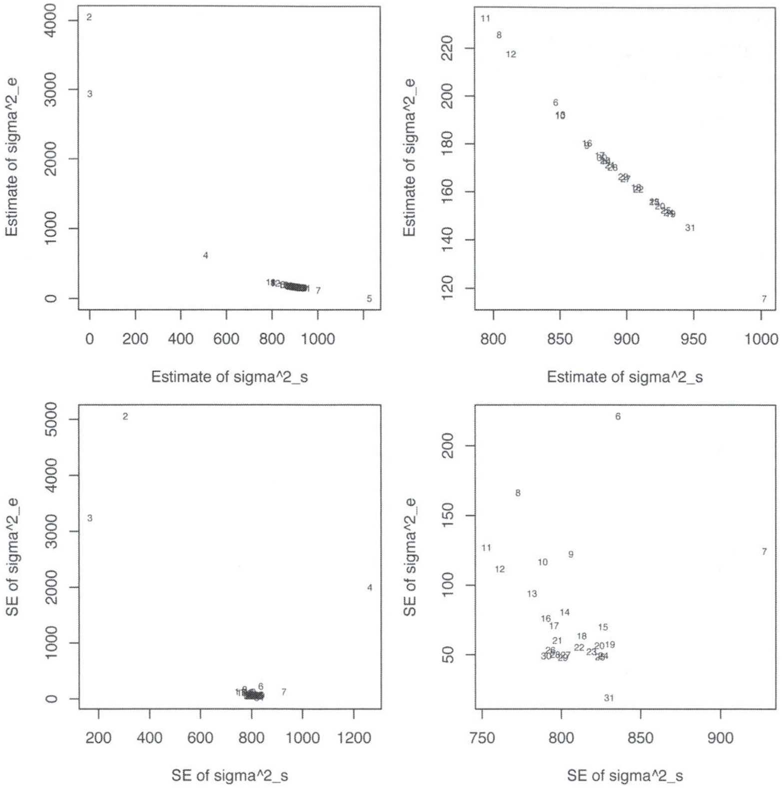

Figure 16.1: Global mean surface temperature data: Penalized spline model, maximizing values and approximate standard errors (SEs) from modified restricted likelihoods. Top row plots show maximizing values, bottom row plots show approximate SEs. Left column plots include all m; right column plots include $m = 6, \ldots, 31$. The plotting character is m.

mostly provide information about σ_s^2 because the range within which $\hat{\sigma}_s^2$ changes in the upper-right panel, for $m \geq 6$, is smaller as a proportion of the full-data $\hat{\sigma}_s^2$ than is the range of changes in $\hat{\sigma}_e^2$, but this is a weak argument. The approximate SEs provide a stronger argument. In the lower-right plot, the approximate SE for $\hat{\sigma}_s^2$ (horizontal axis) varies in a fairly narrow range for $m \geq 8$ and actually tends to increase as m goes from 8 to 14, which suggests that for $m \geq 8$ changes in this approximate standard error are mainly driven by accumulating information about σ_e^2. By contrast, the approximate SE for $\hat{\sigma}_e^2$ declines as m increases, although at a decreasing rate, before dropping from 49 to 19 when the free terms ($m = 31$) are added.

For the simple ICAR fit to the attachment-loss data, Figure 16.2 shows the estimates (top row) and SEs (bottom row) for each m. For $m = 3, \ldots, 9$, the maximizing value of σ_s^2 is zero, so the lower left plot of approximate SEs shows $m \geq 10$. After

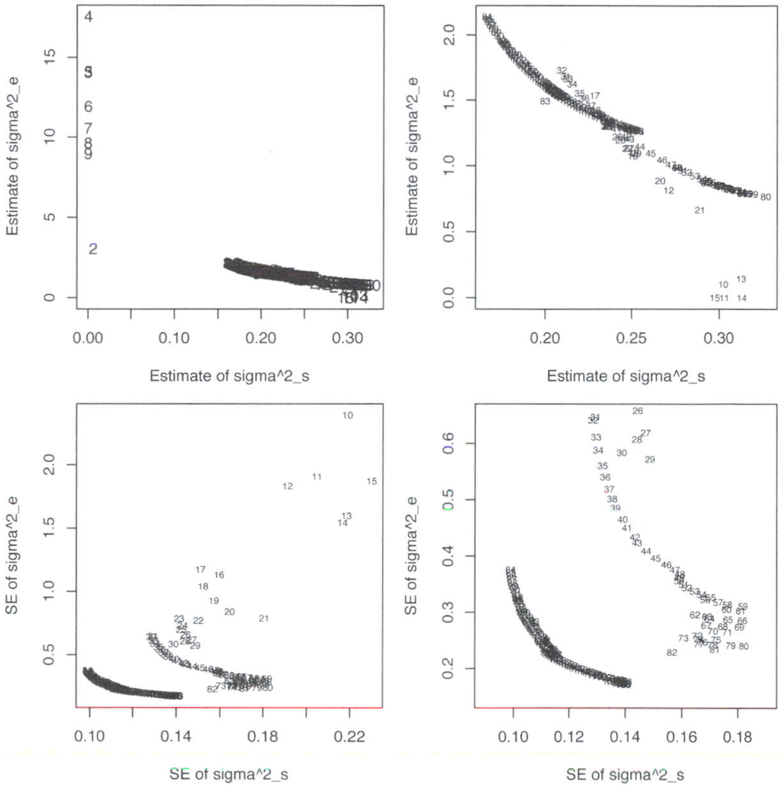

Figure 16.2: Attachment loss data: Simple ICAR model, maximizing values and approximate standard errors (SEs) from modified restricted likelihoods. Top row plots show maximizing values, bottom row plots show approximate SEs. Left column plots include all available m; right column plots reduce the range of the axes to show more detail. The plotting character is m.

jumping about for small m, the maximizing values tend to move together as mixed terms are added. After big jumps for $m = 30, 31$ (upper right panel), adding mixed terms $m = 32$ to 82 tends to decrease the maximizing value of σ_e^2 and increase the maximizing value of σ_s^2. Adding the extreme outliers $j = 83, 84$ yanks the maximizing values in two steps over to the upper left of the plot, after which the following less influential observations coax the modified restricted likelihood back toward the full-data estimates $\hat{\sigma}_s^2 = 0.25$ and $\hat{\sigma}_s^2 = 1.26$. Again, the maximizing values of the modified restricted likelihood provide tepid support for the idea that mixed terms with large a_j mostly provide information about σ_s^2 and those with small a_j mostly provide information about σ_e^2.

Again, however, the approximate SEs provide rather stronger support for this generalization. The approximate SE for $\hat{\sigma}_s^2$ stops systematically decreasing at about

$m = 16$. For larger m, it actually increases as m increases and the approximate SE for $\hat{\sigma}_e^2$ declines, with interruptions at influential $\hat{v}_j$, e.g., $j = 31, 83, 84$, each of which pulls the approximate SE for $\hat{\sigma}_s^2$ to a smaller value, from which it gradually increases to a final value of 0.14 for $m = 166$, little smaller than the approximate SE of 0.16 for $m = 16$. By contrast, the approximate SE for $\hat{\sigma}_e^2$ declines steadily, if at an increasingly slower rate, in the intervals between the jerks administered by the influential $\hat{v}_j$.

16.1.4 Summary and an Important Corollary

Broadly speaking, mixed terms with large a_j mostly provide information about σ_s^2 and mixed terms with small a_j mostly provide information about σ_e^2. This was most striking in the penalized spline example because its a_j are much more differentiated than are the ICAR's. However, the mixed terms were still differentiated for the ICAR: Observations $j = 3, 4$ had a large influence on $\hat{\sigma}_s^2$ even though their scaled residuals were not especially large, while $j = 83, 84$, with huge scaled residuals, had a relatively modest effect on $\hat{\sigma}_s^2$. By contrast, these two pairs of cases had, respectively, relatively small and quite large influences on $\hat{\sigma}_e^2$.

An important corollary emerges from this discussion. Both of these models, but especially the ICAR, are often described as local smoothers because their fits respond to local values of measurements, local to sites for the ICAR and local to predictor values for the spline. However, it is now clear that penalized splines and ICARs are not, in fact, local smoothers when either σ_s^2 or the smoothing ratio $r = \sigma_e^2/\sigma_s^2$ is estimated as part of the analysis. Rather, the degree of smoothing depends on the global properties of the data that determine the estimate or posterior of σ_s^2 or r. For the penalized spline, those global properties are contrasts in the data corresponding to low-order polynomials, especially the roughly cubic and quartic contrasts. For the ICAR model, those global properties are the contrasts in the data that produce the $\hat{v}_j$ for relatively small j. This is, I admit, somewhat vague because the a_j and the resulting influence decline more or less continuously as j increases and decline rather slowly for the ICAR model in particular. Nonetheless, for either model, fitting the model to data with a strong low-order polynomial trend will give a large estimate of σ_s^2 and thus little smoothing. Below, this fact helps explain one of our mysteries.

16.2 Two Mysteries Explained

The mystery in Section 14.1 — why did $\hat{\sigma}_s^2$ increase and $\hat{\sigma}_e^2$ decrease when the two contrasts comparing direct and interproximal sites were changed from random effects to fixed effects? — is easy to explain after the previous section's discussion. These two contrasts have fairly small a_j (0.33) and thus mostly provide information about the error variance σ_e^2. Both contrasts in the data are far larger than expected under the model so when they are changed to fixed effects, $\hat{\sigma}_e^2$ is reduced. It seems the increase in $\hat{\sigma}_s^2$ is mostly secondary to the reduction in $\hat{\sigma}_e^2$. Table 16.3 shows the exact and one-step approximate changes in the two variance estimates from changing one of v_{83} and v_{84} to fixed effects, and from changing both to fixed effects.

Now for the mystery in Section 14.3: Why do the penalized spline and ICAR give such different smooths for the global-mean surface temperature data and why does

Table 16.3: Attachment Loss Data, Simple ICAR Fit: Exact and Approximate Changes from Changing v_{83} and v_{84} from Random to Fixed Effects

	Estimate's Change from			
v_j Changed	$\tilde{\sigma}_e^2 = 1.26$		$\tilde{\sigma}_s^2 = 0.25$	
to Fixed Effects	1-Step	Exact	1-Step	Exact
only $j = 83$	−0.242	−0.253	0.028	0.043
only $j = 84$	−0.312	−0.329	0.036	0.060
$j = 83$ and 84	−0.558	−0.609	0.065	0.147

the ICAR perform so poorly when you force its fit to be as smooth as the spline fit? First I examine the a_j, canonical predictors, case-influence, and modified likelihood for the ICAR model for this dataset as a third example of these tools. Then I use some of those results and another tool, the DF in the fit of each canonical random effect v_j, to explain this mystery.

Figure 16.3 shows some canonical predictors for the simple ICAR with adjacent years as neighbor pairs. The first six canonical predictors are roughly linear, quadratic, and so on up to 6th order polynomials, or alternatively roughly sine curves with 0.5, 1, 1.5, 2, 2.5, and 3 cycles respectively. The last few canonical predictors are like those from the dynamic linear model (Section 15.2.4). Figure 16.3c shows the third from last ($j = 122$); the analogous plots for $j = 123, 124$ show two cycles and one cycle in the amplitude, respectively.

The variance estimates from maximizing the full-data restricted likelihood are $\hat{\sigma}_e^2 = 119.9$ and $\hat{\sigma}_s^2 = 21.8$. (For the spline, the error mean square for an unsmoothed 30-knot fit was 137.4; the ICAR *really* overfits this dataset.) Table 16.4 shows a_j and $\hat{v}_j^2$ for the ICAR fit, among other things, for selected j. (Specifically, the first 20, the last 15, and intermediate j with large scaled residuals. An exercise replicates these computations for all j.) The a_j decline in j at roughly the same pace as for the attachment-loss ICAR model and the DLM, much more slowly than for the penalized spline. The first four $\hat{v}_j^2$ are larger by far than any subsequent $\hat{v}_j^2$ and there are no gross inconsistencies with the ICAR model except arguably for $j = 69$.

Now consider Table 16.4's case-influence measures. For $j = 1,2,3,4$, none of the scaled residuals is especially large despite very large $\hat{v}_j^2$. Nonetheless, these cases have strong influence on $\hat{\sigma}_s^2$, especially $j = 4$. This happens because these cases have large multipliers for σ_s^2, which happens because their a_j are large.[3] These cases have moderate scaled residuals because they are influential; the fit accomodates them.

This example reinforces the generalization that cases with large a_j mostly provide information about σ_s^2 while cases with small a_j mostly provide information about σ_e^2. For example, for $j = 1,\ldots,9$, deletion of any of these cases changes $\hat{\sigma}_s^2$ by at least 1.3 in absolute value even though none of the scaled residuals is large. By contrast, deletion of cases $j = 47$ or 59, which do have large scaled residuals, has a big effect on $\hat{\sigma}_e^2$ but little or none on $\hat{\sigma}_s^2$; the multiplier for σ_s^2 is about 0 for $j = 47$ and just

[3]These multipliers are roughly analogous to leverage in linear models but σ_s^2 and σ_e^2 have different multipliers, while leverage applies to the whole coefficient vector.

Table 16.4: Global Mean Surface Temperature Data, ICAR Fit: Exact and Approximate Changes in Maximum-Restricted-Likelihood Estimates of σ_e^2 and σ_s^2 from Deleting Case j, for Selected j

						Estimate's change from			
			scaled	Multiplier for		$\tilde{\sigma}_e^2 = 119.9$		$\tilde{\sigma}_s^2 = 21.8$	
j	a_j	$\hat{v}_j^2$	resid	σ_e^2	σ_s^2	1-step	exact	1-step	exact
1	1583.2	52872	0.4	2.4	−3.0	0.9	1.2	−1.1	−1.4
2	395.9	2657	−0.5	2.3	−2.9	−1.1	−1.4	1.4	1.7
3	176.0	7388	0.6	2.2	−2.9	1.4	1.9	−1.8	−2.2
4	99.0	7732	1.7	2.1	−2.8	3.6	5.4	−4.7	−6.0
5	63.4	1	−0.7	2.0	−2.7	−1.4	−1.8	1.9	2.2
6	44.1	15	−0.7	1.8	−2.6	−1.3	−1.7	1.8	2.1
7	32.4	57	−0.7	1.6	−2.4	−1.1	−1.5	1.6	1.9
8	24.8	146	−0.6	1.4	−2.3	−0.8	−1.1	1.3	1.5
9	19.6	4	−0.7	1.3	−2.2	−0.9	−1.3	1.5	1.8
10	15.9	494	0.0	1.1	−2.0	0.0	0.1	−0.1	−0.1
11	13.2	240	−0.3	0.9	−1.9	−0.3	−0.4	0.5	0.6
12	11.1	1078	1.4	0.7	−1.8	1.0	1.3	−2.5	−2.7
13	9.5	215	−0.2	0.5	−1.6	−0.1	−0.2	0.4	0.5
14	8.2	66	−0.6	0.4	−1.5	−0.2	−0.4	0.8	1.0
15	7.1	24	−0.6	0.2	−1.4	−0.2	−0.4	0.9	1.1
16	6.3	365	0.3	0.1	−1.3	0.0	0.1	−0.4	−0.4
17	5.6	30	−0.6	0.0	−1.2	0.0	−0.2	0.7	0.9
18	5.0	125	−0.3	−0.2	−1.1	0.1	0.0	0.4	0.4
19	4.5	116	−0.3	−0.3	−1.0	0.1	0.0	0.3	0.4
20	4.0	72	−0.5	−0.4	−0.9	0.2	0.1	0.4	0.5
25	2.6	588	1.6	−0.9	−0.6	−1.4	−1.3	−1.0	−1.1
28	2.1	926	3.2	−1.1	−0.5	−3.4	−3.4	−1.5	−1.6
30	1.8	906	3.3	−1.2	−0.4	−3.9	−3.8	−1.3	−1.3
47	0.8	1214	5.5	−1.7	0.0	−9.6	−9.9	0.0	0.2
59	0.5	964	4.5	−1.9	0.1	−8.5	−9.0	0.5	0.9
69	0.4	1690	8.5	−2.0	0.2	−16.9	−18.8	1.5	2.9
110	0.3	58	−0.4	−2.1	0.3	0.8	0.9	−0.1	−0.1
111	0.3	96	−0.2	−2.1	0.3	0.4	0.4	0.0	−0.1
112	0.3	14	−0.6	−2.1	0.3	1.3	1.4	−0.2	−0.2
113	0.3	50	−0.4	−2.1	0.3	0.9	1.0	−0.1	−0.1
114	0.3	2	−0.7	−2.1	0.3	1.5	1.6	−0.2	−0.2
115	0.3	24	−0.6	−2.1	0.3	1.2	1.3	−0.1	−0.2
116	0.3	2	−0.7	−2.1	0.3	1.5	1.6	−0.2	−0.2
117	0.3	245	0.7	−2.1	0.3	−1.4	−1.5	0.2	0.2
118	0.3	34	−0.5	−2.1	0.3	1.1	1.1	−0.1	−0.2
119	0.3	9	−0.7	−2.1	0.3	1.4	1.5	−0.2	−0.2
120	0.3	78	−0.3	−2.1	0.3	0.6	0.6	−0.1	−0.1
121	0.3	124	0.0	−2.1	0.3	0.0	0.0	0.0	0.0
122	0.3	79	−0.3	−2.1	0.3	0.6	0.6	−0.1	−0.1
123	0.3	515	2.2	−2.1	0.3	−4.7	−5.1	0.6	0.9
124	0.3	28	−0.6	−2.1	0.3	1.2	1.2	−0.1	−0.2

Note: The first 20 and the last 15 j are included; selected other j are between the horizontal lines.

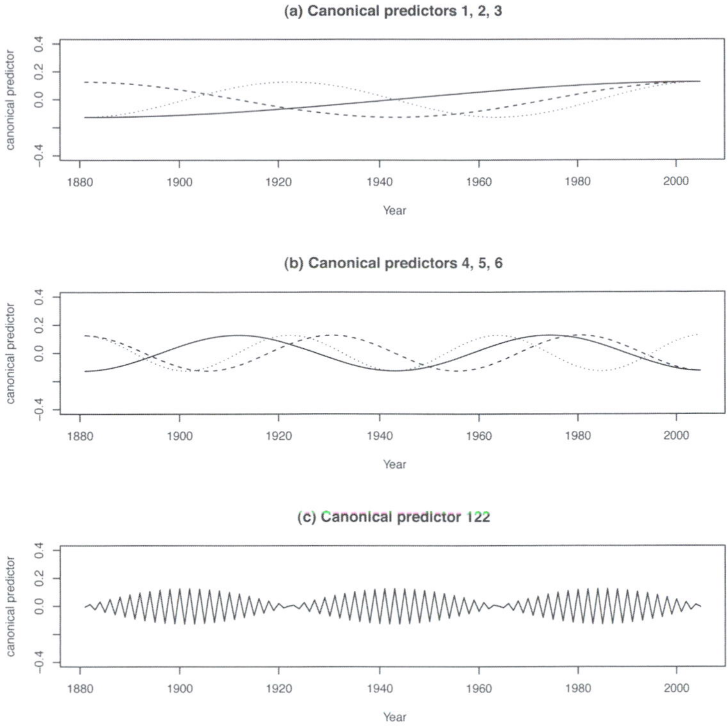

Figure 16.3: Global mean surface temperature data: Simple ICAR model, canonical predictors. The horizontal axis is the row in $\Gamma_Z\mathbf{P}$ (labeled by years), the vertical axis is the value of the predictor for that row. Panel (a): Predictors corresponding to a_1 (solid line), a_2 (dashed line), a_3 (dotted line). Panel (b): Predictors corresponding to a_4 (solid line), a_5 (dashed line), a_6 (dotted line). Panel (c): Predictor corresponding to a_{122}.

0.113 for $j = 59$. By contrast, $j = 69$ has the largest scaled residual in the dataset and by far the largest effect on $\hat{\sigma}_e^2$. It has the second largest exact influence on $\hat{\sigma}_s^2$ (its multiplier is about 50% larger than for $j = 59$) but in the "wrong" direction; several other cases with much larger a_j have influence on $\hat{\sigma}_s^2$ of similar magnitude but in the "right" direction.

Finally, note that cases $j = 1, 2$, and 4 have positive scaled residuals while cases 5 to 9 all have negative scaled residuals. These two groups of influential observations give discordant information about σ_s^2; we'll see below that this has interesting consequences.

Turning to the modified restricted likelihood, Figure 16.4 shows for each m the maximizing values (top row of plots) and approximate standard errors (bottom row of plots). This is similar to the previous two examples in that for m larger than about 11, when one maximizing value becomes larger, the other becomes smaller; the same

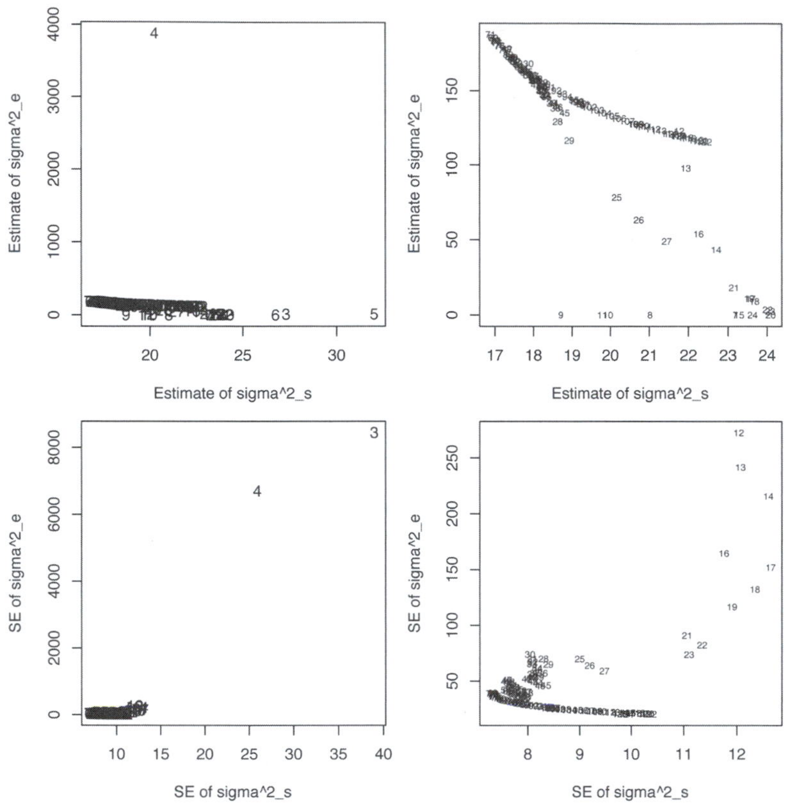

Figure 16.4: Global mean surface temperature data: ICAR model, maximizing values and approximate standard errors (SEs) from modified restricted likelihoods. Top row plots show maximizing values, bottom row plots show approximate SEs. Left column plots include all m; right column plots include $m \geq 7$. The plotting character is m. Approximate SEs were not computed for m = 7 to 11, 15, 20, and 24 because the maximizing value of σ_e^2 was zero.

is true for the approximate SEs for m larger than about 23. The approximate SE of $\hat{\sigma}_s^2$ stops declining systematically by $m = 25$, though it is yanked to smaller values by influential cases and then increases. For $\hat{\sigma}_e^2$, the approximate SE never stops declining. The maximizing values for σ_s^2 have a quite narrow range for $m \geq 7$, between 17 and 24. By contrast, for several low m the modified restricted likelihood is maximized at $\sigma_e^2 = 0$ and even for $m \geq 12$ the maximizing values range from 0 to nearly 190, though they settle into a much narrower range for $m \geq 28$.

So far, there are no surprises here. The j are less differentiated for this ICAR than for the penalized spline in terms of influence of individual j but as yet we have no apparent explanation for our mystery.

We get a hint, however, from considering further the effect of $j = 1,2,3,4$. Cases with large a_j are influential for $\hat{\sigma}_s^2$; for the penalized spline, these j have canonical predictors that are roughly cubic, quartic, etc., while for the ICAR these influential canonical predictors include roughly linear and quadratic predictors as well. So if we modify the ICAR by changing its linear and quadratic canonical predictors from random effects to fixed effects, i.e., by omitting them from the restricted likelihood, will the ICAR's fit look like the penalized spline's fit? The answer is no: As the single-case influence measures suggest, the variance estimates hardly change at all, to $\hat{\sigma}_s^2 = 22.2$ and $\hat{\sigma}_e^2 = 119.5$, and the DF in the new fit actually goes up a bit, to 26.8 DF (3 for fixed effects, 23.8 for random effects). The ICAR and spline behave differently even when the ICAR is modified so the two models have essentially the same fixed and random effects.

We'll return to that tantalizing observation below but first you have to wonder what happens if the ICAR's roughly cubic canonical predictor is changed to a fixed effect. The ICAR still overfits, though a bit less: Now $\hat{\sigma}_s^2 = 19.5$ and $\hat{\sigma}_e^2 = 121.9$, with 25.0 DF (4 for fixed effects, 21.0 for random effects). Well, what if the roughly quartic canonical predictor is changed to a fixed effect? Now we get a big change: $\hat{\sigma}_s^2 = 0$, $\hat{\sigma}_e^2 = 149.5$, and the resulting fit with 5 DF, all in the fixed effects, is *smoother* than the penalized spline (6.7 df, 3 in fixed effects). Perhaps you anticipated this after I pointed out above that the scaled residuals were positive for $j = 1,2,4$ but negative for $j = 5,\ldots,9$. An exercise asks you to reproduce these calculations.

Having answered that tantalizing question, we now solve our mystery by following up on the observation that the ICAR and penalized spline behave differently even after we change the ICAR's first two canonical predictors to fixed effects, so the two models have essentially the same fixed and random effects. With this modification to the ICAR, the only substantial difference between the two models is their a_j. Recall that conditional on (σ_s^2, σ_e^2), the DF in the fitted coefficient v_j of canonical predictor j is $a_j/(a_j + \sigma_e^2/\sigma_s^2)$. Thus, the solution to our mystery lies in the different ways the ICAR and spline allocate the DF in their fits among the canonical predictors.

Table 16.5 shows the DF allocated to each fixed effect and canonical random effect for the penalized spline fit with 6.7 total DF, the ICAR fit using (σ_s^2, σ_e^2) that maximizes the restricted likelihood (26.5 total DF), and the ICAR fit with (σ_s^2, σ_e^2) giving a fit with 6.7 total DF. For the spline, the DF decline quickly as the polynomial order increases, so 3.4 of the 3.7 DF for random effects are allocated to the first 4 canonical predictors. The DF decline much more slowly for the ICAR model. For the ICAR fit with 26.5 DF, canonical predictor $j = 30$ has 0.25 DF; for the spline, canonical predictor $j = 30$ has 0.00002 DF. When we force the ICAR to have 6.7 DF, it gives the cubic and quartic predictors only 0.63 and 0.49 DF respectively, while the spline gives nearly 1 DF to each. Thus the ICAR with 6.7 DF cannot capture the most prominent feature in this data series, the decline-then-rise between about 1940 and 1970. But this ICAR fit is still bumpy in spots because, for example, it gives 0.14 DF — one-seventh of a DF — to the canonical predictor that's a 10th-order polynomial ($j = 10$).

Using the notation of Chapter 5, the simple ICAR model for the GMST series shrinks the first differences of the random effect δ, $\delta_t - \delta_{t-1}$, and thus smooths δ

Table 16.5: Global Mean Surface Temperature Data: How the Penalized Spline and ICAR Allocate DF among Their Fixed Effects and Canonical Random-Effect Predictors

Polynomial Order	Spline DF = 6.7	ICAR DF = 26.5	ICAR DF = 6.7	j
Intercept	1	1	1	FE
Linear	1	0.997	0.94	1
Quadratic	1	0.99	0.80	2
Cubic	0.996	0.97	0.63	3
Quartic	0.95	0.95	0.49	4
Quintic	0.79	0.92	0.38	5
(etc.)	0.49	0.89	0.30	6
	0.24	0.85	0.24	7
	0.11	0.82	0.20	8
	0.05	0.78	0.16	9
	0.03	0.74	0.14	10
	0.0002	0.42	0.04	20
	0.00002	0.25	0.018	30
		0.16	0.011	40
		0.12	0.007	50
		0.09	0.005	60
		0.07	0.004	70
		0.06	0.003	80
		0.05	0.003	90
		0.05	0.003	100
		0.04	0.003	110
		0.04	0.002	120
		0.04	0.002	124

Note: Rows are labeled by the (rough) polynomial order of the fixed effects and canonical random effects. They could also be labeled as 0, 0.5, 1.0, 1.5, etc., the number of cycles of the corresponding approximate sine curves. "FE" means "fixed effect." Below the horizontal line, only selected canonical predictors j are shown.

toward a constant. Thus, if this ICAR is fit to data with a strong low-order polynomial trend, it will undersmooth if it is allowed to fit that trend or it will fit poorly if it is forced to be more smooth. To reiterate the "important corollary" of the previous section: The ICAR model is not a local smoother if σ_s^2 is chosen as part of the fit.

Higher-order ICAR models (Rue & Held 2005, Section 3.4) can avoid the dilemma of the simple ICAR. A 2nd-order ICAR model shrinks the 2nd differences $(\delta_t - \delta_{t-1}) - (\delta_{t-1} - \delta_{t-2}) = \delta_t - 2\delta_{t-1} + \delta_{t-2}$ and thus smooths toward a straight line, where the implicit prior on the line's intercept and slope are flat (improper). A 3rd-order ICAR model shrinks the 3rd differences $\delta_t - 3\delta_{t-1} + 3\delta_{t-2} - \delta_{t-3}$ and thus smooths toward a quadratic with flat priors on its three coefficients.

The 3rd-order ICAR model sounds very much like the penalized spline I've used, which has a truncated-quadratic basis and thus shrinks toward a quadratic with flat priors on its three coefficients. The canonical representation for the restricted likelihood shows that these two models are, indeed, almost identical.[4] Briefly — computing the following results is given as an exercise — the first 30 a_j for the 3rd-order ICAR and the 30 a_j from the penalized spline have correlation very close to 1 (the R function "cor" gives "1"). The 3rd-order ICAR's first 30 a_j are larger than the spline's a_j by a factor of about 1.7×10^6, but the ratios a_1/a_j are very similar for the two models, with the similarity diminishing a bit as j gets close to 30. The DF in canonical random effect v_j is $a_j/(a_j + \sigma_e^2/\sigma_s^2)$, so the ratio of the a_j to each other matters here, not their absolute values. (Multiply all the a_j by a constant and that just changes the scale of σ_s^2.) Thus the 3rd-order ICAR will have essentially the same allocation of DF among the v_j as the penalized spline. Also, the jth canonical predictors for the two models have correlation greater than 0.99 in absolute value for $j \leq 20$ and greater than 0.9 for $j \leq 25$. (The correlations for the last five j have absolute values 0.89, 0.84, 0.75, 0.62, 0.43.) In this sense, the 3rd-order ICAR is almost identical to a penalized spline with a truncated-quadratic basis.

Exercises

Regular Exercises

1. (Section 16.1.2) Prove that the one-step approximation to the changes in the estimates of σ_s^2 and σ_e^2 can be computed using the updating formula from simple linear models.
2. (Section 16.1.2) Derive equations (16.5) and (16.6), the one-step approximation to the change in the maximum restricted likelihood estimate from deleting the "observation" $\hat{v}_j^2$.
3. (Section 16.1.2) Derive the expressions for items (iii) and (iv) given in equations (16.10) and (16.11) respectively.
4. (Section 16.1.2) Compute the items in Table 16.2 for all 166 mixed terms.
5. (Section 16.2) Compute the items in Table 16.4 for all 124 mixed terms.
6. (Section 16.2) For the ICAR model fit to the global-mean surface temperature data, derive the re-expressed restricted likelihood omitting canonical observations with low j and show that you can omit $j = 1,2,3$ without making the fit much smoother but that omitting $j \leq 4$ forces $\hat{\sigma}_s^2 = 0$.
7. (Section 16.2) For the 3rd-order ICAR model, derive the re-expressed restricted likelihood discussed in this Chapter's last paragraph.

Open Questions

1. (Section 16.1.2). Give a better explanation of why a change that increases the estimate of one variance usually also decreases the estimate of the other variance,

[4] This result is also probably in the spline literature somewhere; see the footnote in Section 15.2.2.

and a change that increases the approximate standard error of one variance estimate usually also decreases the approximate standard error of the other variance estimate.

Chapter 17

Extending the Re-expressed Restricted Likelihood

Chapter 15 derived a simple form for the restricted likelihood for two-variance models. This also yielded a vector of known linear functions of the data **y**, the $\hat{v}_j$, with nice properties: They contain all the information about the two variances and conditional on (σ_s^2, σ_e^2) they are independent with mean zero and variance $\sigma_s^2 a_j + \sigma_e^2$, where the a_j are known functions of the design matrices **X** and **Z**. Also, for two-variance models, the re-expressed restricted likelihood is a particular generalized linear model. These facts provided some tools with which we pried open the black box of the restricted likelihood and analyzed it, much as we might analyze a linear model fit.

Unfortunately, this re-expression is not possible for mixed linear models in general although it is for a good-sized subset of them. Section 17.1 proves that the restricted likelihoods implied by two particular models cannot be re-expressed in this simple form. It then describes some classes of models for which the restricted likelihood can be re-expressed simply, though not necessarily as the likelihood from a generalized linear model. It's extremely unlikely this catalog includes all models with re-expressible restricted likelihoods but it gives some sense of what is possible. Section 17.2 then considers expedients for problems for which the restricted likelihood cannot be re-expressed. These expedients are approximate or sloppy, depending on your preference, in that each relies on a choice to ignore some aspect of the restricted likelihood. They do seem to yield useful information so they may suggest ways to supercede or extend the re-expressed restricted likelihood.

17.1 Restricted Likelihoods That Can and Can't Be Re-expressed

Section 17.1.1 shows two models for which the restricted likelihood cannot be re-expressed as in Chapters 15 and 16. The difficulty is the need to diagonalize simultaneously more than two symmetric positive semi-definite matrices. Sections 17.1.2 to 17.1.5 then describe classes of models for which the restricted likelihood can be re-expressed. Each of the latter sections describes the class of models, shows one or two examples, and either proves that the restricted likelihood can be re-expressed or gives it as an exercise (when it seems straightforward). So far I have done very little with these models, so exploration of them is wide open.

17.1.1 Two Restricted Likelihoods That Can't Be Re-expressed

The first example is an additive model with two predictors. For a given spline basis, Chapter 4 showed how this model can be written in mixed-linear-model form as

$$\mathbf{y} = \mathbf{X}\beta + \mathbf{Z}_1\mathbf{u}_1 + \mathbf{Z}_2\mathbf{u}_2 + \varepsilon, \tag{17.1}$$

where $\mathbf{X}\beta$ includes the unshrunk parts of the splines for both predictors and $\mathbf{Z}_j\mathbf{u}_j$ contains the shrunk part of the spline for predictor j, so the random effects design matrix in standard form is $\mathbf{Z} = (\mathbf{Z}_1|\mathbf{Z}_2)$. If the spline for predictor j has K_j knots, then the standard-form random effect $\mathbf{u} = (\mathbf{u}_1', \mathbf{u}_2')'$ has covariance matrix

$$\mathbf{G} = \begin{bmatrix} \sigma_{s1}^2\mathbf{I}_{K_1} & \mathbf{0} \\ \mathbf{0} & \sigma_{s2}^2\mathbf{I}_{K_2} \end{bmatrix}. \tag{17.2}$$

Assume $\mathbf{X}$, $\mathbf{Z}_1$, and $\mathbf{Z}_2$ and $(\mathbf{X}|\mathbf{Z}_1|\mathbf{Z}_2)$ are full rank (this may not be necessary). Define $\Gamma_X, \Gamma_1, \Gamma_2, \Gamma_{12}$ with n rows and s_X, s_1, s_2, and s_{12} columns respectively having the following properties. (Constructing these matrices is an exercise.) Each of these matrices has orthonormal columns, e.g., $\Gamma_X'\Gamma_X = \mathbf{I}_{s_X}$, and each is orthogonal to all the others, e.g., $\Gamma_X'\Gamma_{12} = \mathbf{0}_{s_X \times s_{12}}$. $(\Gamma_X|\Gamma_1|\Gamma_2|\Gamma_{12})$ has the same column space as $(\mathbf{X}|\mathbf{Z}_1|\mathbf{Z}_2)$. The columns of Γ_X are an orthonormal basis for the column space of $\mathbf{X}$; the columns of $(\Gamma_X|\Gamma_1|\Gamma_{12})$ are an orthonormal basis for the column space of $(\mathbf{X}|\mathbf{Z}_1)$; the columns of $(\Gamma_X|\Gamma_2|\Gamma_{12})$ are an orthonormal basis for the column space of $(\mathbf{X}|\mathbf{Z}_2)$. Γ_1's columns are a basis for the part of fitted-value space that is specific to $\mathbf{Z}_1$, i.e., that is confounded with neither the fixed effects $\mathbf{X}$ nor the other random effects $\mathbf{Z}_2$; Γ_2 has the analogous role for $\mathbf{Z}_2$. Γ_{12}'s columns are a basis for the part of fitted-value space that is included in the column spaces of $\mathbf{Z}_1$ and $\mathbf{Z}_2$ but not the column space of $\mathbf{X}$. If $\mathbf{Z}_1$ and $\mathbf{Z}_2$ are orthogonal, e.g., if the predictors are measured on a rectangular grid with equal replication at each grid point, then Γ_{12} is null. Partition the matrix $\mathbf{M}$ satisfying $(\mathbf{X}|\mathbf{Z}_1|\mathbf{Z}_2) = (\Gamma_X|\Gamma_1|\Gamma_2|\Gamma_{12})\mathbf{M}$ as

$$\mathbf{M} = \begin{bmatrix} \mathbf{M}_{XX} & \mathbf{M}_{1X} & \mathbf{M}_{2X} \\ \mathbf{0} & \mathbf{M}_{11} & \mathbf{0} \\ \mathbf{0} & \mathbf{0} & \mathbf{M}_{22} \\ \mathbf{0} & \mathbf{M}_{112} & \mathbf{M}_{212} \end{bmatrix}, \tag{17.3}$$

where $\mathbf{M}$'s column partitions have s_X, K_1, and K_2 columns and $\mathbf{M}$'s row partitions have s_X, s_1, s_2, and s_{12} rows. Finally, define Γ_c with dimension $n \times (n - s_X - s_1 - s_2 - s_{12})$ so the columns of $(\Gamma_X|\Gamma_1|\Gamma_2|\Gamma_{12}|\Gamma_c)$ are an orthonormal basis for real n-space.

The restricted likelihood for $(\sigma_e^2, \sigma_{s1}^2, \sigma_{s2}^2)$ is the likelihood from the model for $\mathbf{w} = \mathbf{K}'\mathbf{y}$, where $\mathbf{K} = (\Gamma_1|\Gamma_2|\Gamma_{12}|\Gamma_c)$. The rest of this derivation is stated with proofs given as an exercise. The restricted likelihood is a function of $(\sigma_e^2, \sigma_{s1}^2, \sigma_{s2}^2)$ only through the matrix

$$\mathbf{W} = \sigma_e^2\mathbf{I}_n + \begin{bmatrix} \sigma_{s1}^2\mathbf{M}_{11}\mathbf{M}_{11}' & \mathbf{0} & \sigma_{s1}^2\mathbf{M}_{11}\mathbf{M}_{112}' & \mathbf{0} \\ \mathbf{0} & \sigma_{s2}^2\mathbf{M}_{22}\mathbf{M}_{22}' & \sigma_{s2}^2\mathbf{M}_{22}\mathbf{M}_{212}' & \mathbf{0} \\ \sigma_{s1}^2\mathbf{M}_{112}\mathbf{M}_{11}' & \sigma_{s2}^2\mathbf{M}_{212}\mathbf{M}_{22}' & \sigma_{s1}^2\mathbf{M}_{112}\mathbf{M}_{112}' + \sigma_{s2}^2\mathbf{M}_{212}\mathbf{M}_{212}' & \mathbf{0} \\ \mathbf{0} & \mathbf{0} & \mathbf{0} & \mathbf{0} \end{bmatrix}. \tag{17.4}$$

To re-express the restricted likelihood for $(\sigma_e^2, \sigma_{s1}^2, \sigma_{s2}^2)$ in the desired simple form, it must be possible to diagonalize $\mathbf{W}$ by pre- and post-multiplying it by a known matrix. However, to diagonalize the 3rd diagonal block $\sigma_e^2 \mathbf{I}_{s_{12}} + \sigma_{s1}^2 \mathbf{M}_{112}\mathbf{M}'_{112} + \sigma_{s2}^2 \mathbf{M}_{212}\mathbf{M}'_{212}$, it must be possible to write $\mathbf{M}_{112}\mathbf{M}'_{112} = \mathbf{P}\mathbf{D}_1\mathbf{P}'$ and $\mathbf{M}_{212}\mathbf{M}'_{212} = \mathbf{P}\mathbf{D}_2\mathbf{P}'$ for diagonal $\mathbf{D}_j$ and $\mathbf{P}$ an orthogonal matrix, and such a $\mathbf{P}$ exists if and only if $\mathbf{M}_{112}\mathbf{M}'_{112}\mathbf{M}_{212}\mathbf{M}'_{212}$ is symmetric (Graybill 1983, Theorem 12.2.12), which is not true in general.

A sufficient condition permitting this model's restricted likelihood to be re-expressed is that $\mathbf{M}_{112}$ and $\mathbf{M}_{212}$ are null. These matrices are either both non-zero or both null; if either is zero, the other is necessarily zero and Γ_{12} is null. This sufficient condition implies a form of balance in the two predictors, e.g., it will hold if the two predictors are observed on a rectangular grid with the same number of replicates at each grid point. Because of $\mathbf{W}$'s non-zero off-diagonals, I suspect this is also a necessary condition but I have not proved it; an exercise examines this question.

The second example of a restricted likelihood that can't be re-expressed is the ICAR model with two classes of neighbor relations (2NRCAR) introduced in Section 14.2. (This proof is taken from Reich et al. 2007.) The n-vector $\mathbf{y}$ is modeled as $\mathbf{y} = \delta + \varepsilon$, where $\varepsilon \sim N_n(\mathbf{0}, \sigma_e^2 \mathbf{I}_n)$ and δ has an improper normal density specified in terms of its precision matrix:

$$f(\delta|\sigma_{s1}^2, \sigma_{s2}^2) \propto \exp\left[-\frac{1}{2}\delta' \left(\mathbf{Q}_1/\sigma_{s1}^2 + \mathbf{Q}_2/\sigma_{s2}^2\right)\delta\right]; \tag{17.5}$$

$\mathbf{Q}_k = (q_{ij,k}), k = 1,2$ encodes class-k neighbor pairs of measurement sites and σ_{sk}^2 controls the similarity of the δ_i induced by class-k neighbor pairs. By Newcomb (1961), there is a non-singular matrix $\mathbf{B}$ such that $\mathbf{Q}_k = \mathbf{B}'\mathbf{D}_k\mathbf{B}$, where $\mathbf{D}_k$ is diagonal with non-negative diagonal elements. $\mathbf{B}$ is non-singular but is orthogonal if and only if $\mathbf{Q}_1\mathbf{Q}_2$ is symmetric (Graybill 1983, Theorem 12.2.12). As noted in Section 14.2, $\mathbf{D}_k$ has I_k zero diagonal elements, where I_k is the number of $\mathbf{Q}_k$'s zero eigenvalues. $I_k \geq I$, where I is the number of zero eigenvalues of $\mathbf{Q}_1 + \mathbf{Q}_2$ and also the number of zero diagonal entries in $\mathbf{D}_1 + \mathbf{D}_2$. Define $\mathbf{D}_{k+}$ as the upper-left $(n-I)\times(n-I)$ submatrix of $\mathbf{D}_k$ and let $d_{kj}, j = 1, \ldots, n-I$, be the diagonal elements of $\mathbf{D}_{k+}$.

To simplify expressions, the rest of this derivation uses precisions instead of variances, so define the precisions $\tau_e = 1/\sigma_e^2$, $\tau_1 = 1/\sigma_{s1}^2$, and $\tau_2 = 1/\sigma_{s2}^2$. The joint posterior (omitting the prior on the τs) is

$$\begin{aligned} &|\tau_e \mathbf{I}_n|^{0.5} \quad \left|\prod_{j=1}^{n-I}\left(d_{1j}\tau_1 + d_{2j}\tau_2\right)\right|^{0.5} \quad \exp\left(-0.5\tau_e(\mathbf{y}-\delta)'(\mathbf{y}-\delta)\right) \\ &\exp\left(-0.5\delta'(\tau_1\mathbf{Q}_1 + \tau_2\mathbf{Q}_2)\delta\right). \end{aligned} \tag{17.6}$$

$|\prod_{j=1}^{n-I}\left(d_{1j}\tau_1 + d_{2j}\tau_2\right)|$ is the determinant of $\mathbf{D}_{1+}\tau_1 + \mathbf{D}_{2+}\tau_2$; it is the normalizing constant I have chosen to use for ICAR models (Section 5.2.1). The restricted likelihood can be obtained by integrating δ out of (17.6) to give

$$|\tau_e \mathbf{I}_n|^{0.5} \quad \left|\prod_{j=1}^{n-I}\left(d_{1j}\tau_1 + d_{2j}\tau_2\right)\right|^{0.5} \quad |\mathbf{H}|^{-0.5} \exp\left(-0.5\tau_e(\mathbf{y}'\mathbf{y} - \tau_e\mathbf{y}'\mathbf{H}^{-1}\mathbf{y})\right) \tag{17.7}$$

where $\mathbf{H} = \tau_e\mathbf{I}_n + \tau_1\mathbf{Q}_1 + \tau_2\mathbf{Q}_2$ (the proof is an exercise).

To re-express this restricted likelihood in the simplified form, **B** must be an orthogonal matrix. As noted, this happens if and only if $\mathbf{Q}_1\mathbf{Q}_2$ is symmetric, which is not true in general. A sufficient condition for re-expressing 2NRCAR and similar models is discussed in Section 17.1.4; again, this condition is a form of balance.

That's the bad news; now for some good news, namely large classes of models giving restricted likelihoods that can be re-expressed in the desired simple form.

17.1.2 Balanced Designs

One such class of models has a particular form of balance, as in "balanced experimental design." To define this class, I first define general balance; the models that yield simple restricted likelihoods are a large subset of models displaying general balance. This discussion draws very heavily on Speed (1983). My understanding of general balance is crude so I suspect it offers more opportunities than I have used.

In mixed-linear-model terms, general balance can be understood as a condition on the random-effects design matrix **Z** and a second condition on the relationship between **Z** and the fixed-effect design matrix **X**. I follow Speed (1983) in defining these conditions using notation that deviates somewhat from the standard form for mixed linear models. First I give some preliminary material, then the conditions defining general balance, then some examples, and finally the main result, which is proved in Section 17.1.2.1.

Consider models in which the n-vector of outcomes **y** has marginal expectation (i.e., marginal to the random effects) $E(\mathbf{y}) = \mathbf{X}\beta$, for **X** a known full-rank $n \times p$ matrix and β unknown. (Full rank is probably not necessary.) Also, **y**'s marginal covariance matrix has the form $\text{cov}(\mathbf{y}) = \sum_{j=1}^{s} \theta_j \mathbf{C}_j$, where the $\mathbf{C}_j$ are known $n \times n$ symmetric positive semi-definite matrices and the θ_j are unknown positive scalars, $j = 1, \ldots, s$. The $\mathbf{C}_j$ could (but do not necessarily) arise from a model of the form

$$\mathbf{y} = \mathbf{X}\beta + \sum_{j=1}^{s} \mathbf{Z}_j \mathbf{u}_j, \quad \text{cov}(\mathbf{u}_j) = \theta_j \mathbf{I}, \tag{17.8}$$

giving $\mathbf{C}_j = \mathbf{Z}_j\mathbf{Z}_j'$, but note that unlike the standard form for mixed linear models, this model does not have an explicit error term ε. Finally, define $\mathbf{P} = \mathbf{X}(\mathbf{X}'\mathbf{X})^{-1}\mathbf{X}'$ to be the (unweighted) orthogonal projection onto the column space of **X**.

The two conditions that define general balance are:

- (C) The matrices $\mathbf{C}_1, \ldots, \mathbf{C}_s$ all commute.
- (GB) The matrices $\mathbf{PC}_1\mathbf{P}, \ldots, \mathbf{PC}_s\mathbf{P}$ all commute.

Here, "commute" means, e.g., $\mathbf{C}_1\mathbf{C}_2 = \mathbf{C}_2\mathbf{C}_1$. Speed (1983) notes that "Condition (C) is equivalent to the matrices $\mathbf{C}_1, \ldots, \mathbf{C}_s$ being simultaneously diagonalizable," so (C) is equivalent to

$$\text{(C)*} \quad \mathbf{C}_j = \mathbf{C}_j^2 = \mathbf{C}_j', \quad \mathbf{C}_{j_1}\mathbf{C}_{j_2} = \mathbf{0} \text{ if } j_1 \neq j_2 \text{ and } \mathbf{C}_1 + \ldots + \mathbf{C}_s = \mathbf{I}_n. \tag{17.9}$$

These conditions imply the existence of an orthogonal matrix Γ and diagonal matrices Δ_j such that $\mathbf{C}_j = \Gamma\Delta_j\Gamma'$ and Δ_j's diagonal entries are all 0 or 1. Thus the columns of Γ can be re-arranged so $\Gamma = (\Gamma_1|\ldots|\Gamma_s)$, the columns of Γ are an orthonormal basis for real n-space, and $\mathbf{C}_j = \Gamma_j\Gamma'_j$.

Similarly, Speed (1983) notes that condition (GB) implies the existence of orthogonal symmetric idempotent $n \times n$ matrices $\mathbf{P}_1,\ldots,\mathbf{P}_K, K \leq p$, and coefficients λ_{jk} such that

$$\text{(GB)*} \quad \mathbf{PC}_j\mathbf{P} = \sum_{k=1}^{K} \lambda_{jk}\mathbf{P}_k, j = 1,\ldots,s, \tag{17.10}$$

with $\lambda_{jk} \in [0,1]$, $\sum_{j=1}^{s} \lambda_{jk} = 1$ for each k, and $\sum_{k=1}^{K} \mathbf{P}_k = \mathbf{P}$. (Proofs of all these assertions are an exercise.) For some designs satisfying (C)* and (GB)*, some λ_{jk} are in the *open* interval (0,1); for any such k, λ_{jk} is in the open interval (0,1) for more than one j because $\sum_{j=1}^{s} \lambda_{jk} = 1$.

A large class of models displays general balance: "[a]lmost all experimental designs ever used ... satisfy the condition of general balance [provided they have no missing observations]. This includes *all* block designs, Latin and Graeco-Latin square designs, and all designs with balanced confounding or proportional replication. ... Exceptions include some multiphase and changeover designs" (Speed 1983, p. 320). Models that do not arise from designed experiments can also display general balance, e.g., the additive model with two predictors discussed above in Section 17.1.1, for which Γ_{12} is null.

To make this rather abstract set-up more concrete, here are two examples taken from Speed (1983, p. 322).[1] First, consider block designs, where the block effects are considered random effects. Suppose we have f blocks of m plots each, so $n = fm$, and that there are p treatments. In the usual mixed-linear-model form, let $\mathbf{Z}$ be the incidence matrix for blocks, i.e., if observation i is in block b, $Z_{ib} = 1$, otherwise $Z_{ib} = 0$. Suppose the error variance is σ_e^2 and the between-blocks variance is σ_s^2 so $\mathbf{G} = \sigma_s^2\mathbf{I}_f$. Then in the notation defined above, $\mathbf{C}_1 = f^{-1}\mathbf{ZZ}'$, $\mathbf{C}_2 = \mathbf{I}_n - f^{-1}\mathbf{ZZ}'$, $\theta_1 = m\sigma_s^2 + \sigma_e^2$, and $\theta_2 = \sigma_e^2$. This design satisfies the conditions of general balance "no matter how the [p] treatments are allocated to plots" (Speed 1983, p. 322), though some allocations give $\lambda_{jk} \in (0,1)$.

For a second example, consider a split-plot design. Suppose there are f blocks of m plots each and that each plot is subdivided into g subplots. Suppose blocks, plots, and subplots have incidence matrices $\mathbf{Z}_1$, $\mathbf{Z}_2$, and $\mathbf{Z}_3 = \mathbf{I}_n$ respectively, with σ_{s1}^2, σ_{s2}^2, and σ_e^2 being respectively the variance between blocks, the variance between sub-plots within blocks, and the error variance. Then $\mathbf{C}_1 = (mg)^{-1}\mathbf{Z}_1\mathbf{Z}'_1$, $\mathbf{C}_2 = g^{-1}\mathbf{Z}_2\mathbf{Z}'_2 - (mg)^{-1}\mathbf{Z}_1\mathbf{Z}'_1$, $\mathbf{C}_3 = \mathbf{I}_n - g^{-1}\mathbf{Z}_2\mathbf{Z}'_2$, $\theta_1 = mg\sigma_{s1}^2 + g\sigma_{s2}^2 + \sigma_e^2$, $\theta_2 = g\sigma_{s2}^2 + \sigma_e^2$, and $\theta_3 = \sigma_e^2$. As in the preceding example, the θ_j are linear functions of the usual variances of a mixed linear model.

Now for the main result. For designs satisfying conditions (C)* and (GB)* *and the additional restriction* that the λ_{jk} are all either 0 or 1, the log restricted likelihood

[1] Speed (1983) presents more general versions of these examples. He clearly has reasons for doing so but I don't know this area well enough to understand them.

for the θ_j is

$$-0.5\sum_j \left(\text{rank}\mathbf{C}_j - \sum_k \lambda_{jk}\text{rank}\mathbf{P}_k\right)\log(\theta_j) - 0.5\sum_j \theta_j^{-1}\mathbf{y}'\mathbf{C}_j\left[\mathbf{I}_n - \mathbf{P}\mathbf{C}_j\mathbf{P}\right]\mathbf{C}_j\mathbf{y}. \tag{17.11}$$

Note that $\mathbf{y}'\mathbf{C}_j\left[\mathbf{I}_n - \mathbf{P}\mathbf{C}_j\mathbf{P}\right]\mathbf{C}_j\mathbf{y}$ is a scalar that can also be written

$$\mathbf{y}'\Gamma_j\left[\mathbf{I}_{\text{rank}\mathbf{C}_j} - \Gamma_j'\mathbf{P}\mathbf{C}_j\mathbf{P}\Gamma_j\right]\Gamma_j'\mathbf{y}. \tag{17.12}$$

Because (17.12) is scalar and each θ_j is a linear function of the usual variances in a mixed linear model, this restricted likelihood has the desired simple form. This re-expressed restricted likelihood is the likelihood from a gamma-errors GLM with identity link and s "observations" indexed by j, given by (17.12). For the jth "observation," the linear predictor — linear in the variances σ_e^2 and σ_{sl}^2 when the model for $\mathbf{y}$ is written in the usual mixed-linear-model form — is θ_j, and the gamma shape parameter ν_j is $\left(\text{rank}\mathbf{C}_j - \sum_k \lambda_{jk}\text{rank}\mathbf{P}_k\right)/2$. Generally, the j correspond to effects in an ANOVA and ν_j is half of j's usual degrees of freedom. (McCullagh & Nelder 1989, p. 287, Section 8.3.5, noted this though not in the detail presented here.)

Designs satisfying the conditions above are *orthogonal designs* (Houtman & Speed 1983, Section 4.1), which includes (White 1985) randomized block designs with equal treatment allocations in each block; Latin squares and certain combinations of them (e.g., Graeco-Latin squares); and any other design that is balanced in the sense that "each level of the first factor occurs as frequently as each level of the second" (White 1985, p. 528), i.e., all designs that are balanced in the simple sense that I knew before I learned about general balance. Another example of an orthogonal design is the additive model with two predictors discussed above in Section 17.1.1, with null Γ_{12}.

Designs that satisfy the conditions of general balance but are not orthogonal include balanced or partially balanced incomplete block designs and randomized block designs with unequal allocations of treatments to blocks, among others. For such designs, which satisfy conditions (C)* and (GB)* and have at least one λ_{jk} that is not 0 or 1, Section 17.1.2.1 shows that the restricted likelihood can be made more explicit than the usual matrix formula but does not have the desired simple form and is not the likelihood arising from a generalized linear model.

We have seen examples of the re-expressed restricted likelihood for an orthogonal designs, namely the balanced one-way random effects model (Section 15.2.1) and the two-crossed-random-effects model (Section 12.1.2). An exercise asks you to express the latter restricted likelihood as the likelihood from a gamma GLM. I show one further example here, the model used to analyze the epidermal nerve density data in Section 14.4. For the present example, I assume the dataset was balanced; the actual dataset was not, though it was close.

In this tidied-up example, we have N subjects, each sampled at 2 locations (calf, thigh), with 2 blisters at each location and 2 images per blister. Location (calf vs. thigh) is a fixed effect. The four unknown variance components σ_{s1}^2, σ_{s2}^2, σ_{s3}^2, and σ_e^2 describe, in order down the design's hierarchy, variation between subjects, variation between subjects in the difference between calf and thigh (the subject-by-location interaction), variation between blisters within subject and location, and "error" (variation between images within blisters). For this design, the log restricted likelihood is

$$\log RL(\sigma_{s1}^2, \sigma_{s2}^2, \sigma_{s3}^2, \sigma_e^2|y) = -0.5 \sum_{j=1}^{4} DF_j \left[\log \theta_j + MS_j/\theta_j\right] \tag{17.13}$$

where DF_j and MS_j are the usual degrees of freedom and mean square for effect j in a standard ANOVA. (The proof is an exercise.) The θ_j are

- subject effect, $\theta_1 = 8\sigma_{s1}^2 + 4\sigma_{s2}^2 + 2\sigma_{s3}^2 + \sigma_e^2$;
- subject-by-location interaction, $\theta_2 = 4\sigma_{s2}^2 + 2\sigma_{s3}^2 + \sigma_e^2$;
- blisters within locations, $\theta_3 = 2\sigma_{s3}^2 + \sigma_e^2$; and
- error, $\theta_4 = \sigma_e^2$.

The θ_j are the expected mean squares from the standard ANOVA table, which is probably true for all orthogonal designs (an exercise incites you to prove this). The mean square MS_j is the sum of squared orthogonal contrasts making up effect j, divided by DF_j; all contrasts corresponding to a given effect have the same θ_j. The analogs to the two-variance model's a_j are now 3-vectors: For $j = 1$, this is $(8,4,2)$; for $j = 2$, it's $(0,4,2)$; etc. The replication provided by images within blisters means this model has free terms for error, the "observation" $j = 4$. This restricted likelihood is the likelihood arising from a gamma-errors generalized linear model with four "observations," the MS_j. An exercise asks you to write down this GLM.

17.1.2.1 Derivations Regarding Balance

Here are four facts I use below.

- (a) Let $\text{cov}(\mathbf{y}) = \mathbf{V} = \sum_{j=1}^{s} \theta_j \mathbf{C}_j$. Then $\mathbf{V}^{-1} = \sum_{j=1}^{s} \theta_j^{-1} \mathbf{C}_j$. This follows because $(\sum_{j=1}^{s} \theta_j \mathbf{C}_j)^{-1} = \Gamma(\sum_{j=1}^{s} \theta_j \Delta_j)^{-1}\Gamma' = \Gamma(\sum_{j=1}^{s} \theta_j^{-1} \Delta_j)\Gamma$, which holds because each Δ_j is diagonal with diagonal entries 0 or 1 and $\sum_{j=1}^{s} \Delta_j = \mathbf{I}_n$.
- (b) $\mathbf{P}$, the orthogonal projection onto the column space of $\mathbf{X}$, can be written $\mathbf{P} = \mathbf{L}\mathbf{L}'$ where $\mathbf{L}$ is $n \times p$ with orthonormal columns (because $\mathbf{X}$ has full rank). The matrices $\mathbf{P}_k$ can be written as $\mathbf{L}_k\mathbf{L}_k'$ where $\mathbf{L}_k$ is $n \times \text{rank}\mathbf{P}_k$. Thus one suitable $\mathbf{L}$ is $(\mathbf{L}_1|\dots|\mathbf{L}_K)$.
- (c) $\mathbf{P}\mathbf{C}_j\mathbf{P} = \sum_{k=1}^{K} \lambda_{jk}\mathbf{P}_k = \sum_{k=1}^{K} \lambda_{jk}\mathbf{L}_k\mathbf{L}_k' = \mathbf{L}\Delta_{pj}\mathbf{L}'$, where Δ_{pj} is a diagonal matrix with diagonal elements $\lambda_{jk(l)}, l = 1,\dots,p$. The indexing function $k(l)$ accounts for the rank of each $\mathbf{P}_k$ or $\mathbf{L}_k$; thus for example if $\text{rank}\mathbf{P}_1 = 3$, $\lambda_{jk(1)} = \lambda_{jk(2)} = \lambda_{jk(3)} = \lambda_{j1}$.
- (d) There exist matrices $\mathbf{D}_X$ $p \times p$ diagonal and $\mathbf{V}_X$ $p \times p$ orthogonal such that $\mathbf{X} = \mathbf{L}\mathbf{D}_X\mathbf{V}_X'$. Why? Because $\mathbf{L}$ is a basis for the column space of $\mathbf{X}$, there exists a $p \times p$ matrix $\mathbf{F}$ such that $\mathbf{X} = \mathbf{L}\mathbf{F}$; but $\mathbf{X}'\mathbf{X} = \mathbf{F}'\mathbf{L}'\mathbf{L}\mathbf{F} = \mathbf{F}'\mathbf{F} = \mathbf{V}_X\mathbf{D}_X^2\mathbf{V}_X$ by the spectral decomposition.

So, begin with the log restricted likelihood as stated in equation (1.16),

$$-0.5\left(\log|\mathbf{V}| + \log|\mathbf{X}'\mathbf{V}^{-1}\mathbf{X}| + \mathbf{y}'\left[\mathbf{V}^{-1} - \mathbf{V}^{-1}\mathbf{X}(\mathbf{X}'\mathbf{V}^{-1}\mathbf{X})^{-1}\mathbf{X}\mathbf{V}^{-1}\right]\mathbf{y}\right), \tag{17.14}$$

where I've omitted the unimportant constant. By an argument like the one for fact (a), $\log|\mathbf{V}| = \sum_{j=1}^{s} \text{rank}\mathbf{C}_j \log(\theta_j)$. As for the other determinant,

$$\begin{aligned}
|\mathbf{X}'\mathbf{V}^{-1}\mathbf{X}| &\propto \left|\mathbf{L}'\left(\sum_{j=1}^{s}\theta_j^{-1}\mathbf{C}_j\right)\mathbf{L}\right| \text{ by facts (a) and (d)} \\
&= \left|\sum_{j=1}^{s}\theta_j^{-1}\mathbf{L}'\mathbf{C}_j\mathbf{L}\right| \\
&= \left|\sum_{j=1}^{s}\theta_j^{-1}\Delta_{pj}\right| \text{ by facts (b) and (c)} \\
&= \prod_{l=1}^{p}\left(\sum_{j=1}^{s}\theta_j^{-1}\lambda_{jk(l)}\right),
\end{aligned} \tag{17.15}$$

which does not simplify further if some $\lambda_{jk(l)} \in (0,1)$.

Now consider $\mathbf{y}'\mathbf{V}^{-1}\mathbf{X}(\mathbf{X}'\mathbf{V}^{-1}\mathbf{X})^{-1}\mathbf{X}\mathbf{V}^{-1}\mathbf{y}$. By fact (d) and some steps like those shown immediately above, this equals

$$\mathbf{y}'\Gamma\left(\mathbf{W}\Gamma'\mathbf{L}\right)\left(\sum_{j=1}^{s}\theta_j^{-1}\Delta_{pj}\right)^{-1}\left(\mathbf{L}'\Gamma\mathbf{W}\right)\Gamma'\mathbf{y} \tag{17.16}$$

for $\mathbf{W} = \sum_{j=1}^{s}\theta_j^{-1}\Delta_j$. At the center of (17.16) is the $p \times p$ matrix

$$\left(\sum_{j=1}^{s}\theta_j^{-1}\Delta_{pj}\right)^{-1} = \text{diag}\left(\sum_{j=1}^{s}\theta_j^{-1}\lambda_{jk(l)}\right)^{-1}, \tag{17.17}$$

which also does not simplify further if some $\lambda_{jk(l)} \in (0,1)$.

So now we consider two cases. First we consider the case in which all λ_{jk} are either 0 or 1, which gives equation (17.11). After that, we assume some λ_{jk} is positive and less than 1 and derive an expression for the restricted likelihood that is more explicit than usual but not really simple.

So assume all λ_{jk} are either 0 or 1. Then the determinant in equation (17.15) simplifies to $|\mathbf{X}'\mathbf{V}^{-1}\mathbf{X}| = \prod_{l=1}^{p}\theta_{j(l)}^{-1}$, where $j(l)$ is the single j with $\lambda_{jk(l)} = 1$; there is exactly one such j because λ_{jk} is either 0 or 1 and $\sum_j \lambda_{jk} = 1$. Then $\prod_{l=1}^{p}\theta_{j(l)}^{-1} = \prod_{j=1}^{s}\theta_j^{-\sum_k \lambda_{jk}\text{rank}\mathbf{P}_k}$, as needed.

Regarding equations (17.16) and (17.17), under the assumption that all λ_{jk} are either 0 or 1,

$$\left(\sum_{j=1}^{s}\theta_j^{-1}\Delta_{pj}\right)^{-1} = \sum_{j=1}^{s}\theta_j\Delta_{pj}. \tag{17.18}$$

To simplify the quadratic form (17.16), note that

$$\mathbf{L}'\Gamma\mathbf{W} = (\theta_1^{-1}\mathbf{L}'\Gamma_1|\ldots|\theta_s^{-1}\mathbf{L}'\Gamma_s), \text{ a } p \times n \text{ matrix.} \tag{17.19}$$

Then $(\mathbf{W}\Gamma'\mathbf{L})\left(\sum_{j=1}^{s}\theta_j^{-1}\Delta_{pj}\right)^{-1}(\mathbf{L}'\Gamma\mathbf{W})$, the central part of equation (17.16), becomes a partitioned matrix with s row and column partitions, and its (j_1, j_2) block is

$$\theta_{j_1}^{-1}\theta_{j_2}^{-1}\Gamma'_{j_1}\left(\sum_j \theta_j \mathbf{L}\Delta_{pj}\mathbf{L}'\right)\Gamma_{j_2} = \theta_{j_1}^{-1}\theta_{j_2}^{-1}\sum_j \theta_j\left(\Gamma'_{j_1}\mathbf{P}\mathbf{C}_j\mathbf{P}\Gamma_{j_2}\right) \text{ by fact (c).} \tag{17.20}$$

In the right-hand part of (17.20), the expression in round brackets is 0 unless $j_1 = j_2$. To see this, note that under the assumption that $\lambda_{jk} = 0$ or 1,

$$(\mathbf{P}\mathbf{C}_{j_1}\mathbf{P})(\mathbf{P}\mathbf{C}_{j_2}\mathbf{P}) = \sum_{k=1}^{K}\lambda_{j_1k}\lambda_{j_2k}\mathbf{P}_k, \tag{17.21}$$

which is zero unless $j_1 = j_2$. But pre-multiplying (17.21) by $\Gamma_{j_1}\mathbf{L}'$ and post-multiplying it by $\mathbf{L}\Gamma_{j_2}$ gives the round-bracketed expression in (17.20).

Therefore,

$$\begin{aligned} & \mathbf{y}'\mathbf{V}^{-1}\mathbf{X}(\mathbf{X}'\mathbf{V}^{-1}\mathbf{X})^{-1}\mathbf{X}\mathbf{V}^{-1}\mathbf{y} \\ = & \left(\mathbf{y}'\Gamma\right)\text{blockdiag}\left[\theta_j^{-1}\Gamma'_j\mathbf{L}\Delta_{pj}\mathbf{L}'\Gamma_j\right]\left(\Gamma'\mathbf{y}\right) \\ = & \sum_{j=1}^{s}\theta_j^{-1}\left[\mathbf{y}'\Gamma_j\Gamma'_j\mathbf{L}\Delta_{pj}\mathbf{L}'\Gamma_j\Gamma'_j\mathbf{y}\right], \end{aligned} \tag{17.22}$$

where the expression inside the square brackets is a scalar. Now $\mathbf{y}'\mathbf{V}^{-1}\mathbf{y} = \sum_{j=1}^{s}\theta_j^{-1}\mathbf{y}'\Gamma_j\Gamma'_j\mathbf{y}$, so the entire quadratic form in (17.14) is as given in equation (17.11).

Now assume some λ_{jk} is positive and less than 1. The determinants in the restricted likelihood cannot be simplified beyond the expressions in (17.15) and just above it. The quadratic form in (17.16) can be made a bit more explicit. Note that

$$\left(\mathbf{L}'\Gamma\mathbf{W}\right)\Gamma'\mathbf{y} = \sum_j \theta_j^{-1}\mathbf{L}'\mathbf{C}_j\mathbf{y}, \tag{17.23}$$

so (17.16) can be written

$$\begin{aligned} & \left(\sum_j \theta_j^{-1}\mathbf{L}'\mathbf{C}_j\mathbf{y}\right)'\text{diag}\left(\sum_j \theta_j^{-1}\lambda_{jk(l)}\right)^{-1}\left(\sum_j \theta_j^{-1}\mathbf{L}'\mathbf{C}_j\mathbf{y}\right) \\ = & \sum_m\sum_i(\theta_i\theta_m)^{-1}\mathbf{y}'\mathbf{C}_m\mathbf{L}\,\text{diag}\left(\sum_j \theta_j^{-1}\lambda_{jk(l)}\right)^{-1}\mathbf{L}'\mathbf{C}_i\mathbf{y}, \end{aligned} \tag{17.24}$$

where the diagonal matrix is $p \times p$ with $l = 1, \ldots, p$ indexing the diagonal elements. I believe this diagonal matrix cannot be simplified further but perhaps you can prove me wrong.

17.1.3 Gaussian Processes Using the Spectral Approximation

This section describes a way to extend the re-expressed restricted likelihood for two-variance models to geostatistical models, specifically Gaussian processes (GPs). At this point, it is barely more than an idea, though it does seem to explain two things that have been widely observed in fitting Gaussian process models to data.

The idea is to use the spectral approximation to turn GPs observed on a rectangular grid into models to which Chapter 15's tools are easily extended. The tools are applied to an approximation and not to the GP itself, but this is still of interest for two reasons. First, if we can develop interesting facts about the approximation, these facts are hypotheses about GPs that can be tested by simulations or mathematical derivations. Second, spectral representations or approximations are of interest in themselves and as an expedient for reducing the computational burden associated with GPs (e.g., Paciorek 2007, Fuentes & Reich 2010), so facts about the spectral representation itself are of interest.

The presentation here draws heavily on Appendices A.1 and A.3 of Paciorek (2007), which built on Wikle (2002). My purpose here is simply to describe the idea and give a sense of how it might help us understand Gaussian processes as data-analytic tools. Thus, I consider only the one-dimensional case, although Paciorek (2007) gives explicit expressions for the two-dimensional case and his R package spectralGP handles arbitrary dimensions. For the present purpose, I ignore the periodicity problem discussed in Appendix A.2 of Paciorek (2007), though a full development will need to face this problem.

For concreteness, suppose we have an updated version of the global-mean surface temperature dataset with $n = 128$ observations; Paciorek (2007) requires a rectangular grid with each dimension's grid size a power of 2. Suppose we want to smooth this series using a one-dimensional Gaussian process without any fixed effects. Some will object on the grounds that this series is clearly not stationary, so it's inappropriate to fit a stationary model like the GP. I find this objection uncompelling. First, in practice GPs are routinely used to smooth data that are obviously not stationary. This may be ill-advised but standard software makes it easy and even sophisticated GP users do it sometimes. Thus, this case has inherent interest. Second, the material to follow is a first step toward analyses in which fixed effects are included. For the global-mean surface temperature data, these might include linear, quadratic, and cubic terms but even if such fixed effects were included, the remaining variation in the series is not stationary, it is simply non-stationary at higher frequencies and with lower amplitude compared to the original series. The question then is how the residual non-stationarity affects estimates of the GP's variance and other unknown parameters, and we can get a sense of that in the material to follow.

Consider, then, a one-dimensional Gaussian process for modeling an n-vector $\mathbf{y}$, where n is a power of 2. The approximation is a mixed linear model with the intercept as the only fixed effect, so $p = 1$, and with an $(n-1)$-dimensional random effect having diagonal covariance matrix $\mathbf{G}$. The equation for an observation

$y_t, t = 1, \ldots, n$ is

$$y_t = \beta_0 + 2 \sum_{m=1}^{\frac{n}{2}-1} (u_{1m} \cos(\omega_m 2\pi t/n) - u_{2m} \sin(\omega_m 2\pi t/n)) + u_{1,n/2} \cos(\omega_{n/2} 2\pi t/n) + \varepsilon, \tag{17.25}$$

where β_0 is the intercept, the terms containing u_j are random effects, and ε is, as usual, an n-vector of iid $N(0, \sigma_e^2)$ error. (I have simplified Paciorek's notation and made it consistent with the rest of the book.) The ω_m are frequencies taking values $m = 1, 2, \ldots, n/2$. Thus the first $2n-2$ columns in the random-effect design matrix $\mathbf{Z}$ are cos/sin pairs where the sin and cos terms share the frequency ω_m but have different random-effect coefficients u_{1m} and u_{2m}. $\mathbf{Z}$'s columns are orthonormal by construction (Paciorek 2007, p. 4). The covariance matrix for the random effects $\mathbf{G}$ is diagonal, as follows. Suppose the spectral density of the GP's covariance function is $\sigma_s^2 \phi(\omega; \theta)$ for frequency ω and unknown parameters θ. Then in this approximation, the variance of $u_{1,n/2}$ is $\sigma_s^2 \phi(\omega_{n/2}; \theta)$ while u_{1m} and u_{2m} have variance $0.5\sigma_s^2 \phi(\omega_m; \theta)$. I examine some specific $\phi(\omega; \theta)$ below.

This approximate GP extends the two-variance model of Chapter 15 by changing the variance of each random effect from σ_s^2 times a constant to σ_s^2 times a function of the frequency ω_m and an unknown parameter θ. Because $\mathbf{Z}$ has orthonormal columns that are orthogonal to $\mathbf{X} = \mathbf{1}_n$, the desired simple form of the restricted likelihood follows immediately. There are no free terms for σ_e^2 and each random effect provides one mixed term to the log restricted likelihood,

$$-0.5 \log(\sigma_s^2 a_j(\theta) + \sigma_e^2) - 0.5 \hat{v}_j^2 / (\sigma_s^2 a_j(\theta) + \sigma_e^2), \tag{17.26}$$

where the $a_j(\theta)$ are functions of the unknowns θ, namely $\phi(\omega_{n/2}; \theta)$ for the mixed term corresponding to $u_{1,n/2}$ and $0.5\phi(\omega_m; \theta)$ for mixed terms corresponding to u_{1m} and u_{2m}. Thus $j = 1, 2$ correspond to the frequency ω_1, $j = 3, 4$ to the frequency ω_2, and so on. The canonical predictors are the columns of $\mathbf{Z}$ and the $\hat{v}_j$ are the coefficients of projections of $\mathbf{y}$ onto those columns. Thus the $\hat{v}_j$ decompose $\mathbf{y}$ into components corresponding to the frequencies ω_m; for each $m < n/2$, the pair of $\hat{v}_j$ corresponding to frequency ω_m are the sin and cos portions of (17.25) for ω_m. Note that the canonical predictors and thus the $\hat{v}_j$ are the same for *all* GPs on a given rectangular grid, so different GPs on a given grid are differentiated *only* by their $a_j(\theta)$, i.e., by their spectral densities.

Because $a_j(\theta)$ is a function of the unknowns in θ, this scalarized restricted likelihood is not a gamma-errors generalized linear model with identity link. However, for a given value of θ, this spectral approximation *is* a two-variance problem and its restricted likelihood is a gamma-errors GLM like the ones in Chapters 15 and 16. In those chapters, the a_j were the key to understanding the restricted likelihood, so an understanding of how a GP works as a data-analysis engine (using the spectral approximation) begins with examination of the $a_j(\theta)$ as a function of θ.

To test this idea's potential, we'll consider two GPs in the Matérn family. Paciorek (2007, equation 5) parameterizes the Matérn family for D-dimensional space so that the spectral density is

$$\sigma_s^2 \phi(\omega;\rho,\nu) = \sigma_s^2 K(\nu,D)\rho^D \left(1+\frac{(\pi\rho)^2}{4\nu}\omega'\omega\right)^{-(\nu+D/2)}, \tag{17.27}$$

where ν and ρ are the Matérn family's smoothness and range parameters, ω is the D-vector of frequencies, and $K(\nu,D)$ is a function of ν and D but not of ρ or ω. The function $\phi(\omega;\rho,\nu)$ has the form of a multivariate t density with dispersion matrix proportional to the identity and scale parameter $\sqrt{2}/\pi\rho$. For the one-dimensional case we are considering, $D=1$ and $\omega'\omega=\omega^2$. In the Matérn model, ρ describes the range at which the correlation between pairs of observations (t_1,t_2) decays to a small value (Paciorek 2007 parameterizes the range somewhat differently than usual).

The parameter ν is usually described as controlling the smoothness of realizations generated from the GP used as a probability model (as distinct from using it as a likelihood) and is usually fixed *a priori* in analyses. The two GPs to be considered fix $\nu=0.5$, the so-called exponential form, and $\nu=\infty$, the so-called squared exponential form. For $\nu=0.5$ and $D=1$, $\phi(\omega;\rho)$ has the form of a Cauchy density,

$$\phi(\omega;\rho) = \frac{1}{\sqrt{2}}\rho\left(1+\frac{(\pi\rho)^2}{2}\omega^2\right)^{-1}. \tag{17.28}$$

For $\nu=\infty$ and $D=1$, $\phi(\omega;\rho)$ has the form of a normal (Gaussian) density,

$$\phi(\omega;\rho) = \frac{\sqrt{\pi}}{2}\rho\exp\left(-\frac{(\pi\rho)^2}{4}\omega^2\right). \tag{17.29}$$

The functions $\phi(\omega;\rho)$ give the $a_j(\rho)$ in the re-expressed restricted likelihood for these two (approximate) GP models. How do these $a_j(\rho)$ decline as the frequency ω_m increases, and how does this differ between $\nu=0.5$ and $\nu=\infty$? Figure 17.1 tells the story. Figure 17.1 shows the $a_j(\rho)$ for two values of ρ, but the pattern is the same for any ρ. The two models ($\nu=0.5$ and $\nu=\infty$) have quite similar $a_j(\rho)$, and these $a_j(\rho)$ are the *only* aspect of the restricted likelihood that differs between these two (approximate) GP models. This explains the common observation that in fitting a GP to data, the data provide hardly any information about ν. The two ν depicted here are the conventional small and large extreme values of ν and they produce restricted likelihoods that barely differ, so it is no wonder that data cannot distinguish more subtle differences in ν.

Figure 17.1 shows that the $a_j(\rho)$ do depend strongly on ρ. Recalling that for fixed ρ this model is a gamma-errors GLM with identity link but that ρ can also be adjusted, we can develop some intuition for how the unknowns ρ, σ_s^2, and σ_e^2 are adjusted to fit the data. In this gamma-errors GLM, the "observations" are the $\hat{v}_j^2$, which have expectation

$$E(\hat{v}_j^2|\sigma_s^2,\rho,\sigma_e^2) = \sigma_s^2 a_j(\rho)+\sigma_e^2 = \sigma_s^2\rho\times Hf(\rho,\omega_m^2)+\sigma_e^2, \tag{17.30}$$

where H is a constant and $f(\rho,\omega_m^2)$ is depicted in Figure 17.1 as a function of ω_m,

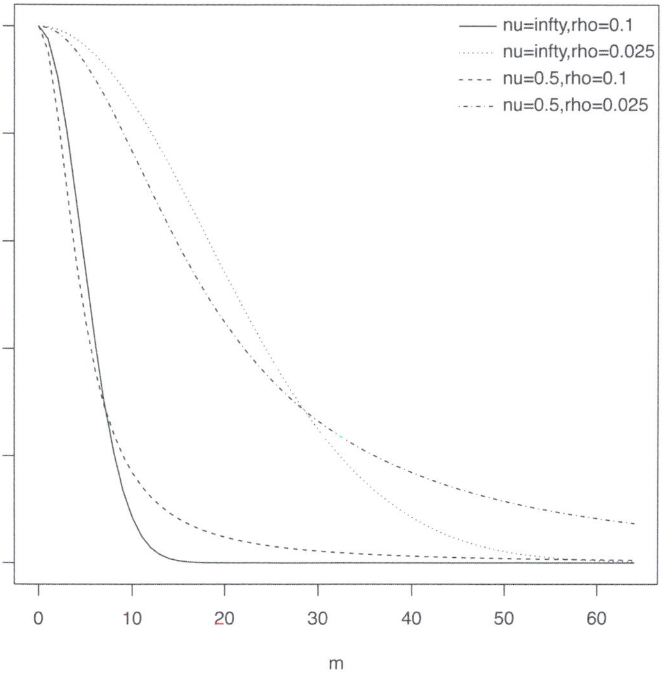

Figure 17.1: For a global-mean surface temperature series with $n = 128$, $a_j(\rho)$ for the GPs with $\nu = 0.5$ and $\nu = \infty$. The horizontal axis is the index of frequencies ω_m, with higher frequencies to the right; recall that $j = 1,2$ share frequency ω_1, etc. The vertical axis is $a_j(\rho)$.

i.e., of j. For a given dataset and ν, ρ is chosen to best fit the decline in $\hat{v}_j^2$ with j. This answers my original question about GPs, namely "what functions of the data determine ρ?" The answer is: The rate of decline in the $\hat{v}_j^2$ as a function of the frequency ω_m. Having determined ρ (speaking very loosely; more on this below), σ_s^2 and σ_e^2 are then chosen to make $\sigma_s^2 \rho \times K f(\rho, \omega_m^2) + \sigma_e^2$ go through the middle of the $\hat{v}_j^2$. For this to happen, σ_s^2 must be large enough so $\sigma_s^2 \rho$ captures the largest $\hat{v}_j^2$ (for small j), while σ_e^2 must describe the level of the $\hat{v}_j^2$ for large j.

With this intuition, it is easy to understand the common observation that σ_s^2 and ρ are poorly identified: The only thing identifying them is $f(\rho, \omega_m^2)$ and if we recall the "noise" in $\hat{v}_j^2$ as a function of j for earlier problems we've considered (Figures 15.2, 15.5, and 15.11 and Tables 16.1, 16.2, and 16.4), it is clear that ρ is not especially well-determined, so σ_s^2 isn't, either.

To suggest how this approximate representation might be used to learn about GPs, I offer some conjectures about how influential $\hat{v}_j^2$ — squared lengths of projections onto particular sine or cosine basis functions — might affect estimates of ρ, σ_s^2, and

σ_e^2. These conjectures, generated by examining the spectral approximation, can be tested for actual GPs using simulation experiments. (An exercise incites you to do so.)

One conjecture is based on the observation that a strong low-frequency trend (e.g., quadratic) will create large $\hat{v}_j^2$ for small j followed by much smaller $\hat{v}_j^2$ for succeeding j. This will force ρ's estimate to be large to capture the sharp decline in the $\hat{v}_j^2$ as j increases; the estimate of σ_s^2 will be adjusted so that for small j, $\sigma_s^2 a_j(\rho) + \sigma_e^2$, which is dominated by $\sigma_s^2 a_j(\rho)$, fits the (large) $\hat{v}_j^2$. However, the resulting fit to $\mathbf{y}$ will not necessarily be smooth. That depends on the level of the $\hat{v}_j^2$ for large j, which determines the estimate of σ_e^2 and thus the DF in the fit of a canonical predictor's coefficient v_j, which is

$$\frac{a_j(\rho)}{a_j(\rho) + \sigma_e^2/\sigma_s^2} = \frac{\sigma_s^2 a_j(\rho)}{\sigma_s^2 a_j(\rho) + \sigma_e^2}. \tag{17.31}$$

This depends on $\sigma_s^2 a_j(\rho)$, both factors of which adjust in fitting the data. If σ_e^2 is estimated to be large relative to $\sigma_s^2 a_j(\rho)$, the fit will be smooth; otherwise, the fit will be rough.

Another conjecture is that an outlier in a high frequency will have little effect on the estimates of σ_s^2 or ρ but will inflate the estimate of σ_e^2 and thus result in a smoother fit.

17.1.4 Separable Models

The restricted likelihood has the simple re-expression for models that are separable in the following sense. Suppose the n-vector $\mathbf{y}$ is modeled as $\mathbf{y} = \delta + \varepsilon$, where $\varepsilon \sim N_n(\mathbf{0}, \sigma_e^2 \mathbf{I}_n)$ and δ has a possibly improper normal density with mean zero and precision matrix of the form $\sum_{k=1}^{K} \tau_k \mathbf{Q}_k$, where the τ_k are scalar precision parameters and the $\mathbf{Q}_k$ are $n \times n$ matrices. The $\mathbf{Q}_k$ have the form $\mathbf{Q}_k = \mathbf{A}_1 \otimes \ldots \otimes \mathbf{A}_M$, where $\otimes$ is the Kronecker product, $\mathbf{A}_l = \mathbf{I}$ for $l \neq k$, $\mathbf{A}_k$ is positive semi-definite, and $\mathbf{A}_l$ has the same dimensions for all $\mathbf{Q}_k$. The proof is straightforward and given as an exercise; the idea is that by properties of the Kronecker product, we just need to simultaneously diagonalize the identity and one positive semi-definite $\mathbf{A}_k$ at a time, and this can always be done.

This class of models includes some 2NRCAR models, for example, those in which the areas form a rectangular grid with neighbors within rows being one class of neighbors and neighbors within columns being the second class (Besag & Higdon 1999). This class of models also includes spatio-temporal ICAR models in which spatial neighbors are one class of neighbors and temporal neighbors are the second class. Spatio-temporal ICAR models with more than one class of spatial neighbor pairs are separable if the spatial part of the model is.

The following two assertions appear to be true in general for separable models as defined above but I have not proved them (proofs are exercises). This form of separability is a type of balance but separable models as defined here do not satisfy the conditions of general balance. Also, models of this form do not produce restricted likelihoods that are likelihoods for generalized linear models.

The intuition for these assertions comes from a separable 2NRCAR model. Regarding balance-but-not-general balance, suppose $\mathbf{B}$ in Section 17.1.1 is an orthogonal matrix. Then referring to the paragraph containing equation (14.2), the 2NRCAR model in mixed-linear-model form is

$$\mathbf{y} = \Gamma_X\beta + \Gamma_Z\mathbf{u} + \varepsilon \tag{17.32}$$

where $\Gamma_X'\Gamma_Z = \mathbf{0}$, the columns $(\Gamma_X|\Gamma_Z)$ are an orthonormal basis for real n-space, and $\mathbf{u}$ has covariance matrix $\mathbf{G} = (\tau_1\mathbf{D}_{1+} + \tau_2\mathbf{D}_{2+})^{-1} = \text{diag}\left[\sigma_{s1}^2\sigma_{s2}^2/(\sigma_{s1}^2 d_{2j} + \sigma_{s2}^2 d_{1j})\right]$. Thus the marginal covariance of $\mathbf{y}$ is

$$\begin{aligned} & \Gamma_Z\text{diag}\left[\sigma_{s1}^2\sigma_{s2}^2/(\sigma_{s1}^2 d_{2j} + \sigma_{s2}^2 d_{1j})\right]\Gamma_Z' + \sigma_e^2\mathbf{I}_n \\ = \; & \Gamma_Z(\text{diag}\left[\sigma_{s1}^2\sigma_{s2}^2/(\sigma_{s1}^2 d_{2j} + \sigma_{s2}^2 d_{1j})\right] + \sigma_e^2\mathbf{I}_n)\Gamma_Z' + \sigma_e^2(\mathbf{I}_n - \Gamma_Z\Gamma_Z'). \end{aligned} \tag{17.33}$$

The conditions defining general balance require that the marginal covariance of $\mathbf{y}$ have the form $\sum_j \theta_j\mathbf{C}_j$ for scalar θ_j and suitable matrices $\mathbf{C}_j$, but (17.33) implies this is not possible. Thus, this form of balance is not general balance, at least for separable 2NRCAR models.

To see why this restricted likelihood cannot be the likelihood for a generalized linear model, suppose again that $\mathbf{B}$ is an orthogonal matrix. Then the restricted likelihood in equation (17.7) is

$$\tau_e^{(n-I)/2}\prod_{j=1}^{n-I}\left(\frac{d_{1j}\tau_1 + d_{2j}\tau_2}{\tau_e + d_{1j}\tau_1 + d_{2j}\tau_2}\right)^{0.5}\exp\left[-0.5\tau_e\sum_{j=1}^{n}(\mathbf{y}'\mathbf{B})_j^2\frac{d_{1j}\tau_1 + d_{2j}\tau_2}{\tau_e + d_{1j}\tau_1 + d_{2j}\tau_2}\right] \tag{17.34}$$

(the proof is an exercise). This still has the form of a gamma-errors model but it cannot be written as a generalized linear model with predictor a linear function of the unknown variances or precisions. I don't have a formal proof of the latter but I've spent a lot of time trying and I'm convinced it's impossible. An exercise lets you see for yourself.

17.1.5 Miscellaneous Other Models

The preceding sections described some large classes of models yielding restricted likelihoods that can be re-expressed in the desired simple form. I am nowhere near a general characterization of models that have this property. Here are some more models that do.

The clustering and heterogeneity model of Section 12.1.1 has a restricted likelihood with the desired simple form. This is the only such example I've seen with three or more unknowns that does not appear to be balanced in some sense, or perhaps this model just has an extreme form of balance in that every area has a parameter for *both* the clustering and heterogeneity effects.

Separable models more complex than those in Section 17.1.4 can also yield simple restricted likelihoods. At least two kinds of extensions are possible. First, Section 17.1.4 assumed $\mathbf{Q}_k = \mathbf{A}_1 \otimes \ldots \otimes \mathbf{A}_M$, where $\mathbf{A}_l = \mathbf{I}$ for $l \neq k$ and $\mathbf{A}_k$ is positive semi-definite. This condition is sufficient but not necessary; it is only necessary to assume each of the Kronecker product's M factors can be diagonalized simultaneously in all the $\mathbf{Q}_k$. This is true for the specification in Section 17.1.4 because a given factor of the Kronecker product is $\mathbf{I}$ for all but one l, but as the discussion of general balance shows, certain more complex structures permit simultaneous diagonalization. I cannot think of any examples of this extension of separable models but someone is probably using one.

Another extension of Section 17.1.4's separable models would include fixed effects, which yield a restricted likelihood with the desired simple form under certain conditions on the $\mathbf{Q}_k$ and $\mathbf{X}$, the fixed-effect design matrix. I don't know what those conditions are so it's possible they exist but are vacuous; an exercise explores this possibility.

17.2 Expedients for Restricted Likelihoods That Can't Be Re-expressed

The re-expressed restricted likelihoods considered in the last three chapters were derived by applying a known linear transformation to the data $\mathbf{y}$ and a known re-parameterization to the mixed linear model. The trick is to figure out the transformation and re-parameterization and as we've seen, they do not exist for some mixed linear models. So what shall we do?

One possibility, of course, is to abandon this approach and pursue a different one. I don't have any ideas, but perhaps you do. Another possibility is to allow either the transformation of the data or the re-parameterization of the mixed linear model to depend on unknowns. Such an approach might permit a scalarized form for the restricted likelihood but with unknowns in the expressions for the $\hat{v}_j^2$. This is somewhat similar to the approach to Gaussian processes described in Section 17.1.3, in which the a_j became functions of the unknown range parameter ρ.

I have not yet considered either of these approaches. The next two subsections show two expedients I have used in problems in which the restricted likelihood could not be scalarized.

17.2.1 Expedient 1: Ignore the Error Variance

This expedient was part of Brian Reich's doctoral dissertation; it was done before the work on two-variance models and prompted the idea for re-expressing the restricted likelihood. Thus it is not a bastardization but an imperfect version of the re-expression, though it was good enough to solve the 2NRCAR mystery described in Section 14.2, to which we now turn.

Reich et al. (2004, 2007) formulated 2NRCAR models not as mixed linear models but in a Bayesian style using precision matrices. I do the same here; an exercise asks you to consider how to extend this expedient to models specified in mixed-linear-model form. I consider a single patient; Reich et al. (2004) considered data for one or three patients while Reich et al. (2007) considered data for 50 patients.

Before we proceed, note that when τ_e has a gamma prior, it is easy to use (17.7) to derive the joint marginal posterior of $r_1 = \tau_1/\tau_e$ and $r_2 = \tau_2/\tau_e$ and thus of

Table 17.1: Definitions of Classifications A, B, and C of Neighbor Pairs

Classification	Type I	Type II	Type III	Type IV	Description
A	1	1	2	2	Sides vs. interproximal
B	1	2	2	2	Direct vs. interproximal
C	2	1	2	2	Type II vs. others

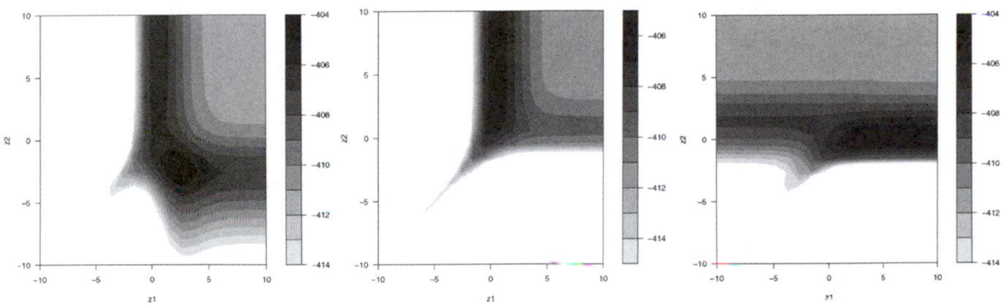

(a) Classification A (b) Classification B (c) Classification C

Figure 17.2: Contour plots of the log marginal posterior of (z_1, z_2) for one person's attachment loss measurements. Panel (a): Classification A; Panel (b): Classification B; Panel (c): Classification C. (From Reich et al. 2007).

$z_1 = \log(r_1)$ and $z_2 = \log(r_2)$. If the prior for the three precisions is $\pi(\tau_e, \tau_1, \tau_2) = \pi(\tau_e)\pi(z_1, z_2)$, where $\pi(\tau_e) \propto \tau_e^{a_e}\exp(-b_e\tau_e)$, then (z_1, z_2) has marginal posterior

$$\pi(z_1, z_2|\mathbf{y}) \quad \propto \quad \pi(z_1, z_2)\prod_{j=1}^{n-I}(e^{z_1}d_{1j} + e^{z_2}d_{2j})^{0.5}\,|\mathbf{I}_n + e^{z_1}\mathbf{Q}_1 + e^{z_2}\mathbf{Q}_2|^{-0.5} \tag{17.35}$$

$$\left[b_e + 0.5(\mathbf{y}'\mathbf{y} - \mathbf{y}'[\mathbf{I}_n + e^{z_1}\mathbf{Q}_1 + e^{z_2}\mathbf{Q}_2]^{-1}\mathbf{y})\right]^{-(n-I)/2-a_e}. \tag{17.36}$$

With this, we can draw contour plots of the marginal posterior $\pi(z_1, z_2|\mathbf{y})$ and see where the posterior mass lies. Figures 17.2a,b show contour plots for a single patient's data using Classifications A and B (reproducing Figure 14.3). Figure 17.2c shows the contour plot arising from the same patient's data analyzed using a third partition of the neighbor pairs into two classes, Classification C. Table 17.1 summarizes how Classifications A, B, and C partition the four types of neighbor pairs (Figure 14.2) into two classes. For all three plots, the prior $\pi(z_1, z_2)$ is flat on the rectangle with each of z_1 and z_2 between -15 and 15, so the plots show the data contribution to this marginal posterior. (Well, not quite: The prior for τ_e had $a_e = b_e = 0.01$.)

Now for the expedient, which is to condition the joint posterior (17.6) on the random effect δ and consider

$$\pi(\tau_1, \tau_2|\delta) \quad \propto \quad \prod_{j=1}^{n-I}(d_{1j}\tau_1 + d_{2j}\tau_2)^{0.5}\exp\left(-0.5\delta'(\tau_1\mathbf{Q}_1 + \tau_2\mathbf{Q}_2)\delta\right)$$

Table 17.2: Counts of **u**-Free and -Mixed Terms for Classifications A, B, and C for the Patient Whose Marginal Posteriors Are Shown in Figure 17.2

		Counts of Islands			Counts of **u**-Free Terms		**u**-Mixed
Classification	n	I	I_1	I_2	for z_1	for z_2	Terms
A	162	3	6	84	81	3	75
B	162	3	54	84	81	51	27
C	162	3	114	3	0	111	48

$$= \prod_{j=1}^{n-I} (d_{1j}\tau_1 + d_{2j}\tau_2)^{0.5} \exp\left(-0.5\mathbf{u}'(\tau_1 \mathbf{D}_{1+} + \tau_2 \mathbf{D}_{2+})\mathbf{u}\right)$$

$$= \prod_{j=1}^{n-I} (d_{1j}\tau_1 + d_{2j}\tau_2)^{0.5} \exp\left(-0.5 \sum_{j=1}^{n-I} u_j^2 (d_{1j}\tau_1 + d_{2j}\tau_2)\right), \quad (17.37)$$

where the $(n-I)$-dimensional **u** is a known linear function of δ defined in Section 14.2. Apart from its use in the present expedient, this conditional distribution is of some interest because it can be used in MCMC algorithms to make draws of (τ_1, τ_2); the marginal posterior of (τ_1, τ_2) is (17.37) integrated against the marginal posterior of δ.

Equation (17.37) has an obvious similarity to the re-expressed restricted likelihood for a two-variance model but with an important difference. In a two-variance model, σ_s^2 and σ_e^2 are inherently asymmetric because σ_e^2 can have free terms but σ_s^2 cannot. By contrast, τ_1 and τ_2 in (17.37) are defined symmetrically — the labels "class 1" and "class 2" are arbitrary — and indeed this expedient's usefulness arises from applying the idea of free and mixed terms to both classes of neighbor pairs.

Define **u**-free terms for τ_1 as those j for which $d_{2j} = 0$, **u**-free terms for τ_2 as those j for which $d_{1j} = 0$, and **u**-mixed terms as those j for which both $d_{1j} > 0$ and $d_{2j} > 0$. Counts of **u**-free terms are easily determined from counts of islands (disconnected regions) in the maps implied by class 1 and class 2 neighbor pairings. If I is the number of islands in a spatial map that ignores the distinction between classes of neighbor pairs and class k neighbor pairings have I_k islands, then $I_k - I$ of the d_{kj} are 0. Thus τ_1 has $I_2 - I$ **u**-free terms, τ_2 has $I_1 - I$ **u**-free terms, and there are $n - I_1 - I_2 + I$ **u**-mixed terms. Table 17.2 gives counts of **u**-free and **u**-mixed terms for Classifications A, B, and C for the patient whose marginal posteriors are shown in Figure 17.2. (This patient was missing a maxillary lateral incisor so his spatial map has $I = 3$ islands. Figure 3 of Reich et al. 2007 shows the data.) Classification B has many **u**-free terms for both z_1 and z_2, while Classification A has few **u**-free terms for z_2 and Classification C has no **u**-free terms for z_1.

We can now explain the odd marginal posteriors in Figure 17.2. The "legs" in Figures 17.2a,b (Classifications A and B) and the strange contours in Figure 17.2c (Classification C) arise from **u**-free terms. Reich et al. (2007, Section 4) presents the argument, which would require a lot of new notation so I won't recapitulate it here. The intuition is that the **u**-free terms for z_1 considered alone have contours parallel to the z_2 axis — like the upward-pointing "leg" in Figures 17.2a,b — because they contain no information about z_2. Similarly, the **u**-free terms for z_2 considered alone have contours parallel to the z_1 axis, as in all three panels of Figure 17.2. When these two sets of **u**-free terms and the mixed terms are multiplied together, they produce the two-legged contour plots for Classifications A and B. Classification C's contour plot has only one "leg" parallel to the z_1 axis because it has no **u**-free terms for z_1.

This leads to a secondary puzzle: If the **u**-free terms dominate the marginal posterior of (z_1, z_2), why don't Figures 17.2a,b have "legs" pointing down and to the right, giving substantial posterior mass to small values of z_2 and z_1 respectively? To understand this, we need to examine the marginal posterior itself, equation (17.35). The intuition is that as either z_1 or z_2 takes increasingly large values — as either class of neighbor pairs is shrunk more stringently — the marginal posterior declines slowly and eventually becomes flat. This happens because past a certain point, neighbors have already been shrunk together as much as they practically can be and a further increase in z_1 or z_2 has no practical effect. Mechanically, it happens because $\left[b_e + 0.5(\mathbf{y}'\mathbf{y} - \mathbf{y}'[\mathbf{I}_n + e^{z_1}\mathbf{Q}_1 + e^{z_2}\mathbf{Q}_2]^{-1}\mathbf{y})\right]^{-(n-I)/2-a_e}$ converges to a finite positive constant while $\prod_{j=1}^{n-I}(e^{z_1}d_{1j} + e^{z_2}d_{2j})^{0.5}$ and $|\mathbf{I}_n + e^{z_1}\mathbf{Q}_1 + e^{z_2}\mathbf{Q}_2|^{-0.5}$ offset each other. This flattening can be seen in the upper right corner of Figures 17.2a,b and the upper half of Figure 17.2c. By contrast, as z_1 takes increasingly negative values, the marginal posterior drops off precipitiously if there are **u**-free terms for z_1 and analogously for z_2. Mechanically, this happens because

$$\prod_{j=1}^{n-I}(e^{z_1}d_{1j} + e^{z_2}d_{2j})^{0.5} \propto e^{n_1 z_1/2} e^{n_2 z_2/2} \prod_{\text{mixed terms}} (e^{z_1}d_{1j} + e^{z_2}d_{2j})^{0.5} \qquad (17.38)$$

where n_k is the number of **u**-free terms for z_k. If one z_k becomes more negative with the other held constant, $e^{n_k z_k/2}$ goes to zero while the other two terms in (17.38) go to finite positive constants.

The counts of **u**-free terms also help us understand why, in Figures 17.2a,b, the two legs differ in mass and width. For Classifications A and B, z_1 has more **u**-free terms than z_2 so in both contour plots the "leg" parallel to the z_2 axis is both higher and narrower than the "leg" parallel to the z_1 axis. For Classification C, z_1 has no free terms so there is no "leg" parallel to the z_2 axis.

17.2.2 Expedient 2: Ignore the Non-zero Off-Diagonals

I developed this expedient as yet another attempt to get some leverage on Michael Lavine's dynamic linear model puzzle (Chapters 6, 9, and 12). I've only used it on this example but it takes advantage of features that are present in other time series and spatial models so it may be useful for other problems.

To present this expedient, I begin with the restricted likelihood for Model 2, which includes signal, mystery (one harmonic), respiration (one harmonic), and heartbeat components having smoothing variances σ^2_{ss}, σ^2_{sm}, σ^2_{sr}, and σ^2_{sh} respectively, with error variance σ^2_e. (In earlier chapters I used the symbol W instead of σ^2.) Model 2's fit was a puzzle because it shrank the signal component to a straight line and shrank the respiration component close to a sine curve while the mystery curve captured almost all the variation that those two components had captured in Model 1's fit.

As in Section 12.2, gather the four components' fixed effects into a single design matrix $\mathbf{X}$ and parameter β but leave their random effects in separate design matrices $\mathbf{Z}_k, k = s,m,r,h$ and random effects $\mathbf{u}_k$. Thus Model 2 for these data is

$$\mathbf{y} = \mathbf{X}\beta + \sum_k \mathbf{Z}_k\mathbf{u}_k + \varepsilon, \tag{17.39}$$

where $\mathbf{X}$ is 650×8, each $\mathbf{Z}_k$ is 650×642, $\text{cov}(\mathbf{u}_k) = \sigma^2_{sk}\mathbf{I}_{642}$, and $\mathbf{R} = \sigma^2_e\mathbf{I}_{650}$. The projection onto the orthogonal complement of $\mathbf{X}$'s column space is $\mathbf{I} - \mathbf{X}(\mathbf{X}'\mathbf{X})^{-1}\mathbf{X}' = \mathbf{K}\mathbf{K}'$ for any 650×642 matrix $\mathbf{K}$ with orthonormal columns spanning the orthogonal complement of $\mathbf{X}$'s column space. The restricted likelihood is then the likelihood arising from the transformed data $\mathbf{K}'\mathbf{y}$, which is

$$|\sum_k \sigma^2_{sk}\mathbf{K}'\mathbf{Z}_k\mathbf{Z}_k'\mathbf{K} + \sigma^2_e\mathbf{I}_{642}|^{-0.5} \exp\left(-0.5\mathbf{y}'\mathbf{K}\left[\sum_k \sigma^2_{sk}\mathbf{K}'\mathbf{Z}_k\mathbf{Z}_k'\mathbf{K} + \sigma^2_e\mathbf{I}_{642}\right]^{-1}\mathbf{K}'\mathbf{y}\right). \tag{17.40}$$

To put this restricted likelihood in the desired simple form, the four $\mathbf{K}'\mathbf{Z}_k\mathbf{Z}_k'\mathbf{K}$ and $\mathbf{I}_{642}$ must be simultaneously diagonalized and that is impossible. In earlier chapters, however, we noted that the canonical re-expressions of quasi-cyclic DLM components have design matrix columns that are, loosely speaking, sinusoidal with frequencies that start just above the nominal frequency of the quasi-cyclic component (e.g., 1/117 cycles per time step for the mystery term) and increase from there. Also, the signal component's canonical re-expression has design matrix columns that (loosely speaking) start with a half-cycle of a sinusoidal curve and increase in frequency from there. This suggests that if we transform the data to diagonalize $\mathbf{I}_{642}$ and $\mathbf{K}'\mathbf{Z}_s\mathbf{Z}_s'\mathbf{K}$ (for the signal component), the $\mathbf{K}'\mathbf{Z}_k\mathbf{Z}_k'\mathbf{K}$ for mystery, respiration, and heartbeat will have mostly small off-diagonals. If so, we can approximate the restricted likelihood (17.40) by ignoring those off-diagonals and this approximation has the desired form. Because the four model components have different nominal frequencies, the components should have large diagonal elements for different canonical predictors. As we'll now see, this worked for the present problem, more or less.

So let $\mathbf{K}'\mathbf{Z}_s\mathbf{Z}_s'\mathbf{K} = \Gamma\mathbf{A}_s\Gamma'$ be the spectral decomposition of $\mathbf{K}'\mathbf{Z}_s\mathbf{Z}_s'\mathbf{K}$. Then the restricted likelihood (17.40) is proportional to

$$|\sum_k \sigma^2_{sk}\mathbf{A}_k + \sigma^2_e\mathbf{I}_{642}|^{-0.5} \exp\left(-0.5\mathbf{y}'\mathbf{K}\Gamma\left[\sum_k \sigma^2_{sk}\mathbf{A}_k + \sigma^2_e\mathbf{I}_{642}\right]^{-1}\Gamma'\mathbf{K}'\mathbf{y}\right), \tag{17.41}$$

where $\mathbf{A}_k = \Gamma'\mathbf{K}'\mathbf{Z}_k\mathbf{Z}_k'\mathbf{K}\Gamma, k = s,m,r,h$. $\mathbf{A}_s$ is diagonal; $\mathbf{A}_m$, $\mathbf{A}_r$, and $\mathbf{A}_h$ are not, but we now approximate each one with a diagonal matrix. Define a_{jk} to be the jth diagonal element of $\mathbf{A}_k$; the approximate restricted likelihood is then

$$\prod_j(\sum_k \sigma^2_{sk}a_{jk} + \sigma^2_e)^{-0.5} \exp\left(-0.5\sum_j \hat{v}^2_j/(\sum_k \sigma^2_{sk}a_{jk} + \sigma^2_e)\right), \tag{17.42}$$

where the canonical observations $\hat{v}_j$ are $\hat{\mathbf{v}} = \Gamma'\mathbf{K}'\mathbf{y}$. This approximate restricted likelihood is the likelihood from a gamma-errors GLM with identity link, "observations" $\hat{v}_j^2$ having expected value $\sum_k \sigma_{sk}^2 a_{jk} + \sigma_e^2$, and shape parameter $\nu_j = 0.5$.

The first question about this approximation is whether the off-diagonals we've discarded are, in fact, small. For the mystery component, almost all of them are. Considering the absolute values of the 205,761 distinct off-diagonals, the mystery term's off diagonals have 99.9th percentile 0.064; the 99.99th is 0.46 and the maximum is 0.72. For the respiration component, the off-diagonals are not quite so small. The 95th percentile is 0.060, the 99th is 0.32, and the maximum is 0.80. For the heartbeat component, the median absolute off-diagonal is 5×10^{-6}, but the percentiles increase quickly from there. The 55th percentile is 0.11, the 75th is 0.46, the 95th is 0.71, and the maximum is 0.82. None of this is surprising, given that in the canonical representation of individual quasi-cyclic components, the progression of design-matrix columns through increasingly high frequencies is good for lower frequencies but breaks down after a point.

The approximation is surprisingly good in some respects. For example, Table 17.3 shows numerical maxima of the exact and approximate restricted likelihoods for four sets of starting values.[2] The apparent global maximum is similar for the two functions (starting values 1), as is the apparent secondary maximum (which has larger respiration variance; starting values 2 for the exact RL, starting values 3 for the approximation). The maximizing values for signal and error appear to vary a lot but these are both zero or very close to it, so the "maximizing" value is an accident of the algorithm. (We will return to this point below.) For two sets of starting values (3 and 4), the exact restricted likelihood became numerically singular (in my naïve implementation) while the approximate restricted likelihood did not.

With this modest justification, we proceed to examine the approximation, for what it's worth.

For two-variance models, the a_j were the key to understanding the restricted likelihood. Here, each of the four smoothing variances has its own a_{jk}. Figure 17.3 shows the $\log(a_{jk})$. Each component's a_{jk} vary over many orders of magnitude. The largest a_{jk} are about the same for the three quasi-cyclic components (mystery, respiration, heartbeat), while the largest a_{jk} for signal are higher than these by about 10 logs. The peaks of the four curves follow the expected order: Signal's a_{js} are maximized at $j = 1$ by construction, mystery's a_{jm} peak at a slightly higher frequency (i.e., slightly larger j), respiration's a_{jr} peak at a somewhat higher frequency (about $j = 70$), and heartbeat's a_{jh} peak far from the others. Each curve is fairly smooth, although with progressively more jiggle for mystery, respiration, and heartbeat.

If we think of the approximate restricted likelihood as the likelihood from a gamma GLM with these a_{jk} as covariates in the linear predictor, Figure 17.3 suggests the covariates for signal and mystery will be highly collinear, that respiration will have some collinearity with signal and mystery, and that heartbeat will not be collinear with the other components. Table 17.4 shows the correlations among the

[2] I used the R function nlminb with default settings, parameterizing the two restricted likelihoods using the logs of the five variances. The starting values given in Table 17.3 are on the raw scale, not the log scale.

Table 17.3: Optical Imaging Data, Model 2: Maximizing Values of the Exact and Approximate Restricted Likelihoods (Upper Table), for Various Starting Values (Lower Table)

Maximizing Values								
Sv	RL	Sig	Myst	Resp	Hb	Error	Maximum	Bomb?
1	exact	1.3(−16)	3.8(−1)	3.8(−3)	1.6(−2)	6.5(−9)	−180.04	No
	approx	3.9(−14)	3.6(−1)	5.0(−3)	1.1(−2)	6.5(−9)	−180.45	No
2	exact	1.8(−47)	1.7(−1)	1.2(−1)	1.7(−2)	2.4(−29)	−185.73	No
	approx	1.8(−22)	3.6(−1)	5.0(−3)	1.1(−2)	6.1(−11)	−180.45	No
3	exact	1.6(−3)	3.1(−1)	2.0(−1)	1.8(−2)	9.6(−8)	1.3(+15)	Yes
	approx	2.3(−5)	3.8(−2)	1.9(−1)	1.2(−2)	9.6(−8)	−183.21	No
4	exact	4.6(−7)	3.2(−1)	4.1(−3)	1.7(−2)	6.5(−2)	3.9(+14)	Yes
	approx	1.8(−14)	3.6(−1)	5.0(−3)	1.1(−2)	1.5(−11)	−180.45	No
Starting Values							Description	
1		3.1(−8)	0.38	0.0039	0.016	6.5(−9)	Max likelihood Model 2	
2		1	1	1	1	1	A common default	
3		0.00158	1	0.234	0.0176	9.6(−8)	Max likelihood Model 1	
4	exact	1.2(−6)	3.8(−1)	3.8(−3)	1.6(−2)	1.3(−1)	"Best" for myst, resp, hb,	
	approx	1.2(−6)	3.6(−1)	5.0(−3)	1.06(−2)	1.3(−1)	"large" for sig, error	

Note: "RL" indicates exact or approximate restricted likelihood. The notation "(−16)" indicates scientific notation with the power of 10 in parentheses. "Sig," "Myst," "Resp," "Hb" are the signal, mystery, respiration, and heartbeat components respectively. The maximum value ("maximum") came from a version of the objective function that ignored numerical singularities; "bomb?" indicates whether a different version stopped because of a numerical singularity.

$\log(a_{jk})$ plotted in Figure 17.3 (second column from left), confirming this impression. But this may be misleading because in the approximate-restricted-likelihood-as-GLM, the predictors are not the $\log(a_{jk})$ but the a_{jk} themselves, and the a_{jk} are not highly correlated at all (Table 17.4, third column from left). But this might not be relevant either because in a GLM, the observations are implicitly weighted by the reciprocals of their variances. The right-most column of Table 17.4 shows the correlation among the a_{jk} after weighting in this manner. This does not support the hypothesis that collinearity in the approximate-restricted-likelihood-as-GLM is causing its strange behavior.

We'll now consider Chapter 16's one-step approximate case-influence measures applied to the approximate restricted likelihood. Figure 17.4 shows the $\hat{v}_j^2$ (triangles) and fitted values (line) from maximizing the approximate restricted likelihood for Model 2. The vertical scale is logarithmic because otherwise the plot would be dominated by $j = 1, 2$. From left to right, the fitted values peak at j for which the a_{jk} peak for mystery, respiration, and heartbeat; there is no peak for signal because $\hat{\sigma}_{ss}^2$ is very close to zero. At the peaks in the fitted values for respiration and heartbeat, the $\hat{v}_j^2$ also show peaks (shifted a bit for heartbeat) but the $\hat{v}_j^2$ have no analogous peak near mystery's fitted-value peak. Also, the $\hat{v}_j^2$ arguably have two un-modeled peaks

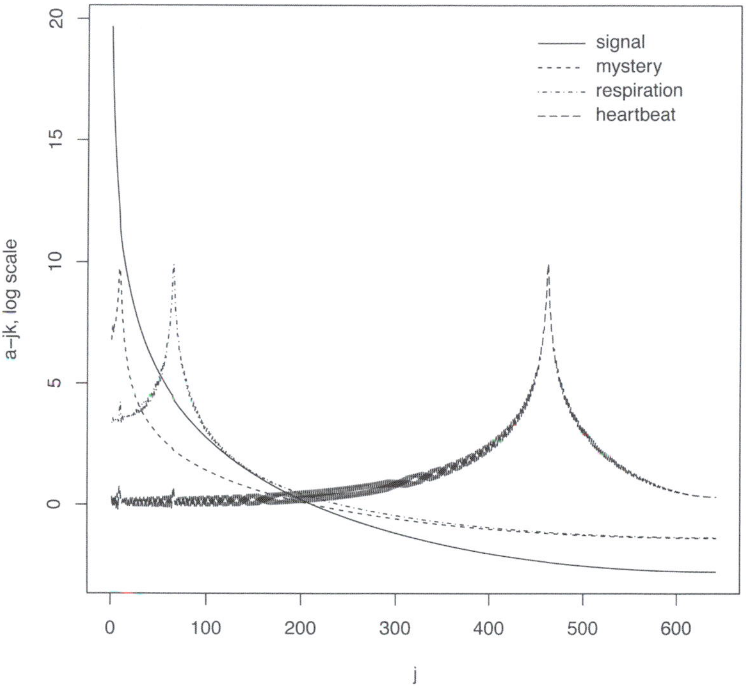

Figure 17.3: Optical imaging data, Model 2: $\log(a_{jk})$ for components k = signal, mystery, respiration, heartbeat.

Table 17.4: Optical Imaging Data, Model 2: Correlations among the a_{jk} for Pairs of Components

	Correlation of		
	$\log(a_{jk})$	a_{jk}	a_{jk}, weighted
sig vs myst	0.98	0.04	0.04
sig vs resp	0.84	–0.004	–0.01
sig vs hb	–0.44	–0.005	–0.04
myst vs resp	0.81	–0.006	–0.17
myst vs hb	–0.42	–0.01	–0.92
resp vs hb	–0.46	–0.01	–0.24

Note: The weighted correlation (right column) is the correlation of $\mathbf{W}^{0.5}\mathbf{B}$; these symbols are defined in the text.

at about j = 140 and 210, which are roughly the second and third harmonics of the respiration component.

Now consider Figure 17.5, which shows the scaled residuals (top panel) and approximate changes in the five variance estimates from deleting each $\hat{v}_j^2$. The changes

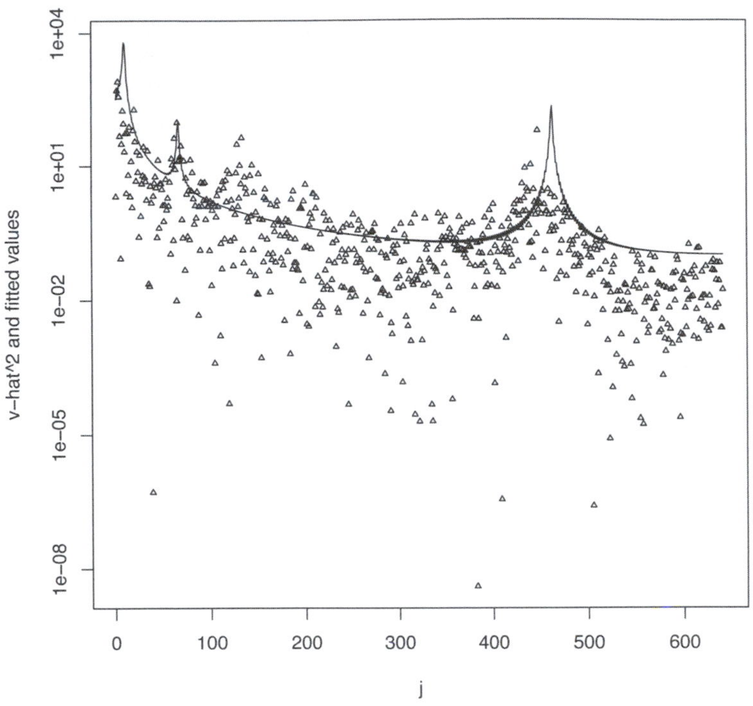

Figure 17.4: Optical-imaging data, Model 2: $\hat{v}_j^2$ (triangles) and fitted values (line) from maximizing the approximate restricted likelihood. The vertical scale is logarithmic.

are scaled because the variances differ so much in magnitude. For mystery, respiration, and heartbeat, Figure 17.5 shows the approximate case-deletion change divided by the full-data estimate. Thus for mystery and respiration, the largest changes are about 20% of the full-data estimate, while for heartbeat the largest change is about 40% of the full-data estimate. When I used the same scaling for signal and error, the scaled changes were enormous. This probably arises because these two variance estimates are defective (too small), the one-step approximation is defective for them, or both. To give reasonable scales to the approximate changes for signal and error, I arbitrarily divided them by numbers chosen so the largest scaled change for each was about 0.20.

The largest scaled residuals are all positive and all near j = 140, 210, or 440. We noticed the first two of these in the plot of fitted values and $\hat{v}_j^2$; they correspond roughly to 2nd and 3rd harmonics of the respiration component. The large scaled residuals around $j = 440$ correspond to the large a_{jh} for the heartbeat component.

The heartbeat component's σ_{sh}^2 is sensitive to $\hat{v}_j^2$ only for j about 440, but it is quite sensitive to three such $\hat{v}_j^2$. The error variance σ_e^2 is also sensitive to these $\hat{v}_j^2$,

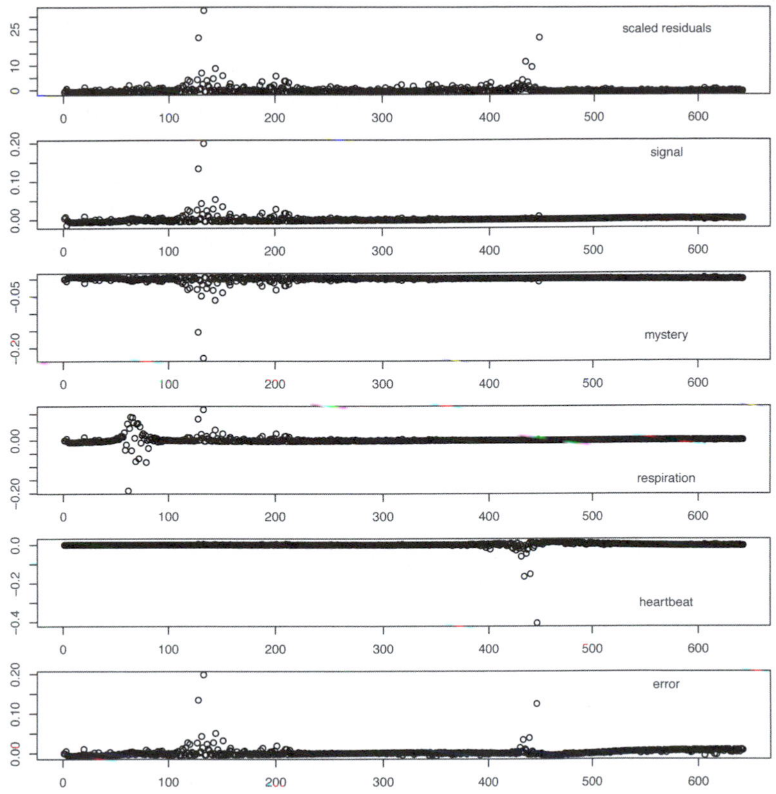

Figure 17.5: Optical imaging data, Model 2: Scaled residuals (top panel) and scaled approximate changes in each variance estimate (next five panels). In all plots, the horizontal axis is j. The scaling is explained in the text.

but the other three variances are not. The respiration component alone is sensitive to j near 70, where its a_{jr} peak. Signal, mystery, respiration, and error are all sensitive to j near 140 and 210. Deleting either of these groups of $\hat{v}_j^2$ reduces mystery's σ_{sm}^2 and increases signal's σ_{ss}^2 and respiration's σ_{sr}^2.

These figures tell us a few things about Model 2's fit. First, for the smallest j, the observed $\hat{v}_j^2$ can only be fit by making the signal variance σ_{ss}^2 or the mystery variance σ_{sm}^2 large. The machinery made σ_{ss}^2 very small and σ_{sm}^2 large even though this created a narrow peak in the fit for small j where there is no analogous peak in the $\hat{v}_j^2$. However, these plots do not tell us why the machinery chose mystery over signal. Below, I explain this by applying this approximate restricted-likelihood analysis to Model 1. Second, it is fairly clear why the heartbeat component is not affected much by changes in the other parts of the model: Information about the heartbeat component comes from $\hat{v}_j^2$ that are largely irrelevant to signal, mystery, and respiration, and vice versa. The data do suggest that the nominal frequency of the heartbeat component might be a bit high, because the fitted peak is a bit to the right of the peak in the $\hat{v}_j^2$.

These figures suggest that instead of adding a second harmonic to the mystery term, it may have been preferable to add second and third harmonics to the respiration term. When I fit such a model using the dlm package (this is given as an exercise), σ_{sr}^2 was reduced compared to Model 2's fit and the fitted respiration component was smoother than for Model 2 because the three-harmonic component needed less evolution error to fit the data. Some DF freed up by reducing σ_{sr}^2 went to the signal component; its fitted curve became reasonable, visually identical to Figure 9.3's fit for Model 3 (which had second harmonics for mystery and respiration). The remaining DF freed up by reducing σ_{sr}^2 went to the mystery component, which was a bit bumpier than in either Model 2 or 3. In Figure 17.5, deleting j corresponding to the second and, to a lesser extent, third harmonics of respiration increased the estimate of σ_{ss}^2 and reduced the estimate of σ_{sm}^2. This is probably why the signal component got some of the DF freed up by adding harmonics to the respiration component, enough to have a reasonable shape.

One small puzzle in this variant model, with three harmonics for respiration, is why none of the freed-up DF went to error. In re-visiting Michael Lavine's models and fitting variants on them, I tried many starting values and while I did find a secondary maximum in the restricted likelihood for Model 2 (Table 17.3), I found no secondary maxima in the likelihoods (using the dlm package) for any of the variant models. It appears that this DLM differs from Chapter 16's two-variance models because with four non-error components in the DLM, some non-error component competes with error to explain $\hat{v}_j^2$ for all j. Based on Figure 17.3 I hypothesize that for these DLMs, the heartbeat component in particular competes with error because heartbeat's a_{jh} do not become small as j becomes large, unlike the other components' a_{jk}. In other words, there is no range of j that is distinctively informative about σ_e^2. (I fit a variant on Model 2 that omitted the heartbeat component. Error absorbed variation previously captured by that omitted component and the rest of Model 2's fit was essentially unchanged.)

The foregoing analysis of Model 2's fit doesn't tell us why adding the mystery component to Model 1 (which had signal, respiration with one harmonic, and heartbeat) effectively eliminated the signal component. To consider this, I approximated the restricted likelihood for Model 1 in the same manner as above. Now the fixed-effect design matrix **X** has six columns, two each for signal, respiration, and heartbeat, so there are 644 $\hat{v}_j$. Figure 17.6 shows that the $\hat{v}_j^2$ differ somewhat from Model 2's but follow the same general pattern. For Model 1, the fitted values are much too high for the smallest j and too low for j about 30–50. Otherwise, the fitted values and $\hat{v}_j^2$ look like their analogs for Model 2, in particular the fitted peaks for respiration and heartbeat fit peaks in the $\hat{v}_j^2$ and respiration appears to have unmodeled second and perhaps third harmonics.

The scaled residuals and scaled case-influence diagnostics (Figure 17.7) confirm these impressions. (In Figure 17.7, each variance estimate's approximate changes are scaled by the variance's full-data estimate.) The scaled residuals differ from Model

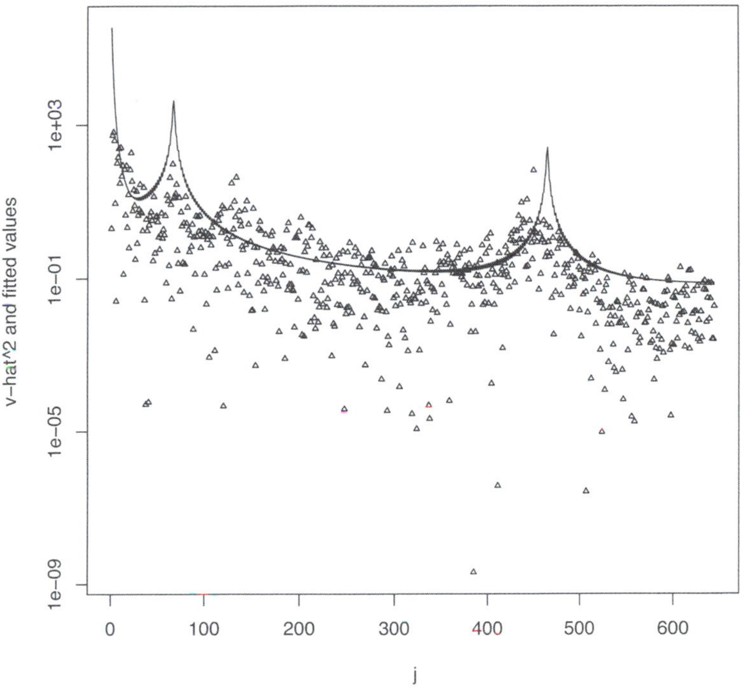

Figure 17.6: Optical-imaging data, Model 1: $\hat{v}_j^2$ (triangles) and fitted values (line) from maximizing the approximate restricted likelihood. The vertical scale is logarithmic.

2's mainly in that they are large for j about 10 to 40. The signal variance σ_{ss}^2 is sensitive to $\hat{v}_j^2$ for j less than about 40: Deleting the smallest j would increase the estimate while deleting j about 20–40 would substantially decrease it. In other words, small j are pulling this estimate down, while j from 20 to 40 are forcing it up.

The signal component was wiped out when the mystery component was added to this model because the signal component fits badly in the range of j that determine its fit and the mystery component's fit. The mystery component's narrow peak for small j fits poorly (Figure 17.4) but that affects few j and otherwise mystery fits much better than signal in the pertinent range of j.

All this suggests that something is wrong in the form of the signal component and that a signal component with a_{js} declining more slowly would fit the data better and perhaps out-compete the mystery component and force it into error. In the repertoire of Petris's (2010) dlm package, the obvious alternative for signal is a higher-order polynomial component for signal. Models 1 through 3 used a 2nd-order (i.e., linear) polynomial model; I hypothesized that this component over-fit the signal in Model 1 because it shrinks toward a straight line and thus needs an overly large σ_{ss}^2 to fit the

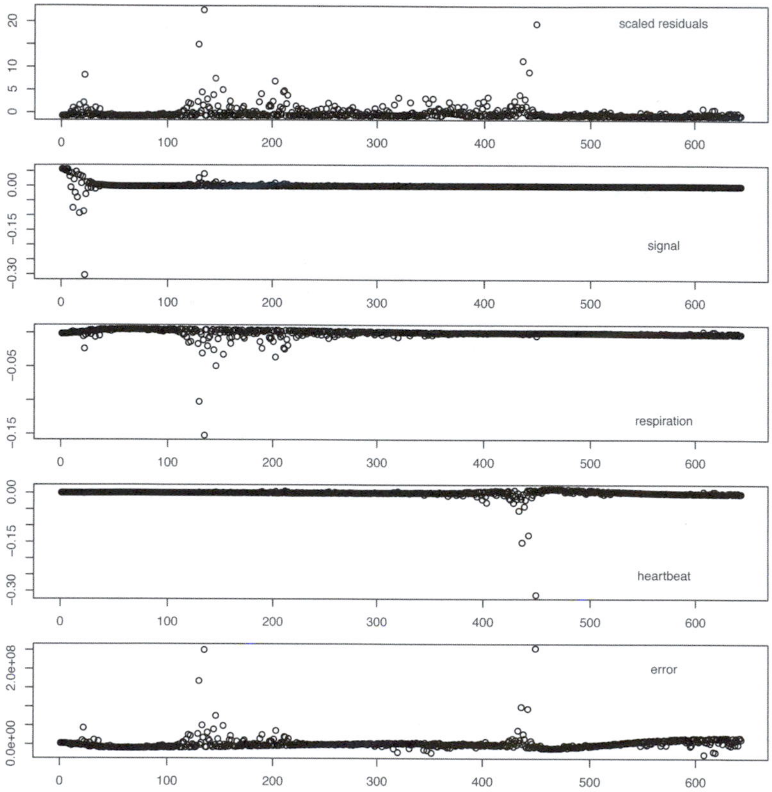

Figure 17.7: Optical imaging data, Model 1: Scaled residuals (top panel) and scaled approximate changes in each variance estimate (next four panels). In all plots, the horizontal axis is j. Approximate changes are scaled by the full-data variance estimates.

data. Therefore, I tried variants on Model 1 (no mystery component) with 3rd- and 4th-order polynomial components (quadratic and cubic) for signal. However, their fits differed little from those of Model 1: They all had bumpy fits for signal, capturing the variation that Model 2 gave to the mystery component. (I also tried fitting the signal component with a 1st-order polynomial, an ICAR model, and it showed spectacularly *more* overfitting, so my intuition was not completely mistaken.)

It's possible that another form of smoother for the signal component might give a more sensible fitted signal without including the unsatisfing mystery component, but I think this is unlikely. The $\hat{v}_j^2$ are substantial for $j \leq 40$ and unless we include the mystery component, any smoother that captures the really low-frequency variation we want in the signal will also capture the moderately low-frequency variation that ended up in the mystery component.

Exercises

Regular Exercises

1. (Section 17.1.1) For the additive model with two predictors, prove that $\Gamma_X, \Gamma_1, \Gamma_2$, and Γ_{12} exist and give a method for constructing them.
2. (Section 17.1.1) For the additive model with two predictors, show that the restricted likelihood is a function of the variances only through the matrix $\mathbf{W}$ and show that $\mathbf{W}$ has the form given in equation (17.4).
3. (Section 17.1.2) Prove all the assertions made in the two paragraphs defining general balance.
4. (Section 17.1.2) For the two-crossed-random-effects model in Section 12.1.2, show that the restricted likelihood, equation (12.18), is the likelihood for a gamma-errors GLM with identity link by identifying the "observations," linear predictor, etc., for that GLM.
5. (Section 17.1.2) Derive the restricted likelihood for the tidied-up epidermal nerve density data, equation (17.13), show that it is the likelihood for a gamma-errors GLM with identity link, and give the "observations," etc., for that GLM.
6. (Section 17.1.4) Show that for the separable models, the restricted likelihood can be re-expressed in the simplified form by diagonalizing all the $\mathbf{Q}_k$ simultaneously.
7. (Section 17.1.4) Derive the restricted likelihood for a separable 2NRCAR model, given in equation (17.34).
8. (Section 17.1.4) Considering the restricted likelihood for a separable 2NRCAR model in equation (17.34), prove that this cannot be re-paramaterized to produce a likelihood from a generalized linear model (or prove that it can).
9. (Section 17.2.2) Fit the variant on Model 2 that adds second and third harmonics to the respiration component.

Open Questions

1. (Section 17.1.1) For the additive model with two predictors, a sufficient condition for re-expressing the restricted likelihood in a simple form is that $\mathbf{M}_{112}$ and $\mathbf{M}_{212}$ are both null. Is this also a necessary condition?
2. (Section 17.1.4) Prove that separable models do not satisfy the conditions of general balance. Prove that for separable models, the restricted likelihood is not the likelihood for any generalized linear model.
3. (Section 17.1.2) For orthogonal designs, are the θ_j always the expected mean squares of the effects?
4. (Section 17.1.3) This section ended with two conjectures about the effects of specific data features on parameter estimates in fitting a one-dimensional Gaussian process. Design and execute simulation experiments to test these conjectures or conjectures of your own.

5. (Section 17.1.5) For the model with separable random effects and fixed effects, derive conditions on $\mathbf{Q}_k$ and the design matrix for the fixed effects under which the resulting restricted likelihood can be re-expressed in the desired simple form. Is the resulting class of models empty, that is, do any actual models satisfy these conditions?

6. (Section 17.2.1) How might you implement this expedient in the mixed-linear-model formulation? The obvious approach is to condition on $\mathbf{u}$, but it is not clear this makes sense.

Chapter 18

Zero Variance Estimates

The literature in this area is almost non-existent. The only papers I know that explicitly address maxima on the boundary of the parameter space are Erkanli (1994) and earlier papers by the same author, which describe a Laplace-type approximation for posterior moments.

Unfortunately, I have little to add so this chapter is brief, intended mainly to suggest directions for development. Section 18.1 uses two examples to make some observations about aspects of the data and design that produce zero variance estimates with a nearly flat restricted likelihood near zero, while Section 18.2 considers some simple tools that I, at least, would find useful. The emphasis here is on "simple" so these tools can be incorporated easily into standard software.

18.1 Some Observations about Zero Variance Estimates

The following observations regarding zero variance estimates and flat restricted likelihoods have implications for study design when it is important to estimate variances; these implications are suggested but not developed in detail.

18.1.1 Balanced One-Way Random Effects Model

As usual, the simplest interesting case is the balanced one-way random effects model, which can be written as $y_{ij} = \mu + u_i + \varepsilon_{ij}$ for $i = 1, \ldots, N$ groups and $j = 1, \ldots, m$ observations per group so $n = Nm$; the u_i are iid $N(0, \sigma_s^2)$ and the ε_{ij} are iid $N(0, \sigma_e^2)$. Several authors have shown that this model's restricted likelihood has a unique maximum. Define the usual sum of squares for error $S_E = \sum_{ij}(y_{ij} - \bar{y}_{i.})^2$, where $\bar{y}_{i.}$ is the average for group i, and define $S_M = \sum_i(\bar{y}_{i.} - \bar{y}_{..})^2$, where $\bar{y}_{..}$ is the overall average (this is $1/m$ times the usual sum of squares for groups). The restricted likelihood is given in equation (15.11) and its maximizing values are

$$
\begin{array}{llll}
\text{if } \frac{S_M}{N-1} & \geq \frac{S_E}{m(n-N)} & \hat{\sigma}_e^2 = & S_E/(n-N) \\
 & & \hat{\sigma}_s^2 = & S_M/(N-1) - \hat{\sigma}_e^2/m \\
 & & & \\
\text{if } \frac{S_M}{N-1} & < \frac{S_E}{m(n-N)} & \hat{\sigma}_e^2 = & (S_E + mS_M)/(n-1) \\
 & & \hat{\sigma}_s^2 = & 0.
\end{array}
\qquad (18.1)
$$

Superficially, $\hat{\sigma}_s^2 = 0$ if variation between groups (S_M) is small relative to variation within groups ($S_E/[m(n-N)]$); we will return to this below.

A convenient way to describe the restricted likelihood near the estimate $\hat{\sigma}_s^2 = 0$ is the first derivative of the log restricted likelihood with respect to σ_s^2 evaluated at $\sigma_s^2 = 0$ and $\hat{\sigma}_e^2$:

$$\frac{N-1}{2}\left(\frac{\hat{\sigma}_e^2}{m}\right)^{-1}\left(\frac{S_M}{N-1}\left(\frac{\hat{\sigma}_e^2}{m}\right)^{-1} - 1\right). \tag{18.2}$$

If $S_M/(N-1) < \hat{\sigma}_e^2/m$, this derivative is negative. It is also small in absolute value — i.e., the restricted likelihood is nearly flat near $\hat{\sigma}_s^2 = 0$ — if either $\hat{\sigma}_e^2/m$ is large or $\hat{\sigma}_e^2/m - S_M/(N-1)$ is close to zero. These two conditions have different implications, which we now consider.

If $\hat{\sigma}_e^2/m$ is large, i.e., if m is small, $\hat{\sigma}_e^2$ is large, or both, we can get both $\hat{\sigma}_s^2 = 0$ and a nearly flat restricted likelihood at $\sigma_s^2 = 0$. In this case, the design and data provide low resolution: Variation within groups is so large that for the given sample sizes N and m, the most likely between-group variance σ_s^2 is zero but a wide interval of positive σ_s^2 have restricted likelihoods near the maximum because the data cannot distinguish among them. The restricted likelihood does provide the information we need but the usual summary (the maximizing value) ignores most of it.

The only way to increase the design's resolution and a positive $\hat{\sigma}_s^2$ is to increase m. If we increase m holding constant N, S_M, and $\hat{\sigma}_e^2$, $\hat{\sigma}_s^2$ eventually becomes positive. Note that increasing N while holding constant $S_M/(N-1)$ and $\hat{\sigma}_e^2/m$ leaves the key condition $S_M/(N-1) < \hat{\sigma}_e^2/m$ unchanged and $\hat{\sigma}_s^2$ remains zero. However, if we increase N in this manner, (18.2) becomes larger in magnitude, i.e., the restricted likelihood declines more steeply at zero.

If $\hat{\sigma}_e^2/m - S_M/(N-1)$ is close to zero, (18.2) is negative but small in magnitude irrespective of $\hat{\sigma}_e^2/m$, because $\sigma_s^2 = 0$ is close to the maximizing value. In this case however the restricted likelihood does not necessarily decline slowly from $\sigma_s^2 = 0$. When $S_M/(N-1) > 0.5\hat{\sigma}_e^2/m$, the second derivative of the log restricted likelihood with respect to σ_s^2 is negative and increases in magnitude as N increases for fixed $\hat{\sigma}_e^2/m$ and $[S_M/(N-1)]/[\hat{\sigma}_e^2/m]$, so the restricted likelihood can drop off quickly despite being nearly flat at $\sigma_s^2 = 0$ (showing this is an exercise).

18.1.2 Balanced ANOVA for the Nerve-Density Example

Consider the tidied-up (i.e., balanced) version of this example discussed in Section 17.1.2 and assume we have $N = 20$ subjects. As before, each subject is sampled at 2 body locations with 2 blisters at each location and 2 images per blister. Location is a fixed effect; the four unknown variance components σ_{s1}^2, σ_{s2}^2, σ_{s3}^2, and σ_e^2 describe variation between subjects, variation between subjects in the difference between locations, variation between blisters within subject and location, and error (variation between images within a blister). The log restricted likelihood is

$$\log RL(\sigma_{s1}^2, \sigma_{s2}^2, \sigma_{s3}^2, \sigma_e^2 | y) = -0.5 \sum_{j=1}^{4} DF_j \left[\log \theta_j + \hat{\theta}_j^u/\theta_j\right], \tag{18.3}$$

where the DF_j are the usual degrees of freedom (for N = 20 these are 19, 19, 40, and 80 for $j = 1,2,3,4$) and $\hat{\theta}_j^u$ are the respective mean squares. The θ_j are $\theta_1 = 8\sigma_{s1}^2 + 4\sigma_{s2}^2 + 2\sigma_{s3}^2 + \sigma_e^2$, $\theta_2 = 4\sigma_{s2}^2 + 2\sigma_{s3}^2 + \sigma_e^2$, $\theta_3 = 2\sigma_{s3}^2 + \sigma_e^2$, and $\theta_4 = \sigma_e^2$, so $\theta_1 \geq \theta_2 \geq \theta_3 \geq \theta_4$. The $\hat{\theta}_j^u$ maximize (18.3) and do not necessarily follow the ordering $\hat{\theta}_1^u \geq \hat{\theta}_2^u \geq \hat{\theta}_3^u \geq \hat{\theta}_4^u$; they are thus unconstrained estimates of the θ_j (hence the superscript "u"). If the $\hat{\theta}_j^u$ do follow this ordering, they are also the maximum restricted-likelihood estimates of the θ_j; the corresponding estimates of the variances σ_{sj}^2 and σ_e^2 are obtained by a linear transformation and are all positive. However, if $\hat{\theta}_j^u < \hat{\theta}_{j+1}^u$ for $j = 1,2$, or 3, then these unconstrained estimates differ from the constrained estimates $\hat{\theta}_j^c$, which maximize the restricted likelihood subject to $\hat{\theta}_1^c \geq \hat{\theta}_2^c \geq \hat{\theta}_3^c \geq \hat{\theta}_4^c$.

The interesting question is how exactly these constrained estimates behave as functions of the mean squares $\hat{\theta}_j^u$. Table 18.1 shows what happens when we begin with $\hat{\theta}_1^u \geq \hat{\theta}_2^u \geq \hat{\theta}_3^u \geq \hat{\theta}_4^u$ and then steadily increase $\hat{\theta}_4^u$, the error mean square. The first row in Table 18.1 corresponds to a hypothetical dataset with mean squares $\hat{\theta}_j^u =$ 15, 7, 3, and 1 for $j = 1,2,3,4$ respectively, so the maximum restricted likelihood estimates of the four variances (2nd through 5th columns from the left) are all 1. The maximum restricted likelihood estimates of the θ_j (the right-most four columns) are, of course, 15, 7, 3, and 1. Each subsequent row in Table 18.1 changes the hypothetical dataset by increasing the error mean square ($\hat{\theta}_4^u$, left-most column) by 1. With the first such increment, $\hat{\theta}_4^c$ and $\hat{\sigma}_e^2$ increase from 1 to 2; $\hat{\theta}_3^c$ stays at 3 but $\hat{\sigma}_{s3}^2$ decreases from 1.0 to 0.5, while $\hat{\sigma}_{s1}^2$ and $\hat{\sigma}_{s2}^2$ are unchanged. When $\hat{\theta}_4^u$ is increased again to 3, $\hat{\theta}_3^c$ stays at 3 but $\hat{\sigma}_{s3}^2$ is reduced to 0. When $\hat{\theta}_4^u$ is increased to 4, $\hat{\theta}_4^c$ no longer increases by 1, and it now equals $\hat{\theta}_3^c$. Also, $\hat{\sigma}_{s2}^2$ is now reduced below 1 while $\hat{\theta}_2^c$ remains at 7. Further increases in $\hat{\theta}_4^u$ increase $\hat{\sigma}_e^2$ and decrease $\hat{\sigma}_{s2}^2$ with $\hat{\sigma}_{s3}^2$ stuck at zero. After $\hat{\sigma}_{s2}^2$ is reduced to zero, $\hat{\sigma}_{s1}^2$ then begins declining until it too is reduced to zero.

Several things are interesting here. First, increasing the error mean square reduces $\hat{\sigma}_{s3}^2$, $\hat{\sigma}_{s2}^2$, and $\hat{\sigma}_{s1}^2$ to zero in order up the hierarchy of this experimental design. In the terms used in the previous subsection, Table 18.1 shows that if we hold the design fixed and reduce its resolution by increasing the error mean square, the reduced resolution obscures higher-level components of variation in order of their proximity to error in the hierarchy. Thus for example increasing the error mean square can only obscure σ_2^2 if it also obscures σ_3^2. The analogous thing happens if we hold $\hat{\theta}_4^u$ constant and instead increase $\hat{\theta}_3^u$. It is not obvious why this should happen. It seems this is a consequence not of the design's hierarchy but rather of the coefficients of each variance in the θ_j, so I conjecture similar behavior occurs in non-nested designs. This question is given as an exercise.

Second, when $\hat{\theta}_1^u > \hat{\theta}_2^u > \hat{\theta}_4^u > \hat{\theta}_3^u$, the restricted likelihood is maximized subject to the order constraint by making $\hat{\theta}_4^u > \hat{\theta}_4^c = \hat{\theta}_3^c > \hat{\theta}_3^u$. This is easy to prove (the proof is an exercise) and the mechanics are simple. However, if we see this restricted likelihood as the likelihood arising from a generalized linear model, then it is not clear why this GLM gives these estimates while, for example, $\hat{\theta}_1^c = \hat{\theta}_1^u$ is unchanged until the error mean square increases to 10.

Table 18.1: Nerve Density Example: Hypothetical Data Showing How the Variance Estimates Go to Zero as the Error Mean Square Is Increased

$\hat{\theta}_4^u$	$\hat{\sigma}_{s1}^2$	$\hat{\sigma}_{s2}^2$	$\hat{\sigma}_{s3}^2$	$\hat{\sigma}_e^2$	$\hat{\theta}_1^c$	$\hat{\theta}_2^c$	$\hat{\theta}_3^c$	$\hat{\theta}_4^c$
1	1.00	1.00	1.00	1.00	15.00	7.00	3.00	1.00
2	1.00	1.00	0.50	2.00	15.00	7.00	3.00	2.00
3	1.00	1.00	0.00	3.00	15.00	7.00	3.00	3.00
4	1.00	0.83	0.00	3.67	15.00	7.00	3.67	3.67
5	1.00	0.67	0.00	4.33	15.00	7.00	4.33	4.33
6	1.00	0.50	0.00	5.00	15.00	7.00	5.00	5.00
7	1.00	0.33	0.00	5.67	15.00	7.00	5.67	5.67
8	1.00	0.17	0.00	6.33	15.00	7.00	6.33	6.33
9	1.00	0.00	0.00	7.00	15.00	7.00	7.00	7.00
10	0.93	0.00	0.00	7.58	15.00	7.58	7.58	7.58
11	0.86	0.00	0.00	8.15	15.00	8.15	8.15	8.15
12	0.78	0.00	0.00	8.73	15.00	8.73	8.73	8.73
13	0.71	0.00	0.00	9.30	15.00	9.30	9.30	9.30
14	0.64	0.00	0.00	9.88	15.00	9.88	9.88	9.88
15	0.57	0.00	0.00	10.45	15.00	10.45	10.45	10.45
16	0.50	0.00	0.00	11.03	15.00	11.03	11.03	11.03
17	0.42	0.00	0.00	11.60	15.00	11.60	11.60	11.60
18	0.35	0.00	0.00	12.18	15.00	12.18	12.18	12.18
19	0.28	0.00	0.00	12.76	15.00	12.76	12.76	12.76
20	0.21	0.00	0.00	13.33	15.00	13.33	13.33	13.33
21	0.14	0.00	0.00	13.91	15.00	13.91	13.91	13.91
22	0.06	0.00	0.00	14.48	15.00	14.48	14.48	14.48
23	0.00	0.00	0.00	15.05	15.05	15.05	15.05	15.05
24	0.00	0.00	0.00	15.56	15.56	15.56	15.56	15.56
25	0.00	0.00	0.00	16.06	16.06	16.06	16.06	16.06

Note: $N = 20$. The mean squares are 15, 7, and 3 for θ_1, θ_2, and θ_3 respectively; the mean square for error ($\hat{\theta}_4^u$) is in the left-most column. The 2nd through 5th columns from the left are the constrained estimates of the variances; the right-most four columns are the constrained estimates of the θ_j. Horizontal lines indicate when a $\hat{\sigma}_{sj}^2$ reaches zero.

If $\hat{\theta}_1^u > \hat{\theta}_2^u > \hat{\theta}_4^u > \hat{\theta}_3^u$, so that $\hat{\sigma}_{s3}^2 = 0$, under what conditions is the restricted likelihood flat near $\sigma_{s3}^2 = 0$? As above we consider this using derivatives of the log restricted likelihood but first note that in this case

$$\begin{aligned} \hat{\sigma}_e^2 &= \hat{\theta}_3^c = \hat{\theta}_4^c = \frac{DF_3 \hat{\theta}_3^u + DF_4 \hat{\theta}_4^u}{DF_3 + DF_4} \\ \text{and } \hat{\sigma}_{s3}^2 &= 0 \end{aligned} \tag{18.4}$$

(the proof is an exercise). Now differentiate the log restricted likelihood with respect to σ^2_{s3}, giving

$$\frac{\partial \log RL}{\partial \sigma^2_{s3}} = -\sum_{j=1}^{3} DF_j \left(\frac{1}{\theta_j} - \frac{MS_j}{\theta_j^2} \right) \tag{18.5}$$

(the derivation is an exercise). Evaluated at the $\hat{\theta}^c_j$, i.e., at $\hat{\sigma}^2_{s3} = 0$, this is $-DF_3[1/\hat{\theta}^c_3 - MS_3/(\hat{\theta}^c_3)^2] < 0$ because $MS_3 = \hat{\theta}^u_3 < \hat{\theta}^c_3$. As in the balanced one-way random effect model, this derivative is small in magnitude if $\hat{\theta}^c_3 = \hat{\sigma}^2_e$ is large or if $MS_3/\hat{\theta}^c_3$ is just barely less than 1 and again, these two possibilities have different implications, as we now discuss.

If $\hat{\theta}^c_3 = \hat{\sigma}^2_e$ is large, the design has low resolution and the restricted likelihood declines slowly from $\sigma^2_{s3} = 0$ as a function of σ^2_{s3}. To see this, consider the second derivative with respect to σ^2_{s3},

$$\frac{\partial^2 \log RL}{\partial (\sigma^2_{s3})^2} = 4\sum_{j=1}^{3} \frac{DF_j}{\theta_j^2} \left(0.5 - \frac{MS_j}{\theta_j} \right) \tag{18.6}$$

(the derivation is an exercise). To allow consideration of the design's effect, let N be the number of subjects as before and let m be the number of blisters per subject/location. Then this second derivative, evaluated at the $\hat{\theta}^c_j$, is

$$N\left[-\frac{2(N-1)}{N} \left(\frac{1}{(\hat{\theta}^c_1)^2} + \frac{1}{(\hat{\theta}^c_2)^2} \right) + \frac{4m}{(\hat{\theta}^c_3)^2} \left(0.5 - \frac{MS_3}{\hat{\theta}^c_3} \right) \right]. \tag{18.7}$$

If $\hat{\theta}^c_3 = \hat{\sigma}^2_e$ is large relative to MS_3, then $\hat{\theta}^c_1$ and $\hat{\theta}^c_2$ are even larger so (18.7) is small in magnitude (it can even be positive). Thus if $\hat{\sigma}^2_e$ is large, the restricted likelihood is flat near the maximum with $\sigma^2_{s3} = 0$. If $MS_3 > 0.5\hat{\theta}^c_3$, so this second derivative is necessarily negative, the only way to make the restricted likelihood less flat while holding the $\hat{\theta}^u_j$ and $\hat{\theta}^c_j$ fixed is to increase N, the number of subjects. If $MS_3 < 0.5\hat{\theta}^c_3$ and the second derivative is positive, even that doesn't necessarily shorten the likelihood's upper tail in σ^2_{s3}.

On the other hand, if $MS_3/\hat{\theta}^c_3$ is just barely less than 1, then the second derivative (18.7) is necessarily negative and can be increased in magnitude for fixed $\hat{\theta}^u_j$ and $\hat{\theta}^c_j$ by increasing either N, the number of subjects, or m, the number of blisters per subject/location.

18.2 Some Thoughts about Tools

My goal here is modest and utterly pragmatic: When the restricted likelihood is maximized by a zero variance estimate, I want to know whether the data are consistent with values of that variance that differ much from zero, in other words, whether the restricted likelihood has a flat left tail in that variance. At this point I have only a few thoughts about this problem.

For a conventional analysis, the obvious form for this information is a one-sided confidence interval for the variance that has a zero estimate. The preceding section suggests that a diagnostic for a flat left tail might instead be based on derivatives of the log restricted likelihood evaluated at the maximum, i.e., with the variance in question at zero. If the first and second derivatives of the log restricted likelihood are small in magnitude, this suggests a flat left tail. However, it is not clear (to me, at least) how to determine what magnitudes qualify as "small," though this may be possible. By contrast, a simple confidence interval, if one exists, avoids this calibration problem.

For a Bayesian analysis, such a confidence interval would be informative because if the restricted likelihood's left tail is fairly flat, the posterior's left tail is largely determined by the prior. This is true because the non-prior part of the marginal posterior for ϕ, the unknowns in **G** and **R**, is identical to the restricted likelihood if the prior on β is flat, or is identical to the restricted likelihood for a slightly modified model and dataset if the prior on β is a proper normal distribution. Obviously, you could see the posterior's left tail directly by doing a Bayesian analysis with a flat prior on the variance in question (restricted to a closed interval to ensure a proper posterior). But as of today, Bayesian software requires rather more coding than, say, SAS's MIXED procedure, so for many problems at least it would be much faster and simpler to do a conventional analysis to diagnose the left tail, if a simple confidence interval were available. (The INLA package for R does not use MCMC, so perhaps a simple addition to INLA could give a suitable diagnostic.)

Assuming, then, that a one-sided confidence interval is desirable, I would argue that in the current situation — in which standard software gives as a confidence interval either (0,0) or (missing,missing) — it would be a real advance to have a simple confidence interval even if its coverage was often below nominal. If such a confidence interval gave low coverage when it failed to give near-nominal coverage, then when the interval was wide, it would support an *a fortiori* argument that the restricted likelihood is consistent with large values of the variance in question. That is, in the current dismal state of affairs, we need not be too fussy.

An obvious candidate for such a one-sided confidence interval would use the profile log restricted likelihood for σ_s^2, i.e., that function of σ_s^2 obtained by maximizing, for each value of σ_s^2, the log restricted likelihood in the other unknowns in ϕ. The upper end of the interval would be the value of σ_s^2 at which the profile log restricted likelihood is smaller than the maximum by a critical value c. Some software already computes the profile restricted likelihood and with such a routine in hand it should not be too difficult to compute the upper end of this interval.

The hard question, in view of Ruppert et al.'s (2003) sobering advice about the mostly poor quality of asymptotic approximations for these models, is whether any such c can be chosen to give adequate coverage for a large collection of models. This can only be determined by large simulation experiments. Such experiments would need to include several classes of models (splines, DLMs, ANOVA models, etc.) and a designed collection of variants within each class. One complication in designing such simulation experiments is that for some simulated datasets, the restricted likelihood will not be maximized by a zero variance. Thus, to test a proposed one-sided confidence interval, we have two choices: Devise a one-sided confidence interval that is applied whether or not the restricted likelihood is maximized at $\sigma_s^2 = 0$, or define a confidence procedure that gives a one-sided interval when the restricted likelihood is maximized at $\sigma_s^2 = 0$ and a different, two-sided interval (e.g., Satterthwaite's interval) when the restricted likelihood is maximized at $\sigma_s^2 > 0$. An exercise asks you to explore this idea and the associated simulation experiments.

(An alternative interval would use a simple distributional form to approximate the restricted likelihood as a function of the variance estimated at zero, fixing the other unknowns in ϕ at their maximizing values. This is quite crude but that's OK in the present state of affairs. I considered doing this approximation by matching the first two derivatives of the log restricted likelihood to the same derivatives of the log density of some standard distribution and then computing an approximate interval using a percentile of that distribution. To my surprise, no standard univariate distribution on the positive real numbers has a log density with first and second derivatives having enough flexibility to match the log restricted likelihood's derivatives, even for the balanced one-way random-effects model. Thus, this idea was a non-starter.)

Exercises

Regular Exercises

1. (Section 18.1.1) For the balanced one-way random effects model, derive the second derivative of the log restricted likelihood with respect to σ_s^2. Show that if $S_M/(N-1) > 0.5\hat{\sigma}_e^2/m$, this second derivative is negative and its magnitude increases as N increases for fixed $\hat{\sigma}_e^2/m$ and $[S_M/(N-1)]/[\hat{\sigma}_e^2/m]$.
2. (Section 18.1.2) For the nerve-density example, prove that if $\hat{\theta}_1^u > \hat{\theta}_2^u > \hat{\theta}_4^u > \hat{\theta}_3^u$, then $\hat{\theta}_4^u > \hat{\theta}_4^c = \hat{\theta}_3^c > \hat{\theta}_3^u$. A suggestion: I did this in three steps, each using the first derivatives of the log restricted likelihood. First, prove that if $\theta_3 = \theta_4 = \hat{\theta}_4^u$, the restricted likelihood can be increased by reducing $\theta_3 = \theta_4$. Second, prove that if $\theta_3 = \theta_4 = \hat{\theta}_3^u$, the restricted likelihood can be increased by increasing $\theta_3 = \theta_4$. Finally, prove that if $\hat{\theta}_4^u > \theta_3 > \theta_4 > \hat{\theta}_3^u$, the restricted likelihood can be increased by reducing θ_3 or increasing θ_4.
3. (Section 18.1.2) Derive the estimates in equation (18.4).
4. (Section 18.1.2) Derive the derivatives in equations (18.5) and (18.6).

Open Questions

1. (Section 18.1.2) In the nerve-density example, increasing the error mean square caused the estimates of the other variances to go to zero in a specific order. This order corresponds to the hierarchy in this particular design, but the hierarchy *per se* appears irrelevant. Rather, it seems that the key is the ordering of the coefficients of the variances in the θ_j. Is this also true, for example, in a balanced design with two crossed random effects (Section 12.1.2) and different numbers of levels for the row and column effects?
2. (Section 18.2) Develop and test the idea of a simple one-sided interval based on the log profile restricted likelihood. Publish it and become famous, or at least frequently cited.

Chapter 19

Multiple Maxima in the Restricted Likelihood and Posterior

This chapter's subject has some literature, though not much. Henn & Hodges (2013) reviewed the literature on multiple maxima in likelihoods, restricted likelihoods, and posterior distributions, emphasizing mixed linear models and the unknowns in ϕ. This chapter summarizes points of particular relevance to this book.

As my students and I have learned the hard way, the mathematics of multiple maxima are messy and difficult. Accordingly, most papers in this area are about simple problems, e.g., two groups or very small numbers of observations. Exceptions include Carriquiry & Kliemann (2007) on the marginal posterior of the fixed effects and Oliveira & Ferreira (2011) on Gaussian Markov random fields.

However, some generalizations seem clear. Multiple local maxima occur more readily in the likelihood for a mixed linear model than in the restricted likelihood. Because the likelihood is of secondary interest, interested readers are referred to Henn & Hodges (2013) for details. Section 19.1 below discusses multiple local maxima in the restricted likelihood using the re-expression developed in preceding chapters. If the marginal posterior for ϕ is understood as the restricted likelihood multiplied by a prior, you might expect that the marginal posterior is more prone to multiple local maxima than the restricted likelihood, and it is. Section 19.2 discusses this using the balanced one-way random effects model and the hierarchical model for the HMO data in Chapter 8. It summarizes some odd and disturbing aspects of posterior multimodality reported in Liu & Hodges (2003) and Henn & Hodges (2103), the most inconvenient being that the number of posterior modes can change with a change of variables from variances to precisions or vice versa.

19.1 Restricted Likelihoods with Multiple Local Maxima

Henn & Hodges (2013) found exactly one report in the literature of multiple local maxima in the restricted likelihood, by Welham & Thompson (2009) for a penalized spline model. These authors expressed the restricted likelihood in a scalarized form like the one used in earlier chapters of this book. Henn & Hodges (2013) used this scalarized form to explore multiple local maxima in restricted likelihoods more generally and to show how to manufacture new examples. The remainder of this section summarizes their results.

Section 15.1 derived this scalarized form for the restricted likelihood for a two-variance model (equation 15.4):

$$\begin{aligned}\log RL(\sigma_s^2, \sigma_e^2|\mathbf{y}) &= B - \frac{n - s_X - s_Z}{2}\log(\sigma_e^2) - \frac{1}{2\sigma_e^2}\mathbf{y}'\Gamma_c\Gamma_c'\mathbf{y} \\ &\quad - \frac{1}{2}\sum_{j=1}^{s_Z}\left[\log(\sigma_s^2 a_j + \sigma_e^2) + \frac{\hat{v}_j^2}{\sigma_s^2 a_j + \sigma_e^2}\right], \qquad (19.1)\end{aligned}$$

for B an unimportant constant, s_X the rank of $\mathbf{X}$, and $s_Z = \text{rank}(\mathbf{X}|\mathbf{Z}) - \text{rank}\mathbf{Z}$. The first piece, $-(n - s_X - s_Z)\log(\sigma_e^2)/2 - \mathbf{y}'\Gamma_c\Gamma_c'\mathbf{y}/2\sigma_e^2$, was called the free terms for σ_e^2 and the rest of (19.1) was called the mixed terms. Section 15.1 noted that although (19.1) is the likelihood for a gamma-errors generalized linear model (GLM) with the identity link, (19.1) also has the form of a likelihood from a gamma-errors GLM with identity link (the mixed terms only) multiplied by a prior distribution for σ_e^2 (the free terms). Obviously the free terms are not a prior distribution; they're part of the restricted likelihood and merely have the same form as a particular prior distribution. However, thinking of the free terms this way reminds us that for many prior-likelihood combinations (though not all; Lucas 1993), a posterior can have a second mode if the prior and likelihood provide very different information, which suggests that restricted likelihoods with multiple local maxima can arise if the free and mixed terms provide very different information about σ_e^2. Henn & Hodges (2013, Section 4) used this approach to construct an example, an ICAR model with repeat measurements in some regions. Thus conflict between the free and mixed terms provides a recipe for manufacturing examples of restricted likelihoods with multiple maxima. An exercise invites you to manufacture your own examples.

Can the restricted likelihood have multiple maxima without conflict between the free and mixed terms? Wedderburn (1976) showed that the likelihood for a gamma-errors GLM with identity link is not necessarily a strictly concave function, so a unique maximum is not assured. Welham & Thompson's (2009) restricted likelihood is an example where this uniqueness fails. Although the example in Henn & Hodges (2013, Section 4) was constructed by imagining that the free terms were a prior distribution, the resulting restricted likelihood, free terms and mixed terms, is the likelihood from a gamma-errors GLM with identity link and thus provides a second example in which uniqueness fails. This raises the possibility that a restricted likelihood — even for a two-variance model — could have three local maxima, if the mixed terms (a gamma-errors GLM with identity link) had two maxima and the free terms conflicted with the mixed terms sufficiently. Henn & Hodges (2013) noted this possibility but could not produce an example. An exercise asks whether you can.

19.2 Posteriors with Multiple Modes

This section concludes the book by trying to frighten you with strange but true results about bimodality in posterior distributions for two simple problems. The point is that multimodal posteriors probably happen far more often than is generally known or acknowledged, so it would be helpful to develop a collection of examples as a basis for further research.

19.2.1 Balanced One-Way Random Effect Model

Consider the simplest interesting case, the balanced one-way random effects model, examined in detail in Liu & Hodges (2003). Once again, this model can be written as $y_{ij} = \mu + u_i + \varepsilon_{ij}$ for $i = 1,\ldots,N$ groups and $j = 1,\ldots,m$ observations per group, so $n = Nm$; u_i iid $N(0,\sigma_s^2)$ and ε_{ij} iid $N(0,\sigma_e^2)$. Many authors have shown that this model's restricted likelihood necessarily has a single maximum. The intuition, from equation (15.11), is that the restricted likelihood has free terms and mixed terms with a single distinct a_j, so a single maximum must occur at the (σ_s^2,σ_e^2) that maximizes both the free and mixed terms. Because the restricted likelihood has one maximum, multimodal posteriors necessarily arise from conflict between the restricted likelihood and prior, but it turns out this conflict is far from simple.

Liu & Hodges (2003) mostly considered posteriors for this model when the two variances have independent inverse gamma priors. For this prior, they gave conditions under which the full joint posterior of all the unknowns $(\mu, u_1,\ldots,u_N,\sigma_s^2,\sigma_e^2)$, the marginal posterior of $(\mu,\sigma_s^2,\sigma_e^2)$, and the marginal posterior of (σ_s^2,σ_e^2) are unimodal or bimodal. They showed these three posteriors can have no more than two modes and do not necessarily have the same modality (O'Hagan 1976 showed the latter in rather greater generality). For each posterior, the stationary points are solutions to a cubic equation that depends on the two priors, the error and between-group sums of squares, and the two sample sizes N and m. The derivations are entirely lacking in intuition so I will show just a few results obtained using them.[1]

Figure 19.1 (Figure 2d in Liu & Hodges 2003) shows typical results, in this case for the full joint posterior with N = 10 groups, m = 10 observations per group, and both inverse gamma priors having parameters (0.001,0.001). (A similar plot could be drawn for either of the marginal posteriors mentioned above.) Any dataset is represented on this plot by its error mean square (horizontal axis) and between-groups mean square (vertical axis); datasets below and above the thick line give unimodal and bimodal posteriors respectively. Thus, for the case shown in Figure 19.1, for a given error mean square, when the between-groups mean square is near zero the posterior has a single mode corresponding to considerable shinkage ("much shrinkage"). The joint posterior is unimodal until the between-groups mean square reaches the thick line, at which point a second mode arises corresponding to little shrinkage of the group means toward the overall mean ("little shrinkage"). In other words, the posterior is bimodal when the F-statistic is larger than a threshold depending on the sample sizes, the two mean squares, and the prior parameters. As the between-groups mean square continues to increase, the difference in height between the two modes changes. The thin lines in Figure 19.1 indicate between-group mean squares at which the difference in height, "little shrinkage" mode minus "much shrinkage" mode, reaches a given value in log units. Obviously a mode's probability mass is not determined solely by its height but also by its breadth; nonetheless, this plot shows that as the between-groups mean square increases, mass shifts from the "much

[1] Two people have independently reported that they cannot reproduce the analyses of the peak discharge data in Liu & Hodges (2003, Section 4), so I assume these results contain an error. No other errors in this paper have been reported to me.

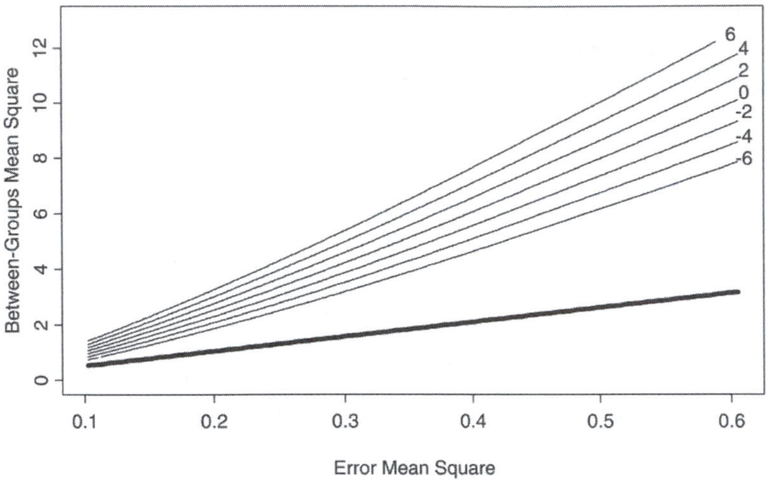

Figure 19.1: Balanced one-way random effects model: Datasets with a bimodal full joint posterior are those above the thick line; datasets below the thick line have unimodal posteriors. The other lines are contours for the difference in height between the two modes. Further details are in the text.

shrinkage" mode to the "little shrinkage" mode. The "much shrinkage" mode represents the prior's contribution to the posterior, while the "little shrinkage" mode represents the data's contribution (i.e., the restricted likelihood). The key observation is that the single mode that's present when the between-groups mean square is small does not shift as the between-groups mean square increases; rather, a distant second mode arises and mass moves from one mode to the other.

The obvious question is how often the posterior is bimodal in actual datasets. When Jiannong Liu did this research, I had 67 balanced one-way datasets from my statistical practice representing a variety of subject matter and data types. (Sufficient statistics for these 67 datasets are in Liu 2001, Table 4.6.) With inverse gamma (0.001,0.001) priors for the variances, 47 datasets (70%) gave bimodal marginal posteriors for $(\mu, \sigma_s^2, \sigma_e^2)$; for 4 of these (6%) the two modes differed in height by 4 natural-log units or less. Noteworthy bimodality was not common but also not rare.

Here are a few other facts from Liu & Hodges (2003), ranging from inconvenient to truly strange. As noted, if the posterior is unimodal and the between-groups mean square is increased with all else held fixed, the posterior eventually becomes bimodal. Thus, no prior in this conjugate class guarantees a unimodal posterior (that is, without cheating by using the data in the prior). Further, if the posterior is bimodal and the number of groups N is increased while holding constant m, the error mean square, and the between-groups mean square, the posterior does not necessarily become unimodal. Finally, it is not always the case that, as in Figure 19.1, a simple straight line separates datasets that produce uni- and bimodal posteriors. If the two variances have inverse gamma priors with parameters (0.001,0.001), $N = 15$, $m = 20$, and the within-group sum of squares is fixed at 5.2, the posterior is unimodal when the between-groups sum of squares is 0.9, bimodal when it's 1.1, unimodal again when it's 1.5, and bimodal again when the between-groups sum of squares reaches 300. This particular oddity appears to be rare and arises from details of the cubic equations that determine posterior modality, about which I have no intuition at all.

No conjugate prior on the variances guarantees a unimodal posterior; does any prior? (The following results are from Liu 2001, Chapter 6.) The uniform shrinkage prior — a uniform prior on $\xi = \sigma_e^2/(\sigma_e^2 + m\sigma_s^2) \in [0,1]$ — with an inverse gamma or flat prior on σ_e^2 guarantees unimodal marginal posteriors for $(\mu, \sigma_s^2, \sigma_e^2)$ and (σ_s^2, σ_e^2). However, the full joint posterior has an infinite spine at $\xi = 1$ for any $\sigma_e^2 > 0$ and if the F-statistic is large enough, it has a second mode. If ξ has a beta prior with parameters (η, κ), the full joint posterior has a single mode if $\kappa \geq N/2+1$.

What are we to think of all this? A member of Dr. Liu's dissertation committee argued, citing de Finetti, that if this model and prior truly capture Your beliefs (capitalizing *à la* de Finetti), then a posterior with a mixture of extreme beliefs is the coherent (i.e., correct) result. But one could argue with equal force, citing de Finetti again, that this mixture of extremes merely shows hidden strength in a specification made almost automatic by habit.[2] The message, perhaps, is that we understand this specification less well than we would like to think — and this is the simplest nontrivial mixed linear model.

19.2.2 Two-Level Model: The HMO Data (Example 9) Revisited

Section 14.5 reported Wakefield's (1998) re-analysis of the HMO data using a model with two state-level predictors. The marginal posterior for the between-state and within-state variances was bimodal; Wakefield concluded "there are two competing explanations for the observed variability in the data" (p. 525). Henn & Hodges (2013) re-examined this analysis; the figures below are taken from that paper and use its notation, τ^2 for the between-state variance (usually σ_s^2) and σ^2 for the within-state variance (usually σ_e^2). Figures 19.2a,b are contour plots of the log restricted likelihood and log posterior respectively of the variances τ^2 and σ^2. For the between-groups variance τ^2, the log restricted likelihood (Figure 19.2a) has a single peak at about 100. The tail of small τ^2 is flat because for small τ^2, the state-level errors are shrunk almost to zero and the restricted likelihood cannot distinguish tiny differences in shrinkage. The log posterior (Figure 19.2b) has two distinct peaks; the peak from the restricted likelihood, with τ^2 about 100, is rather lower than the peak from the prior, with τ^2 just under 0.01. With the logarithmic vertical scale, the peak with small τ^2 appears more massive but on the original scale it is quite narrow while the peak with large τ^2 is broad. So the "two competing explanations" do not, in fact, both come from the data; the restricted likelihood provides fairly weak information about τ^2 and the prior almost overpowers it.

While examining this problem, Lisa Henn changed variables in Wakefield's posterior from variances to precisions, giving the log posterior with contours shown in

[2]De Finetti actually said this about the Central Limit Theorem.

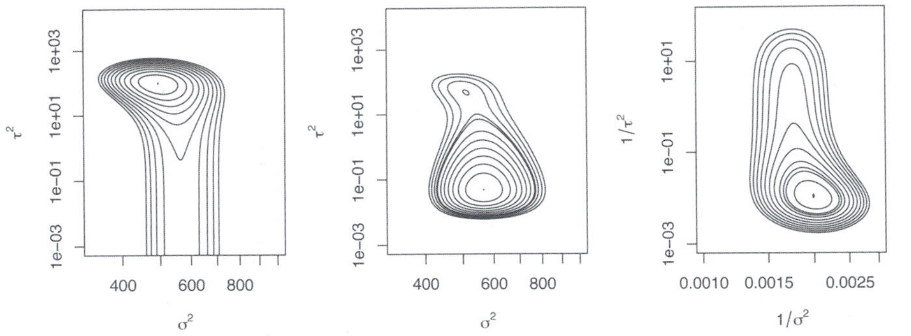

(a) Restricted likelihood (b) Marginal posterior, variances (c) Marginal posterior, precisions

Figure 19.2: HMO data, model with two state-level predictors, Wakefield's analysis revisited: Panel a: Log restricted likelihood for the between-state variance τ^2 and within-state variance σ^2; Panel b: Log marginal posterior for the variances; Panel c: Log marginal posterior for the precisions. Horizontal and vertical scales are logarithmic. Contours are 1-log increments with an extra contour in Panel b to show the secondary maximum.

Figure 19.2c. Simply changing variables in this posterior changed a bimodal posterior to a unimodal posterior. This happens because the mode for τ^2 created by the prior is spread out over a wide range of $1/\tau^2$ and becomes a shoulder in O'Hagan's (1985) terms rather than a local maximum.

One might reasonably argue that the posterior in Figure 19.2c is *almost* bimodal, so the change of modality from a mere change of variables should not be too disturbing. But having found this one example, Ms. Henn easily devised another that cannot be dismissed. Changing Wakefield's analysis simply by changing the prior on the error precision σ^2 to a gamma with parameters $\alpha = 3$ and $\beta = 1$ (so the prior mean is $3/1$) gives marginal posteriors for the precisions and variances shown in Figures 19.3a,b. The posterior for the precisions has two clear modes, while the posterior for the variances has a single mode with no shoulder.

Ms. Henn stumbled on the first example but produced the second with little further effort, so either she was unusually lucky or it is not hard to manufacture such examples. The latter is more plausible, in which case we should be concerned that modality not uncommonly depends on whether we consider variances or precisions. An exercise incites you to look for more examples.

Until our theory of richly parameterized models has developed to the point where we can easily determine posterior modality for a given dataset, model, and prior, it is reasonable to ask how to protect yourself against surprises like this. Henn & Hodges (2013, Section 6) gave some advice based on the theory developed in that paper, but collectively we are nowhere near understanding when multiple local maxima are present in restricted likelihoods or marginal posteriors.

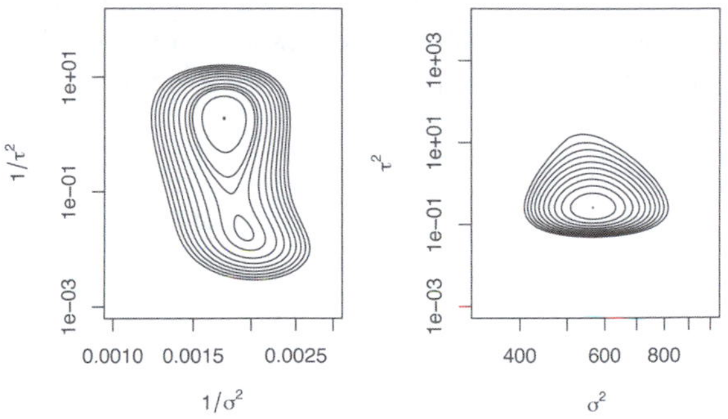

(a) Marginal posterior, precisions (b) Marginal posterior, variances

Figure 19.3: HMO data, model with two state-level predictors, Wakefield's analysis modified by adding a Gamma(3,1) prior for the error variance σ^2. Panel a: Log marginal posterior for the precisions; Panel b: Log marginal posterior for the variances. Horizontal and vertical scales are logarithmic. Contours are 1-log increments with an extra contour in Panel a to show the secondary maximum.

Exercises

Regular Exercises

1. (Section 19.1) By setting the free terms in conflict with the mixed terms, see if you can produce a restricted likelihood with two local maxima for a model other than the ICAR model. The balanced one-way random effect model's restricted likelihood necessarily has one mode, but is the same true for the unbalanced one-way random effect model?
2. (Section 19.2.2) Using the HMO dataset or other datasets, look for more examples in which changing variables from variances to precisions or vice versa changes the posterior's modality.

Open Questions

1. (Section 19.1) Produce an example of a restricted likelihood with three or more modes or prove that is impossible. Welham and Thompson (2009) described their restricted likelihood with two local maxima as arising from conflict between components that have small eigenvalues in **G**, the random-effects covariance matrix, and components with large eigenvalues in **G**. In terms of the scalarized restricted likelihood, this suggests grouping the mixed terms j into two groups, one with large a_j and one with small a_j, and constructing $\hat{v}_j^2$ for the two groups that convey very different information about σ_s^2. If this produces mixed terms with two local maxima, a third local maximum might be obtained by constructing free terms with information about σ_e^2 that is inconsistent with both local maxima in the mixed terms. Alternatively, it may be possible to prove that a two-variance model's

restricted likelihood can have no more than two local maxima, analogous to the proof in Liu & Hodges (2003) that posteriors from a balanced one-way random effect model with inverse gamma priors on the variances have at most two modes.

References

Adams JL, Hammitt JK, Hodges JS (2000). Periodicity in the global mean temperature series? In Morton SC, Rolph JE, eds., *Public Policy and Statistics: Case Studies from RAND*. New York: Springer-Verlag, 75–93.

Assunçao R, Krainski E (2009). Neighborhood dependence in Bayesian spatial models. *Biometrical Journal*, **51**:851–869.

Atkinson AC (1985). *Plots, Transformations, and Regression*. Oxford: Clarendon Press.

Banerjee S, Carlin BP, Gelfand AE (2004). *Hierarchical Modeling and Analysis for Spatial Data*. Boca Raton, FL: Chapman&Hall/CRC Press.

Barnard J, McCulloch R, Meng X-L (2000). Modeling covariance matrices in terms of standard deviations and correlations, with applications to shrinkage. *Statistica Sinica*, **10**:1281–1311.

Bates DM, DebRoy S (2004). Linear mixed models and penalized least squares. *Journal of Multivariate Analysis*, **91**:1–17.

Bayarri MJ, DeGroot MH (1992). Difficulties and ambiguities in the definition of a likelihood function. *Journal of the Italian Statistical Society*, **1**:1–15.

Bayarri MJ, DeGroot MH, Kadane JB (1988). What is the likelihood function? *Statistical Decision Theory and Related Topics IV, Vol. 1*, eds. Gupta SS, Berger JO, Springer Berlin, Heidelberg, 3–16.

Berger JO (1992). Discussion of a paper by JS Hodges. In Bernardo JM, Berger JO, Dawid AP, and Smith AFM, eds., *Bayesian Statistics 4*. Oxford: Clarendon, 256–258.

Berger JO (1993). Contributed discussion. In C Gatsonis, JS Hodges, RE Kass, ND Singpurwalla, eds., *Case Studies in Bayesian Statistics*. New York: Springer, 302–303.

Berkson J (1980). Minimum chi-square, not maximum likelihood! *Annals of Statistics*, **8**:457–487.

Bernardo JM, Smith AFM (1994). *Bayesian Theory*. New York: Wiley.

Besag J (1974). Spatial interaction and the statistical analysis of lattice systems (with discussion). *Journal of the Royal Statistical Society, Series B*, **36**:192–236.

Besag J, Higdon D (1999). Bayesian analysis of agricultural field experiments (with discussion). *Journal of the Royal Statistical Society, Series B*, **61**:691–746.

Besag J, York JC, and Mollié A (1991). Bayesian image restoration, with two applications in spatial statistics (with discussion). *Annals of the Institute of Statistical Mathematics*, **43**:1–59.

Bondell HD, Reich BJ (2009). Simultaneous factor selection and collapsing of levels in ANOVA. *Biometrics*, **69**:169–177.

Bradlow ET, Zaslavsky AM (1997). Case influence analysis in Bayesian inference. *Journal of Computational and Graphical Statistics*, **6**:314–331.

Breiman L (2001). Statistical modeling: The two cultures (with discussion). *Statistical Science*, **16**:199–231.

Browne W, Goldstein H, Rasbash J (2001). Multiple membership multiple classification (MMMC) models. *Statistical Modelling*, **1**:103–124.

Carlin BP, Louis TA (2008). *Bayesian Methods for Data Analysis*, third edition, Boca Raton, FL: Chapman and Hall/CRC Press.

Celeux G, Forbes F, Robert CP, Titterington DM (2001). Deviance information criteria for missing data models (with discussion). *Bayesian Analysis*, **1**:651–706.

Chaloner K (1994). Residual analysis and outliers in Bayesian hierarchical models. In *Aspects of Uncertainty*, eds. AFM Smith, PR Freeman, Chichester: Wiley, 153–161.

Chang MC, Ko CC, Liu CC, Douglas WH, DeLong R, Seong WJ, Hodges J, An KN (2003). Elasticity of alveolar bone near dental implant-bone interfaces after one month's healing. *Journal of Biomechanics*, **36**:1209–1214.

Chen Z, Dunson DB (2003). Random effects selection in mixed linear models. *Biometrics*, **59**:762–769.

Christensen, R (2011). *Plane Answers to Complex Questions: The Theory of Linear Models*, fourth edition. New York: Springer.

Clayton DG, Bernardinelli L, Montomoli C (1993). Spatial correlation in ecological analysis. *International Journal of Epidemiology*, **22**:1193–1201.

Cook RD (1986). Assessment of local influence (with discussion). *Journal of the Royal Statistical Society*, Series B, **48**:133–169.

Cook RD, Weisberg S (1982). *Residuals and Influence in Regression*. New York: Chapman and Hall.

Cowles MK, Carlin BP (1996). Markov chain Monte Carlo convergence diagnostics: A comparative review. *Journal of the American Statistical Association*, **91**:883–904.

Cressie NAC (1991). *Statistics for Spatial Data*. New York: Wiley.

Critchley F (1998). Discussion of Hodges (1998). *Journal of the Royal Statistical Society, Series B*, **60**:528–529.

Cui Y (2008). Smoothing analysis of variance for general designs and partitioning degrees of freedom in hierarchical and other richly parameterized models. Unpublished doctoral dissertation, Division of Biostatistics, University of Minnesota.

Cui Y, Hodges JS (2009). Smoothing analysis of variance for general designs. Manuscript, available from this book's web site.

Cui Y, Hodges JS, Kong X, Carlin BP (2010). Partitioning degrees of freedom in hierarchical and other richly parameterized models. *Technometrics*, **52**:124–136.

Daniels M (1999). A prior for the variance in hierarchical models. *Canadian Journal of Statistics*, **27**:567–578.

Daniels MJ, Kass RE (1999). Nonconjugate Bayesian estimation of covariance matrices and its use in hierarchical models. *Journal of the American Statistical Association*, **94**:1254–1263.

Daniels MJ, Kass RE (2001). Shrinkage estimators for covariance matrices. *Biometrics*, **57**:1773–1184.

Davison AC, Tsai C-L (1992). Regression model diagnostics. *International Statistical Review*, **60**:337–353.

de Finetti B (1974–75). *Theory of Probability*, English translation by Machi A, Smith AFM. New York: Wiley.

Derksen S, Keselman HJ (1992). Backward, forward and stepwise automated subset selection algorithms: Frequency of obtaining authentic and noise variables. *British Journal of Mathematics and Statistics in Psychology*, **45**:265–282.

Diggle PJ, Liang K-Y, Zeger SL (1994). *Analysis of Longitudinal Data*. New York: Oxford.

Dixon DO, Simon R (1991). Bayesian subset analysis. *Biometrics*, **47**:871–881.

Draper DC, Hodges JS, Mallows CL, Pregibon D (1993). Exchangeability and data analysis (with discussion). *Journal of the Royal Statistical Society, Series A*, **156**:9–37.

Duncan DB, Horn SD (1972). Linear dynamic recursive estimation from the viewpoint of regression analysis. *Journal of the American Statistical Association*, **67**:815–821.

Dunson DB, Chen Z, Harry J (2003). A Bayesian approach for joint modeling of cluster size and subunit-specific outcomes. *Biometrics*, **59**:521–530.

Eaton ML, Freedman DA (2004). Dutch book against some "objective" priors. *Bernoulli*, **10**:861–872.

Eaton ML, Sudderth WD (2004). Properties of right Haar predictive inference. *Sankhyā*, **66**, Part 3:487–512.

Erkanli A (1994). Laplace approximations for posterior expectations when the mode occurs at the boundary of the parameter space. *Journal of the American Statistical Association*, **89**:250–258.

Fahrmeir L, Tutz G (2010). *Multivariate Statistical Modeling Based on Generalized Linear Models*, second edition. New York: Springer-Verlag.

Fellner WH (1986). Robust estimation of variance components. *Technometrics*, **28**:51–60.

Flegal JM, Jones GL (2011). Implementing Markov chain Monte Carlo: Estimating with confidence. In *Handbook of Markov Chain Monte Carlo*, eds. Brooks S, Gelman A, Jones G, Meng X-L, Boca Raton, FL: CRC Press.

Flegal JM, Haran M, Jones GL (2008). Markov chain Monte Carlo: Can we trust the third significant figure? *Statistical Science*, **23**:250–260.

Freedman DA (1983). A note on screening regression equations. *The American Statistician*, **37**:152–155.

Freedman DA, Lane D (1983). A nonstochastic interpretation of reported significance levels. *Journal of Business and Economic Statistics*, **1**:292–298.

Fuller WA (1980). The use of indicator variables in computing predictions. *Journal of Econometrics*, **12**:231–243.

Fuentes M, Reich BJ (2010). Spectral domain. Chapter 5 in *Handbook of Spatial Statistics*, eds. Gelfand AE, Diggle PJ, Fuentes M, Guttorp P, Boca Raton, FL: Chapman & Hall/CRC.

Gelfand AE, Dey DK, Chang H (1992). Model determination using predictive distributions with implementation via sampling-based methods. In *Bayesian Statistics 4*, eds. Bernardo JH, Berger JO, Dawid AP, Smith AFM, Cambridge: Oxford, 147–167.

Gelfand AE, Sahu SK (1999). Identifiability, improper priors, and Gibbs sampling for generalized linear models. *Journal of the American Statistical Association*, **94**:247–253.

Gelfand AE, Smith AFM (1990). Sampling-based approaches to calculating marginal densities. *Journal of the American Statistical Association*, **85**:398–409.

Gelman A (2005a). Analysis of variance — why it is more important than ever (with discussion). *Annals of Statistics*, **33**:1–53.

Gelman A (2005b). Why I don't use the term "fixed and random effects." Blog entry, January 25, 2005, URL http://www.stat.columbia.edu/~cook/movabletype/archives/2005/01/why_i_dont_use.html

Gelman A (2006). Prior distributions for variance parameters in hierarchical models. *Bayesian Analysis*, **1**:515–534.

Gelman A, Carlin JB, Stern HB, Rubin DB (2004). *Bayesian Data Analysis*, second edition, Boca Raton, FL: Chapman and Hall.

Geman S, Geman D (1984). Stochastic relaxation, Gibbs distributions and the Bayesian restoration of images. *IEEE Transactions on Pattern Analysis and Machine Intelligence*, **6**:721–741.

Gilks WR, Richardson S, Spiegelhalter DJ (1996). *Markov Chain Monte Carlo in Practice*. New York: Chapman & Hall.

Gill S, Loprinzi CL, Sargent DJ, Thome SD, Alberts SR, Haller DG, Benedetti J, Francini G, Shepherd LE, Seitz JF, Labianca R, Chen W, Cha SS, Heldebrant MP, Goldberg RM (2004). Pooled analysis of fluorouracil-based adjuvant therapy for stage II and III colon cancer: Who benefits and by how much? *Journal of Clinical Oncology*, **22**:1–10.

Graybill FA (1983). *Matrices with Applications in Statistics*, second edition. Pacific Grove, CA: Wadsworth & Brooks/Cole.

Green PJ (2002). Discussion of Spiegelhalter et al. (2002). *Journal of the Royal Statistical Society, Series B*, **64**:627–628.

Green PJ, Silverman BW (1994). *Nonparametric Regression and Generalized Linear Models: A Roughness Penalty Approach*. New York: Chapman & Hall.

Hastie TJ, Tibshirani RJ (1990). *Generalized Additive Models*. New York: Chapman & Hall.

He Y, Hodges JS (2008). Point estimates for variance-structure parameters in Bayesian analysis of hierarchical models. *Computational Statistics and Data Analysis*, **52**:2560–2577.

He Y, Hodges JS, Carlin BP (2007). Re-considering the variance parameterization in multiple precision models. *Bayesian Analysis*, **2**:529–556.

Henderson CR, Kempthorne O, Searle SR, von Krosigk CN (1959). Estimating of environmental and genetic trends from records subject to culling. *Biometrics*, **15**:192–218.

Henn L, Hodges JS (2013). Multiple local maxima in restricted likelihoods and posterior distributions for mixed linear models. *International Statistical Review*, to appear.

Hilden-Minton JA (1995). Multilevel diagnostics for mixed and hierarachical linear models. Unpublished PhD dissertation, University of California, Los Angeles.

Hodges JS (1998). Some algebra and geometry for hierarchical models, applied to diagnostics (with discussion). *Journal of the Royal Statistical Society, Series B*, **60**:497–536.

Hodges JS, Carlin BP, Fan Q (2003). On the precision of the conditionally autoregressive prior in spatial models. *Biometrics*, **59**:317–322.

Hodges JS, Clayton MK (2010). Random effects old and new. Manuscript, available on this book's web site.

Hodges JS, Cui Y, Sargent DJ, Carlin BP (2007). Smoothing balanced single-error-term analysis of variance. *Technometrics*, **49**:12–25.

Hodges JS, Reich BJ (2010). Adding spatially correlated errors can mess up the fixed effect you love. *The American Statistician*, **64**:325–334.

Hodges JS, Sargent DJ (2001). Counting degrees of freedom in hierarchical and other richly parameterized models. *Biometrika*, **88**:367–379.

Hoffman EB, Sen PK, Weinberg CR (2001). Within-cluster resampling. *Biometrika*, **88**:1121–1134.

Houtman AM, Speed TP (1983). Balance in designed experiments with orthogonal block structure. *Annals of Statistics*, **11**:1069–1085.

Hughes J, Haran M (2013). Dimension reduction and alleviation of confounding for spatial generalized linear mixed models. *Journal of the Royal Statistical Society, Series B*, to appear.

Johnson AA, Jones GL (2010). Gibbs sampling for a Bayesian hierarchical general linear model. *Electronic Journal of Statistics*, **4**:313–333.

Johnson LT, Geyer CJ (2012). Variable transformation to obtain geometric ergodicity in the random-walk Metropolis algorithm. *Annals of Statistics*, **40**:3050–3076.

Jones GL, Haran M, Caffo BS, Neath R (2006). Fixed-width output analysis for Markov chain Monte Carlo. *Journal of the American Statistical Association*, **101**:1537–1547.

Kass F (2007). Survival of full coverage restorations in primary maxillary incisors: A retrospective study. Unpublished Master's thesis, Division of Pediatric Dentistry, School of Dentistry, University of Minnesota.

Kass RE (2011). Statistical inference: The big picture (with discussion). *Statistical Science*, **26**:1–20.

Kass RE, Raftery AE (1995). Bayes factors. *Journal of the American Statistical Association*, **90**:773–795.

Kass RE, Tierney L, Kadane JB (1989). Approximate methods for assessing influence and sensitivity in Bayesian analysis. *Biometrika*, **76**:663–674.

Carriquiry AL, Kliemann W (2007). The modes of posterior distributions for mixed linear models. *Proyecciones*, **26**:327–354.

Kendall KA (2009). High-speed laryngeal imaging compared with videostroboscopy in healthy subjects. *Archives of Otolaryngology–Head and Neck Surgery*, **135**:274–281.

Ko CC, Douglas WH, DeLong R, Rohrer MD, Swift JQ, Hodges JS, An KN, Ritman EL (2003). Effects of implant healing time on crestal bone loss of a controlled-load dental implant. *Journal of Dental Research*, **82**:585–591.

Lavine M, Haglund MM, Hochman DW (2011). Dynamic linear model analysis of optical imaging data acquired from the human neocortex. *Journal of Neuroscience Methods*, **199**:346–362.

Lavine M, Hodges JS (2012). On rigorous specification of ICAR models. *The American Statistician*, **66**:42–49.

Leamer EE (1978). *Specification Searches*. New York: Wiley.

Lee Y (2007). Review of Jiang J, *Linear and Generalized Linear Models and Their Applications*, *Biometrics*, **63**:1297–1298.

Lee Y, Nelder JA (1996). Hierarchical generalized linear models (with discussion). *Journal of the Royal Statistical Society, Series B*, **58**:6719–678.

Lee Y, Nelder JA (2009). Likelihood inference for models with unobservables: Another view (with discussion). *Statistical Science*, **24**:255–302.

Lee Y, Nelder JA, Pawitan Y (2006). *Generalized Linear Models with Random Effects*. Boca Raton, FL: Chapman and Hall/CRC Press.

Lindgren F, Rue H, Lindström J (2011). An explicit link between Gaussian fields and Gaussian Markov random fields: The stochastic partial differential equation approach (with discussion). *Journal of the Royal Statistical Society, Series B*, **73**:423–498.

Liu J (2001). Characterizing modality of the posterior for hierarchical models. Unpublished doctoral dissertation, Division of Biostatistics, University of Minnesota.

Liu J, Hodges JS (2003). Posterior bimodality in the balanced one-way random effects model. *Journal of the Royal Statistical Society, Series B*, **65**:247–255.

Longford NT (1998). Comment on Hodges (1998). *Journal of the Royal Statistical Society, Series B*, **60**:527.

Lu H, Hodges JS, Carlin BP (2007). Measuring the complexity of generalized linear hierarchical models. *Canadian Journal of Statistics*, **35**:69–87.

Lucas TW (1993). When is conflict normal? *Journal of the American Statistical Association*, **88**:1433–1437.

MacEachern SN, Peruggia M (2000). Importance link function estimation for Markov chain Monte Carlo methods. *Journal of Computational and Graphical Statistics*, **9**:99–121.

Marron JS (1996). A personal view of smoothing and statistics. In M.G. Schimek (Ed.), *Statistical Theory and Computational Aspects of Smoothing*, Heidelberg: Physica-Verlag, 1–9.

McCaffrey DF, Lockwood JR, Koretz D, Louis TA, Hamilton L (2004). Models for value-added modeling of teacher effects. *Journal of Behavioral and Educational Statistics*, **29**:67–101.

McCullagh P, Nelder JA (1989). *Generalized Linear Models*, second edition, Boca Raton, FL: Chapman and Hall.

Meng XL (2009). Decoding the h-likelihood (discussion of Lee & Nelder 2009). *Statistical Science*, **24**:280–293.

Milliken GA, Johnson DE (1992). *Analysis of Messy Data, Volume 1: Designed Experiments*, Boca Raton, FL: Chapman and Hall/CRC Press.

Newcomb RW (1961). On the simultaneous diagonalization of two semi-definite matrices. *Quarterly Journal of Applied Mathematics*, **19**:144–146.

Nobile A, Green PJ (2000). Bayesian analysis of factorial experiments by mixture modelling. *Biometrika*, **87**:15–35.

Nychka DW (1988). Confidence intervals for smoothing splines. *Journal of the American Statistical Association*, **83**:1134–1143.

O'Hagan A (1976). On posterior joint and marginal modes. *Biometrika*, **63**:329–333.

O'Hagan A (1985). Shoulders in hierarchical models. In *Bayesian Statistics 2*, eds. Bernardo JM, DeGroot MH, Lindley CV, Smith AFM, Second Valencia International Meeting: Elsevier Science Publishers, 697–710.

O'Malley AJ, Marsden PV (2008). The analysis of social networks. *Health Services Outcomes Research Methodology*, **8**:222–269.

Oliveira VD, Ferreira MAR (2011). Maximum likelihood and restricted maximum likelihood estimation for a class of gaussian markov random fields. *Metrika*, **74**:167–183.

Paciorek CJ (2007). Bayesian smoothing with Gaussian processes using Fourier basis functions in the spectralGP package. *Journal of Statistical Software*, **19(2)**:1–38.

Paciorek CJ (2010). The importance of scale for spatial-confounding bias and precision of spatial regression estimators. *Statistical Science*, **25**:107–125.

Paciorek CJ, Schervish MJ (2006). Spatial modelling using a new class of nonstationary covariance functions. *Environmetrics*, **17**:483–506.

Panoutsopoulou IG, Wendelshafer-Crabb G, Hodges JS, Kennedy WR (2009). Skin blister and skin biopsy to quantify epidermal nerves: a comparative study. *Neurology*, **72**:1205–1210.

Peruggia M (1997). On the variability of case-deletion importance sampling weights in the Bayesian linear model. *Journal of the American Statistical Association*, **92**:199–207.

Pesun IJ, Hodges JS, Lai JH (2002). Effect of finishing and polishing procedures on the gap width between a denture base resin and two long-term resilient denture liners. *Journal of Prosthodontic Dentistry*, **87**:311–318.

Peterson C, Simon M, Hodges J, Mertens P, Egelman E, Anderson D (2001). Composition and mass of the bacteriophage ϕ29 prohead and virion. *Journal of Structural Biology*, **135**:18–25.

Petris G (2010). An R package for dynamic linear models. *Journal of Statistical Software*, **36**(12):1–16.

Pinheiro JC, Bates DM (2000). *Mixed-Effects Models in S and S-PLUS*. New York: Springer.

Plummer M (2008). Penalized loss functions for Bayesian model comparison. *Biostatistics*, **9**:523–539.

Pourahmadi M (2007). Cholesky decompositions and estimate of a covariance matrix: orthogonality of variance-correlation parameters. *Biometrika*, **94**:1006–1013.

Raftery AE, Madigan D, Hoeting J (1993). Model selection and accounting for model uncertainty in linear regression models. Technical Report 262, University of Washington Department of Statistics.

Rappold AG, Lavine M, Lozier S (2007). Subjective likelihood for the assessment of trends in the ocean's mixed-layer depth. *Journal of the American Statistical Association*, **102**:771–780.

Reich BJ, Hodges JS (2008a). Identification of the variance components in the general two-variance linear model. *Journal of Statistical Planning and Inference*, **138**:1592–1604.

Reich BJ, Hodges JS (2008b). Modeling longitudinal spatial periodontal data: A spatially adaptive model with tools for specifying priors and checking fit. *Biometrics*, **64**:790–799.

Reich BJ, Hodges JS, Carlin BP (2004). Spatial analysis of periodontal data using conditionally autoregressive priors having two types of neighbor relations. Division of Biostatistics, University of Minnesota, Research Report RR2004-004. Available from this book's web site.

Reich BJ, Hodges JS, Carlin BP (2007). Spatial analysis of periodontal data using conditionally autoregressive priors having two classes of neighbor relations. *Journal of the American Statistical Association*, **102**:44–55.

Reich BJ, Hodges JS, Zadnik V (2006). Effects of residual smoothing on the posterior of the fixed effects in disease-mapping models. *Biometrics*, **62**:1197–1206. Errata available from JS Hodges.

Ripley BD (2004). *Spatial Statistics*. Hoboken, NJ: Wiley.

Robinson GK (1991). That BLUP is a good thing: The estimation of random effects (with discussion). *Statistical Science*, **6**:15–51.

Román JC, Hobert JP (2012). Convergence analysis of the Gibbs sampler for Bayesian general linear mixed models with improper priors. *Annals of Statistics*, **40**:2823–2849.

Rue H, Held L (2005). *Gaussian Markov Random Fields: Theory and Applications*. Boca Raton, FL: Chapman & Hall/CRC.

Rue H, Martino S, Chopin N (2009). Approximate Bayesian inference for latent Gaussian models by using integrated nested Laplace approximations (with discussion). *Journal of the Royal Statistical Society, Series B*, **71**:319–392.

Ruppert D, Wand MP, Carroll RJ (2003). *Semiparametric Regression*. New York: Cambridge.

Salkever DS (1976). The use of dummy variables to compute predictions, prediction errors, and confidence intervals. *Journal of Econometrics*, **4**:393–397.

Salkowski NJ (2008). Using the SemiPar package. Manuscript, available on this book's web site.

Sargent DJ, Goldberg RM, Jacobson SD, Macdonald JS, Labianca R, Haller DG, Shepherd LE, Seitz JF, Francini G (2001). A pooled analysis of adjuvant chemotherapy for resected colon cancer in elderly patients. *New England Journal of Medicine*, **345**:1091–1097.

Schabenberger O, Gotway CA (2004). *Statistical Methods for Spatial Data Analysis*. Boca Raton, FL: Chapman & Hall/CRC Press.

Scheffé H (1959). *The Analysis of Variance*. New York: Wiley.

Schott JR (1997). *Matrix Analysis for Statistics*. New York: Wiley.

Searle SR, Casella G, McCulloch CE (1992). *Variance Components*. New York: Wiley.

Seber GAF (1984). *Multivariate Observations*. New York: Wiley.

Simpson D, Lindgren F, Rue H (2012). In order to make spatial statistics computationally feasible, we need to forget about the covariance function. *Environmetrics*, **23**:65–74.

Smith AFM (1973). A general Bayesian linear model. *Journal of the Royal Statistical Society, Series B*, **35**:67–75.

Smith AFM (1986). Some Bayesian thoughts on modelling and model choice. *The Statistician*, **35**:97–102.

Snijders TAB, Bosker RJ (2012). *Multilevel Analysis: An Introduction to Basic and Advanced Multilevel Modeling*, second edition. Los Angeles: SAGE.

Speed TP (1983). General balance. In *Encyclopedia of Statistical Sciences*, first edition, Vol. 3, eds. Kotz S, Johnson NL III, Read CB, New York: Wiley, 320–326.

Speed TP (1987). What is an analysis of variance? (with discussion). *Annals of Statistics*, **15**:885–941.

Spiegelhalter DJ, Best NG, Carlin BP, van der Linde A (2002). Bayesian measures of model complexity and fit (with discussion). *Journal of the Royal Statistical Society, Series B*, **64**:583–639.

Stein C (1956). Inadmissibility of the usual estimator for the mean of a multivariate normal distribution. *Proceedings of the Third Berkeley Symposium on Mathematical Statistics and Probability*, Vol. 1, University of California Press, 197–206.

Stein ML (1999). *Interpolation of Spatial Data: Some Theory For Kriging*. New York: Springer-Verlag.

Steinberg DM, Bursztyn D (2004). Data analytic tools for understanding random field regression models. *Technometrics*, **46**:411–420.

Sun J, Loader CR (1994). Simultaneous confidence bands for linear regression and smoothing. *Annals of Statistics*, **22**:1328–1345.

Szpiro AA, Paciorek CJ (2014). Measurement error in two-stage analyses, with application to air pollution epidemiology (with discussion). *Environmetrics*, to appear.

Szpiro AA, Sheppard L, Lumley T (2011). Efficient measurement error correction with spatially misaligned data. *Biostatistics*, **12**:610–623.

Theil H (1971). *Principles of Econometrics*. New York: Wiley.

Tibshirani R (1996). Regression shrinkage and selection via the lasso. *Journal of the Royal Statistical Society, Series B*, **58**:267–288.

Tierney L (1994). Markov chains for exploring posterior distributions (with discussion). *Annals of Statistics*, **22**:1701–1762.

Vaida F, Blanchard S (2005). Conditional Akaike information criterion for mixed effects models. *Biometrika*, **92**:351–370.

Verbeke G, Molenberghs G (1997). *Linear Mixed Models in Practice: A SAS-Oriented Approach*. New York: Springer.

Wakefield J (1998). Discussion of Hodges (1998). *Journal of the Royal Statistical Society, Series B*, **60**:523–526.

Wakefield J (2007). Disease mapping and spatial regression with count data. *Biostatistics*, **8**:158–183.

Wall MM (2004). A close look at the spatial structure implied by the CAR and SAR models. *Journal of Statistical Planning and Inference*, **121**:311–324.

Waller LA, Gotway CA (2004). *Applied Spatial Statistics for Public Health Data*. New York: Wiley.

Ward K (2010). Serum creatinine and cystatin C based GFR estimating models in kidney transplant recipients. MS thesis, Division of Biostatistics, University of Minnesota.

Wei P, Pan W (2012). Bayesian joint modeling of multiple gene networks and diverse genomic data to identify target genes of a transcription factor. *Annals of Applied Statistics*, **6**:334–355.

Weisberg S (1980). *Applied Linear Regression*, first edition. New York: Wiley.

Welham SJ, Thompson R (2009). A note on bimodality in the log-likelihood function for penalized spline mixed models. *Computational Statistics and Data Analysis*, **53**:920–931.

West M, Harrison J (1997). *Bayesian Forecasting and Dynamic Models*, second edition. New York: Springer.

White LV (1985). Orthogonal designs. In *Encyclopedia of Statistical Sciences*, first edition, Vol. 6, eds. Kotz S, Johnson NL III, Read CB, New York: Wiley, 528–530.

Whittaker J (1998). Discussion of Hodges (1998). *Journal of the Royal Statistical Society, Series B*, **60**:533.

Wikle C (2002). Spatial modeling of count data: A case study in modelling breeding bird survey data on large spatial domains. In *Spatial Cluster Modeling*, eds. Lawson A, Denison D, Boca Raton, FL: Chapman & Hall, 199–209.

Williams DA (1987). Generalized linear model diagnostics using the deviance and single case deletion. *Applied Statistics*, **36**:181–191.

Wonder S (1973). Superstition. *Talking Book*. Detroit, MI: Tamla.

Ye J (1998). On measuring and correcting the effects of data mining and model selection. *Journal of the American Statistical Association*, **93**:120–131.

Yuan Y, Johnson VE (2012). Goodness-of-fit diagnostics for Bayesian hierarchical models. *Biometrics*, **68**:156–164.

Zadnik V, Reich BJ (2006). Analysis of the relationship between socioeconomic factors and stomach cancer incidence in Slovenia. *Neoplasma*, **53**:103–110.

Zhang H (2004). Inconsistent estimation and asymptotically equal interpolations in model-based geostatistics. *Journal of the American Statistical Association*, **99**:250–261.

Zhang Y (2009). Extending smoothed analysis of variance to new problems. Unpublished doctoral dissertation, Division of Biostatistics, University of Minnesota.

Zhang Y, Hodges JS, Banerjee S (2010). Smoothed ANOVA with spatial effects as a competitor to MCAR in multivariate spatial smoothing. *Annals of Applied Statistics*, **3**:1805–1830.

Author Index

Subject Index

This index should be used in conjunction with the Table of Contents, which has detailed headings for sections and subsections, and the List of Examples, which lists the sections in which each example is used.